Lecture Notes in Artificial Intelligence 6379

Edited by R. Goebel, J. Siekmann, and W. Wahlster

Subseries of Lecture Notes in Computer Science

Amol Deshpande Anthony Hunter (Eds.)

Scalable Uncertainty Management

4th International Conference, SUM 2010
Toulouse, France, September 27-29, 2010
Proceedings

Springer

Series Editors

Randy Goebel, University of Alberta, Edmonton, Canada
Jörg Siekmann, University of Saarland, Saarbrücken, Germany
Wolfgang Wahlster, DFKI and University of Saarland, Saarbrücken, Germany

Volume Editors

Amol Deshpande
University of Maryland, Department of Computer Science
College Park, MD 20910, USA
E-mail: amol@cs.umd.edu

Anthony Hunter
University College London, Department of Computer Science
Gower Street, London WC1E 6BT, UK
E-mail: a.hunter@cs.ucl.ac.uk

Library of Congress Control Number: 2010934755

CR Subject Classification (1998): I.2, H.4, H.3, H.5, C.2, H.2

LNCS Sublibrary: SL 7 – Artificial Intelligence

ISSN 0302-9743
ISBN-10 3-642-15950-8 Springer Berlin Heidelberg New York
ISBN-13 978-3-642-15950-3 Springer Berlin Heidelberg New York

springer.com

Printed in Germany

Typesetting: Camera-ready by author, data conversion by Scientific Publishing Services, Chennai, India
Printed on acid-free paper 06/3180

Preface

Managing uncertainty and inconsistency has been extensively explored in Artificial Intelligence over a number of years. Now with the advent of massive amounts of data and knowledge from distributed heterogeneous, and potentially conflicting, sources, there is interest in developing and applying formalisms for uncertainty and inconsistency widely in systems that need to better manage this data and knowledge.

The annual International Conference on Scalable Uncertainty Management (SUM) has grown out of this wide-ranging interest in managing uncertainty and inconsistency in databases, the Web, the Semantic Web, and AI. It aims at bringing together all those interested in the management of large volumes of uncertainty and inconsistency, irrespective of whether they are in databases, the Web, the Semantic Web, or in AI, as well as in other areas such as information retrieval, risk analysis, and computer vision, where significant computational efforts are needed. After a promising First International Conference on Scalable Uncertainty Management was held in Washington DC, USA in 2007, the conference series has been successfully held in Napoli, Italy, in 2008, and again in Washington DC, USA, in 2009.

This volume contains the papers presented at the Fourth International Conference on Scalable Uncertainty Management (SUM 2010), which was held in Toulouse, France, during 27-29 September 2010. It contains 26 technical papers, which were selected out of 32 submitted papers. The volume also contains abstracts of the two invited talks. To encourage discussion among the conference participants, we invited several experts in topics related to uncertainty management as *discussants*. The discussants were also invited to contribute to the proceedings; we include six such contributions in the proceedings. We wish to thank all authors who submitted papers and all conference participants for fruitful discussions. We are also very grateful to the invited speakers (Christoph Koch, Torsten Schaub), and the discussants (Salem Benferhat, Didier Dubois, Lluis Godo, Eyke Hüllermeier, Ander de Keijzer, Amedeo Napoli, Odile Papini, Olivier Strauss) for their talks. We would like to thank all the Program Committee members and external referees for their timely expertise in carefully reviewing the submissions. Special thanks are due to Florence Boué, Didier Dubois, and Henri Prade, the local organizers of SUM 2010, for their invaluable help.

September 2010

Amol Deshpande
Anthony Hunter

Organization

Local Organization

Florence Boué	IRIT, CNRS/Université Paul Sabatier, Toulouse
Didier Dubois	IRIT, CNRS/Université Paul Sabatier, Toulouse
Henri Prade	IRIT, CNRS/Université Paul Sabatier, Toulouse

Program Chairs

Amol Deshpande	University of Maryland, USA
Anthony Hunter	University College London, UK

Program Committee

Leila Amgoud	IRIT, France
Chitta Baral	Arizona State University, USA
Salem Benferhat	University of Artois, France
Leopoldo Bertossi	Carleton University, Canada
Isabelle Bloch	ENST, France
Reynold Cheng	University of Hong Kong, China
Carlos Chesñevar	Universidad Nacional del Sur, Argentina
Laurence Cholvy	ONERA, France
Jan Chomicki	SUNY Buffalo, USA
Anish Das Sarma	Yahoo Research, USA
Thierry Denœux	Université de Technologie de Compiègne, France
Jürgen Dix	TU Clausthal, Germany
Francesco Donini	University of Tuscia, Italy
Scott Ferson	Applied Biomathematics, USA
Fabio Gagliardi Cozman	University of Sao Paulo, Brazil
Lluis Godo	IIIA, Spain
Nikos Gorogiannis	University College London, UK
John Grant	Towson University, USA
Sergio Greco	Università della Calabria, Italy
Jon Helton	Sandia, USA
Ihab Ilyas	University of Waterloo, Canada

Vladik Kreinovich	University of Texas, USA
Rudolf Kruse	Universität Magdeburg, Germany
Jonathan Lawry	University of Bristol, UK
Weiru Liu	Queen's University Belfast, UK
Peter Lucas	University of Nijmegen, The Netherlands
Thomas Lukasiewicz	Oxford University, UK; TU Vienna, Austria
Tommie Meyer	Meraka Institute, South Africa
Serafin Moral	Universidad de Granada, Spain
Dan Olteanu	Oxford University, UK
Jeff Z. Pan	Aberdeen University, UK
Bijan Parsia	University of Manchester, UK
Simon Parsons	City University New York, USA
Gabriella Pasi	University of Milan-Bicocca, Italy
Sunil Prabhakar	Purdue University, USA
Andrea Pugliese	Università della Calabria, Italy
Guilin Qi	Southeast University, China
Chris Re	University of Wisconsin at Madison, USA
Thomas Roelleke	Queen Mary University of London, UK
Prithviraj Sen	Yahoo Research, India
Prakash Shenoy	University of Kansas, USA
Guillermo Simari	Universidad Nacional del Sur, Argentina
Umberto Straccia	ISTI-CNR, Italy
Sunil Vadera	University of Salford, UK
Jef Wijsen	Université de Mons, Belgium
Nic Wilson	University College Cork, Ireland
Ronald R. Yager	Iona College, USA

External Referees

Silvia Calegari
Stephane Herbin
Wei Hu
Livia Predoiu

Sponsoring Institutions

This conference was partially supported by the Université Paul Sabatier, Toulouse.

Table of Contents

Invited Talks

Markov Chain Monte Carlo and Databases (Abstract) 1
Christoph Koch

Answer Set Programming, the Solving Paradigm for Knowledge Representation and Reasoning 2
Torsten Schaub

Discussant Contributions

Graphical and Logical-Based Representations of Uncertain Information in a Possibility Theory Framework 3
Salem Benferhat

Probabilistic Data: A Tiny Survey 7
Ander de Keijzer

The Role of Epistemic Uncertainty in Risk Analysis 11
Didier Dubois

Uncertainty in Clustering and Classification 16
Eyke Hüllermeier

Information Fusion 20
Odile Papini

Use of the Domination Property for Interval Valued Digital Signal Processing 24
Olivier Strauss

Regular Contributions

Managing Lineage and Uncertainty under a Data Exchange Setting 28
Foto N. Afrati and Angelos Vasilakopoulos

A Formal Analysis of Logic-Based Argumentation Systems 42
Leila Amgoud and Philippe Besnard

Handling Inconsistency with Preference-Based Argumentation 56
Leila Amgoud and Srdjan Vesic

A Possibility Theory-Oriented Discussion of Conceptual Pattern Structures 70
Zainab Assaghir, Mehdi Kaytoue, and Henri Prade

DK-BKM: Decremental K Belief K-Modes Method 84
Sarra Ben Hariz and Zied Elouedi

On the Use of Fuzzy Cardinalities for Reducing Plethoric Answers to Fuzzy Queries 98
Patrick Bosc, Allel Hadjali, Olivier Pivert, and Grégory Smits

From Bayesian Classifiers to Possibilistic Classifiers for Numerical Data 112
Myriam Bounhas, Khaled Mellouli, Henri Prade, and Mathieu Serrurier

Plausibility of Information Reported by Successive Sources 126
Laurence Cholvy

Combining Semantic Web Search with the Power of Inductive Reasoning 137
Claudia d'Amato, Nicola Fanizzi, Bettina Fazzinga, Georg Gottlob, and Thomas Lukasiewicz

Evaluating Trust from Past Assessments with Imprecise Probabilities: Comparing Two Approaches 151
Sebastien Destercke

Range-Consistent Answers of Aggregate Queries under Aggregate Constraints 163
Sergio Flesca, Filippo Furfaro, and Francesco Parisi

Characterization, Propagation and Analysis of Aleatory and Epistemic Uncertainty in the 2008 Performance Assessment for the Proposed Repository for High-Level Radioactive Waste at Yucca Mountain, Nevada 177
Clifford W. Hansen, Jon C. Helton, and Cédric J. Sallaberry

Comparing Evidential Graphical Models for Imprecise Reliability 191
Wafa Laâmari, Boutheina Ben Yaghlane, and Christophe Simon

Imprecise Bipolar Belief Measures Based on Partial Knowledge from Agent Dialogues 205
Jonathan Lawry

Kriging with Ill-Known Variogram and Data 219
Kevin Loquin and Didier Dubois

Event Modelling and Reasoning with Uncertain Information for Distributed Sensor Networks 236
Jianbing Ma, Weiru Liu, and Paul Miller

Uncertainty in Decision Tree Classifiers 250
Matteo Magnani and Danilo Montesi

Efficient Policy-Based Inconsistency Management in Relational Knowledge Bases 264
Maria Vanina Martinez, Francesco Parisi, Andrea Pugliese, Gerardo I. Simari, and V.S. Subrahmanian

Modelling Probabilistic Inference Networks and Classification in Probabilistic Datalog 278
Miguel Martinez-Alvarez and Thomas Roelleke

Handling Dirty Databases: From User Warning to Data Cleaning — Towards an Interactive Approach 292
Olivier Pivert and Henri Prade

Disjunctive Fuzzy Logic Programs with Fuzzy Answer Set Semantics ... 306
Emad Saad

Cost-Based Query Answering in Action Probabilistic Logic Programs ... 319
Gerardo I. Simari, John P. Dickerson, and V.S. Subrahmanian

Clustering Fuzzy Data Using the Fuzzy EM Algorithm 333
Benjamin Quost and Thierry Denœux

Combining Multi-resolution Evidence for Georeferencing Flickr Images 347
Olivier Van Laere, Steven Schockaert, and Bart Dhoedt

A Structure-Based Similarity Spreading Approach for Ontology Matching 361
Ying Wang, Weiru Liu, and David A. Bell

Risk Modeling for Decision Support 375
Ronald R. Yager

Author Index 389

Markov Chain Monte Carlo and Databases
(Abstract)

Christoph Koch

École Polytechnique Fédérale de Lausanne
CH-1015 Lausanne, Switzerland

Several currently ongoing research efforts aim to combine Markov Chain Monte Carlo (MCMC) with database management systems. The goal is to scale up the management of uncertain data in contexts where only MCMC is known to be applicable or where the range and flexibility of MCMC provides a compelling proposition for powerful and interesting systems. This talk surveys recent work in this area and identifies open research challenges.

The talk starts with a discussion of applications that call for the combination of MCMC with ideas from database management. This is followed by a brief discussion of the now somewhat maturing field of probabilistic databases not based on MCMC, and what can be learned from these. Next, the architecture of an MCMC-based database management system is sketched, and key technical and algorithmic challenges are discussed. For efficient MCMC, it is key to be able to quickly evaluate queries on a sequence of many sample databases among which consecutive samples differ only moderately. The talk discusses techniques for efficiently solving this problem by aggressive incremental query evaluation. The locality of changes between consecutive samples is also key to scaling MCMC beyond state sizes that fit conveniently into a computer's main memory.

The second part of the talk addresses query languages beyond industry-standard languages such as SQL, which have limited appeal in the context of the scientific applications of MCMC. Computational problems to which MCMC is applied are often best expressed in terms of iteration and fixpoints. Database research knows languages centered around these principles, and it is interesting to understand how iteration as a query language construct interacts with MCMC sampling. The talk presents recent results in this space, including considerations of complexity and expressive power of query languages specifically designed for MCMC.

A. Deshpande and A. Hunter (Eds.): SUM 2010, LNAI 6379, p. 1, 2010.

Answer Set Programming, the Solving Paradigm for Knowledge Representation and Reasoning

Torsten Schaub*

University of Potsdam, Germany
torsten@cs.uni-potsdam.de

Abstract. Answer Set Programming (ASP; [1,2,3,4]) is a declarative problem solving approach, combining a rich yet simple modeling language with high-performance solving capacities. ASP is particularly suited for modeling problems in the area of Knowledge Representation and Reasoning involving incomplete, inconsistent, and changing information. From a formal perspective, ASP allows for solving all search problems in NP (and NP^{NP}) in a uniform way (being more compact than SAT). Applications of ASP include automatic synthesis of multiprocessor systems, decision support systems for NASA shuttle controllers, reasoning tools in systems biology, and many more. The versatility of ASP is also reflected by the ASP solver `clasp` [5,6,7], developed at the University of Potsdam, and winning first places at ASP'09, PB'09, and SAT'09.

The talk will give an overview about ASP, its modeling language, solving methodology, and portray some of its applications.

References

1. Gelfond, M., Lifschitz, V.: The stable model semantics for logic programming. In: Proceedings of the Fifth International Conference and Symposium of Logic Programming (ICLP 1988), pp. 1070–1080. The MIT Press, Cambridge (1988)
2. Niemelä, I.: Logic programs with stable model semantics as a constraint programming paradigm. Annals of Mathematics and Artificial Intelligence 25(3-4), 241–273 (1999)
3. Baral, C.: Knowledge Representation, Reasoning and Declarative Problem Solving. Cambridge University Press, Cambridge (2003)
4. Gelfond, M.: Answer sets. In: Lifschitz, V., van Hermelen, F., Porter, B. (eds.) Handbook of Knowledge Representation, pp. 285–316. Elsevier, Amsterdam (2008)
5. Gebser, M., Kaufmann, B., Neumann, A., Schaub, T.: Conflict-driven answer set solving. In: Proceedings of the Twentieth International Joint Conference on Artificial Intelligence (IJCAI 2007), pp. 386–392. AAAI Press/The MIT Press (2007)
6. Gebser, M., Kaufmann, B., Schaub, T.: The conflict-driven answer set solver clasp: Progress report. In: Erdem, E., Lin, F., Schaub, T. (eds.) LPNMR 2009. LNCS, vol. 5753, pp. 509–514. Springer, Heidelberg (2009)
7. Potassco, the Potsdam Answer Set Solving Collection, http://potassco.sourceforge.net/

* Affiliated with Simon Fraser University, Canada, and Griffith University, Australia.

A. Deshpande and A. Hunter (Eds.): SUM 2010, LNAI 6379, p. 2, 2010.

Graphical and Logical-Based Representations of Uncertain Information in a Possibility Theory Framework

Salem Benferhat

Université Lille-Nord de France
Artois, F-62307 Lens, CRIL, F-62307 Lens
CNRS UMR 8188, F-62307 Lens
benferhat@cril.fr

1 Introduction

Developing efficient approaches for reasoning under uncertainty is an important issue in many applications. Several graphical [3] and logical-based methods have been proposed to reason with incomplete information in various uncertainty theory frameworks. This paper focuses on possibility theory which is a convenient uncertainty theory framework to represent different kinds of prioritized pieces of information. It provides a brief overview of main compact representation formats, and their associated inference tools, that exist in a possibility theory framework. In particular, we discuss :

- logical-based representations, by means of possibilistic knowledge bases, which naturally extend propositional logic. Possibilistic knowledge bases gather propositional formulas associated with degrees belonging to a linearly ordered scale. These degrees reflect certainty or priority, depending if the formulas encode pieces of beliefs or goals to be pursued.
- conditional-based representations that allow to deal with generic rules having exceptions of the form "generally, if α is true then β is true" and
- possiblistic graphical models that can be viewed as counterparts of probabilistic bayesian networks.

We point out different connections that exist between these different knowledge representation formats. We also analyses various extensions that has been proposed to cope for instance with partially ordered information or multiple-source information.

2 Possibility Distribution

We consider a finite set of propositional variables V and a propositional language built from V, $\{\top, \bot\}$ and the connectives $\wedge, \vee, \neg, \rightarrow, \leftrightarrow$ in the usual way. Formulas, i.e., elements of PL_V are denoted by Greek letters.

One of the basic object of possibility theory is the concept of *possibility distribution*, which is a mapping from the set of classical interpretations Ω to the

A. Deshpande and A. Hunter (Eds.): SUM 2010, LNAI 6379, pp. 3–6, 2010.

interval [0,1]. A possibility distribution π represents the available knowledge about what the real world is. By convention, $\pi(\omega) = 1$ means that it is totally possible for ω to be the real world, $\pi(\omega) > 0$ means that ω is only somewhat possible, while $\pi(\omega) = 0$ means that ω is certainly not the real world. π induces two mappings grading respectively the possibility and the certainty of a formula:

- The possibility measure: $\Pi(\alpha) = max\{\pi(\omega) : \omega \models \alpha\}$ which evaluates to what extent α is consistent with the available knowledge expressed by π.
- The certainty (or necessity) measure: measure $N(\alpha) = 1 - \Pi(\neg\alpha)$ which evaluates to what extent α is entailed by the knowledge expressed by π.

Other uncertainty measures have been proposed such as the notion of guaranteed possibility measure [4], denoted by Δ. Intuitively, if α encodes an agent's goal then $\Delta(\alpha) \geq a$ means that any solution satisfying the goal Δ is satisfactory to a degree at least equal to a.

3 Knowledge Representation Formats

3.1 Possibilistic Logic and Its Extensions

One of the well-used and developed compact representation of a possibility distribution is the concept of a possibilistic knowledge base. Possibilistic logic provides a simple format that turns to be useful for handling qualitative uncertainty, exceptions or preferences. A possibilistic logic knowledge base is a set of possibilistic logic formulas where a possibilistic logic formula is a pair made of a classical logic formula ψ and a weight $a \in (0, 1]$ expressing the corresponding certainty. The weight a is interpreted as a lower bound of $N(\psi)$, i.e., the possibilistic logic expression (ψ, a) is understood as $N(\psi) \geq a$.

Possibilistic knowledge bases are compact representations of possibility distributions. Indeed, each possibilistic knowledge base induces a unique possibility distribution such that $\forall \omega \in \Omega$ and $\forall i$ such that $(\psi_i, a_i) \in \Sigma$:

$$\pi_\Sigma(\omega) = \begin{cases} 1 & if \omega \models \psi_i \\ 1 - max\{a_i : \omega \not\models \psi_i\} & \text{otherwise} \end{cases} \tag{1}$$

where $\models$ is propositional logic entailment.

Recently, several approaches(e.g., [2]) have been proposed to reason from partially pre-ordered belief bases using possibilistic logic. Partially pre-ordered belief bases offer much more flexibility in order (compared to totally pre-ordered bases) to efficiently represent incomplete knowledge and to avoid comparing unrelated pieces of information. Reasoning from partially pre-ordered belief bases also comes down to reason with a family of "compatible" possibilistic knowledge base. A compatible base represents a possible completion (i.e., by relating incomparable formulas) of a partially pre-ordered belief base.

Other extensions of possibilistic logic have been proposed, for instance, to cope with multiple-source information or to cope with temporal information.

3.2 Conditional Knowledge Bases

A possibility distribution allows to expresse what is the normal situation given any context. This can compactly be represented by a conditional knowledge base T which is a set of default rules having exceptions, simply called here conditional assertions. The possibilistic handling of conditional bases consists in viewing each conditional assertion $\alpha \to \beta$ as a constraint expressing that the situation where α and β is true has a greater possibility than the one where α and $\neg\beta$ is true. This statement is expressed in a possibility theory framework by $\Pi(\alpha \wedge \beta) > \Pi(\alpha \wedge \neg\beta)$. Hence, a conditional base T can be viewed as a restricting a family $\prod(T)$ of possibility distributions satisfying constraints induced by T. Considering the whole family $\prod(T)$ is equivalent to System P (P as Preferential) which is a set of postulates encoded by a reflexivity axiom and five inference rules namely Left Logical Equivalence, Right Weakening, Or, Cautious Monotony and Cut.

Selecting a possibility distribution from $\prod(T)$ using the minimum specificity principle is equivalent to System Z.

3.3 Possibilistic Graphical Models

Graphical models provide a simple representation of cause-and-effect relationships among key variables. In possibility theory there are two main ways to define the counterpart of Bayesian networks. This is due to the existence of two definitions of possibilistic conditioning: product-based and min-based conditioning. When we use the product form of conditioning, we get a possibilistic network close to the probabilistic one sharing the same features and having the same theoretical and practical results. However, this is not the case with min-based networks.

A possibilistic network over a set of variables V, denoted by ΠG_{min} is composed of:

- a *graphical component* that is a DAG (Directed Acyclic Graph) where nodes represent variables and edges encode the links between the variables. The parent set of a node X_i is denoted by U_i.
- a *numerical component* that quantifies different links. For every root node X_i ($U_i = \emptyset$.), uncertainty is represented by the a priori possibility degree $\Pi(x_i)$ of each instance $x_i \in D_{X_i}$, such that $max_{x_i} \Pi(x_i) = 1$. For the rest of the nodes ($U_i \neq \emptyset$.) uncertainty is represented by the conditional possibility degree $\Pi(x_i|u_i)$ of each instances $x_i \in D_{X_i}$ and $u_i \in D_{U_i}$. These conditional distributions satisfy: $max_{x_i} \Pi(x_i|u_i) = 1$, for any u_i.

The set of a priori and conditional possibility degrees in a possibilistic network induce a unique joint possibility distribution defined by:

$$\pi_{\otimes}(X_1, .., X_N) = \otimes_{i=1..N} \ \ \Pi(X_i \mid U_i), \tag{2}$$

where the $\otimes$ can be either a minimum or a product depending on the used conditioning.

Guaranteed possibilistic networks can also be easily defined by replacing in the above equation the possibility measure Π by the guaranteed possibility measure Δ and $\otimes$ by max. Guaranteed possibilistic networks may be very useful for representing preferences.

4 Concluding Discussions

This paper briefly presented different compact representation formats of a same possibility distribution. Each of this format has its merits from a knowledge representation point of view. For instance, guaranteed possibilistic knowledge bases are useful for representing agents' preferences while possibilistic networks are more appropriate for representing independence information, non-binary variables, causal information and interventions. Several equivalent transformations have been proposed from one representation format to another. These transformation procedures are important for merging heterogeneous multiple-source information. The transformation procedures from possibilistic networks to possibilistic logic are often achieved in a polynomial time while the converse is not tractable. The transformation procedure from a conditional knowledge base to a possibilistic logic base needs N (N is the size of the conditional knowledge base) calls to the satisfiability problem (SAT problem). There are also, from very particular situations, some linear transformations from conditional knowledge bases to partially ordered belief bases since both of them use the concept of compatible bases or distributions.

From the inference point of view, the methods usually proposed in graphical models differ from the one proposed in the possibilistic logic framework. The inference in possibilistic logic is basically based on propositional satisfiability task. In fact, the computational complexity of possibilistic logic is the one of classical logic multiplied by the logarithm of the number of distinct levels used in the base. In graphical model, the inference (called propagation) is more achieved using compilation approaches, by transforming initial graphs to possibilistic trees from which inference can be achieved in a linear. Recently, different CNF encodings (e.g., [1]) has been proposed for both possibilistic networks and possibilistic knowledge bases. These encodings, taking advantages of propositional knowledge compilation, will offer alternative inference tools which may be useful in many online applications such as access control in computer security.

References

1. Ayachi, R., Ben Amor, N., Benferhat, S., Haenni, R.: Compiling possibilistic networks: Alternative approaches to possibilistic inference. In: UAI 2010 (2010)
2. Benferhat, S., Lagrue, S., Papini, O.: Reasoning with partially ordered information in a possibilistic framework. Fuzzy Sets and Systems 144, 25–41 (2004)
3. Darwiche, A.: Modeling and Reasoning with Bayesian Networks. Cambridge University Press, New York (2009)
4. Dubois, D., Hajek, P., Prade, H.: Knowledge-driven versus data-driven logics. Journal of Logic, Language, and Information 9, 65–89 (2000)

Probabilistic Data: A Tiny Survey

Ander de Keijzer

University of Twente
MIRA - Institute for Biomedical Technology and Technical Medicine
PO.Box 217, 7500AE Enschede, The Netherlands

In this survey, we will visit existing projects and proposals for uncertain data, all supporting probabilistic handling of confidence scores.

1 Relational Data

Several models for uncertain data have been proposed over the years. Initial efforts all focused on relational data [3] and also currently efforts are being made in the relational setting [10,4,5,7,2]. With relational data models, two methods to associate confidences with data are commonly used. The first method associates these confidence scores with entire tuples (Type-1) [4], whereas the second method associates the confidence scores with individual attributes (Type-2) [3], Table 1 shows examples of uncertain relational data using the two types of uncertainty. The first table uses attribute level uncertainty, whereas the second table uses tuple level uncertainty. Omitted confidence scores in the tables indicate a score of 1. Both tables contain address book information on persons named John and Amy and both capture uncertainty about their room and phone number. Table 1(a) uses Type-2 uncertainty and captures the fact that John either occupies room 3035 (with probability 40%), or 3037 (with probability 60%), but certainly has phone number 1234. Amy, in this table, either occupies room 3122 (with probability 60%), or room 3120 (with probability 40%) and *independently* of the room has phone number 4321 (with probability 60%) or 5678 (with probability 40%). Table 1(b) uses Type-1 uncertainty and contains the same choices for room numbers and phone numbers for both persons, but in this case the room number and phone number for Amy are *dependent* on each other. If Amy occupies room 3122, then her phone number is 4321 analogously, if she occupies

Table 1. Attribute and Tuple level uncertainty

(a) Attribute level uncertainty

name	room	phone
John	3035 [.4]	1234
	3037 [.6]	
Amy	3122 [.6]	4321 [.6]
	3120 [.4]	5678 [.4]

(b) Tuple level uncertainty

name	room	phone	
John	3035	1234	.4
	3037		.6
Amy	3122	4321	.6
	3120	5678	.4

A. Deshpande and A. Hunter (Eds.): SUM 2010, LNAI 6379, pp. 7–10, 2010.

room 3120, then her room number is 5678. Observe that with tuple level uncertainty the expressiveness is larger, since dependencies between attributes can be expressed. This is impossible in the case of attribute level uncertainty. In the case of type-1 uncertainty it is, of course, possible to express the situation where both attributes are independent by enumerating all possibilities.

2 Semistructured Data

Semistructured data, and in particular XML has also been used as a data model for uncertain data [8,1]. As with the relational based models, there are two basic strategies. The first strategy is *event based* uncertainty, where choices for particular alternatives are based on specified events [1,8]. The occurence of an event validates a certain part of the tree and invalidates the alternatives. Using these events, possible worlds are created. Each combination for all events selects one of the possible worlds. In event based models, the events are independent of each other. The other strategy for semistructured models is the *choice point based* uncertainty [9]. With this strategy, at specific points in the tree a choice between the children has to be made. Choosing one child node, and as a result an entire subtree, invalidates the other child nodes. As with the event based strategy, possible worlds can be selected by choosing specific child nodes of choice points. The model presented in this thesis is based on the choice point strategy.

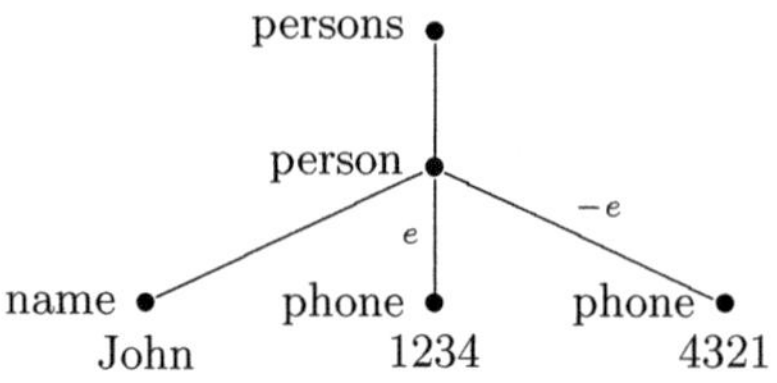

event	prob
e	0.3

(a) Fuzzy Tree representation

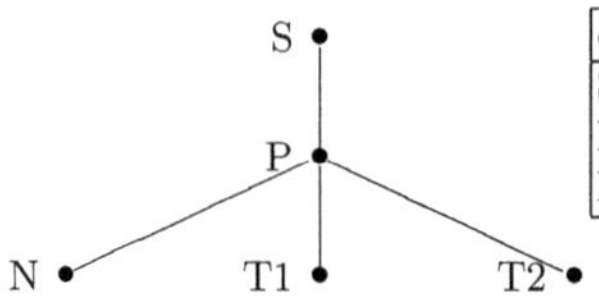

o	l	$lch(o,l)$
S	person	{P}
P	name	{N}
P	phone	{T1, T2}

o	l	$card(o,l)$
S	person	[1, 1]
P	name	[1, 1]
P	phone	[1, 1]

$c \in PC(P)$	$\wp(P)(c)$
{T1}	0.3
{T2}	0.7

(b) PXML representation

Fig. 1. Semistructured event based documents

Figure 2 contains two XML documents containing identical information. The first document (Figure 1(a)) is a Fuzzy Tree [1], whereas the second tree (Figure 1(b)) is a probabilistic XML document according to the PXML model of [8]. Both XML documents are event based. Both documents contain address book

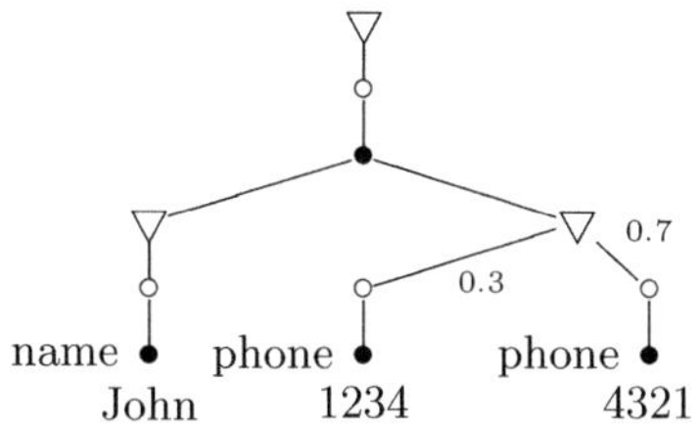

Fig. 2. Semistructured choice point based document

information for a person named John. For this person, only a phone number, either 1234, or 4321 is stored. Figure 1(a) contains one event, called e. The name in the document is independent of the event and therefore, the name element is always present. In other words, the name element is associated with event *true* and therefore always present. If e is true, then the phone number is 1234, otherwise phone number is 4321. The likelihood of e being true is 30%. The same information captured in a choice point based model is presented in Figure 2. At each choice point, indicated by $\bigtriangledown$ one of the child elements can be chosen. The probability of each of the child nodes is given at the edge to that child node.

In Figure 1(b) the PXML of [8], an event based model, is shown. In addition to the tree, the functions *lch*, *card* and $\wp$ are provided. Function *lch* shows the child nodes of any given node o in the tree and associates a label l with the edge. Here, node S *has* a person node P. Function *card* gives the cardinality interval for each of the nodes in the tree, based on the labels of the edges. In this case, all cardinalities are exactly one. For node P this means that there is exactly one name edge, as well as exactly one phone edge. The final function $\wp$ provides probabilities for nodes that are uncertain. In this case, only T1 and T2 are uncertain. Since the cardinality constraint dictates that T1 and T2 are mutually exclusive, the probabilities for T1 and T2 add up to 1.

Confidence Scores

With the probabilistic paradigm, all confidence scores are regarded as probabilities and are propagated as such. The result is, that at any given time, the total probability mass, or the sum of all probabilities, can't exceed 1. When calculating this probability mass, several things have to be taken into account, such as local vs. global probabilities and dependencies. Type-1 probabilities, for example, are global probabilities when no joins are used. Type-2 probabilities on the other hand, are local to the tuple and only when alternatives for all of the attributes in a tuple are chosen, can the global, Type-1 probability be calculated. Most data models and systems using probabilities assume independency among the tuples, but queries can create dependencies. If these dependencies are not taken into account, the calculated probability is incorrect. Systems using the probabilistic approach are MystiQ [5] and Trio [11].

Besides discrete probability distributions, continuous distributions are another possibility for storing uncertainty about data. Here, the distribution itself also represents the data value of an attribute. Continuous uncertainty is supported by the ORION system [7,6]. Consider, for example, a sensor application, that stores the data coming from a temperature sensor. Most producers of such sensors state that the sensor can report a temperature with a predefined uncertainty. We assume for this particular example that the actual temperature is normal distributed with the reported temperature as its mean and a static maximum deviation of $1°C$.

References

1. Abiteboul, S., Senellart, P.: Querying and updating probabilistic information in XML. In: Ioannidis, Y., Scholl, M.H., Schmidt, J.W., Matthes, F., Hatzopoulos, M., Böhm, K., Kemper, A., Grust, T., Böhm, C. (eds.) EDBT 2006. LNCS, vol. 3896, pp. 1059–1068. Springer, Heidelberg (2006)
2. Antova, L., Koch, C., Olteanu, D.: Maybms: Managing incomplete information with probabilistic world-set decompositions. In: ICDE, pp. 1479–1480. IEEE, Los Alamitos (2007)
3. Barbará, D., Garcia-Molina, H., Porter, D.: A probabilistic relational data model. In: Bancilhon, F., Tsichritzis, D.C., Thanos, C. (eds.) EDBT 1990. LNCS, vol. 416, pp. 60–74. Springer, Heidelberg (1990)
4. Benjelloun, O., Sarma, A.D., Hayworth, C., Widom, J.: An introduction to uldbs and the trio system. IEEE Data Eng. Bull. 29(1), 5–16 (2006)
5. Boulos, J., Dalvi, N.N., Mandhani, B., Mathur, S., Re, C., Suciu, D.: MYSTIQ: a system for finding more answers by using probabilities. In: Proceedings of SIGMOD, Baltimore, Maryland, USA, pp. 891–893 (2005)
6. Cheng, R., Prabhakar, S.: Sensors, Uncertainty Models and Probabilistic Queries. In: Encyclopedia of Database Technologies and Applications. Idea Group Publishing, USA (2005)
7. Cheng, R., Singh, S., Prabhakar, S.: U-DBMS: A database system for managing constantly-evolving data. In: Proceedings of VLDB, Trondheim, Norway, pp. 1271–1274 (2005)
8. Hung, E., Getoor, L., Subrahmanian, V.S.: PXML: A probabilistic semistructured data model and algebra. In: Proceedings of ICDE, Bangalore, India, pp. 467–478 (2003)
9. van Keulen, M., de Keijzer, A., Alink, W.: A probabilistic xml approach to data integration. In: Proceedings of ICDE, Tokyo, Japan, pp. 459–470 (2005)
10. Lakshmanan, L.V.S., Leone, N., Ross, R., Subrahmanian, V.S.: ProbView: a flexible probabilistic database system. ACM Transactions on Database Systems 22(3), 419–469 (1997)
11. Widom, J.: Trio: A system for integrated management of data, accuracy, and lineage. In: CIDR, pp. 262–276 (2005)

The Role of Epistemic Uncertainty in Risk Analysis

Didier Dubois

Institut de Recherche en Informatique de Toulouse,
118 Route de Narbonne
31062 Toulouse Cedex 9, France
dubois@irit.fr

The notion of uncertainty has been a controversial issue for a long time. In particular the prominence of probability theory in the scientific arena has blurred some distinctions that were present from its inception, namely between uncertainty due to the variability of physical phenomena, and uncertainty due to a lack of information. The Bayesian school claims that whatever its origin, uncertainty can be modeled by single probability distributions [18]. This assumption has been questioned in the last thirty years or so. Indeed the use of unique distributions so as to account for incomplete information leads to paradoxical uses of probability theory. One well-known major flaw is that the unique distribution representation is scale-sensitive: a probability distribution supposedly representing the absence of information on one scale may be changed into an informative distribution on another scale. Besides, empirical findings by decision scientists indicate that decision-makers do not follow the theory of expected utility with respect to a unique subjective probability distribution when information is missing: they rather make decisions as if they were using a set of distributions as potential priors, selecting an appropriate distribution likely to protect them against the lack of information each time they have to compare two acts [14]. In fact, there is not a one-to-one correspondence between epistemic states and probability distributions, that is, several individuals having distinct epistemic states may be led to propose the same betting rates, whose meaning is then ambiguous (viz. uniform distributions may express known full-fledged randomness or total ignorance).

Risk can be defined as the combination of the likelihood of occurrence of an undesirable event and the severity of the damage that can be caused by this event. In the area of risk analysis, especially concerning environmental matters, it is crucial to account for variability and incomplete information separately, even if conjointly, in uncertainty propagation techniques [12,16]. Indeed, it should be clear at the decision level what is the part of uncertainty due to partial ignorance (hence reducible by collecting more information) from uncertainty due to variability (to be faced with concrete actions). New uncertainty theories have emerged [16,11], which have the potential to meet this challenge, and where the unique distribution is replaced by a convex set of probabilities [21], this set being all the larger as less information is present. Generally, computing with probability sets is a very burdensome task. Special cases of such representations, which enable efficient calculation methods, are based on random sets [19] and possibility theory [10,5] (using fuzzy sets of possible values [22]).

A. Deshpande and A. Hunter (Eds.): SUM 2010, LNAI 6379, pp. 11–15, 2010.

The distinction between epistemic and aleatory uncertainties leads to a significant alteration of the traditional risk analysis methodology relying on the Bayesian credo that any state of information can be rendered by means of a unique probability distribution. Any risk analysis methodology includes the following steps [6,8]:

1. Information collection on input parameters and their representation.
2. Propagation of uncertainty through a mathematical model.
3. Extraction of useful information.
4. Decision.

In step 1, one adopts a *faithfulness* principle : choosing the type of representation in agreement with the amount of available information, so as to reflect as much as possible information gaps. Simple representations (possibility distributions, generalized p-boxes) naturally capture expert interval information along with confidence levels, quantiles, mean, median, mode, etc.

- If variability prevails and enough statistical information is available, unique probability distributions can be used.
- If there is incomplete information on some value whether constant or not: intervals, possibility distributions (fuzzy intervals, understood as nested intervals with various confidence levels [9]) can be used. In the case of ill-known variability, probabilistic inequalities provide the right way to account for known properties of the variability distribution (symmetry, mode, mean and variance especially) [7]. Median and quantile information rather lead to random set representations.
- For a parameterized model with ill-known parameters, a p-box (a pair of cumulative distribution functions one stochastically dominating the other) is the most natural representation.

An example of elicitation procedure designed to query an expert, collect the available information, and find the appropriate representation is in [15,8]. Appropriate independence or dependence assumptions also have to be made. Interestingly the enlarged uncertainty framework enables the distinction between epistemic (in)dependence (between sources of information) and stochastic independence (between physical random variables) to be made. Computation schemes have been devised [2] to account for situations such as dependent sources of incompleteness and independence between variables, or independent sources *and* variables, or on the contrary, no assumption of independence (the latter is computationally more difficult).

The propagation step is often carried out by means of Monte-Carlo style simulations. In the presence of epistemic uncertainty, the joint use of Monte-Carlo methods and interval analysis tools (possibly fuzzy intervals [9]) is needed [13]. It presupposes that all ill-known parameters can be represented by means of random intervals, an umbrella common to most simple representations of epistemic uncertainty (see [16] and the whole issue of the journal). Interestingly, such simple epistemic representations like possibility distributions and p-boxes are not preserved through propagation, which results in a discrete random set or even a fuzzy random variable [3].

This kind of result is hard to interpret, more difficult than the outcome of a traditional Bayesian approach where a single distribution is almost always obtained. Hence step 3 cannot be reduced to the extraction of a mean and a variance. More parameters are needed to describe the results [3]. For instance one can compute the average level of imprecision of the result, and the variance of this imprecision, or on the contrary the degree of apparent variability, or an interval-valued variance that contains the possible values of the variance one could have obtained if the lack of information had be removed. Moreover, one can extract p-boxes and possibility distributions from the outputs. Contrary to the Bayesian case, these are just partial summaries of the available information. The choice between presenting a p-box and a possibility distribution depends on the question of interest motivating the study. If the issue is to check that the output does not violate a safety threshold, then a p-box is the right answer. Precise results with high variability are encoded by a pair of close distributions with high variance, while a significant lack of knowledge leads to a pair of remote distributions. If the question of interest is whether the output remains close to to a prescribed value or stays within prescribed bounds, then a pair of possibility distributions corresponding to nested intervals with upper and lower probability bounds is a more appropriate response.

In any case, the enlarged uncertainty setting for risk analysis leads to computing upper and lower probability of occurrence of a certain risky event. This kind of result poses a difficulty at the decision step because it puts the decision-maker in front of his or her lack of knowledge, if any. That is, the aim of the risk analysis process is no longer just to inform the decision-maker whether there is actual risk or not in a given phenomenon. If it can be done, so much the better. However the above risk analysis methodology also informs the decision-maker about the amount of available knowledge. It is then up to the decision-maker to consider whether the information is sufficient to decide between acting so as to circumvent the risk or not acting, or on the contrary, to collect additional data likely to improve the informativeness of the propagation step. Aven [1] argues that imprecise probability approaches only deal with aleatory uncertainty, interpreting intervals as mere information gaps, hence provide no help to the decision-maker. However, as intervals represent epistemic uncertainty, one may argue that they are as subjective representations as probability, since they are attached to an observer [6]. Moreover the whole approach to imprecise probabilities due to Walley [21] is an extension of subjective probability.

A consensus regarding the best way of making decisions in the imprecise probability setting does not exist yet. A number of new decision criteria have been proposed, both in economics (see [4] for a survey) and in connection with Walley's imprecise probability theory, following pioneering works by Isaac Levi (see Troffaes [20]). There are basically two schools of thought:

- Comparing set-valued utility estimations under more or less strict conditions. These decision rules usually do not result in a total ordering of decisions, and some scholars may consider that the problem is not fully solved then. Nevertheless they provide rationality constraints on the final decision.

- Comparing point-valued estimations through the selection of a "reasonable" utility value within the computed bounds. This approach leads to clear-cut best decisions but the responsibility of the choice of the representative value then relies on the decision-maker.

The second approach is sometimes based on a generalization of the (pessimistic) maximin criterion of Wald proposed by Gilboa and Schmeidler [14]. It comes down to deciding on the basis of lower probability bounds induced by a family of priors, which may sound overpessimistic. An alternative is to get inspiration from Hurwicz criterion involving the degree of pessimism of the decision-maker[17,8] in order to select a more reasonable cumulative distribution inside a p-box, for supporting the decision step.

In conclusion, the distinction between epistemic and aleatory uncertainty looks essential in risk analysis, so as to provide more relevant decision support. New uncertainty theories basically complement the usual probabilistic approach by their capability to lay bare this distinction. There is some hope that new uncertainty theories eventually get better recognition. At the turn of this century, Lindley [18] insisted again that

> measurements of uncertainty must obey the rules of the probability calculus. Other rules, like those of fuzzy logic or possibility theory, dependent on maxima and minima, rather than sums and products, are out.

He also recalls that first principles of comparative probability

> ... lead to probability being the only satisfactory expression of uncertainty.

However he pursues as follows :

> The last sentence is not strictly true. A fine critique is Walley, who went on to construct a system with a pair of numbers... instead of the single probability. The result is a more complicated system. My position is that the complication seems unnecessary.

Recent papers in imprecise probability clearly show that on the contrary, possibility theory is part of the imprecise probability landscape. Moreover, the use of simplified representations like p-boxes, possibility distributions and the like in imprecise probability methods for risk analysis makes uncertainty management under incomplete probabilistic information more and more scalable.

References

1. Aven, T.: On the need for restricting the probabilistic analysis in risk assessments to variability. Risk Analysis 30(3), 354–360 (2010)
2. Baudrit, C., Dubois, D.: Comparing methods for joint objective and subjective uncertainty propagation with an example in a risk assessment. In: Proc. Fourth International Symposium on Imprecise Probabilities and Their Application (ISIPTA 2005), Pittsburg, USA, pp. 31–40 (2005)

3. Baudrit, C., Guyonnet, D., Dubois, D.: Joint propagation and exploitation of probabilistic and possibilistic information in risk assessment. IEEE Trans. Fuzzy Systems 14, 593–608 (2006)
4. Chateauneuf, A., Cohen, M.: Cardinal extensions of the EU model based on Choquet integral. In: Bouyssou, D., Dubois, D., Pirlot, M., Prade, H. (eds.) Decision-Making Process- Concepts and Methods, ch. 3, pp. 401–433. ISTE & Wiley, London (2009)
5. Dubois, D.: Possibility theory and statistical reasoning. Computational Statistics & Data Analysis 51, 47–69 (2006)
6. Dubois, D.: Representation, Propagation, and Decision Issues in Risk Analysis Under Incomplete Probabilistic Information. Risk Analysis 30(3), 361–368 (2010)
7. Dubois, D., Foulloy, L., Mauris, G., Prade, H.: Probability-possibility transformations, triangular fuzzy sets, and probabilistic inequalities. Reliable Computing 10, 273–297 (2004)
8. Dubois, D., Guyonnet, D.: Risk-informed decision-making in the presence of epistemic uncertainty. Int. J. General Systems (to appear, 2010)
9. Dubois, D., Kerre, E., Mesiar, R., Prade, H.: Fuzzy interval analysis. In: Dubois, D., Prade, H. (eds.) The Handbook of Fuzzy Sets. Fundamentals of Fuzzy Sets, vol. I, pp. 483–581. Kluwer Academic Publishers, Dordrecht (2000)
10. Dubois, D., Prade, H.: Possibility Theory: An Approach to Computerized Processing of Uncertainty. Plenum Press, New York (1988)
11. Dubois, D., Prade, H.: Formal representations of uncertainty. In: Bouyssou, D., Dubois, D., Pirlot, M., Prade, H. (eds.) Decision-Making Process- Concepts and Methods, ch. 3, pp. 85–156. ISTE & Wiley, London (2009)
12. Ferson, S., Ginzburg, L.R.: Different methods are needed to propagate ignorance and variability. Reliability Engineering and System Safety 54, 133–144 (1996)
13. Ferson, S., Tucker, T.W.: Sensitivity analysis using probability bounding. Reliability Engineering and System Safety 91, 1435–1442 (2006)
14. Gilboa, I., Schmeidler, D.: Maxmin expected utility with a non-unique prior. Journal of Mathematical Economics 18, 141–153 (1989)
15. Guyonnet, D., Bellenfant, G., Bouc, O.: Soft methods for treating uncertainties: Applications in the field of environmental risks. In: Dubois, D., Lubiano, M.A., Prade, H., Gil, M.A., Grzegorzewski, P., Hryniewicz, O. (eds.) SMPS. Advances in Soft Computing, vol. 48, pp. 16–26. Springer, Heidelberg (2008)
16. Helton, J.C., Oberkampf, W.L.: Alternative representations of epistemic uncertainty. Reliability Engineering and System Safety 85, 1–10 (2004)
17. Jaffray, J.Y.: Linear utility theory for belief functions. Operations Research Letters 8, 107–112 (1989)
18. Lindley, D.V.: The philosophy of statistics. The Statistician 49 (Part 3), 293–337 (2000)
19. Shafer, G.: A mathematical theory of evidence. Princeton University Press, Princeton (1976)
20. Troffaes, M.: Decision making under uncertainty using imprecise probabilities. Int. J. Approx. Reasoning 45(1), 17–29 (2007)
21. Walley, P.: Statistical reasoning with imprecise probabilities. Chapman and Hall, Boca Raton (1991)
22. Zadeh, L.A.: Fuzzy sets as a basis for a theory of possibility. Fuzzy Sets and Systems 1, 3–28 (1978)

Uncertainty in Clustering and Classification

Eyke Hüllermeier

Department of Mathematics and Computer Science
University of Marburg, Germany
eyke@mathematik.uni-marburg.de

Clustering and classification are among the most important problem tasks in the realm of data analysis, data mining and machine learning. In fact, while clustering can be seen as the most popular representative of *unsupervised learning*, classification (together with regression) is arguably the most frequently considered task in *supervised learning*. Even though the literature on clustering and classification abounds, the interest in these topics seems to be unwaning, both from a research and application point of view.

Learning from data, whether supervised or unsupervised, is inseparably connected with uncertainty. This is largely due to the fact that learning, understood as generalizing beyond the observed data, is necessarily based on a process of *induction*. Inductive inference replaces specific observations by general models of the data generating process, but these models are always hypothetical and, therefore, afflicted with uncertainty. Indeed, observed data can generally be explained by more than one candidate theory, which means that one can never be sure of the truth of a particular model (and the predictions it implies).

Apart from the uncertainty inherent in inductive inference, additional sources of uncertainty exist. Indeed, essentially all "ingredients" of the inference process can be afflicted with uncertainty, including

- the observed data (e.g., points in $\mathbb{R}^d$ in clustering, or feature vectors associated with a discrete class label in classification);
- background knowledge including underlying model assumptions (e.g., the assumption that the data is generated by a mixture of Gaussians, the assumption that two classes can be separated by a linear decision boundary, or the prior distribution on the model space in Bayesian inference);
- the induction principle (e.g., the maximum likelihood principle or structural risk minimization);
- a learning algorithm realizing this induction principle (e.g., the Expectation-Maximization algorithm or support vector machines for classification).

Traditionally, all sorts of uncertainty in clustering and classification, like in data analysis in general, have been modeled in a probabilistic way, and indeed, probability theory has always been considered as the ultimate tool for uncertainty handling in fields like statistics and machine learning. The situation has started to change in recent years, in which "data mining" at large has attracted a lot of interest in other research communities as well. Being less dogmatic, some of these communities have also looked at alternative uncertainty formalisms, including generalizations of classical probability theory. Indeed, without questioning the importance of

A. Deshpande and A. Hunter (Eds.): SUM 2010, LNAI 6379, pp. 16–19, 2010.

probability, one may well argue that not all types of uncertainty relevant to learning from data are of a probabilistic nature and, therefore, that other uncertainty formalisms can complement probability theory in a reasonable way.

Uncertainty Modeling in Clustering and Classification

As mentioned earlier, uncertainty may in principle concern all aspects ("ingredients") of a learning process and, correspondingly, alternative uncertainty formalisms are potentially useful for all inputs and outputs of this process. For example, a possibilistic variant of the heuristic "Occam's razor" induction principle, i.e., a formalization of this principle in the framework of possibility theory, has been proposed in [1]. In the following, however, we shall focus on the modeling of the observed data and the representation of predictions (directly related to the representation of models), since these aspects have received the most attention so far.

Uncertain Data and Observations

Alternative uncertainty formalisms can be used for different reasons, in particular (i) to *complement* probabilistic representations in order to capture non-probabilistic types of uncertainty, such as imprecision, vagueness or gradedness; (ii) to *generalize* probabilistic representations, thereby allowing to model partial ignorance in a more adequate way. What has been studied quite intensively, for example, is the use of Dempster-Shafer's theory for evidence for modeling uncertain data. Indeed, a number of clustering and classification algorithms have been extended correspondingly, including, e.g., fuzzy c-means clustering [2] and nearest-neighbor estimation [3,4]. Likewise, representations on the basis of possibility theory have been used [5,6,7].

An increasing number of publications is devoted to the learning of models from "fuzzy data", where observations are modeled in terms of fuzzy subsets of the original output space [8,9]. Again, this idea requires the extension of corresponding learning algorithms. Unfortunately, this is often done without clarifying the actual meaning of a fuzzy observation and the interpretation of membership functions, even though different interpretations obviously require different extensions. Apart from fuzzy sets, other types of uncertainty representations have been used for modeling training data, both in clustering and classification as well as extended settings such as multi-label classification [10].

Uncertainty can also be represented in a purely qualitative way. This is a key idea in the emerging field of *preference learning* [11]. For example, consider the case of ordinal classification, where a total order is defined on the set of classes (e.g., hotel categories ranging from 1 to 5 stars). Even though the true class membership might be unknown for training instances (hotels) A and B, it may still be known that the class of A is higher than the class of B. This can be seen as a kind of indirect supervision, or indirect information that can be exploited by a learning algorithm in order to induce a ranking or classification function [12].

Uncertain Predictions

As mentioned above, a model that has been learned from data is always afflicted with uncertainty, and so are the predictions produced by such a model. The question of how to characterize this uncertainty in a proper way has received increasing attention in machine learning in general. Topics that have recently been addressed in this regard include calibration methods for turning classifier scores into proper probability estimates [13]; the learning of reliable classification models that allow for abstaining in cases of uncertainty [14] and guarantee a certain level of confidence when making a prediction [15], the prediction of a set of candidate classes instead a single class [16,17,18]; qualitative representations of uncertainty, e.g., by predicting a ranking of classes ranging from the most likely to the least likely candidate [12].

Uncertainty can already be addressed on the level of the model itself, not only on the prediction level. Generally, this means considering a set of candidate models instead of inducing a single model, for example, making inference on the basis of the full posterior distribution over the hypothesis space instead of the MAP prediction in Bayesian analysis. This approach will normally come with a high computational complexity and can often be implemented only approximately, for example using sampling techniques [19]. In the case of clustering, the idea to produce multiple (or even all) reasonable clustering structures instead of only a single one has recently been addressed in [20].

In [21,22], the authors propose to distinguish two types of uncertainty in classification, called *conflict* and *ignorance*. Roughly speaking, given a query instance to be classified, a conflict occurs if the observed data, in conjunction with the underlying model assumptions, provides evidence in favor of more than one of the classes, whereas ignorance means that none of the classes is sufficiently supported. These two types of uncertainty are arguably difficult to distinguish in probability theory. Instead, the authors develop a formalization within the framework of fuzzy preference relations.

Scalability

Modeling and handling uncertainty in a proper way normally comes with an increased computational complexity—storing a single value for an attribute and computing with this value is cheaper than doing the same with a probability distribution over the whole domain, which in turn is cheaper than using a belief function. Since data sets to be analyzed tend to increase in size and computationally intense, resource-bounded frameworks such as mining data streams [23] become increasingly relevant, the aspects of computational complexity and scalability should always be kept in mind when developing methods for uncertainty handling in data analysis. Indeed, the aspect of scalability in general and its interplay with uncertainty in particular are important topics on ongoing research [24].

References

1. Hüllermeier, E.: Possibilistic induction in decision tree learning. In: Elomaa, T., Mannila, H., Toivonen, H. (eds.) ECML 2002. LNCS (LNAI), vol. 2430, pp. 173–184. Springer, Heidelberg (2002)
2. Masson, M., Denoeux, T.: ECM: An evidential version of the fuzzy c-means algorithm. Pattern Recognition 41(4), 1384–1397 (2008)
3. Denoeux, T.: A k-nearest neighbor classification rule based on Dempster-Shafer theory. In: Yager, R., Liu, L. (eds.) Classic Works of the Dempster-Shafer Theory of Belief Functions. Springer, Heidelberg (2008)
4. Younes, Z., Abdallah, F., Denoeux, T.: An evidence-theoretic k-nearest neighbor rule for multi-label classification. In: Godo, L., Pugliese, A. (eds.) SUM 2009. LNCS, vol. 5785, pp. 297–308. Springer, Heidelberg (2009)
5. Hüllermeier, E.: Possibilistic instance-based learning. Artificial Intelligence 148(1-2), 335–383 (2003)
6. Haouari, B., Amor, A.B., Elouedi, Z., Mellouli, K.: Naïve possibilistic network classifiers. Fuzzy Sets and Systems 160(22), 3224–3238 (2009)
7. Jenhani, I., Amor, N.B., Elouedi, Z.: Decision trees as possibilistic classifiers. Int. J. Approx. Reasoning 48(3), 784–807 (2008)
8. Diamond, P.: Fuzzy least squares. Information Sciences 46(3), 141–157 (1988)
9. Yang, M., Ko, C.: On a class of fuzzy c-numbers clustering procedures for fuzzy data. Fuzzy Sets and Systems 84(1), 49–60 (1996)
10. Cheng, W., Dembczynski, K., Hüllermeier, E.: Graded multi-label classification: The ordinal case. In: Proc. ICML 2010, Haifa, Israel (2010)
11. Fürnkranz, J., Hüllermeier, E.: Preference Learning. Springer, Heidelberg (2010)
12. Fürnkranz, J., Hüllermeier, E., Vanderlooy, S.: Binary decomposition methods for multipartite ranking. In: Proc. ECML/PKDD 2009, Bled, Slovenia (2009)
13. Zadrozny, B., Elkan, C.: Transforming classifier scores into accurate multiclass probability estimates. In: Proc. KDD 2002, pp. 694–699 (2002)
14. Yuan, M., Wegkamp, M.: Classification methods with reject option based on convex risk minimization. J. Machine Learning Research 11, 111–130 (2010)
15. Campi, M.: Classification with guaranteed probability of error. Machine Learning 80(1) (2010)
16. Hüllermeier, E.: Credible case-based inference using similarity profiles. IEEE Transactions on Knowledge and Data Engineering 19(5), 847–858 (2007)
17. Corani, G., Zaffalon, M.: Learning reliable classifiers from small or incomplete data sets: The naive credal classifier 2. J. Machine Learning Research 9, 581–621 (2008)
18. del Coz, J., Diez, J., Bahamonde, A.: Learning nondeterministic classifiers. J. Machine Learning Research 10, 2273–2293 (2009)
19. Minka, T.: A family of algorithms for approximate Bayesian inference. PhD thesis, MIT (2001)
20. Niu, D., Dy, J., Jordan, M.: Multiple non-redundant spectral clustering views. In: Proc. ICML 2010, Haifa, Israel (2010)
21. Hüllermeier, E., Brinker, K.: Learning valued preference structures for solving classification problems. Fuzzy Sets and Systems 159(18), 2337–2352 (2008)
22. Hühn, J., Hüllermeier, E.: FR3: A fuzzy rule learner for inducing reliable classifiers. IEEE Transactions on Fuzzy Systems 17(1), 138–149 (2009)
23. Gaber, M., Zaslavsky, A., Krishnaswamy, S.: ACM SIGMOD Record 34(2), 18–26 (2005)
24. Laurent, A., Lesot, M. (eds.): Scalable Fuzzy Algorithms for Data Management and Analysis: Methods and Design. IGI Global, Hershey (2009)

Information Fusion

Odile Papini

LSIS-CNRS,
Université de Méditerranée, ESIL,
163 av. de Luminy - 13288 Marseille Cedex 09. France
papini@esil.univmed.fr

Merging information coming from different sources is an important issue in various domains of computer science like knowledge representation for artificial intelligence, decision making or databases. The aim of fusion is to obtain a global point of view, exploiting the complementarity between sources, solving different existing conflicts, reducing the possible redundancies.

When focusing on merging, one has to pay attention to the nature of the targeted information to be merged: beliefs, generic knowledge, goals or preferences, laws or regulations since the kind of fusion deeply depends on the nature of information provided by the sources [1]. Beliefs are factual information, they represent agent's perceptions or observations and can be false. A belief base is an agent's description of the world according to its perceptions and the fusion of belief bases expresses the beliefs of a group of agents on the basis of the invidual beliefs. On constrast, generic knowledge is unquestionable information. When merging generic knowledge coming from different sources, the only acceptable fusion method is the conjunction of the information provided by the sources and this conjunction has to be consistent. Moreover, beliefs and generic knowledge represent the world as it is assumed to be, however goals or preferences represent the world as it should evolve for the agent and merging goal bases or preference bases amounts to find which goals a group of agents should converge to in order to best satisfy the group. This is related to preference aggregation. Regulations, laws, specifications describe the world as it should be ideally. The aim of this kind of fusion is to provide a consistent base of regulations from several initial bases that could conflict.

Among the various approaches of multi-sources information merging, symbolic approaches gave rise to increasing interest within the artificial intelligence community [2,3,4,5,6] the last decade. Belief bases merging has received much attention and most of the approaches have been defined within the framework of classical logic, more often propositional.

Postulates characterizing the rational behavior of fusion operations have been proposed [7] which capture the following basic assumptions. The sources are mutually independant and no implicit link between the information from the different sources are assumed. All sources have the same level of importance and provide consistent belief bases. All information from a source have the same level of reliability or priority. Due to the non-constructive nature of these postulates, the core problem is the definition of fusion operations. Several merging operations have been proposed that can be divided into two families. The semantic

A. Deshpande and A. Hunter (Eds.): SUM 2010, LNAI 6379, pp. 20–23, 2010.

(or model-based) ones which select interpretations that are the "closest" to the original belief bases [7,8,9,10,11,12,13] and the syntactic (or formula-based) ones which select some formulas from the initial bases [14,15,16,17,18].

In some situations, the belief bases are not flat and the beliefs are stratified or equipped with priority levels, in other cases the belief bases are flat but the sources are not equally reliable and there exists a preference relation between sources. In such cases, prioritized merging consists of combining belief bases taking into account the stratification of the belief bases or the preference relation [19]. Prioritized merging has been studied within the framework of propositional logic [10,15,20] as well as within the possibilistic logic one [18,21]. The links between iterated revision and prioritized merging has been discussed and rational postulates for prioritized merging have been proposed in [10], [22].

The worst case computational complexity of semantic approaches generally is at the second level of the polynomial hierarchy and few implementations have been proposed. Model-based merging operations reformulated in terms of dilation have been proposed [23] with an implementation stemming from Binary Decision Diagrams (BDDs). The syntactic Removed Set Fusion (RSF) has been implemented thanks to Answer Set Programming (ASP) [24]. Both approaches conducted experimental studies that are difficult to compare.

Among the numerous open issues for information fusion some promising research directions, without exhaustivity, are listed below.

Concerning belief bases fusion within a classical logic framework,

- tractability: although some complexity results are known like model-based approaches stemming from the Hamming distance [12], there is no systematic study on the theoretical complexity of fusion operations and an interesting direction could be to provide restrictive assumptions in order to decrease the complexity and to find tractable classes of fusion problems.
  Moreover, concerning the practical complexity, there is a lack of benchmarks for testing and comparing fusion approaches and a challenge could be to construct such a set of of benchmarks in the same spirit of the ones used to implement the SAT problem.

- extension to partially preordered information: when belief bases are not flat they are equipped with total preorders and the various sources are assumed commensurable. However, in some applications, an agent has not always a total preorder between situations at his disposal, but is only able to define a partial preorder between them, particularly in case of partial ignorance and incomplete information. Moreover, in some cases the sources are incommensurable. In such cases, merging of partially preorderd belief bases has to be investigated. Recently, the fusion of incommensurable sources has been addressed and several operations have been proposed [25].

From now on, most of the approaches focus on belief bases merging, or on preference merging however mixing information of different nature could be another interesting research direction that could probably lead to extending fusion to other logical frameworks:

- description logics: in the context of Semantic Web, ontologies are designed to represent generic knowledge, they are written in expressive description logics. Investigating to which extent logical approaches of fusion could be used seems to be a promising issue.

- non-monotonic frameworks: defining fusion operations for non-monotonic frameworks also seems a promising research direction. A preliminary approach has been proposed [26] where belief bases are represented in logic programming with Answer Sets Semantics.

References

1. Grégoire, E., Konieczny, S.: Logic-based approaches to information fusion. Inf. Fusion 7(1), 4–18 (2006)
2. Baral, C., Kraus, S., Minker, J., Subrahmanian, V.S.: Combining knowledge bases consisting of first order theories. In: Raś, Z.W., Zemankova, M. (eds.) ISMIS 1991. LNCS, vol. 542, pp. 92–101. Springer, Heidelberg (1991)
3. Revesz, P.Z.: On the semantics of theory change: arbitration between old and new information. In: 12th ACM SIGACT-SGMIT-SIGART symposium on Principes of Databases, pp. 71–92 (1993)
4. Lin, J.: Integration of weighted knowledge bases. Artif. Intell. 83, 363–378 (1996)
5. Revesz, P.Z.: On the semantics of arbitration. Journ. of Alg. and Comp. 7(2), 133–160 (1997)
6. Cholvy, L.: Reasoning about merging information. Handbook of DRUMS 3, 233–263 (1998)
7. Konieczny, S., Pérez, R.P.: On the logic of merging. In: Proc. of KR 1998, pp. 488–498 (1998)
8. Lafage, C., Lang, J.: Logical representation of preferences for group decision making. In: Proc. of KR 2000, pp. 457–468. Morgan Kaufmann, San Francisco (2000)
9. Konieczny, S.: On the difference between merging knowledge bases and combining them. In: Proc of KR 2000, pp. 135–144. Morgan Kaufmann, San Francisco (2000)
10. Delgrande, J., Dubois, D., Lang, J.: Iterated revision as prioritized merging. In: Proc. of KR 2006, pp. 210–220 (2006)
11. Fagin, R., Kuper, G.M., Ullman, J.D., Vardi, M.Y.: Updating logical databases. In: Advances in Computing Research, pp. 1–18 (1986)
12. Konieczny, S., Lang, J., Marquis, P.: Distance-based merging: A general framework and some complexity results. In: Proc. of KR 2002, pp. 97–108 (2002)
13. Bloch, I., Lang, J.: Towards mathematical morpho-logics. In: Technologies for constructing intelligent systems: tools, pp. 367–380. Springer, Heidelberg (2002)
14. Meyer, T., Ghose, A., Chopra, S.: Syntactic representations of semantic merging operations. In: Proc. of Workshop IDK, IJCAI 2001, pp. 36–42 (2001)
15. Yue, A., Liu, W., Hunter, A.: Approaches to constructing a stratified merged knowledge base. In: Mellouli, K. (ed.) ECSQARU 2007. LNCS (LNAI), vol. 4724, pp. 54–65. Springer, Heidelberg (2007)

16. Hue, J., Papini, O., Würbel, E.: Syntactic propositional belief bases fusion with removed sets. In: Mellouli, K. (ed.) ECSQARU 2007. LNCS (LNAI), vol. 4724, pp. 66–77. Springer, Heidelberg (2007)
17. Dubois, D., Lang, J., Prade, H.: Possibilistic Logic. In: Handbook of Logic in Artificial Intelligence and Logic Programming, vol. 3, pp. 439–513 (1994)
18. Benferhat, S., Dubois, D., Kaci, S., Prade, H.: Possibilistic Merging and Distance-based Fusion of Propositional Information. AMAI 34(1-3), 217–252 (2002)
19. Benferhat, S., Dubois, D., Prade, H.: Reasoning in inconsistent stratified knowledge bases. In: Proc. of ISMVL 1996, pp. 184–189 (1996)
20. Hunter, A., Liu, W.: Knowledge base stratification and merging based on degree of support. In: Sossai, C., Chemello, G. (eds.) ECSQARU 2009. LNCS, vol. 5590, pp. 383–395. Springer, Heidelberg (2009)
21. Benferhat, S., Kaci, S.: Logical representation and fusion of prioritized information based on guaranteed possibility measures: Application to the distance-based merging of classical bases. Artif. Intell. 148(1-2), 291–333 (2003)
22. Hue, J., Papini, O., Würbel, E.: Implementing prioritized merging with asp. In: Hüllermeier, E., Kruse, R., Hoffmann, F. (eds.) Computational Intelligence for Knowledge-Based Systems Design. LNCS (LNAI), vol. 6178, Springer, Heidelberg (2010)
23. Gorogiannis, N., Hunter, A.: Implementing semantic merging operators using binary decision diagrams. Int. J. Approx. Reasoning 49(1), 234–251 (2008)
24. Hue, J., Papini, O., Würbel, E.: Removed sets fusion: Performing off the shelf. In: Proc. of ECAI 2008. Frontiers in Art. Intel. and Appli., pp. 94–98. IOS Press, Amsterdam (2008)
25. Benferhat, S., Lagrue, S., Rossit, J.: An analysis of sum-based incommensurable belief base merging. In: Godo, L., Pugliese, A. (eds.) SUM 2009. LNCS, vol. 5785, pp. 55–67. Springer, Heidelberg (2009)
26. Hue, J., Papini, O., Würbel, E.: Merging belief bases represented by logic programs. In: Sossai, C., Chemello, G. (eds.) ECSQARU 2009. LNCS, vol. 5590, pp. 371–382. Springer, Heidelberg (2009)
27. Yue, A., Liu, W., Hunter, A.: Approaches to constructing a stratified merged knowledge base. In: Mellouli, K. (ed.) ECSQARU 2007. LNCS (LNAI), vol. 4724, pp. 54–65. Springer, Heidelberg (2007)
28. Bloch, I., Lang, J.: Towards mathematical morpho-logics. In: Technologies for Constructing Intelligent Systems: Tools, pp. 367–380. Springer, Heidelberg (2002)
29. Konieczny, S., Pérez, R.P.: Merging with integrity constraints. In: Hunter, A., Parsons, S. (eds.) ECSQARU 1999. LNCS (LNAI), vol. 1638, pp. 233–244. Springer, Heidelberg (1999)
30. Lin, J., Mendelzon, A.O.: Merging databases under constraints. Int. Journ. of Coop. Inf. Sys. 7(1), 55–76 (1998)
31. Qi, G.: A model-based approach for merging prioritized knowledge bases in possibilistic logic. In: Proc. of AAAI 2007, pp. 471–476 (2007)
32. Benferhat, S., Cayrol, C., Dubois, D., Lang, J., Prade, H.: Inconsistency management and prioritized syntax-based entailment. In: Proc. of IJCAI 1993, pp. 640–647 (1993)
33. Seinturier, J., Papini, O., Drap, P.: A reversible framework bases merging. In: Hunter, T., Dix., J. (eds.) Proc. of NMR 2006, pp. 490–496 (2006)
34. Benferhat, S., Kaci, S., Berre, D.L., Williams, M.A.: Weakening conflicting information for iterated revision and knowledge integration. Artif. Intell. 153(1-2), 339–371 (2004)

Use of the Domination Property for Interval Valued Digital Signal Processing

Olivier Strauss

LIRMM Université Montpellier II, 161 rue Ada, 34392 Montpellier cedex 5, France
Olivier.Strauss@lirmm.fr

Abstract. Imprecise probability framework is usually dedicated to decision processes. In recent work, we have shown that this framework can also be used to compute an interval-valued signal containing all outputs of processes involving a coherent family of conventional linear filters. This approach is based on a very straightforward extension of the expectation operator involving appropriate concave capacities.

Keywords: Convex capacities, real intervals, linear filtering.

1 Introduction

Digital signal processing (DSP) is a significant issue in many applications (automatic control, image processing, speech recognition, monitoring, radar signal analyze, etc.). DSP is mainly dedicated to filtering, analyzing, compressing, storing or transmitting real world analog signals or sampled measurements. When used to mimic real world signal processing, converting the input signal is required from an analog to a digital form, i.e. a sequence of numbers. This conversion is achieved in two steps: sampling and quantization. Sampling consists of estimating the analog signal value associated with discrete values of the reference space (time, spatial localization). Quantization means associating an integer value to a class of real values. For example, in digital image processing, the reference space is a box in $\mathbb{R}^2$, the sampled space is an interval $[1, n] \times [1, m]$ of $\mathbb{N}^2$, the signal is the projected illumination (or any other activity e.g. radioactivity) and the integer range value is the interval $[0, 255]$ if the grey level is coded on 8 bits.

Within the classical approach, the digital signal to be processed is assumed to be composed of precise real valued quantities associated with precisely known values of the reference domain. Naturally, this is not true. Converting an analog signal into a digital signal transforms the information contained in the signal to be processed. Therefore, the classical approach consisting of mimicking analog signal processing by arithmetical operations leads to unquantified computation errors.

One of the most natural ways to represent the loss of information due to quantization is replacing the real precise valued number associated with each sample by an interval valued number. This representation not only solve the problem of representation of real numbers on a digital scale but is also a suitable way for representing the expected fluctuations in the sampled value due to noise

A. Deshpande and A. Hunter (Eds.): SUM 2010, LNAI 6379, pp. 24–27, 2010.

or error in measurement. This approach leads to bounded error estimation [7] when using interval generalization of the involved arithmetic operations (see also [1], [2], [5]). Different interpretations are possible for interval valued data, e.g. a range in which one could have a certain level of confidence of finding the true value of the observed variable [14], a range of values that the real data can take when the measurement process involves quantization and/or sampling [7], [8], or a representation of the known detection limits, sensitivity or resolution of a sensor [4], etc. Within any interval-based signal processing application, there is a strong need for a reliable representation of the variability domain of each involved observation. An important issue is the meaning of the interval and the consistency of this meaning with respect to the tools used for further analysis or processing.

There are still three weaknesses in digital signal processing that are difficult to account for by using classical tools which are: a lack of knowledge of the sampling process, the use of linear digital equations to approximately model non-linear continuous processes, an imprecise knowledge of a filtering process.

In recent papers (see e.g. [9], [10], [6], [13]), we have proposed to use the ability of imprecise probability framework to define family of functions to cope with these weaknesses. Consider a process based on a function for which you have partial information. A way to account for this lack of knowledge is to replace this imprecisely known function by a set of functions that is coherent with your knowledge on the suitable function to be used. Moreover, such a model leads to a new interpretation of interval valued data.

2 Linear Signal Processing and Impulse Response

In signal processing, filtering consists of modifying a real input signal by blocking pre-specified particular components (usually frequency components). Finite impulse response (FIR) filters are the most popular type of filters. They are usually defined by their responses to the individual frequency components that constitute the input signal. In this context, the mathematical manipulation consists of convolving the input samples with a particular digital signal called the impulse response of the filter. This impulse response is simply the response of the digital filter to a Kroenecker impulse input.

Let $X = (X_n)_{n=1,\ldots,N}$ be a set of N digital samples of a signal. Let $\rho = (\rho_i)_{i\in\mathbb{Z}}$ be the impulse response of the considered filter. The computation of Y_k, the kth component of Y the filter output, is given by $Y_k = \sum_{n=1}^{N} \rho_{k-n} X_n$.

When the impulse response is positive and has a unitary gain ($\forall i \in \mathbb{Z}$, $\rho_i \geq 0$ and $\sum_{i\in\mathbb{Z}} \rho_i = 1$), it can be considered as a probability distribution inducing a probability measure P on each subset A of $\mathbb{Z}$ by $P(A) = \sum_{i\in A} \rho_i$. This special case of impulse response is often called *summative kernels* [10], or simply *kernels*, when used to ensure interplay between continuous and discrete domains. Thus, computing Y_k is equivalent to computing a discrete expectation operator involving a probability measure P_k induced by $(\rho_{k-n})_{n\in\mathbb{Z}}$, the probability distribution obtained by translating the probability distribution ρ in k: $Y_k = \sum_{n=1}^{N} \rho_{k-n} X_n = E_{P_k}(X)$.

When the impulse response is not positive or has not a unitary gain then it can be expressed as a linear combination of, at most, two summative kernels in the following way. Let $\varphi = (\varphi_i)_{i\in\mathbb{Z}}$ be the real finite impulse response of a discrete filter such that $\sum_{i\in\mathbb{Z}} \varphi_i < \infty$. Let $\varphi_i^+ = \max(0, \varphi_i)$ and $\varphi_i^- = \max(0, -\varphi_i)$. Let $A^+ = \sum_{i\in\mathbb{Z}} \varphi_i^+$ and $A^- = \sum_{i\in\mathbb{Z}} \varphi_i^-$. Let $\rho_i^+ = \frac{\varphi_i^+}{A^+}$ and $\rho_i^- = \frac{\varphi_i^-}{A^-}$. By construction, ρ_i^+ and ρ_i^- are summative kernels and $\varphi_i = \rho_i^+ A^+ - \rho_i^- A^-$.

Thus, any discrete linear filtering operation can be considered as a weighted sum of, at most, two expectation operations. Let P_k^+ (rsp.P_k^-) be the probability measure based on the summative kernel ρ_{k-i}^+ (rsp.ρ_{k-i}^-), X an input signal and Y the corresponding output signal, then $Y_k = A^+ E_{P_k^+}(X) - A^- E_{P_k^-}(X)$.

The decomposition of φ into ρ^+ , ρ^-, A^+ and A^- is called its canonical decomposition and is denoted as $\{A^-, A^+, \rho^-, \rho^+\}$.

3 Extension of Linear Filtering via a Pair of Two Conjugate Capacities

Let us consider a pair of capacities ν^+ and ν^- such that $P^+ \in \text{core}(\nu^+)$ and $P^- \in \text{core}(\nu^-)$. By translating the confidence measures, we also define ν_k^+ and ν_k^- such that $P_k^+ \in \text{core}(\nu_k^+)$ and $P_k^- \in \text{core}(\nu_k^-)$. It is thus easy to extend linear filtering to a convex set of impulse responses defined by ν^+ and ν^- by: $[Y_k] = A^+ \overline{\underline{E}}_{\nu_k^+}(X) \ominus A^- \overline{\underline{E}}_{\nu_k^-}(X)$, with $\ominus$ being the Minkowski sum [11], $\overline{\underline{E}}_\nu$ the extension of expectation to concave capacities [12] and $[Y_k]$ the kth component of the output of the imprecise filter. Due to the domination properties [3] it verifies: $Y_k = A^+ E_{P_k^+}(X) - A^- E_{P_k^-}(X) \in [Y_k]$.

It can also be extended by considering two real intervals $[I^+]$ and $[I^-]$ and using $\otimes$, the extension of the multiplication to interval valued quantities:

$$[Y_k] = ([I^+] \otimes \overline{\underline{E}}_{\nu_k^+}(X)) \ominus ([I^-] \otimes \overline{\underline{E}}_{\nu_k^-}(X)). \tag{1}$$

Thus every filter with an impulse response φ whose canonical decomposition $\{A^-, A^+, \rho^-, \rho^+\}$ is such that $P^+ \in \text{core}(\nu^+)$ and $P^- \in \text{core}(\nu^-)$ and $A^+ \in [I^+]$ and $A^- \in [I^-]$ has an output that belongs to $[Y]$ the output of the interval-valued filter defined by Equation 1.

4 Discussion and Conclusion

The new method we propose is an extension of the conventional signal filtering approach that enables us to handle imperfect knowledge about the impulse response of the filter to be used. It mostly consist in replacing the usual single precise impulse response by a set of impulse responses that is consistent with the user's expert knowledge. It can be perceived as a surprising way of using the imprecise probability framework. It allows a new interpretation and a new way of computing the imprecision associated with an observed value. According to this interpretation, the imprecision of an observation can be due to the observation process but also to poor knowledge on the proper post-processing to be used to filter the raw measured signal. Defining the pair of convex capacities also defines the convex set of impulse responses. In our recent papers, we have shown different approaches to define pair of capacities that able to handle with a lack of knowledge of the sampling process and an imprecise knowledge of a filtering process (see e.g. [12]). It also ables the propagation of input random noise level to the output filtered value [9] and thus use this information for automatically define thresholds in image analysis processes [6].

Our actual approach only considers a precise signal input. It thus would be useful extend our work to an imprecise signal input, whose imprecision could come from a previous imprecise filtering or be due to pre-calibration of the expected signal error. This could be a way to deal with the measurement uncertainty that is only indirectly taken into account within our approach.

References

1. Boukezzoula, R., Galichet, S.: Optimistic arithmetic operators for fuzzy and gradual intervals. In: Hüllermeier, E., Kruse, R., Hoffmann, F. (eds.) IPMU 2010. LNCS, vol. 6178, pp. 440–450. Springer, Heidelberg (2010)
2. Brito, P.: Modelling and analysing interval data. In: Advances in Data Analysis, pp. 197–208. Springer, Heidelberg (2007)
3. Denneberg, D.: Non-Additive Measure and Integral. Kluwer Academic Publishers, Dordrecht (1994)
4. Kreinovich, V.: et al. Interval versions of statistical techniques with applications to environmental analysis, bioinformatics, and privacy in statistical databases. J. Comput. Appl. Math. 199(2), 418–423 (2007)
5. Gardenes, E., Sainz, M.A., Jorba, I., Calm, R., Estela, R., Mielgo, H., Trepat, A.: Modal intervals. Reliable Computing 7(2), 77–111 (2001)
6. Jacquey, F., Loquin, K., Comby, F., Strauss, O.: Non-additive approach for gradient-based detection. In: ICIP 2007, San Antonio, Texas, pp. 49–52 (2007)
7. Jaulin, L., Kieffer, M., Didrit, O., Walter, E.: Applied Interval Analysis with Examples in Parameter and State Estimation, Robust Control and Robotics. Springer, Heidelberg (2001)
8. Jaulin, L., Walter, E.: Set inversion via interval analysis for nonlinear bounded-error estimation. Automatica 29(4), 1053–1064 (1993)
9. Strauss, O., Loquin, K.: Noise quantization via possibilistic filtering. In: Proc. 6th Int. Symp. on Imprecise Probability:Theories and Application, Durham, United Kingdom, July 2009, pp. 297–306 (2009)
10. Loquin, K., Strauss, O.: On the granularity of summative kernels. Fuzzy Sets and Systems 159(15), 1952–1972 (2008)
11. Moore, R.E.: Interval analysis. Prentice-Hall, Englewood Cliffs (1966)
12. Rico, A., Strauss, O.: Imprecise expectations for imprecise linear filtering. International Journal of Approximate Reasoning (to appear, 2010)
13. Rico, A., Strauss, O., Mariano-Goulart, D.: Choquet integrals as projection operators for quantified tomographic reconstruction. Fuzzy Sets and Systems 160(2), 198–211 (2009)
14. Zhu, Y., Li, B.: Optimal interval estimation fusion based on sensor interval estimates with confidence degrees. Automatica 42, 101–108 (2006)

Managing Lineage and Uncertainty under a Data Exchange Setting

Foto N. Afrati and Angelos Vasilakopoulos

National Technical University of Athens
afrati@softlab.ece.ntua.gr, avasila@central.ntua.gr

Abstract. We present a data exchange framework that is capable of exchanging uncertain data with lineage and give meaningful certain answers on queries posed on the target schema. The data are stored in a *database with uncertainty and lineage (ULDB)* which represents a set of possible instances that are *databases with lineage (LDBs)*. Hence we need first to revisit all the notions related to data exchange for the case of LDBs. Producing all possible instances of a ULDB, like the semantics of certain answers would indicate, is exponential. We present a more efficient approach: a u-chase algorithm that extends the known chase procedure of traditional data exchange and show that it can be used to correctly compute certain answers for conjunctive queries in PTIME for a set of weakly acyclic tuple generating dependencies. We further show that if we allow equality generating dependencies in the set of constraints then computing certain answers for conjunctive queries becomes NP-hard.

Keywords: Data Uncertainty, Lineage, Data Exchange.

1 Introduction

Data exchange is the problem of translating data that is described in a source schema to a different target schema. The relation between source and target schemas is typically defined by schema mappings. Recently the data exchange problem has been widely investigated, even for various uncertain frameworks, e.g. probabilistic [9]. One aspect of the problem is finding procedures that compute in polynomial-time target instances that represent adequate (usually for query answering purposes) information. A challenging problem that has received considerable attention is the problem of giving meaningful semantics and computing answers of queries posed on the target schema of a data exchange setting [1,3,9]. In [10] it was shown that computing certain answers for conjunctive queries over ordinary relational databases can be done with polynomial data complexity if the constraints satisfy specific conditions (are *weakly acyclic*). Query answering for data exchange was also studied for aggregate queries [2] and queries with arithmetic comparisons [3]. To the best of our knowledge the data exchange problem and query answering has not been studied for models of databases that incorporate both uncertainty and lineage. On Uncertainty-Lineage Databases (ULDBs)

A. Deshpande and A. Hunter (Eds.): SUM 2010, LNAI 6379, pp. 28–41, 2010.

an uncertain database represents a set of *possible instances - PIs* which are certain databases with lineage (LDBs). The semantics for possible instances of an uncertain instance are that only one of them "captures the truth", but we do not know which. Thus we have two kinds of uncertainty: i) uncertainty about the possible instances that this source represents (which one is the "true") and ii) uncertainty that arises due to heterogeneous source and data schemas. Many modern applications like data extraction from the web, scientific databases, sensors and even data exchange systems of certain sources often contain uncertain data. These applications may require recording the origin of data, which is modeled through *lineage* or *provenance.* Databases that track down the provenance of certain data have been extensively studied [5,6,12].

The Trio model [4] combines and supports both lineage and uncertainty in ULDBs. This model was proven to be complete for succinctly representing any set of databases with lineage (LDBs) [4]. In this model a) each uncertain tuple (x-tuple) is a set of traditional tuples, called "alternatives", that represent its possible values (i.e. we have uncertainty as to which of them is the "true" one) and b) each alternative comes with its lineage. One of the reasons ULDB model was introduced is that it would be important for data exchange [4]. Computing conjunctive queries, that is SPJ queries, on ULDBs was shown to require polynomial-time [4]. Trio implements ULDBs in a database system build on top of a traditional SQL system with necessary encoding [4].

In general a data exchange problem for a data model consists of a source instance represented in this model and of a set Σ of constraints that consists of source-to-target constraints Σ_{st} and target constraints Σ_t ($\Sigma = \Sigma_{st} \cup \Sigma_t$). Given a finite source instance I, the data exchange problem is to find a finite target instance J such that $< I, J >= I \cup J$ satisfies Σ_{st} and J satisfies Σ_t. Such a J is called *a solution for I* or, simply *a solution* if the source instance I can be easily understood [10]. The query answering problem on a data exchange setting is which target instance to materialize to be used for obtaining meaningful answers (called *certain answers*) of a query. In this paper we investigate the *data exchange* problem for the ULDB model. We consider source-to-target (s-t) and target tuple-generating dependencies (tgds) as constraints. In addition we consider a ULDB source whose lineage is "well-behaved", a constraint defined in [4] and in practice is true for most databases. The important property of "well-behaved" lineage for our results is that the lineage of a tuple, even if expanded, cannot refer to itself (lineage transitive closure does not contain cycles).

We investigate the data exchange problem of the Trio model and address the relevant query answering problem for Conjunctive Queries (CQs). Our contributions are: i) We present natural semantics for ULDB certain answers. ii) We give *u-chase* algorithm which extends the well-known chase algorithm for certain databases. iii) We use u-chase to show that if the source is a well-behaved ULDB, the source to target constraints consists of tgds and the target of a set of weakly acyclic tgds then we can compute ULDB certain answers in PTIME. iv) We prove that if we incorporate equality generating dependencies (egds) in the set of target dependencies then computing certain answers for conjunctive queries on a ULDB data exchange setting becomes NP-hard.

Intuitively the semantics of ULDB certain answers are the following: It is a ULDB that represents a set of possible LDB instances. We want those instances to be the same as if we had considered first the LDB possible instances of the source ULDB and computed certain answers at each one of them. An obvious way to tackle this problem is to follow the direction that semantics indicate: if the ULDB source has n possible LDB instances then produce n LDB certain answers computed from n LDB data exchange problems. But the number of the possible instances of a (source) ULDB can be exponential on the size of the data [4]. As a result this procedure would be computationally very expensive and would not be suitable for large data sets. In contrast our proposed u-chase algorithm computes certain answers in polynomial time for weakly acyclic tgds. Another reason which shows that moving from the well-studied data exchange problem for ordinary databases to databases with uncertainty and lineage is not trivial is the following: In the former case computing certain answers with egds in the target dependencies remains polynomial, while in our case we prove that this problem becomes NP-hard. The correct meaning of *ULDB certain answer* semantics is illustrated in Example 1. For simplicity it does not include any target constraints.

Example 1. Suppose that a local police department contains a database **A** with two relations: `Saw(caseID,witness,car)` and `Drives(person,car)`. (We have borrowed the source schema from [4]). The department has a list of drivers along with their cars that contains the information that Hank drives a Honda, Jimmy a Mazda and Billy a Toyota car. All this information is certain and stored in relation `Drives(person,car)`. The source ULDB relation `Drives` is shown in Figure 1. For every crime if a witness reports that he/she saw a car near the crime scene then the local department stores in relation `Saw` the witness, the car-make that the witness saw along with a `caseID` for the relevant case. Suppose now that we have two cases with CaseIDs: *Case*1 and *Case*2. In the first case a witness named *Cathy* saw a car near the crime scene but she was not sure whether it was a *Honda* or a *Mazda*. In the second case a witness named *Amy* saw a car but she was uncertain whether it was a *Toyota* or a *Honda*. Note that in Figure 1 empty lineage of the source data is omitted.

A ULDB relation consists of x-tuples instead of tuples. Each x-tuple has a unique identifier and a multiset of "alternative values", separated with || symbol. The semantics are that only one alternative (or zero if we have symbol '?') from each x-tuple can be true in a possible instance [4]. For example ULDB relation *Saw* on Figure 1 has 4 LDB possible instances due to the 2 choices of alternatives for x-tuples 11 and 12. If *Saw* had n x-tuples each with 2 alternatives then it would represent 2^n possible instances (exponential in the number of alternatives).

Now a private investigator owns a database **B** that has a single relation: `Files&Suspects(caseID,suspect,date)`. Suppose that the private investigator wants to transfer the information from the database of the local police department to his own database. But he only wants to store the caseID, suspects and date for each crime since he is not interested about witness' names or car information. The following source-to-target tgd ξ models this example: $\texttt{Saw(caseID, witness,car), Drives(p,car)} \rightarrow \exists D\, \texttt{Files\&Suspects(caseID, p, D)}$. The date attribute is not present in the source schema and is represented as an "unknown" - "null" value in the target. Even in certain data exchange the heterogeneity between the source and the target schema gave rise to "null" values that appear as distinct variables in a database instance [10]. Hence a materialized target instance that we expect due to tgd ξ is shown in Figure 2.

ξ: $\texttt{Saw(caseID, witness, car), Drives(p, car)} \rightarrow \exists D\, \texttt{Files\&Suspects(caseID, p, D)}$
Q: $\texttt{Q(suspect) : -Files\&Suspects(caseID, suspect, date)}$

ID	**Saw**(caseID,witness,car)
11	Case1,Cathy,Honda || Case1,Cathy,Mazda
12	Case2,Amy,Toyota || Case2,Amy,Honda

ID	**Drives**(person,car)
21	Hank, Honda
22	Jimmy, Mazda
23	Billy, Toyota

Fig. 1. ULDB **source** database **A**

ID	**Files&Suspects** (caseID,suspect,date)
31	Case1, Hank,d_1 ?
32	Case1, Jimmy,d_2 ?
33	Case2, Billy,d_3 ?
34	Case2, Hank,d_4 ?

$\lambda(31) = \{(11, 1), (21, 1)\}$
$\lambda(32) = \{(11, 2), (22, 1)\}$
$\lambda(33) = \{(12, 1), (23, 1)\}$
$\lambda(34) = \{(12, 2), (21, 1)\}$

Fig. 2. Expected materialized **target** ULDB instance

ID	**Q**(suspect)
41	Hank ?
42	Jimmy ?
43	Billy ?
44	Hank ?

$\lambda_B(41) = \{(11, 1), (21, 1)\}$
$\lambda_B(42) = \{(11, 2), (22, 1)\}$
$\lambda_B(43) = \{(12, 1), (23, 1)\}$
$\lambda_B(44) = \{(12, 2), (21, 1)\}$

Fig. 3. Expected Certain Answer of Q posed on **target** database **B**

Suppose now that the private investigator wants to ask about all the names of all the suspects. This query posed over his target database **B** is the following conjunctive query Q: $\texttt{Q(suspect) :- Files\&Suspects(caseID, suspect, date)}$. Query Q is a projection of attribute suspect of Files&Suspects. Since the source instance is a database with uncertainty, the target database **B** will also be uncertain and have lineage pointing to source data. It will represent a set of certain *possible instances* which are certain databases with lineage. Accordingly the answer of the query Q posed on B will represent a set of possible instances. Intuitively we expect the following:

i) *Hank* should exist as a suspect in the certain answers of Q if the car that *Cathy* saw was in reality a *Honda* car. So in the relation Q the lineage will record the pointer back to the fact that *Cathy* saw a *Honda* (alternative $11, 1$). Answer *Hank* also comes from the certain fact that *Hank* drives a *Honda* (alternative $21, 1$). ii) But *Hank* can also be a suspect in the certain answers of Q if the car that *Amy* saw was in reality a *Honda* car (this time he will be a suspect for *Case*2). Thus the suspect *Hank* will be recorded twice in the answers but with different lineage. As we see we can now have that tuples with same data can appear many times (we have a multiset/bag of tuples) if their lineage is different.

iii) Similarly we expect *Jimmy* to exist as a suspect in the certain answers of Q if the car that *Cathy* saw was in reality a *Mazda* car, so with lineage that contradicts the lineage $(11, 1)$ of the first tuple with *Hank* data.

As a result we expect the certain answer of Q to be a ULDB as shown in Figure 3. With $s(i, j)$ we denote the j-th alternative of x-tuple with identifier i. *Base data* of a ULDB instance is its source data with empty lineage. If two alternatives point to the same base data we say that they have the same *base lineage* which is lineage extended and containing only base data. In Figure, 3 $\lambda_B(s(i, j))$ denotes the *base lineage* of an alternative $s(i, j)$. We note that lineage does not just track where data comes from, but also poses logical restrictions to the possible LDB instances that a ULDB represents. An alternative of an x-tuple can be true in a possible instance if its lineage is true. For example Jimmy cannot appear in all possible instances but only to the two ones that have selected alternative $(11, 2)$ to be true. Symbol '?' indicate that there exists a possible instance that does not contain any alternative from this x-tuple (such a tuple is called a "maybe" x-tuple [4]). For more information about ULDB possible instances we refer to [4].

2 Preliminaries

We have that the *possible instances (PIs)* of a ULDB are LDBs [4]. Since we do not know which one captures the "truth", intuitively all the LDB possible instances of a ULDB solution must satisfy data exchange constraints. So we need to revisit all the data exchange relevant definitions of the certain case for an LDB data exchange setting. The data exchange problem for databases with lineage has the same setting as in the certain case. The difference now is that the tuples in the source instance and the target instance (to be materialized) have lineage information. The model of LDB extends the relational model in the way that every tuple apart from its data has also a unique identifier attached and a lineage function. Definition of LDBs is presented in [4]. In general a tuple t^{LDB} in an LDB, belonging in a set $\bar{R}$ of relation names, consists of three things: i) its identifier symbol denoted as $ID(t)$ and belonging to a set S of symbols, ii) its data t and iii) its lineage $\lambda(ID(t))$. We have the following definition:

Definition 1 (LDB tuple). *A tuple t^{LDB} of an LDB $D = (\bar{R}, S, \lambda)$ is a triple $\big(ID(t),\ t,\ \lambda(ID(t))\big)$ where: i) $t \in \bar{R}$, ii) $ID(t) \in S$ and iii) $\lambda(ID(t)) \in \lambda$.*

As we already discussed in the Introduction, heterogeneity between the source and the target schema can give rise to "null" values that appear as distinct variables in a target instance. When tuples have values from constants then we say that it is a *ground database instance with lineage (ground LDB instance).* When tuples have values from constants and variables (where these two sets are disjoint) then we say that we have a *database instance with lineage (LDB instance).*

The lineage function is empty for base relations. For derived relations it points to the identifiers of the tuples from which it was derived. If a tuple is derived as an answer of a query, lineage points to the tuples that were used in order to get this tuple in the answer relation. In [4] an algorithm for computing conjunctive queries (CQs) over ULDB was presented.

In our work we focus on computing certain answers and correct lineage. Correct lineage for a tuple that is derived as an answer of a CQ has been already defined in CQ answering algorithm of [4]. But now we also have to compute semantically correct lineage for a tuple that is derived from a data exchange setting. The important difference between LDB conjunctive query computing and CQ computing on ordinary databases is the following: Now a tuple can exist in the answer of a query more than once if it is derived from different tuples. So two or more tuples can have the same data if they have different lineage. In a similar way the semantics of tuple generating dependencies should also be slightly different. We will refer to the right hand side (left hand side respectively) of a tgd with *rhs* or *body* (*lhs* or *head* respectively). It is natural to define that a tuple generating dependency rule d is LDB-satisfied if it is satisfied in the traditional sense and, moreover, the lineage of the image of the lhs tgd atoms is related to the lineage of the image of the rhs tgd atoms. The formal definition follows:

Definition 2 (LDB satisfaction of a tgd). *Let D be an LDB and d be a tgd of the form:*
$\forall \mathbf{x}(R_1(\mathbf{x_1}), R_2(\mathbf{x_2}), \ldots, R_n(\mathbf{x_n}) \rightarrow \exists \mathbf{y}(T_1(\mathbf{x'_1}, \mathbf{y_1}), T_2(\mathbf{x'_2}, \mathbf{y_2}), \ldots, T_m(\mathbf{x'_m}, \mathbf{y_m})))$
We will say that LDB D l-satisfies d, if for each homomorphism h that maps $R_1(\mathbf{x})$, $R_2(\mathbf{x}), \ldots,$, $R_n(\mathbf{x})$ to tuples: $R_1(t_1)$ with $ID = ID_1$, $R_2(t_2)$ with $ID = ID_2, \ldots,$ $R_n(t_n)$ with $ID = ID_n$ in D, then there exists a homomorphism h' that is an extension of h that maps the rhs of d to tuples $T_1(t'_1), T_2(t'_2), \ldots, T_m(t'_m)$ in D with: $\lambda(t'_1) = \{ID_1, ID_2, \ldots, ID_n\}$, $\lambda(t'_2) = \{ID_1, ID_2, \ldots, ID_n\}$, $\ldots$, $\lambda(t'_m) = \{ID_1, ID_2, \ldots, ID_n\}$.

The identifiers in the answer of a CQ apart from the fact that need to be unique and different than the existing ones, their exact value is not important. For example a tuple with identifier 41, data value $Hank$ and lineage $\lambda(41) = \{11, 21\}$ says that $Hank$ exists in the answer due to facts with ids $\{11, 21\}$. At this point the newly created answer-identifier 41 could have been as well 51: the important fact is that $Hank$ is in the answer and due to tuples $\{11, 21\}$. We will have in mind this difference in order to define l-certain answers: For a tuple t^{LDB} of an LDB we can "drop" its identifier information and thus take a pair $(t, \lambda(ID(t)))$, which we denote by $ID_{drop}(t^{LDB})$. We extend ID_{drop} to apply to databases:

Definition 3. *Given LDB D, we define $ID_{drop}(D)$ to be derived from D by changing each tuple t^{LDB} in each relation of D to $ID_{drop}(t^{LDB})$.*

Since we have no alternatives, LDB tuples are always present and their unique identifiers can be single numbers and not pairs referring to alternatives like in ULDBs. When a tuple's identifier is clear from the context, we may abuse the above notation and refer to an LDB tuple only with its identifier ID (i.e. denote its lineage as $\lambda(ID)$). Alternatively, we may say that we want to have an LDB tuple with data t and lineage $\lambda(t)$ present in our database, when confusion does not arise. Also we can extract only relations of target schema J using polynomial extraction algorithm found in [4]. It extracts a subset of relations of an LDB or a ULDB along with all their base lineage information. Data from tuples not appearing in retained relations is not being kept. Only their lineage and only if it is referred through the base lineage of remaining tuples. From now on when we refer to *extracting* a relation we will mean using extraction algorithm of [4]. We can now define LDB certain answers and LDB solutions:

Definition 4 (LDB Solutions, Certain Answers). *Let* $(\mathbf{S}, \mathbf{T}, \mathbf{\Sigma_{st}}, \mathbf{\Sigma_t})$ *be an LDB data exchange setting. An LDB* solution *to this data exchange setting is an LDB* J *of target schema* $\mathbf{T}$ *such that together with* I *l-satisfies the tgds in* Σ.

Let q *be a conjunctive query over the target schema* T *and* I *an LDB source instance. For each solution* J *we compute* $q(J)$ *and drop the identifiers producing* $ID_{drop}(q(J))$. *Then* $l\text{-}cert_{drop}(q, I) = \cap\{ID_{drop}(q(J))\}$. *Finally the LDB* certain answers *(l-certain answers) are produced by attaching a fresh distinct identifier to each tuple of* $l\text{-}cert_{drop}(q, I)$ *and produce* $l\text{-}cert(q, I)$.

Universal solutions of a certain data exchange setting are target relations, such that there exists a homomorphism from them to each other solution [10]. In order to define LDB Universal Solutions we first have to revisit the notion of homomorphism. We keep the "data part" of definitions as it is in the certain case and for the lineage part we require the mapped tuples to have the same base lineage. The usefulness of such a definition is more obviously seen when we use it in the context of ULDB data exchange query answering. For now we only note that the base data is the one which defines: i) the number of possible instances of a ULDB and ii) which derived tuples with well-behaved lineage will be also appearing (apart from the base ones) at each instance. This property is proved in [4]. So base data is the one which exclusively determines (through base lineage) the LDB possible instances of a ULDB. We can now define LDB homomorphism (denoted as l-homomorphism) and LDB Universal Solution.

Definition 5 (LDB homomorphism h_l). *Let* D_1, D_2 *be two LDB instances over the same schema with values in* $Const \cup Var$. *We say that* D_1 *l-maps to* D_2 *(*$D_1 \rightarrow_l D_2$*) if there exists a homomorphism* h *from variables and constants of* D_1 *to variables and constants of* D_2 *such that: i) it maps every base fact* t_i^1 *of* D_1 *to a base fact* t_j^2 *of* D_2 *with* $ID(h(t_i^1)) = ID(t_j^2)$ *and ii) for every non base fact* $R_i(t)$ *of* D_1 *we have that* $R_i(h(t))$ *is a fact of* D_2 *with* $\lambda_B(R_i(h(t)) = \lambda_B(R_i(t))$, *where* λ_B *is the lineage extended back to and containing only base data.*

Definition 6 (LDB Universal Solution). *Consider a data exchange setting* $(\mathbf{S}, \mathbf{T}, \mathbf{\Sigma_{st}}, \mathbf{\Sigma_t})$. *If* I *is a source LDB instance, then an l-universal solution for* I *is a solution* J *such that for every solution* J' *we have that there exists an LDB homomorphism* h_l *such that:* $J \rightarrow_{h_l} J'$.

3 Computing LDB Certain Answers

Now that we have all the necessary notions from the previous sections, we can define l-chase. L-chase will be used in the next section in order to compute ULDB certain answers in PTIME. Again the "data part" of l-chase will be similar to the certain definition. Certain chase is a procedure that makes sure that its result will satisfy a given tgd. We will similarly also need to extend it with a "lineage part", according to our definition of l-homomorphism in order for its result to l-satisfy a tgd:

Definition 7 (l-chase step). *Let K be an LDB instance. Let d be a tgd of the form: $\forall \mathbf{x}(R_1(\mathbf{x_1}), R_2(\mathbf{x_2}), \ldots, R_n(\mathbf{x_n}) \rightarrow \exists \mathbf{y}(T_1(\mathbf{x'_1}, \mathbf{y_1}), T_2(\mathbf{x'_2}, \mathbf{y_2}), \ldots, T_m(\mathbf{x'_m}, \mathbf{y_m})))$. Let h be a homomorphism from $R_1(\mathbf{x}), R_2(\mathbf{x}), \ldots, R_n(\mathbf{x})$ to tuples $R_1(t_1)$, $R_2(t_2), \ldots, R_n(t_n)$ in K with IDs: $\{ID_1, ID_2, \ldots, ID_n\}$ such that there exist no homomorphism h' that is an extension of h that maps the rhs of d to tuples $T_1(t'_1), T_2(t'_2), \ldots, T_m(t'_m)$ in K with: $\lambda_B(t'_1) = \lambda_B(t_1) \cup \lambda_B(t_2) \cup \ldots \cup \lambda_B(t_n)$, $\lambda_B(t'_2) = \lambda_B(t_1) \cup \lambda_B(t_2) \cup \ldots \cup \lambda_B(t_n)$, $\ldots$, $\lambda_B(t'_m) = \lambda_B(t_1) \cup \lambda_B(t_2) \cup \ldots \cup \lambda_B(t_n)$. We say that d can be l-applied to K with homomorphism h (in this case l-chase "fires").*

Let LDB K' be the union of K with the set of facts obtained by: (a) extending h to h' such that each variable in $\mathbf{y}$ is assigned a fresh labeled null, followed by (b) taking the image of the atoms of $\mathbf{y}$ under h' and adding them to the relations of the rhs of d along with a new non-used identifier (c) set as lineage of each atom of (b) the set $\{ID_1, ID_2, \ldots, ID_n\}$. We say that the result of applying d to K with h is K' , and write $K \xrightarrow{d,h}_l K'$.

Our l-chase is a finite terminating sequence of l-chase steps. It is an extension of the well known chase procedure for traditional databases. In fact we have proved that termination of l-chase can be determined for the same set of tgds as for (certain) chase. The result of l-chase can be used in order to polynomially compute l-certain answers for weakly acyclic tgds as it is stated in the following Theorem:

Theorem 1 (Computing ldb certain answers). *Consider an LDB data exchange setting with an LDB source instance I with schema S, T be the target schema and Σ be a set of source to target tgds and weakly acyclic target tgds. Let q be a conjunctive query over the target schema T. We can compute $l\text{-}cert(q, I)$ in time polynomial to the size of LDB I.*

4 Data Exchange for Uncertain Databases with Lineage (ULDBs)

For ordinary databases we had the following definition for certain answers: given a query q posed over a certain instance I with null values, then: $certain(q, I) = \{ \cap(J) \mid \text{where } J \text{ is a solution } \}$. However, for a ULDB data exchange setting, if we take the intersection of the answers to the query over all possible instances, it is easy to see that we will mostly derive an empty set of answers. Moreover, this does not agree with the intuition that only one of the possible instances is the correct one. If so, then taking the intersection between facts appearing in the one "true" instance and all other "false" possible instances is meaningless. Thus, in order to semantically define certain answers in a ULDB data exchange setting we think as follows (similar intuition was applied in the definition of certain answers in [3]): Each possible instance D_i of a ULDB source instance I is an LDB and gives rise to an LDB data exchange problem with D_i as its source instance and the same set of dependencies Σ. Since we do not know which possible instance is the one that captures the "truth" it is natural to consider them all but "separately". We note that since possible instances are LDBs, they are allowed in them duplicates of data if they have different lineage. We retain this property, i.e. we do not perform any kind of duplicate elimination.

As a result when we pose a query to any of the solutions of all LDB data exchange problems that arise from the D_i possible instances of a source ULDB we expect to get the corresponding certain answer. We thus have the following definition:

Definition 8 (ULDB Certain answers). *Consider a data exchange setting* $(\mathbf{S}, \mathbf{T}, \mathbf{\Sigma} = \mathbf{\Sigma_{st}} \cup \mathbf{\Sigma_t})$. *If* I *is a source ULDB instance with* $PI(I) = D_1, \ldots, D_n$ *and* q *a query over target schema* $\mathbf{T}$ *then the certain answers of* q *with respect to* I*, denoted as uldb-cert*(q, I) *is a complete ULDB* C *(with no nulls) with* $PI(C) = D''_1, \ldots, D''_n$ *such that:*
each D''_i *is an l-certain answer of the LDB data exchange problem with* D_i *as its LDB source instance and* Σ *as its set of constraints.*

Figure 4 illustrates the notion of ULDB certain answers. We want to give an appropriate procedure to compute *uldb-cert*(q, I) which will be based on l-chase for databases with lineage. We do not to explicitly produce the possible instances of the source instance, which is computationally expensive, but rather compute in polynomial time directly a ULDB that will be used for the computation of *uldb-cert*(q, I). On the other hand the semantics of the result of this procedure should be the following: the possible instances of the ULDB that will be the result of our uncertain-chase (which we will denote by u-chase) procedure should be

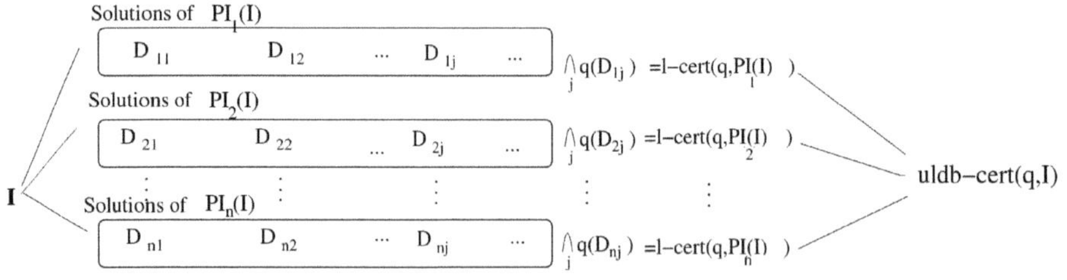

Fig. 4. ULDB Certain Answers

the same with the possible instances that we would have if we first took the possible instances of the ULDB instance and apply l-chase to each one of them.

We will make use of a way to transform a ULDB into a "pseudo-LDB" called its "Horizontal Relation". A similar approach was used in an algorithm in [4] to correctly compute answers of CQs over ULDBs in PTIME without producing all possible instances. Horizontal relations were first defined in [8] in a variation without lineage. Intuitively in order to take the Horizontal database of a ULDB we "flatten" each alternative so that it will become a new x-tuple (with no other alternatives), but retain the lineage information (now referring to tuples and not to alternatives). As is stated in [4], we have that since the size of D_H is the same as the size of ULDB relations R_1, $R_2, \ldots, R_n$, complexity does not increase when we compute a horizontal relation of a ULDB.

Definition 9 (Horizontal Relation, Horizontal Database). *Let R be an relation of a ULDB. We define the* Horizontal relation *of R and denote it with R_H the LDB (with no uncertainty -no alternatives- but with lineage):*
$R_H = \{$ *tuples* $s(i,j) \mid s(i,j)$ *is an alternative in* $R\}$.

Let D be a ULDB with x-relations $\{R_1, \ldots, R_n\}$. We define the Horizontal database of D and denote it with D_H the LDB: $D_H = R_{1H}, \ldots, R_{nH}$ such that $\forall k$, $k \in [1,n]$: $R_{kH} = \{$ tuples $s(i,j) \mid s(i,j)$ is an alternative in $R_k\}$

Definition 10 (U-Chase sequence). ***input:*** *A source schema* **S**, *a target schema* **T**, *a source ULDB D with x-relations $\{R_1, \ldots, R_n\}$ and Σ a set of source-to-target and target tgds.* ***output:*** *a ULDB D'. Steps:*
1. From ULDB source D create the horizontal database D_H. 2. Since D_H is an LDB we can now apply an l-chase sequence with D_H as the source LDB instance and constraints Σ and produce D'_H. 3. Create a ULDB D' that is the union of i) the source ULDB relations $\{R_1, \ldots, R_n\}$ and of ii) target ULDB relations that are created from the target horizontal relations of D'_H by treating each LDB tuple as a maybe x-tuple with one alternative (so with symbol '?') and retaining the lineage relationship of D'_H. 4. Return D'.
A terminating u-chase *sequence is the one that applies a terminating l-chase sequence.*

Note that in the u-chase result we do not regroup alternatives of target relations to x-tuples, like ULDB conjunctive query algorithm in [4] does. The possible instances of our ULDB output will be the same even without regrouping due to the constraints posed by lineage information. In addition when we compute certain answers of a CQ (which is the reason we defined u-chase for), ULDB conjunctive query algorithm will then regroup alternatives to x-tuples.

4.1 U-Chase Result for Our Example

Example 2. Let us consider the ULDB data exchange problem of our running Example 1: The source ULDB I is shown in Figure 1. We follow the steps of u-chase: We first create the horizontal relations of the source database **A**. Since horizontal relations are LDBs we apply l-chase to them and get an intermediate

LDB Files&Suspects$_H$. In our example, this is the result of a single l-chase step since our tgd will not l-fire again.

We form a ULDB from the LDB result of l-chase: Each LDB horizontal tuple becomes a ULDB x-tuple with one alternative and we retain all lineage connections, now referring to alternatives. We have to add '?' at each x-tuple because we no longer have an LDB: In LDBs all tuples are certain (since LDBs have one possible instance). Some of those '?' symbols might be extraneous, but we can remove them in polynomial time, as it proven in [4]. The extracted Files&Suspects target relation of U-chase result is the same as the one we expected in Example 1, shown in Figure 2. Let us denote with J' this materialized target instance of relation Files&Suspects which is the target result of U-chase. This result can be used in order to compute ULDB certain answers (*uldb-cert*(Q, I)): $Q(J'_{\downarrow})$ are the extracted certain answers of query Q of Example 1. Indeed the result of $Q(J'_{\downarrow})$ is the same with the ULDB certain answers of Q which we intuitively expected, shown in Figure 3. So for each possible instance of the source database **A** we get the l-certain answer that would be real if we had a data exchange problem with that instance. The retained lineage will logically restrict the possible instances of the target ULDB. For example tuples having $11, 1$ in their base lineage will never coexist in a possible instance with tuples having $11, 2$ in their lineage, even though the "data part" of source x-tuple 11 would have been removed. This is one of the aims of data exchange: to be able to compute certain answers only from target database when the source data in no longer available. We are now going to prove that the result of u-chase result can always be used in order to compute ULDB certain answers.

4.2 Complexity of Computing ULDB Certain Answers

First we prove that any u-chase sequence on a well-behaved ULDB U that creates a ULDB U' is equivalent to applying an l-chase sequence to each of the possible instances of U. Then in Lemma 1 we prove that a terminating u-chase on U produces a ULDB with possible instances which are the result of a terminating l-chase applied on each of the possible instances of U. Finally we have the last Theorem 3 which states that for a well-behaved source ULDB instance and a set of weakly acyclic tgds we can use the result of u-chase in order to compute ULDB certain answers of conjunctive queries in polynomial time. Its proof will make use of Theorem 2 and Lemma 1:

Theorem 2. *Suppose we start with a well-behaved ULDB D_0 and after a u-chase sequence steps we arrive at a ULDB D_j. Suppose that the possible instances of D_0 are: $D_0^1, D_0^2, \ldots, D_0^i$. Suppose that the possible instances of D_j are: $D_j^1, D_j^2, \ldots, D_j^{i'}$. Then the following holds: i) $i = i'$ and ii) every D_j^k comes from a $D_0^{k'}$ after a sequence of l-chase steps.*

Lemma 1. *Suppose that we have a data exchange problem with a well-behaved ULDB source D_0 with $PI(D_0) = D_0^1, D_0^2, \ldots, D_0^n$ and a set Σ of tgds. Then the result of a terminating u-chase with Σ is a ULDB D_j with $PI(D_j) =$*

$D_j^1, D_j^2, \ldots, D_j^n$ such that for every $i = 1 \ldots n$ we have that D_j^i is an l-universal solution (if we extract target relations) of the LDB data exchange problem with D_0^i as the source and Σ as constraints.

Theorem 3 (ULDB Certain Answers). *Consider a ULDB data exchange setting with a well-behaved ULDB source instance I with schema S, T be the target schema and Σ be a set of source to target tgds and weakly acyclic target tgds. Let q be a conjunctive query over the target schema T. We can compute uldb-cert(q, I) in time polynomial to the size of I using the result of u-chase.*

4.3 Complexity of a ULDB Data Exchange Problem That Includes Egds in Its Set of Constraints Σ

Let us modify a little our running example by adding one new attribute `signature` to relation `Saw` which contains the name that appears in the signature of a witness. Suppose that *Cathy* is not certain if the car she saw was a *Honda* or a *Mazda* but we always have her name in the signature. On the other hand *Amy* saw a *Toyota* car but we have uncertainty if the witness report has a signature with name *Amy* or *Annie* (for example due to a not clear signature). Now we suppose that the private investigator also stores information about the name of the witness and her signature. Hence target `Files&Suspects` now has schema: {`caseID,witness,signature,suspect,date`}. So we change the source to target tgd to the following ξ_1: `Saw(caseID,witness,signature,car)`, `Drives(p,car)` $\rightarrow \exists D$ `Files&Suspects(caseID,witness,signature,p,D)`. Now let us consider that we also have the target egd ξ_2: `Files&Suspects (caseID,witness,signature,suspect,date)` $\rightarrow$ `(witness = signature)`.

Since we do not know which instance captures the "truth" we would want all the possible instances to satisfy the egd. Intuitively the ULDB result of u-chase with ξ_2 should represent only the two possible instances that do not contain any result derived from the possible instances that contain in `Saw` a tuple with name *Amy* and signature *Annie*. The reason is that any possible instance of the source containing tuple $Saw(Case2, Amy, Annie, Toyota)$ will give a failing l-chase. But we prove in the following theorem that even asking for a solution in a ULDB data exchange problem that contains egds in its constraints is an NP-hard problem. The proof will be a reduction from 3-coloring. A UDLB solution will be a target instance that satisfies the constraints. Satisfaction for ULDBs of a constraint d (egd or tgd) is formally defined in Definition 12 which uses the following Definition 11 for egds and LDBs. Finally we note that for an LDB data exchange setting with weakly acyclic tgds, we can include egds and still compute l-certain answers in polynomial time. Due to lack of space we do not include details.

Definition 11 (LDB satisfaction of an egd). *Let D be an LDB and d be an egd of the form: $\forall \mathbf{x}(R_1(\mathbf{x_1}), R_2(\mathbf{x_2}), \ldots, R_n(\mathbf{x_n}) \rightarrow (x_1 = x_2))$. We will say that LDB D l-satisfies d, if for each homomorphism h that maps $R_1(\mathbf{x})$, $R_2(\mathbf{x}), \ldots, R_n(\mathbf{x})$ to tuples: $R_1(t_1)$ with $ID = ID_1$, $R_2(t_2)$ with $ID = ID_2$, $\ldots$, $R_n(t_n)$ with $ID = ID_N$ in D, then $h(x_1) = h(x_2)$.*

Definition 12 (ULDB satisfaction of an egd/tgd). *Let D be a ULDB instance. Suppose that the possible instances of D are LDBs: $D^1, D^2, \ldots, D^n$. Let d be a tgd or an egd constraint. Then D will ULDB satisfy (u-satisfy) d if for every i the following holds: D^i l-satisfies d.*

Theorem 4. *Consider a UDLB data exchange setting with a well-behaved ULDB source I and a set of dependencies Σ that can contain egds as well as tgds and a CQ Q. Then the following problems are an NP-hard: i) "Is there a solution of the data exchange setting I and Σ?", ii) "Is a tuple t in uldb-cert(Q, I)?".*

Proof. (sketch) i) By reduction from 3-coloring: Let G be a graph. Consider a ULDB relation `Color(vertex,coloring)` that for each vertex v_i of G has one x-tuple of the form: $Color(v_i, blue) \,||\, Color(v_i, red) \,||\, Color(v_i, green)$. The ULDB source (with empty lineage) I has relation `Color` and one more relation `edge(x,y)`. Relation `edge` contains an x-tuple with one alternative (v_k, v_l) if there exists an edge in G that connects vertices v_k and v_l and an x-tuple with one alternative (v_i, v_i) for every vertex v_i of G. Consider the following 2 source-to-target (copy) tgds: $Color(x, y) \rightarrow Color_2(x, y)$ and $edge(x, y) \rightarrow e_2(x, y)$ and the following 3 target egds: $e_2(x, y), Color_2(x, blue), Color_2(y, blue) \rightarrow x = y$, $e_2(x, y), Color_2(x, red), Color_2(y, red) \rightarrow x = y$ and $e_2(x, y), Color_2(x, green), Color_2(y, green) \rightarrow x = y$. Then it is easy to see that there exists a solution of this ULDB data exchange problem if and only if there exists a 3-coloring for the graph that is represented by the source relations.
ii) Consider the data exchange problem in the proof of part (i). Let $Q(x, y)$:- $e_2(x, y)$ and let $t \in Q(x, y)$ be an arbitrary arc of G. Then due to the data exchange constraints $t \in$ *uldb-cert*(Q, I) if and only if t is an arc of a graph G that has a 3-coloring.

5 Conclusion and Related Work

We investigated the problem of query answering in a data exchange setting in which the source data that is to be exchanged has uncertainty and lineage. A straightforward way to compute such ULDB certain answers is to first compute all the possible LDB instances of the source and compute LDB certain answers for each one. Even though LDB certain answering of CQs is polynomial for a set of weakly acyclic tgds, the number of the possible instances of a ULDB may be exponential, making this approach computationally expensive and unsuitable for large data sets. In contrast we presented a u-chase procedure that can be used in order to polynomially compute ULDB certain answers of CQs for a set of weakly acyclic tgds and a well-behaved ULDB source. U-chase will create a "pseudo-LDB", use our l-chase procedure on it and finally return a ULDB. Finally we showed that computing certain answers for CQs is no longer polynomial if we allow egds in our dependencies, contrary to what happens in certain data exchange.

Other data models that represent a set of possible worlds to capture incomplete information are presented in [13]. The conditional tables of [13] are complete for representing possible instances that are certain databases but do not support lineage tracking. In [11] peer data exchange for LDBs is considered that uses mappings of various trustworthiness and focuses on filtering out the untrusted answers. In their lineage (provenance) model, they record more detailed information than the LDB model we consider in our work here.

A similar with data exchange problem is data integration. In [7] data integration with uncertain mappings is considered but for certain sources. The procedure that is presented produces first all possible certain data integration problems that are represented due to uncertain mappings and then computes ordinary certain answers separately. In [14] the sources can be uncertain (but with no lineage) and several properties of uncertain data integration (a problem generally more complicated than data exchange) are formalized and discussed. The procedure that is used for certain query answering first produces all possible instances without uncertainty.

References

1. Afrati, F.N., Li, C., Pavlaki, V.: Data exchange: Query answering for incomplete data sources. In: InfoScale 2008, pp. 1–10. ICST (2008)
2. Afrati, F.N., Kolaitis, P.G.: Answering aggregate queries in data exchange. In: PODS, pp. 129–138 (2008)
3. Afrati, F.N., Li, C., Pavlaki, V.: Data exchange in the presence of arithmetic comparisons. In: EDBT, pp. 487–498 (2008)
4. Benjelloun, O., Sarma, A.D., Halevy, A.Y., Theobald, M., Widom, J.: Databases with uncertainty and lineage. VLDB J. 17(2), 243–264 (2008)
5. Buneman, P., Khanna, S., Tan, W.C.: Why and where: A characterization of data provenance. In: Van den Bussche, J., Vianu, V. (eds.) ICDT 2001. LNCS, vol. 1973, pp. 316–330. Springer, Heidelberg (2000)
6. Cui, Y., Widom, J.: Lineage tracing for general data warehouse transformations. VLDB J. 12(1), 41–58 (2003)
7. Das Sarma, A., Dong, L., Halevy, A.: Uncertainty in data integration. In: Managing and Mining Uncertain Data. Springer, Heidelberg (2009)
8. Das Sarma, A., Ullman, J.D., Widom, J.: Schema design for uncertain databases. In: AMW (2009)
9. Fagin, R., Kimelfeld, B., Kolaitis, P.: Probabilistic data exchange. In: ICDT 2010 (to appear, 2010)
10. Fagin, R., Kolaitis, P.G., Miller, R.J., Popa, L.: Data exchange: semantics and query answering. Theor. Comput. Sci. 336(1), 89–124 (2005)
11. Green, T.J., Karvounarakis, G., Ives, Z.G., Tannen, V.: Update exchange with mappings and provenance. In: VLDB, pp. 675–686 (2007)
12. Green, T.J., Karvounarakis, G., Tannen, V.: Provenance semirings. In: PODS, pp. 31–40 (2007)
13. Imielinski, T., Lipski Jr., W.: Incomplete information in relational databases. J. ACM 31(4), 761–791 (1984)
14. Magnani, M., Montesi, D.: Towards relational schema uncertainty. In: SUM, pp. 150–164 (2009)

A Formal Analysis of Logic-Based Argumentation Systems

Leila Amgoud and Philippe Besnard

IRIT-CNRS, 118, route de Narbonne,
31062 Toulouse Cedex 4 France
{amgoud,besnard}@irit.fr

Abstract. Dung's abstract argumentation model consists of a set of arguments and a binary relation encoding attacks among arguments. Different acceptability semantics have been defined for evaluating the arguments. What is worth noticing is that the model completely abstracts from the applications to which it can be applied. Thus, it is not clear what are the results that can be returned in a given application by each semantics. This paper answers this question. For that purpose, we start by plunging the model in a real application. That is, we assume that we have an inconsistent knowledge base (KB) containing formulas of an *abstract monotonic logic*. From this base, we show how to define arguments. Then, we characterize the different semantics in terms of the subsets of the KB that are returned by each extension. We show a full correspondence between maximal consistent subbases of a KB and maximal conflict-free sets of arguments. We show also that stable and preferred extensions choose *randomly* some consistent subbases of a base. Finally, we investigate the results of three argumentation systems that use well-known attack relations.

1 Introduction

Argumentation has become an Artificial Intelligence keyword for the last fifteen years, especially for handling inconsistency in knowledge bases (e.g. [2,6,16]), for decision making under uncertainty (e.g. [4,7,13]), and for modeling interactions between agents (e.g. [3,14,15]).

One of the most abstract argumentation systems was proposed in [12]. This system consists of a set of arguments and a binary relation encoding attacks among arguments. Different acceptability semantics were also defined for evaluating the arguments. A semantics defines the conditions under which a given set of arguments is declared acceptable. What is worth noticing is that the system completely abstracts from the applications to which it can be applied. Thus:

1. The origin and the structure of both arguments and the attack relation are not specified. This is seen in the literature as an advantage of this model. However, some of its instantiations lead to undesirable results. This is due to a lack of *methodology* for defining these two main components of the system.

A. Deshpande and A. Hunter (Eds.): SUM 2010, LNAI 6379, pp. 42–55, 2010.

2. The different acceptability semantics capture mainly different properties of the graph associated to the system. While it is clear that those properties are nice and meaningful, it is not clear whether they really make sense in concrete applications. It is not even clear what are the results that can be returned, in a given application, by each semantics.

In [1], we proposed an extension that fills in the gap between Dung's system and the applications. The idea was to consider all the ingredients involved in an argumentation problem. We started with an abstract monotonic logic which consists of a set of formulas and a consequence operator. We have shown how to build arguments from a knowledge base using the consequence operator of the logic, and how to choose an appropriate attack relation.

Starting from this class of logic-based argumentation systems, this paper characterizes the different semantics in terms of subsets of the knowledge base that are returned by each extension. The results we got show that there is a full correspondence between maximal consistent subbases of a KB and maximal conflict-free sets of arguments. They also show that stable and preferred extensions choose *randomly* some consistent subbases of a base, which are not necessarily maximal in case of preferred extensions. Finally, the paper studies the properties of three well-known argumentation systems, namely the ones that use resp. rebut, strong rebut and undercut as attack relations. We show that the system that is based on undercut returns sound results at the cost of redundancy: duplicating many arguments with exactly the same support.

The paper is organized as follows: Section 2 recalls our extension of Dung's system. Section 3 presents an analysis of acceptability semantics. Section 4 investigates the properties of particular systems.

2 An Extension of Dung's Abstract System

In [1], we proposed an extension of Dung's abstract argumentation system. That extension defines all the ingredients involved in argumentation ranging from the logic used to define arguments to the set of conclusions to be inferred from a given knowledge base. A great advantage of this extension is that it is abstract since it is grounded on an *abstract monotonic logic*. According to [17], such logic is defined as a pair $(\mathcal{L}, \mathsf{CN})$ where members of $\mathcal{L}$ are called *well-formed formulas*, and CN is a *consequence operator*. CN is any function from $2^{\mathcal{L}}$ to $2^{\mathcal{L}}$ that satisfies the following axioms:

1. $X \subseteq \mathsf{CN}(X)$ **(Expansion)**
2. $\mathsf{CN}(\mathsf{CN}(X)) = \mathsf{CN}(X)$ **(Idempotence)**
3. $\mathsf{CN}(X) = \bigcup_{Y \subseteq_f X} \mathsf{CN}(Y)$ **(Finiteness)**
4. $\mathsf{CN}(\{x\}) = \mathcal{L}$ for some $x \in \mathcal{L}$ **(Absurdity)**
5. $\mathsf{CN}(\emptyset) \neq \mathcal{L}$ **(Coherence)**

$Y \subseteq_f X$ means that Y is a finite subset of X. Intuitively, $\mathsf{CN}(X)$ returns the set of formulas that are logical consequences of X according to the logic in question.

It can easily be shown from the above axioms that CN is a closure operator, and satisfies *monotonicity* (i.e. for $X, X' \subseteq \mathcal{L}$, if $X \subseteq X'$ then $\mathsf{CN}(X) \subseteq \mathsf{CN}(X')$). Note that a wide variety of logics can be viewed as special cases of Tarski's notion of an abstract logic (classical logic, intuitionistic logic, modal logic, temporal logic, ...). Formulas of $\mathcal{L}$ encode both defeasible and undefeasible information while the consequence operator CN is used for defining arguments.

Definition 1 (Consistency). *Let $X \subseteq \mathcal{L}$. X is* consistent *in logic $(\mathcal{L}, \mathsf{CN})$ iff $\mathsf{CN}(X) \neq \mathcal{L}$. It is* inconsistent *otherwise.*

In simple English, this says that X is consistent iff its set of consequences is not the set of all formulas. The coherence requirement (absent from Tarski's original proposal but added here to avoid considering trivial systems) forces $\emptyset$ to always be consistent - this makes sense for any reasonable logic as saying emptiness should intuitively be consistent.

We start with an abstract logic $(\mathcal{L}, \mathsf{CN})$ from which the notions of argument and attacks between arguments are defined. More precisely, arguments are built from a knowledge base, say Σ, containing formulas of the language $\mathcal{L}$.

Definition 2 (Argument). *Let Σ be a knowledge base. An* argument *is a pair (X, x) such that:*

1. $X \subseteq \Sigma$
2. *X is consistent*
3. $x \in \mathsf{CN}(X)$
4. *$\nexists X' \subset X$ s.t. X' satisfies the three above conditions*

Notation: $\mathtt{Supp}$ and $\mathtt{Conc}$ denote respectively the *support* X and the *conclusion* x of an argument (X, x). For $\mathcal{S} \subseteq \Sigma$, $\mathtt{Arg}(\mathcal{S})$ denotes the set of all arguments that can be built from $\mathcal{S}$ by means of Definition 2.

Since CN is monotonic, argument construction is a monotonic process (i.e. $\mathtt{Arg}(\Sigma) \subseteq \mathtt{Arg}(\Sigma')$ whenever $\Sigma \subseteq \Sigma' \subseteq \mathcal{L}$).

We have shown in [1] that in order to satisfy the consistency rationality postulate proposed in [9], the attack relation should be chosen in an appropriate way. Otherwise, unintended results may be returned by the argumentation system. An appropriate relation, called *valid*, should ensure that the set of formulas used in arguments of any non-conflicting (called also *conflict-free*) set of arguments is consistent.

Notation: Let $\mathcal{B} \subseteq \mathtt{Arg}(\Sigma)$. $\mathtt{Base}(\mathcal{B}) = \bigcup_{a \in \mathcal{B}} \mathtt{Supp}(a)$.

Definition 3 (Valid attack relation). *Let $\mathtt{Arg}(\Sigma)$ be a set of arguments built from a knowledge base Σ. An attack relation $\mathcal{R} \subseteq \mathtt{Arg}(\Sigma) \times \mathtt{Arg}(\Sigma)$ is* valid *iff $\forall \mathcal{B} \subseteq \mathtt{Arg}(\Sigma)$, if $\mathcal{B}$ is conflict-free, then $\mathtt{Base}(\mathcal{B})$ is consistent.*

In [1], we have investigated the properties of a valid attack relation. Namely, we have shown that it should depend on the *minimal conflicts* contained in Σ, and also sensitive to them.

Definition 4. *Let* $\mathtt{Arg}(\Sigma)$ *be the set of arguments built from* Σ, *and* $\mathcal{R} \subseteq \mathtt{Arg}(\Sigma) \times \mathtt{Arg}(\Sigma)$.

- $C \subseteq \Sigma$ *is a* minimal conflict *iff i)* C *is inconsistent, and ii)* $\forall x \in C$, $C \backslash \{x\}$ *is consistent.*
- $\mathcal{R}$ *is* conflict-dependent *iff* $\forall a, b \in \mathtt{Arg}(\Sigma)$, $(a, b) \in \mathcal{R}$ *implies that there exists a minimal conflict* $C \in \mathcal{C}_\Sigma$[1] *s.t.* $C \subseteq \mathtt{Supp}(a) \cup \mathtt{Supp}(b)$.
- $\mathcal{R}$ *is* conflict-sensitive *iff* $\forall a, b \in \mathtt{Arg}(\Sigma)$ *s.t. there exists a minimal conflict* $C \in \mathcal{C}_\Sigma$ *with* $C \subseteq \mathtt{Supp}(a) \cup \mathtt{Supp}(b)$, *then either* $(a, b) \in \mathcal{R}$ *or* $(b, a) \in \mathcal{R}$.

In [1], we have shown that when the attack relation is conflict-dependent, from any consistent subset Σ, a conflict-free set of arguments is built.

Dung's abstract system is refined as follows.

Definition 5 (Argumentation system). *Given a knowledge base* Σ, *an* argumentation system *over* Σ *is a pair* $(\mathtt{Arg}(\Sigma), \mathcal{R})$ *s.t.* $\mathcal{R} \subseteq \mathtt{Arg}(\Sigma) \times \mathtt{Arg}(\Sigma)$ *is a valid attack relation.*

Among all the arguments, it is important to know which arguments to rely on for inferring conclusions from a base Σ. In [12], different acceptability semantics have been proposed. The basic idea behind these semantics is the following: for a rational agent, an argument is acceptable if he can defend this argument against all attacks on it. All the arguments acceptable for a rational agent will be gathered in a so-called *extension*. An extension must satisfy a consistency requirement (i.e. conflict-free) and must defend all its elements. Recall that a set $\mathcal{B} \subseteq \mathtt{Arg}(\Sigma)$ *defends* an argument a iff $\forall b \in \mathtt{Arg}(\Sigma)$, if $(b, a) \in \mathcal{R}$, then $\exists c \in \mathcal{B}$ such that $(c, b) \in \mathcal{R}$. The different acceptability semantics defined in [12] are recalled below. Let $\mathcal{B}$ be a conflict-free set of arguments:

- $\mathcal{B}$ is an *admissible* extension iff $\mathcal{B}$ defends all its elements.
- $\mathcal{B}$ is a *preferred* extension iff it is a maximal (for set inclusion) admissible extension.
- $\mathcal{B}$ is a *stable* extension iff it is a preferred extension that attacks any argument in $\mathtt{Arg}(\Sigma) \setminus \mathcal{B}$.

3 Relating Acceptability Semantics to the Knowledge Base

The aim of this section is to understand the underpinnings of the different acceptability semantics introduced in [12]. The idea is to analyze the results returned by each semantics in terms of subsets of the knowledge base at hand.

The first result shows that when the attack relation is conflict-dependent and conflict-sensitive, then from a maximal consistent subbase of Σ, it is possible to build a *unique* maximal (wrt set inclusion) conflict-free set of arguments.

---
[1] Let $\mathcal{C}_\Sigma$ denote the set of all minimal conflicts of Σ.

Proposition 1. *Let Σ be a knowledge base, and $(\mathcal{S}_i)_{i\in\mathcal{I}}$ be its maximal consistent subsets. If $\mathcal{R}$ is conflict-dependent and conflict-sensitive, then:*

1. *For all $i \in \mathcal{I}$, $\texttt{Arg}(\mathcal{S}_i)$ is a maximal (wrt set $\subseteq$) conflict-free subset of $\texttt{Arg}(\Sigma)$.*
2. *For all $i, j \in \mathcal{I}$, if $\texttt{Arg}(\mathcal{S}_i) = \texttt{Arg}(\mathcal{S}_j)$ then $\mathcal{S}_i = \mathcal{S}_j$.*
3. *For all $i \in \mathcal{I}$, $\mathcal{S}_i = \texttt{Base}(\texttt{Arg}(\mathcal{S}_i))$.*

Similarly, we show that each maximal conflict-free subset of $\texttt{Arg}(\Sigma)$ is built from a unique maximal consistent subbase of Σ. However, this result is *only* true when the attack relation is chosen in an "appropriate" way, i.e., when it is valid.

Proposition 2. *Let $(\texttt{Arg}(\Sigma), \mathcal{R})$ be an argumentation system over Σ, and $(\mathcal{E}_i)_{i\in\mathcal{I}}$ be the maximal (wrt set $\subseteq$) conflict-free subsets of $\texttt{Arg}(\Sigma)$. If $\mathcal{R}$ is conflict-dependent and valid, then:*

1. *For all $i \in \mathcal{I}$, $\texttt{Base}(\mathcal{E}_i)$ is a maximal (wrt set $\subseteq$) consistent subbase of Σ.*
2. *For all $i, j \in \mathcal{I}$, if $\texttt{Base}(\mathcal{E}_i) = \texttt{Base}(\mathcal{E}_j)$ then $\mathcal{E}_i = \mathcal{E}_j$.*
3. *For all $i \in \mathcal{I}$, $\mathcal{E}_i = \texttt{Arg}(\texttt{Base}(\mathcal{E}_i))$.*

Propositions 1 and 2 provide a full correspondence between maximal consistent subbases of Σ and maximal conflict-free subsets of $\texttt{Arg}(\Sigma)$.

Corollary 1. *Let $(\texttt{Arg}(\Sigma), \mathcal{R})$ be an argumentation system over Σ. If $\mathcal{R}$ is conflict-dependent and valid, then the maximal conflict-free subsets of $\texttt{Arg}(\Sigma)$ are exactly the $\texttt{Arg}(\mathcal{S})$ where $\mathcal{S}$ ranges over the maximal (wrt set inclusion) consistent subbases of Σ.*

This result is very surprising since, apart from stable semantics, all the other acceptability semantics (e.g. preferred and admissible) are not about maximal conflict-free sets of arguments, but rather subsets of them. This means that those semantics do not necessarily yield maximal subsets of the knowledge base Σ.

3.1 Stable Semantics

The idea behind stable semantics is that a set of arguments is "acceptable" if it attacks any argument that is outside the set. This condition makes stable semantics very strong and the existence of stable extensions not guaranteed for every argumentation system. From the results obtained in the previous subsection on the link between maximal conflict-free subsets of $\texttt{Arg}(\Sigma)$ and maximal consistent subbases of Σ, it seems that stable semantics is not adequate. Let us illustrate this issue on the following example.

Example 1. *Let Σ be such that $\texttt{Arg}(\Sigma) = \{a, b, c\}$. Also, let the attack relation be as depicted in the figure below. This relation is assumed to be conflict-dependent and valid.*

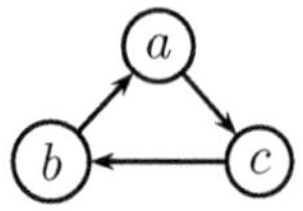

There are three maximal conflict-free sets of arguments. From Corollary 1, there is a full correspondence between the maximal (wrt set $\subseteq$*) consistent subsets of* Σ *and the maximal (wrt set* $\subseteq$*) conflict-free sets of* $\texttt{Arg}(\Sigma)$.

1. $\mathcal{E}_1 = \{a\}$ — $\mathcal{S}_1 = \texttt{Base}(\mathcal{E}_1)$
2. $\mathcal{E}_2 = \{b\}$ — $\mathcal{S}_2 = \texttt{Base}(\mathcal{E}_2)$
3. $\mathcal{E}_3 = \{c\}$ — $\mathcal{S}_3 = \texttt{Base}(\mathcal{E}_3)$

Indeed, the base Σ *has "stable sets of formulas"* $\mathcal{S}_1, \mathcal{S}_2$ *and* $\mathcal{S}_3$ *while these are not captured by the stable semantics since the argumentation system* $(\texttt{Arg}(\Sigma), \mathcal{R})$ *has no stable extension. To make the example more illustrative, let us mention that an instance of the example arises from a logic of set-theoretic difference.*

The above example shows that stable semantics can miss "stable sets of formulas", i.e. maximal consistent subsets of the knowledge base at hand. However, we show that when an argumentation system $(\texttt{Arg}(\Sigma), \mathcal{R})$ has stable extensions, those extensions capture maximal consistent subsets of Σ.

Proposition 3. *If* $\mathcal{R}$ *is conflict-dependent and valid, then any stable extension* $\mathcal{E}$ *of* $(\texttt{Arg}(\Sigma), \mathcal{R})$, $\texttt{Base}(\mathcal{E})$ *is a maximal (wrt set* $\subseteq$*) consistent subset of* Σ.

The fact that stable extensions "return" maximal consistent subsets of Σ does not mean that all of them are returned. The following example illustrates this issue, and shows that Dung's system somehow picks some maximal consistent subbases of Σ.

Example 2. *Let* Σ *be such that* $\texttt{Arg}(\Sigma) = \{a, b, c, d, e, f, g\}$ *and the attack relation is as depicted in the figure below. Assume also that this relation is both conflict-dependent and valid.*

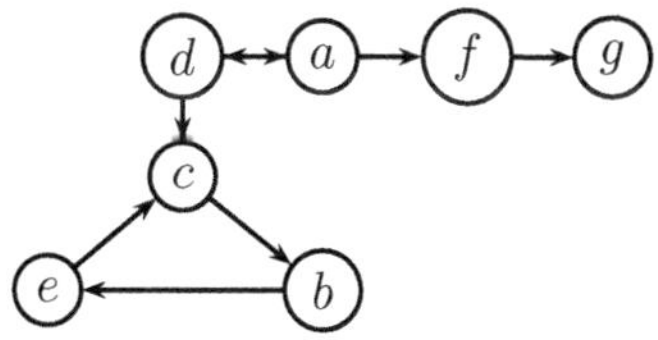

There are 5 maximal conflict-free subsets of $\texttt{Arg}(\Sigma)$*:*

1. $\mathcal{E}_1 = \{d, e, f\}$ — $\mathcal{S}_1 = \texttt{Base}(\mathcal{E}_1)$
2. $\mathcal{E}_2 = \{b, d, f\}$ — $\mathcal{S}_2 = \texttt{Base}(\mathcal{E}_2)$
3. $\mathcal{E}_3 = \{a, c, g\}$ — $\mathcal{S}_3 = \texttt{Base}(\mathcal{E}_3)$
4. $\mathcal{E}_4 = \{a, e, g\}$ — $\mathcal{S}_4 = \texttt{Base}(\mathcal{E}_4)$
5. $\mathcal{E}_5 = \{a, b, g\}$ — $\mathcal{S}_5 = \texttt{Base}(\mathcal{E}_5)$

Since the attack relation is both conflict-dependent and valid, then each set $\mathcal{E}_i$ *returns a maximal consistent subbase* $\mathcal{S}_i$ *of* Σ*. From Corollary 1, it is clear that the maximal consistent subbases of* Σ *are exactly the five* $\texttt{Base}(\mathcal{E}_i)$*. The argumentation system* $(\texttt{Arg}(\Sigma), \mathcal{R})$ *has only one stable extension which is* $\mathcal{E}_2$*. Thus, Dung's approach picks only one maximal subbase, i.e.,* $\mathcal{S}_2$ *among the five. Such* *a choice is clearly counter-intuitive since there is no additional information that allows us to elicit* $\mathcal{S}_2$ *against the others. There is a huge literature on handling inconsistency in knowledge bases. None of the approaches can make a choice between the above five subbases* $\mathcal{S}_i$ *if no additional information is available, like priorities between formulas of* Σ.

The above result shows that from a maximal consistent subbase of Σ, it is not always the case that a stable extension exists as its counterpart in the argumentation framework. For instance, $\mathcal{S}_1$ is a maximal consistent subbase of Σ, but its corresponding set of arguments, $\mathcal{E}_1$, is not a stable extension. This depends broadly on the attack relation that is considered in the argumentation system. The above example shows that when the attack relation is asymmetric then a full correspondence between stable extensions and maximal consistent subbases of Σ is not guaranteed.

Let us now analyze the case of a symmetric attack relation. The following result shows that each maximal consistent subbase of Σ "returns" a stable extension, provided that the attack relation is conflict-dependent and conflict-sensitive.

Proposition 4. *Let* $\mathcal{S}$ *be a maximal (wrt set inclusion) subset of* Σ*. If* $\mathcal{R}$ *is conflict-dependent, conflict-sensitive and symmetric, then* $\mathtt{Arg}(\mathcal{S})$ *is a stable extension of* $(\mathtt{Arg}(\Sigma), \mathcal{R})$.

From the above result, it follows that an argumentation system based on a knowledge base Σ always has stable extensions unless the knowledge base contains only inconsistent formulas.

Corollary 2. *Let* $(\mathtt{Arg}(\Sigma), \mathcal{R})$ *be an argumentation system based on a knowledge base* Σ *such that* $\mathcal{R}$ *is symmetric, conflict-dependent and conflict-sensitive. If there exists* $x \in \Sigma$ *s.t.* $\{x\}$ *is consistent, then* $(\mathtt{Arg}(\Sigma), \mathcal{R})$ *has at least one stable extension.*

In order to have a full correspondence between the stable extensions of $(\mathtt{Arg}(\Sigma), \mathcal{R})$ and the maximal consistent subsets of Σ, the attack relation should be symmetric, conflict-dependent and valid.

Corollary 3. *Let* $(\mathtt{Arg}(\Sigma), \mathcal{R})$ *be an argumentation system based on a knowledge base* Σ *such that* $\mathcal{R}$ *is symmetric, conflict-dependent and valid. Each maximal conflict-free subset of* $\mathtt{Arg}(\Sigma)$ *is exactly* $\mathtt{Arg}(\mathcal{S})$ *where* $\mathcal{S}$ *ranges over the maximal (wrt set inclusion) consistent subsets of* Σ.

However, in [1], we have shown that when the attack relation is symmetric, it is not valid in general. Namely, when the knowledge base contains n-ary (other then binary) minimal conflicts, then the rationality postulate on consistency [9] is violated due to symmetric relations, that are thereby ruled out. What happens in this case is that the argumentation system not only "returns" stable extensions that capture all the maximal consistent subsets of Σ but also other stable extensions corresponding to inconsistent subsets of Σ.

3.2 Other Acceptability Semantics

Preferred semantics has been introduced in order to palliate the limits of the stable semantics. Indeed, in [12], it has been shown that each argumentation system has at least one preferred extension. Preferred extensions are maximal sets of arguments that defend themselves against all attacks. The following example shows that preferred semantics may fail to capture consistent subbases of a knowledge base.

Example 3 (Example 1 cont). *The argumentation system* $(\texttt{Arg}(\Sigma), \mathcal{R})$ *of Example 1 has one preferred extension which is the empty set. The corresponding base is thus the empty set. However, the knowledge base* Σ *has three maximal consistent subbases.*

Let us now analyze the case where the argumentation system $(\texttt{Arg}(\Sigma), \mathcal{R})$ has non-empty preferred extensions. When the attack relation is valid, each preferred extension of $(\texttt{Arg}(\Sigma), \mathcal{R})$ "returns" a consistent subbase of Σ. However, this does not mean that these subbases are maximal wrt set inclusion as shown in the following example.

Example 4 (Example 2 cont). *The argumentation system of Example 2 has two preferred extensions:* $\mathcal{E}_2 = \{b, d, z\}$ *and* $\mathcal{E}_6 = \{a, y\}$. *The set* $\mathcal{E}_6$ *is a subset of* $\mathcal{E}_3$, $\mathcal{E}_4$ *and* $\mathcal{E}_5$ *whose bases are maximal for set inclusion. Therefore,* $\texttt{Base}(\mathcal{E}_6)$ *is not maximal, whereas* $\texttt{Base}(\mathcal{E}_2)$ *is maximal.*

As for stable extensions, preferred extensions rely on a seemingly random pick. In the above example, it is unclear why $\texttt{Base}(\mathcal{E}_6)$ is picked instead of another consistent subset of Σ. Similarly, why is $\texttt{Base}(\mathcal{E}_2)$ picked instead of $\texttt{Base}(\mathcal{S}_1)$ for instance?

It is worth mentioning that the remaining acceptability semantics (i.e. admissible, grounded and complete) also "return" somehow *randomly* some (but not necessarily all) consistent subbases of Σ *even* when the attack relation is defined carefully. Let us consider the following simple example.

Example 5. *Let* Σ *be such that* $\texttt{Arg}(\Sigma) = \{a, b, c\}$. *Also, let the attack relation be as depicted in the figure below. This relation is assumed to be conflict-dependent and valid.*

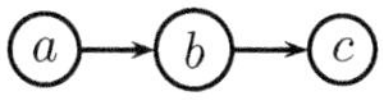

There are two maximal conflict-free sets of arguments: $\mathcal{E}_1$ *and* $\mathcal{E}_2$.

1. $\mathcal{E}_1 = \{a, c\}$ $\quad \mathcal{S}_1 = \texttt{Base}(\mathcal{E}_1)$
2. $\mathcal{E}_2 = \{b\}$ $\quad \mathcal{S}_2 = \texttt{Base}(\mathcal{E}_2)$

From Corollary 1, there is a full correspondence between the maximal (wrt set inclusion) consistent subbases of Σ *and the maximal (wrt set inclusion) conflict-free sets of* $\texttt{Arg}(\Sigma)$. *Thus, the knowledge base* Σ *has exactly two maximal consistent subbases* $\mathcal{S}_1$ *and* $\mathcal{S}_2$. *The argumentation system* $(\texttt{Arg}(\Sigma), \mathcal{R})$ *has one grounded extension which is* $\mathcal{E}_1$. *Thus, conclusions supported by* a *and* c *will be inferred from* Σ. *Here again it is not clear why the base* $\mathcal{S}_2$ *is discarded.*

Recently, several new acceptability semantics have been introduced in the literature. In [8], a weaker version of stable semantics has been defined. The new semantics, called *semi-stable*, is between stable and preferred semantics. It has been shown that each stable extension is a semi-stable one, and each semi-stable extension is a preferred one. When the attack relation has desirable features then, semi-stable extensions return "some" consistent bases which are not necessarily maximal wrt set inclusion. Example 1 has been handled in [8]. It has been shown that the empty set is the only semi-stable extension of the system. However, we have shown that the knowledge base has three maximal consistent subsets.

Corollary 4. *For any semi-stable extension* $\mathcal{E}$ *of* $(\texttt{Arg}(\Sigma), \mathcal{R})$*, where* $\mathcal{R}$ *is valid,* $\texttt{Base}(\mathcal{E})$ *is consistent.*

In [5], six new acceptability semantics have been proposed. That work was merely motivated by the fact that Dung's semantics treat differently odd-length cycles and even-length cycles. Thus, they suggested semantics that handle the two cases in a similar way. Those semantics yield extensions that are not necessarily maximal conflict-free subsets of the original set of arguments. This means that their corresponding bases are not maximal for set inclusion.

The very last acceptability semantics that has been proposed in the literature is the so-called *ideal* semantics [11]. This semantics always gives a single extension. The latter is an admissible extension that is contained in every preferred extension. To show that this semantics suffers from the problems discussed here, it is sufficient to consider Example 5. Indeed, in that example, there is one ideal extension which is $\{a, c\}$. We have shown that this result is ad hoc since it allows to infer formulas from a seemingly randomly chosen consistent subset of the knowledge base Σ.

4 Particular Argumentation Systems

In this section we will study three argumentation systems. These systems use the three attack relations that are commonly used in the literature: rebut, strong rebut and assumption attack (called here "undercut"). Although these relations have originally been defined in a propositional logic context, we will generalize them to any logic in the sense of Tarski.

The rebut relation captures the idea that two arguments with contradictory conclusions are conflicting. This relation is thus defined as follows:

Definition 6 (Rebut). *Let* $a, b \in \texttt{Arg}(\Sigma)$*. The argument* a rebuts b *iff the set* $\{\texttt{Conc}(a),\ \texttt{Conc}(b)\}$ *is inconsistent.*

It is easy to show that this relation is conflict-dependent.

Property 1. The relation "Rebut" is conflict-dependent.

It can be checked that this relation is *not conflict-sensitive*. Let us, for instance, consider the following two arguments built from a propositional base: $(\{x \wedge y\}, x)$ and $(\{\neg y \wedge z\}, z)$. It is clear that the set $\{x \wedge y, \neg y \wedge z\}$ is a minimal conflict, but none of the two arguments is rebutting the other. Note also that the rebut relation violates most of the properties studied in [1]. In particular, it is not homogeneous. Since the rebut relation is symmetric, it is not valid in the case that the knowledge base contains n-ary (other than binary) minimal conflicts. Worse yet, this relation is not valid even in the binary case. This can be seen from the following conflict-free set of arguments, $\{(\{x, x \rightarrow y\}, y), (\{\neg x, \neg x \rightarrow z\}, z)\}$, whose corresponding base is inconsistent. This means that the argumentation system $(\mathtt{Arg}(\Sigma), \text{Rebut})$ violates consistency. Consequently, this system is not recommended for inferring conclusions from a knowledge base Σ.

Property 2. If Σ is such that $\exists C = \{x_1, \ldots, x_n\} \subseteq \mathcal{C}_\Sigma$ where either $n > 2$ or $n = 2$ and $\exists x'_1 \in \mathsf{CN}(\{x_1\}, \exists x'_2 \in \mathsf{CN}(\{x_2\}$ s.t. $\{x'_1, x'_2\}$ is inconsistent then $(\mathtt{Arg}(\Sigma), \mathit{Rebut})$ violates extension consistency.

Let us now consider the second symmetric relation, called "Strong Rebut". This relation is defined as follows:

Definition 7 (Strong Rebut). *Let $a, b \in \mathtt{Arg}(\Sigma)$. The argument a* strongly rebuts *b iff $\exists x \in \mathsf{CN}(\mathtt{S}upp(a))$ and $\exists y \in \mathsf{CN}(\mathtt{S}upp(b))$ such that the set $\{x, y\}$ is inconsistent.*

The following result shows that "Strong Rebut" satisfies interesting properties.

Property 3. "Strong Rebut" is conflict-dependent and conflict-sensitive.

Unfortunately, since this relation is symmetric, then it is not valid in the general case, i.e. when n-ary (other than binary) minimal conflicts occur in the knowledge base Σ.

Property 4. Let Σ be such that $\exists C \in \mathcal{C}_\Sigma$ such that $|C| > 2$. An argumentation system $(\mathtt{Arg}(\Sigma), \text{Strong Rebut})$ violates consistency.

In the particular case where all the minimal conflicts of the knowledge base base Σ are binary, we have shown in [1] that conflict-sensitive attack relations are valid. Since "Strong Rebut" is conflict-sensitive, then it is a valid relation.

Property 5. Let Σ be such that all the minimal conflicts in $\mathcal{C}_\Sigma$ are binary. Any argumentation system $(\mathtt{Arg}(\Sigma), \text{Strong Rebut})$ satisfies consistency.

Finally, we show that when all the minimal conflicts of the knowledge base Σ are binary, then there is a full correspondence between the maximal consistent subsets of Σ and the stable extensions of $(\mathtt{Arg}(\Sigma), \text{Strong Rebut})$. Remember that, since "Strong Rebut" is symmetric, then all the semantics coincide as they all give the maximal conflict-free subsets of $\mathtt{Arg}(\Sigma)$.

Property 6. Let $(\texttt{Arg}(\Sigma), \text{Strong Rebut})$ be an argumentation system over a knowledge base Σ whose minimal conflicts are all binary. For any stable extension $\mathcal{E}$ of $(\texttt{Arg}(\Sigma), \text{Strong Rebut})$, $\texttt{Base}(\mathcal{E})$ is a maximal consistent subset of Σ. For any maximal consistent subset $\mathcal{S}$ of Σ, $\texttt{Arg}(\mathcal{S})$ is a stable extension of $(\texttt{Arg}(\Sigma), \text{Strong Rebut})$.

Let us now investigate the results of an argumentation system that is built upon the attack relation "Undercut". Recall that the idea behind this relation is that an argument may undermine a premise of another argument. Formally:

Definition 8 (Undercut). *Let $a, b \in \texttt{Arg}(\Sigma)$. The argument a* undercuts *b iff $\exists x \in \texttt{Supp}(b)$ such that the set $\{\texttt{Conc}(a), x\}$ is inconsistent.*

The next property shows the main feature of this attack relation.

Property 7. The relation "Undercut" is conflict-dependent.

It can be checked that this relation is unfortunately not conflict-sensitive. For instance, neither the argument $(\{x \wedge y\}, x)$ undercuts the argument $(\{\neg y \wedge z\}, z)$ nor $(\{\neg y \wedge z\}, z)$ undercuts $(\{x \wedge y\}, x)$. However, the set $\{x \wedge y, \neg y \wedge z\}$ is clearly a minimal conflict. Due to this limitation, undercut is *not valid.*

Property 8. The relation "Undercut" is not valid.

Somewhat surprisingly, the argumentation system $(\texttt{Arg}(\Sigma)$, Undercut) satisfies extension consistency. Indeed, the base of any admissible extension of this argumentation system is consistent. The reason is that the arguments that do not behave correctly w.r.t. Undercut, are defended by other arguments.

Property 9. If $\mathcal{E}$ is an admissible extension of an argumentation system $(\texttt{Arg}(\Sigma)$, Undercut), then $\texttt{Base}(\mathcal{E})$ is consistent.

Unfortunately, the fact that the extensions (under a given acceptability semantics) of $(\texttt{Arg}(\Sigma)$, Undercut) satisfy consistency does not mean that "Undercut" is a good attack relation. This relation works only when "all" the arguments that can be built from a knowledge base are extracted and taken into account. In many applications, like dialogue between two or more agents, this is inadequate. Indeed, in a dialogue, agents seldom utter all the possible arguments.

In [10], it has been shown that when the knowledge base Σ is encoded in *propositional logic*, that is, $(\mathcal{L}, \mathsf{CN})$ is propositional logic, there is a correspondence between the stable extensions of the argumentation system $(\texttt{Arg}(\Sigma), \text{Undercut})$ based on Σ and the maximal consistent subsets of Σ. The following result shows that this property holds for any logic in the sense of Tarski.

Proposition 5. *Let $(\texttt{Arg}(\Sigma)$, Undercut) be an argumentation system over a knowledge base Σ.*

1. *For all stable extension $\mathcal{E}$ of $(\texttt{Arg}(\Sigma), \textit{Undercut})$, $\texttt{Base}(\mathcal{E})$ is a maximal (wrt set inclusion) consistent subset of Σ.*

2. *For all stable extension* $\mathcal{E}$ *of* ($\texttt{Arg}(\Sigma)$, *Undercut*), $\mathcal{E} = \texttt{Arg}(\texttt{Base}(\mathcal{E}))$.

The following result shows that each maximal (wrt set inclusion) consistent subset of Σ "returns" exactly one stable extension of ($\texttt{Arg}(\Sigma)$, Undercut).

Proposition 6. *Let* ($\texttt{Arg}(\Sigma)$, *Undercut*) *be an argumentation system over a knowledge base* Σ.

1. *For all maximal (wrt set inclusion) consistent subset* $\mathcal{S}$ *of* Σ, $\texttt{Arg}(\mathcal{S})$ *is a stable extension of* ($\texttt{Arg}(\Sigma)$, *Undercut*).
2. *For all maximal (wrt set inclusion) consistent subset* $\mathcal{S}$ *of* Σ, $\mathcal{S} = \texttt{Base}(\texttt{Arg}(\mathcal{S}))$.

From the above results, a full correspondence between the stable extensions of ($\texttt{Arg}(\Sigma)$, Undercut) and the maximal consistent subsets of Σ is established.

Corollary 5. *Let* ($\texttt{Arg}(\Sigma)$, *Undercut*) *be an argumentation system over a knowledge base* Σ. *The stable extensions of* ($\texttt{Arg}(\Sigma)$, *Undercut*) *are exactly* $\texttt{Arg}(\mathcal{S})$ *where* $\mathcal{S}$ *ranges over the maximal (wrt set inclusion) consistent subsets of* Σ.

Since each maximal consistent subset of Σ "returns" a stable extension of the argumentation system ($\texttt{Arg}(\Sigma)$, Undercut), then the latter always has stable extensions unless the knowledge base Σ contains only inconsistent formulas.

Corollary 6. *Let* ($\texttt{Arg}(\Sigma)$, *Undercut*) *be an argumentation system over a knowledge base* Σ. *If there exists* $x \in \Sigma$ *s.t.* $\{x\}$ *is consistent, then* ($\texttt{Arg}(\Sigma)$, *Undercut*) *has at least one stable extension.*

The results presented in this section are disappointing. They show that the three classes of argumentation systems that are that have been given extensive attention in the literature suffer from collapse. The argumentation system ($\texttt{Arg}(\Sigma)$, Rebut) should not be used since it violates the postulate on consistency, while the system ($\texttt{Arg}(\Sigma)$, Strong Rebut) should only be used in the particular case that all minimal conflicts of Σ are binary. The third argumentation system ($\texttt{Arg}(\Sigma)$, Undercut) is misleading inasmuch as it satisfies extension consistency while the "Undercut" relation is not valid. However, its stable extensions "return" all and only maximal consistent subbases of the knowledge base at hand. Thus, it can be used specifically under the stable semantics. Anyway, we should keep in mind that the "Undercut" relation satisfies extension consistency at the cost of needing to take into account arguments that have different conclusions but the same support: so much redundancy! The results show that the system based on the "Undercut" relation guarantees extension consistency (at a prohibitive cost) but "Undercut" does not enjoy nice properties.

5 Conclusion

The paper has analyzed the different acceptability semantics introduced in [12] in terms of the subbases that are returned by each extension. The results of the analysis show that there is a complete correspondence between maximal consistent subbases of a KB and maximal conflict-free sets of arguments. Regarding stable extensions, we have shown that, when the attack relation is chosen in an appropriate way, then they return maximal consistent subbases of KB. However, they don't return all of them. A choice is made in an unclear way. The case of preferred extensions is worse since they compute consistent subbases but not necessarily maximal ones. When the attack relation is considered symmetric, inconsistent subbases are returned by conflict-free sets of arguments when ternary or more minimal conflicts are available in a KB. This is due to the binary character of the attack relation.

The paper studies three argumentation systems that rely on well-known attack relations. The results show that the system based on the "Undercut" relation guarantees extension consistency (at a prohibitive cost) but "Undercut" does not enjoy nice properties.

References

1. Amgoud, L., Besnard, P.: Bridging the gap between abstract argumentation systems and logic. In: Proceedings of the 3rd International Conference on Scalable Uncertainty, pp. 12–27 (2009)
2. Amgoud, L., Cayrol, C.: Inferring from inconsistency in preference-based argumentation frameworks. International Journal of Automated Reasoning 29(2), 125–169 (2002)
3. Amgoud, L., Parsons, S., Maudet, N.: Arguments, dialogue, and negotiation. In: Proceedings of the 14th European Conference on Artificial Intelligence (ECAI 2000), pp. 338–342. IOS Press, Amsterdam (2000)
4. Amgoud, L., Prade, H.: Using arguments for making and explaining decisions. Artificial Intelligence Journal 173, 413–436 (2009)
5. Baroni, P., Giacomin, M., Guida, G.: Scc-recursiveness: a general schema for argumentation semantics. Artificial Intelligence Journal 168, 162–210 (2005)
6. Besnard, P., Hunter, A.: Elements of Argumentation. MIT Press, Cambridge (2008)
7. Bonet, B., Geffner, H.: Arguing for decisions: A qualitative model of decision making. In: Proceedings of International Conference on Uncertainty in Artificial Intelligence (UAI 1996), pp. 98–105 (1996)
8. Caminada, M.: Semi-stable semantics. In: Proceedings of the 1st International Conference on Computational Models of Argument, pp. 121–130 (2006)
9. Caminada, M., Amgoud, L.: On the evaluation of argumentation formalisms. Artificial Intelligence Journal 171(5-6), 286–310 (2007)
10. Cayrol, C.: On the relation between argumentation and non-monotonic coherence-based entailment. In: Proceedings of the 14th International Joint Conference on Artificial Intelligence, pp. 1443–1448 (1995)
11. Dung, P., Mancarella, P., Toni, F.: Computing ideal skeptical argumentation. Artificial Intelligence Journal 171, 642–674 (2007)
12. Dung, P.M.: On the acceptability of arguments and its fundamental role in nonmonotonic reasoning, logic programming and n-person games. Artificial Intelligence Journal 77, 321–357 (1995)

13. Kakas, A., Moraitis, P.: Argumentation based decision making for autonomous agents. In: Proceedings of the 2nd International Joint Conference on Autonomous Agents and Multi-Agents systems (AAMAS 2003), pp. 883–890 (2003)
14. Kraus, S., Sycara, K., Evenchik, A.: Reaching agreements through argumentation: a logical model and implementation. Journal of Artificial Intelligence 104, 1–69 (1998)
15. Prakken, H.: Coherence and flexibility in dialogue games for argumentation. Journal of Logic and Computation 15, 1009–1040 (2005)
16. Simari, G., Loui, R.: A mathematical treatment of defeasible reasoning and its implementation. AIJ 53, 125–157 (1992)
17. Tarski, A.: On Some Fundamental Concepts of Metamathematics. Logic, Semantics, Metamathematic. Edited and translated by J. H. Woodger. Oxford Uni. Press, Oxford (1956)

Handling Inconsistency with Preference-Based Argumentation

Leila Amgoud and Srdjan Vesic

IRIT-CNRS, 118, route de Narbonne,
31062 Toulouse Cedex 4 France
{amgoud,vesic}@irit.fr

Abstract. Argumentation is a promising approach for handling inconsistent knowledge bases, based on the justification of plausible conclusions by arguments. Due to inconsistency, arguments may be attacked by counterarguments. The problem is thus to evaluate the arguments in order to select the most acceptable ones.

The aim of this paper is to make a bridge between the argumentation-based and the coherence-based approaches for handling inconsistency. We are particularly interested by the case where priorities between the formulas of an inconsistent knowledge base are available. For that purpose, we will use the rich preference-based argumentation framework (PAF) we have proposed in an earlier work. A rich PAF has two main advantages: i) it overcomes the limits of existing PAFs, and ii) it encodes two different roles of preferences between arguments (handling critical attacks and refining the evaluation of arguments). We show that there exist full correspondences between particular cases of these PAF and two well known coherence-based approaches, namely the preferred sub-theories and the democratic as well.

1 Introduction

An important problem in the management of knowledge-based systems is the handling of inconsistency. Inconsistency may be present for mainly three reasons:

- The knowledge base includes default rules. Let us consider for instance the general rules 'birds fly', 'penguins are birds' and the specific rule 'penguins do not fly'. If we add the fact 'Tweety is a penguin', we may conclude that Tweety does not fly because it is a penguin, and also that Tweety flies because it is a bird.
- In model-based diagnosis, a knowledge base contains a description of the normal behavior of a system, together with observations made on this system. Failure detection occurs when observations conflict with the normal functioning mode of the system and the hypothesis that the components of the system are working well; that leads to diagnose which component fails;
- Several consistent knowledge bases pertaining to the same domain, but coming from different sources of information, are available. For instance, each source is a reliable specialist in some aspect of the concerned domain but is less reliable in other aspects. A straightforward way of building a global base Σ is to concatenate the knowledge bases Σ_i provided by each source. Even if each base Σ_i is consistent, it is unlikely that their concatenation will be consistent also.

A. Deshpande and A. Hunter (Eds.): SUM 2010, LNAI 6379, pp. 56–69, 2010.

Classical logic has many appealing features for knowledge representation and reasoning, but unfortunately when reasoning with inconsistent information, i.e. drawing conclusions from an inconsistent knowledge base, the set of classical consequences is trivialized. To solve this problem, two kinds of approaches have been proposed. The first one, called *coherence-based* approach and initiated in [10], proposes to give up some formulas of the knowledge base in order to get one or several consistent subbases of the original base. Then plausible conclusions may be obtained by applying classical entailment on these subbases. The second approach accepts inconsistency and copes with it. Indeed, it retains all the available information but prohibits the logic from deriving trivial conclusions. Argumentation is one of these approaches. Its basic idea is that each plausible conclusion inferred from the knowledge base is justified by some reason(s), called *argument*(s), for believing in it. Due to inconsistency, those arguments may be attacked by other arguments (called counterarguments). The problem is thus to evaluate the arguments in order to select the most acceptable ones.

In [7], it has been shown that the results of the coherence-based approach proposed in [10] can be recovered within Dung's argumentation framework [9]. Indeed, there is a full correspondence between the maximal consistent subbases of a given inconsistent knowledge base and the stable extensions of the argumentation system built over the same base. In [10], the formulas of the knowledge base are assumed to be equally preferred. This assumption has been discarded in [6] and in [8]. Indeed, in the former work, a knowledge base is equipped with a total preorder. Thus, instead of computing the maximal consistent subbases, *preferred sub-theories* are computed. These sub-theories are consistent subbases that privilege the most important formulas. In [8], the knowledge base is rather equipped with a partial preorder. The idea was to define a preference relation, called *democratic relation*, between the consistent subbases. The best subbases, called *democratic sub-theories*, wrt this relation are used for inferring conclusions from the knowledge base.

The aim of this paper is to investigate whether it is possible to recover the results of these two works within an argumentation framework. Since priorities are available, it is clear that we need a preference-based argumentation framework (PAF). Recently, we have shown in [3] that existing PAFs (developed in [2,4]) are not appropriate since they may return unintended results, especially when the attack relation is asymmetric. Moreover, their results are not optimal since they may be refined by the available preferences between arguments. Consequently, we have proposed in the same paper (i.e. [3]) a new family of PAFs, called *rich PAF*, that encodes two distinct roles of preferences between arguments: handling critical attacks (that is an argument is stronger than its attacker) and refining the result of the evaluation of arguments using acceptability semantics. In this paper, we show that there is a full correspondence between the preferred sub-theories proposed in [6] and the stable extensions of an instance of this rich PAF, and also a full correspondence between the democratic sub-theories developed in [8] and another instance of the rich PAF. The two correspondences are obtained by choosing appropriately the main components of a rich PAF: the definition of an argument, the attack relation, the preference relation between arguments and the preference relation between subsets of arguments.

The paper is organized as follows: Sections 2 and 3 recall respectively the rich PAF in [3] and the two works of [6,8]. Section 4 shows how instances of the rich PAF compute preferred and democratic sub-theories of a knowledge base. The last section is devoted to some concluding remarks.

2 Preference-Based Argumentation Frameworks

In [9], Dung has developed the most abstract argumentation framework in the literature. It consists of a set of arguments and an attack relation between them.

Definition 1 (Argumentation framework [9]). *An* argumentation framework *(AF) is a pair $\mathcal{F} = (\mathcal{A}, \mathcal{R})$, where $\mathcal{A}$ is a set of arguments and $\mathcal{R}$ is an attack relation ($\mathcal{R} \subseteq \mathcal{A} \times \mathcal{A}$). The notation $a\mathcal{R}b$ means that the argument a* attacks *the argument b.*

In the above definition, the arguments and attacks are abstract entities since Dung's framework completely abstracts from the application. However, the two components can be defined as follows when handling inconsistency in a *propositional* knowledge base Σ.

Definition 2 (Argument - Undercut). *Let Σ be a propositional knowledge base.*

- *An* argument *is a pair $a = (H, h)$ s.t.*
  - $H \subseteq \Sigma$
  - *H is consistent*
  - $H \vdash h$
  - *$\nexists H' \subset H$ such that H' is consistent and $H' \vdash h$.*
- *An argument (H, h)* undercuts *an argument (H', h') iff $\exists h'' \in H'$ s.t. $h \equiv \neg h''$.*

Notations: Let $a = (H, h)$ be an argument (in the sense of Definition 2). The functions Supp and Conc return respectively the support H and the conclusion h of the argument a. For $\mathcal{S} \subseteq \Sigma$, $\texttt{Arg}(\mathcal{S}) = \{(H, h) \mid (H, h)$ is an argument in the sense of Definition 2 and $H \subseteq \mathcal{S}\}$. Thus, $\texttt{Arg}(\Sigma)$ denotes the set of all the arguments that can be built from the whole knowledge base Σ.

Example 1. Let $\Sigma = \{x, \neg y, x \to y\}$ be a propositional knowledge base. The following arguments are built from this base:

$a_1 : (\{x\}, x)$	$a_2 : (\{\neg y\}, \neg y)$
$a_3 : (\{x \to y\}, x \to y)$	$a_4 : (\{x, \neg y\}, x \wedge \neg y)$
$a_5 : (\{\neg y, x \to y\}, \neg x)$	$a_6 : (\{x, x \to y\}, y)$

The figure below depicts the attacks wrt "undercut".

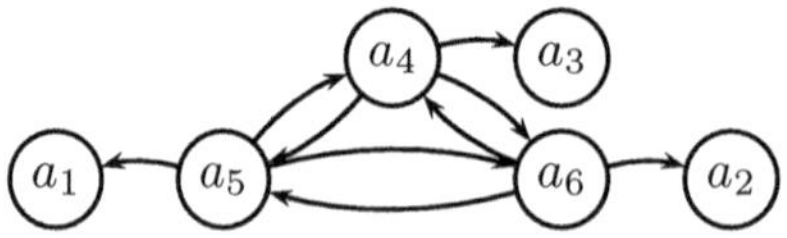

Different *acceptability semantics* for evaluating arguments have been proposed in the same paper [9]. Each semantics amounts to define sets of acceptable arguments, called *extensions*. For the purpose of our paper, we only need to recall stable semantics.

Definition 3 (Conflict-free, Stable semantics [9]). *Let $\mathcal{F} = (\mathcal{A}, \mathcal{R})$ be an AF, $\mathcal{B} \subseteq \mathcal{A}$.*

- *$\mathcal{B}$ is* conflict-free *iff $\nexists$ a, $b \in \mathcal{B}$ such that $a\mathcal{R}b$.*
- *$\mathcal{B}$ is a* stable *extension iff it is conflict-free and attacks any element in $\mathcal{A} \setminus \mathcal{B}$.*

Example 1 (Cont): The argumentation framework of Example 1 has three stable extensions: $\mathcal{E}_1 = \{a_1, a_2, a_4\}$, $\mathcal{E}_2 = \{a_2, a_3, a_5\}$ and $\mathcal{E}_3 = \{a_1, a_3, a_6\}$.

The attack relation is the backbone of any acceptability semantics in [9]. An attack from an argument b towards an argument a always wins unless b is itself attacked by another argument. However, this assumption is very strong because some attacks cannot always 'survive'. Especially when the attacked argument is stronger than its attacker.

Throughout the paper, the relation $\geq \subseteq \mathcal{A} \times \mathcal{A}$ is assumed to be a preorder (reflexive and transitive). For two arguments a and b, writing $a \geq b$ (or $(a, b) \in \geq$) means that a is at least as strong as b. The relation $>$ is the strict version of $\geq$. Indeed, $a > b$ iff $a \geq b$ and not $(b \geq a)$. Examples of such relations are those based on the certainty level of the formulas of a propositional knowledge base Σ. The base Σ is equipped with a total preorder $\trianglerighteq$. For two formulas x and y, writing $x \trianglerighteq y$ means that x is at least as certain as y. In this case, the base Σ is stratified into $\Sigma_1 \cup \ldots \cup \Sigma_n$ such that formulas of Σ_i have the same certainty level and are more certain than formulas in Σ_j where $j > i$. The stratification of Σ enables to define a certainty level of each subset $\mathcal{S}$ of Σ. It is the highest number of stratum met by this subset. Formally:

$$\texttt{Level}(\mathcal{S}) = \max\{i \mid \exists\, x \in \mathcal{S} \cap \Sigma_i\} \text{ (with } \texttt{Level}(\emptyset) = 0).$$

The above certainty level is used in [5] in order to define a total preorder on the set of arguments that can be built from a stratified knowledge base. The preorder is defined as follows:

Definition 4 (Weakest link principle [5]). *Let $\Sigma = \Sigma_1 \cup \ldots \cup \Sigma_n$ be a propositional knowledge base and $(H, h), (H', h') \in \texttt{Arg}(\Sigma)$. The argument (H, h) is preferred to (H', h'), denoted by $(H, h) \geq_{WLP} (H', h')$, iff $\texttt{Level}(H) \leq \texttt{Level}(H')$.*

Example 1 (Cont): Assume that $\Sigma = \Sigma_1 \cup \Sigma_2$ with $\Sigma_1 = \{x\}$ and $\Sigma_2 = \{x \to y, \neg y\}$. It holds that $\texttt{Level}(\{x\}) = 1$ while $\texttt{Level}(\{\neg y\}) = \texttt{Level}(\{x \to y\}) = \texttt{Level}(\{x, \neg y\}) = \texttt{Level}(\{\neg y, x \to y\}) = \texttt{Level}(\{x, x \to y\}) = 2$. Thus, $a_1 \geq_{WLP} a_2, a_3, a_4, a_5, a_6$ while the five other arguments are all equally preferred.

In [2,4], Dung's argumentation framework has been extended by preferences between arguments. The idea behind those works is to remove *critical* attacks[1] and to apply Dung's semantics on the remaining attacks. Unfortunately, this solution does not work, in particular, when the attack relation is asymmetric. It returns extensions which are

[1] An attack $(b, a) \in \mathcal{R}$ is *critical* iff $a > b$ (i.e. $a \geq b$ and not $(b \geq a)$).

not necessarily conflict-free wrt the attack relation. This leads to undesirable results as illustrated by the following example.

Example 1 (Cont): The classical approaches of PAFs remove the critical attack from a_5 to a_1 (since $a_1 >_{WLP} a_5$) and get $\{a_1, a_2, a_3, a_5\}$ as a stable extension. Note that this extension, which intends to support a *coherent point of view*, is conflicting since it contains both a_1 and a_5 and support thus both x and $\neg x$.

The approach followed in [2,4] suffers from another problem. Its results may need to be *refined* by preferences between arguments as shown by the following example.

Example 2. Let us consider the AF depicted in the figure below.

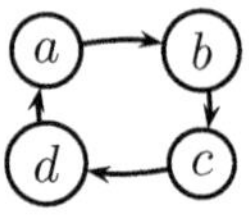

Assume that $a > b$ and $c > d$. The corresponding PAF has two stable extensions: $\{a, c\}$ and $\{b, d\}$. Note that any element of $\{b, d\}$ is weaker than at least one element of the set $\{a, c\}$. Thus, it is natural to consider $\{a, c\}$ as better than $\{b, d\}$. Consequently, we may conclude that the two arguments a and c are "more acceptable" than b and d.

What is worth noticing is that a refinement amounts to *compare* subsets of arguments. In Example 2, the so-called *democratic* relation, $\succeq_d$, can be used for comparing the two sets $\{a, c\}$ and $\{b, d\}$. This relation is defined as follows:

Definition 5 (Democratic relation). *Let Δ be a set of objects and $\geq \subseteq \Delta \times \Delta$ be a partial preorder. For $\mathcal{X}, \mathcal{X}' \subseteq \Delta$, $\mathcal{X} \succeq_d \mathcal{X}'$ iff $\forall x' \in \mathcal{X}' \setminus \mathcal{X}$, $\exists x \in \mathcal{X} \setminus \mathcal{X}'$ such that $x > x'$ (i.e. $x \geq x'$ and not$(x' \geq x)$).*

In [3], we have proposed a novel approach which palliates the limits of the existing ones. It follows two steps:

1. To repair the critical attacks by computing a new attack relation $\mathcal{R}_r$.
2. To refine the results of the framework $(\mathcal{A}, \mathcal{R}_r)$ by comparing its extensions using a refinement relation.

The idea behind the first step is to modify the graph of attacks in such a way that, for any critical attack, the preference between the arguments is taken into account and the conflict between the two arguments of the attack is represented. For this purpose, we *invert* the arrow of the critical attack. For instance, in Example 1, the arrow from a_5 to a_1 is replaced by another arrow emanating from a_1 towards a_5. The intuition behind this is that an attack between two arguments represents in some sense two things: i) an incoherence between the two arguments, and ii) a kind of preference determined by the direction of the attack. Thus, in our approach, the direction of the arrow represents a "real" preference between arguments. Moreover, the conflict is kept between the two arguments. Dung's acceptability semantics are then applied on the modified graph.

Definition 6 (PAF [3]). *A* preference-based argumentation framework *(PAF) is a tuple $\mathcal{T} = (\mathcal{A}, \mathcal{R}, \geq)$ where $\mathcal{A}$ is a set of arguments, $\mathcal{R} \subseteq \mathcal{A} \times \mathcal{A}$ is an attack relation and $\geq$ is a (partial or total) preorder on $\mathcal{A}$. The extensions of $\mathcal{T}$ under a given semantics are the* *extensions of the argumentation framework* $(\mathcal{A}, \mathcal{R}_r)$, *called* repaired framework, *under the same semantics with:* $\mathcal{R}_r = \{(a,b)|(a,b) \in \mathcal{R} \text{ and not } (b > a)\} \cup \{(b,a)|(a,b) \in \mathcal{R} \text{ and } b > a\}$.

This approach does not suffer from the drawback of the existing one. Indeed, it delivers conflict-free extensions of arguments.

Property 1. Let $\mathcal{T} = (\mathcal{A}, \mathcal{R}, \geq)$ be a PAF and $\mathcal{E}_1, \ldots, \mathcal{E}_n$ its extensions under a given semantics. For all $i = 1, \ldots, n$, $\mathcal{E}_i$ is conflict-free wrt $\mathcal{R}$.

At the second step, the result of the above PAF is refined using a refinement relation. The two steps are captured in an abstract framework, called *rich preference-based argumentation framework*.

Definition 7 (Rich PAFs [3]). *A* rich PAF *is a tuple* $\mathcal{T} = (\mathcal{A}, \mathcal{R}, \geq, \succeq)$ *where* $\mathcal{A}$ *is a set of arguments,* $\mathcal{R} \subseteq \mathcal{A} \times \mathcal{A}$ *is an attack relation,* $\geq \subseteq \mathcal{A} \times \mathcal{A}$ *is a (partial or total) preorder and* $\succeq \subseteq \mathcal{P}(\mathcal{A}) \times \mathcal{P}(\mathcal{A})$[2] *is a refinement relation. The extensions of* $\mathcal{T}$ *under a given semantics are the elements of* $\mathtt{Max}(\mathcal{S}, \succeq)$[3] *where* $\mathcal{S}$ *is the set of extensions (under the same semantics) of the PAF* $(\mathcal{A}, \mathcal{R}, \geq)$.

Example 3. Let us consider the argumentation framework depicted in the left side of the following figure.

Assume that $a > b$, $c > d$ and $b > c$. The repaired framework corresponding to $(\mathcal{A}, \mathcal{R}, \geq)$ is depicted in the right side of the above figure. This latter has two stable extensions $\{a, c\}$ and $\{b, d\}$. According to the democratic relation $\succ_d$, it is clear that the first extension is better than the second one. Thus, the set $\{a, c\}$ is the stable extension of the rich PAF $(\mathcal{A}, \mathcal{R}, \geq, \succeq_d)$.

In [3], we have studied deeply the properties of the rich PAF. However, for the purpose of this paper we do not need to recall them.

3 Coherence-Based Approach for Handling Inconsistency

The coherence-based approach for handling inconsistency in a propositional knowledge base Σ follows two steps: At the first step, some subbases of Σ are chosen. In [10], these subbases are the maximal (for set inclusion) consistent ones. At the second step, an inference mechanism is chosen. This later defines the inferences to be made from Σ. An example of inference mechanism is the one that infers a formula if it is a classical conclusion of all the chosen subbases.

[2] $\mathcal{P}(\mathcal{A})$ denotes the power set of the set $\mathcal{A}$.

[3] $\mathtt{Max}(\mathcal{S}, \succeq) = \{s \in \mathcal{S} \mid \nexists s' \in \mathcal{S} \text{ s.t. } s' \succeq s \text{ and not } (s \succeq s')\}$.

Several works have been done on choosing the subbases, in particular when Σ is equipped with a (total or partial) preorder $\trianglerighteq$ ($\trianglerighteq \subseteq \Sigma \times \Sigma$). Recall that when $\trianglerighteq$ is total, Σ is stratified into $\Sigma_1 \cup \ldots \cup \Sigma_n$ such that $\forall i, j$ with $i \neq j$, $\Sigma_i \cap \Sigma_j = \emptyset$. Moreover, Σ_1 contains the most important formulas while Σ_n contains the least important ones.

In [6], the knowledge base Σ is equipped with a total preorder. The chosen subbases privilege the most important formulas.

Definition 8 (Preferred sub-theory [6]). *Let Σ be stratified into $\Sigma_1 \cup \ldots \cup \Sigma_n$. A* preferred sub-theory *is a set $\mathcal{S} = \mathcal{S}_1 \cup \ldots \cup \mathcal{S}_n$ such that $\forall k \in [1, n]$, $\mathcal{S}_1 \cup \ldots \cup \mathcal{S}_k$ is a maximal (for set inclusion) consistent subbase of $\Sigma_1 \cup \ldots \cup \Sigma_k$.*

Example 1 (Cont): The knowledge base $\Sigma = \Sigma_1 \cup \Sigma_2$ with $\Sigma_1 = \{x\}$ and $\Sigma_2 = \{x \to y, \neg y\}$ has two preferred sub-theories: $\mathcal{S}_1 = \{x, x \to y\}$ and $\mathcal{S}_2 = \{x, \neg y\}$.

It can be shown that the preferred sub-theories of a knowledge base Σ are maximal (wrt set inclusion) consistent subbases of Σ.

Property 2. Each preferred sub-theory of a knowledge base Σ is a maximal (for set inclusion) consistent subbase of Σ.

In [8], the above definition has been extended to the case where Σ is equipped with a partial preorder $\trianglerighteq$. The basic idea was to define a preference relation on the power set of Σ. The best elements according to this relation are the preferred theories , called also *democratic sub-theories.* The relation that generalizes preferred sub-theories is the democratic relation (see Definition 5). In this context, Δ is Σ and $\geq$ is the relation $\trianglerighteq$. In what follows, $\triangleright$ denotes the strict version of $\trianglerighteq$. Thus:

$$\text{Let } \mathcal{S}, \mathcal{S}' \subseteq \Sigma.\ \mathcal{S} \succeq_d \mathcal{S}' \text{ iff } \forall x' \in \mathcal{S}' \setminus \mathcal{S}, \exists x \in \mathcal{S} \setminus \mathcal{S}' \text{ such that } x \triangleright x'.$$

Definition 9 (Democratic sub-theory [8]). *Let Σ be propositional knowledge base and $\trianglerighteq \subseteq \Sigma \times \Sigma$ be a partial preorder. A* democratic sub-theory *is a set $\mathcal{S} \subseteq \Sigma$ such that $\mathcal{S}$ is consistent and $(\nexists \mathcal{S}' \subseteq \Sigma)$ s.t. $\mathcal{S}'$ is consistent and $\mathcal{S}' \succeq_d \mathcal{S}$.*

Example 4. Let $\Sigma = \{x, \neg x, y, \neg y\}$ be such that $\neg x \trianglerighteq y$ and $\neg y \trianglerighteq x$. Let $\mathcal{S}_1 = \{x, y\}$, $\mathcal{S}_2 = \{x, \neg y\}$, $\mathcal{S}_3 = \{\neg x, y\}$, and $\mathcal{S}_4 = \{\neg x, \neg y\}$. The three subbases $\mathcal{S}_2$, $\mathcal{S}_3$ and $\mathcal{S}_4$ are the democratic sub-theories of Σ. However, $\mathcal{S}_1$ is not a democratic sub-theory since $\mathcal{S}_4 \succeq_d \mathcal{S}_1$.

It is easy to show that the democratic sub-theories of a knowledge base Σ are maximal (for set inclusion) consistent.

Property 3. Each democratic sub-theory of a knowledge base Σ is a maximal (for set inclusion) consistent subbase of Σ.

4 Computing Sub-theories with Argumentation

This section shows how two instances of the rich PAF presented in Section 2 compute the preferred and the democratic sub-theories of a propositional knowledge base Σ.

The two instances use all the arguments that can be built from Σ using Definition 2 (i.e. the set $\mathtt{Arg}(\Sigma)$). Similarly, they both use the attack relation "Undercut" given also in Definition 2. However, as we will see next, they are grounded on distinct preference relations between arguments. The last component of a rich PAF is a preference relation on the power set of $\mathtt{Arg}(\Sigma)$. Both instances will use the democratic relation $\succeq_d$. Thus, for recovering preferred and democratic sub-theories, we will use two instances of the rich PAF $(\mathtt{Arg}(\Sigma), \text{Undercut}, \geq, \succeq_d)$.

It can be shown that when the preference relation $\geq$ is a total preorder, then the stable extensions of the PAF $(\mathtt{Arg}(\Sigma), \text{Undercut}, \geq)$ are all incomparable wrt the democratic relation $\succeq_d$.

Property 4. Let $\mathcal{T} = (\mathtt{Arg}(\Sigma), \text{Undercut}, \geq)$ be a PAF. For all stable extensions $\mathcal{E}$ and $\mathcal{E}'$ of $\mathcal{T}$ with $\mathcal{E} \neq \mathcal{E}'$, if $\geq$ is a total preorder, then $\neg(\mathcal{E} \succeq_d \mathcal{E}')$.

From the previous property, it follows that the stable extensions of $(\mathtt{Arg}(\Sigma)$, Undercut, $\geq)$ coincide with those of the rich PAF $(\mathtt{Arg}(\Sigma)$, Undercut, $\geq$, $\succeq_d)$.

Property 5. If $\geq$ is a total preorder, then the stable extensions of $(\mathtt{Arg}(\Sigma)$, Undercut, $\geq$, $\succeq_d)$ are exactly the stable extensions of $(\mathtt{Arg}(\Sigma)$, Undercut, $\geq)$.

Notation: For $\mathcal{B} \subseteq \mathtt{Arg}(\Sigma)$, $\mathtt{Base}(\mathcal{B}) = \bigcup \mathtt{Supp}(a)$ where $a \in \mathcal{B}$.

The following result summarizes some useful properties of the two functions: Arg and Base.

Property 6.
- For any consistent subbase $\mathcal{S} \subseteq \Sigma$, $\mathcal{S} = \mathtt{Base}(\mathtt{Arg}(\mathcal{S}))$.
- The function Base is surjective but not injective.
- For any $\mathcal{E} \subseteq \mathtt{Arg}(\Sigma)$, $\mathcal{E} \subseteq \mathtt{Arg}(\mathtt{Base}(\mathcal{E}))$.
- The function Arg is injective but not surjective.

Another property that is important for the rest of the paper relates the notion of consistency of a set of formulas to that of conflict-freeness of a set of arguments.

Property 7. A set $\mathcal{S} \subseteq \Sigma$ is consistent *iff* $\mathtt{Arg}(\mathcal{S})$ is conflict-free.

The following example shows that the previous property does not hold for an arbitrary set of arguments.

Example 5. Let $\mathcal{E} = \{(\{x\}, x), (\{x \to y\}, x \to y), (\{\neg y\}, \neg y)\}$. It is obvious that $\mathcal{E}$ is conflict-free while $\mathtt{Base}(\mathcal{E})$ is not consistent.

Assumption: In the rest of this paper, we assume that a knowledge base Σ contains only consistent formulas.

4.1 Recovering the Preferred Sub-Theories

In this section, we will show that there is a full correspondence between the preferred sub-theories of a knowledge base Σ and the stable extensions of the PAF $(\texttt{Arg}(\Sigma),$ Undercut, $\geq_{WLP})$. Recall that the relation $\geq_{WLP}$ is based on the weakest link principle and privileges the arguments whose less important formulas are more important than the less important formulas of the other arguments. This relation is a total preorder and is defined over a knowledge base that is itself equipped with a total preorder. According to Property 5, the stable extensions of $(\texttt{Arg}(\Sigma),$ Undercut, $\geq_{WLP})$ coincide with those of $(\texttt{Arg}(\Sigma),$ Undercut, $\geq_{WLP}, \succeq_d)$.

The first result shows that from a preferred sub-theory, it is possible to build a unique stable extension of the PAF $(\texttt{Arg}(\Sigma),$ Undercut, $\geq_{WLP})$.

Theorem 1. *Let Σ be a stratified knowledge base. For all preferred sub-theory $\mathcal{S}$ of Σ, it holds that:*

- $\texttt{Arg}(\mathcal{S})$ *is a stable extension of* $(\texttt{Arg}(\Sigma),$ *Undercut,* $\geq_{WLP})$
- $\mathcal{S} = \texttt{Base}(\texttt{Arg}(\mathcal{S}))$

Similarly, we show that each stable extension of $(\texttt{Arg}(\Sigma),$ Undercut, $\geq_{WLP})$ is built from a unique preferred sub-theory of Σ.

Theorem 2. *Let Σ be a stratified knowledge base. For all stable extension $\mathcal{E}$ of $(\texttt{Arg}(\Sigma),$ Undercut, $\geq_{WLP})$, it holds that:*

- $\texttt{Base}(\mathcal{E})$ *is a preferred sub-theory of* Σ
- $\mathcal{E} = \texttt{Arg}(\texttt{Base}(\mathcal{E}))$

The next theorem shows that there exists a one-to-one correspondence between preferred sub-theories of Σ and stable extensions of $(\texttt{Arg}(\Sigma),$ Undercut, $\geq_{WLP})$.

Theorem 3. *Let $\mathcal{T} = (\texttt{Arg}(\Sigma),$ Undercut, $\geq_{WLP})$ be a PAF over a stratified knowledge base Σ. The stable extensions of $\mathcal{T}$ are exactly the $\texttt{Arg}(\mathcal{S})$ where $\mathcal{S}$ ranges over the preferred sub-theories of Σ.*

From the above result, it follows that the PAF $(\texttt{Arg}(\Sigma),$ Undercut, $\geq_{WLP})$ has at least one stable extension unless the formulas of Σ are all inconsistent.

Corollary 1. *The PAF $(\texttt{Arg}(\Sigma),$ Undercut, $\geq_{WLP})$ has at least one stable extension.*

Example 1 (Cont): Figure 1 shows the two preferred sub-theories of Σ as well as the two stable extensions of the corresponding PAF.

4.2 Recovering the Democratic Sub-theories

Recall that the democratic sub-theories of a knowledge base Σ generalize the preferred sub-theories when Σ is equipped with a partial preorder $\trianglerighteq$. Thus, in order to capture the democratic sub-theories, we will use the generalized version of the preference relation $\geq_{WLP}$ which is defined in [1] as follows:

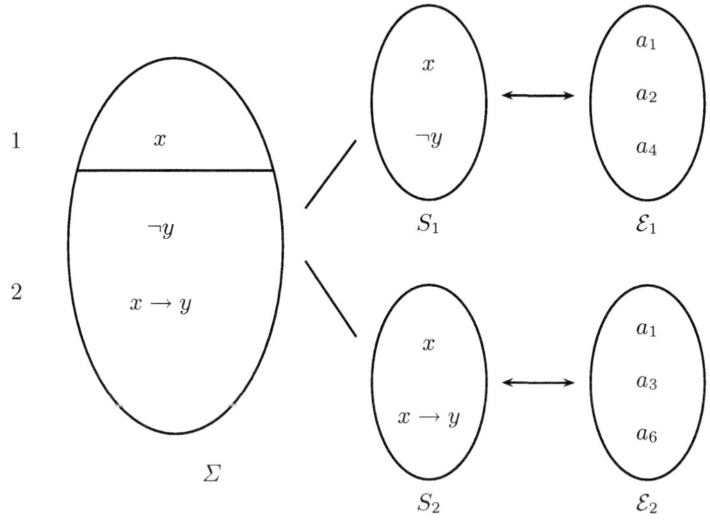

Fig. 1. Preferred sub-theories of Σ + Stable extensions of $(\texttt{Arg}(\Sigma), \text{Undercut}, \geq_{WLP})$

Definition 10 (Generalized weakest link principle [1]). *Let Σ be a knowledge base which is equipped with a partial preorder $\trianglerighteq$. For two arguments $(H, h), (H', h') \in \texttt{Arg}(\Sigma)$, $(H, h) \geq_{GWLP} (H', h')$ iff $\forall k \in H$, $\exists k' \in H'$ such that $k \triangleright k'$ (i.e. $k \trianglerighteq k'$ and not $(k' \trianglerighteq k)$).*

It can be shown that from each democratic sub-theory of a knowledge base Σ, a stable extension of $(\texttt{Arg}(\Sigma), \text{Undercut}, \geq_{GWLP})$ can be built.

Theorem 4. *Let Σ be a knowledge base which is equipped with a partial preorder $\trianglerighteq$. For all democratic sub-theory $\mathcal{S}$ of Σ, it holds that $\texttt{Arg}(S)$ is a stable extension of $(\texttt{Arg}(\Sigma), \text{Undercut}, \geq_{GWLP})$.*

The following result shows that each stable extension of the PAF $(\texttt{Arg}(\Sigma), \text{Undercut}, \geq_{GWLP})$ returns a maximal consistent subbase of Σ.

Theorem 5. *Let Σ be a knowledge base which is equipped with a partial preorder $\trianglerighteq$. For all stable extension $\mathcal{E}$ of $(\texttt{Arg}(\Sigma), \text{Undercut}, \geq_{GWLP})$, it holds that:*

- $\texttt{Base}(\mathcal{E})$ *is a maximal (for set inclusion) consistent subbase of Σ.*
- $\mathcal{E} = \texttt{Arg}(\texttt{Base}(\mathcal{E}))$.

The following example shows that the stable extensions of $(\texttt{Arg}(\Sigma), \text{Undercut}, \geq_{GWLP})$ do not necessarily return democratic sub-theories.

Example 4 (Cont): Recall that $\Sigma = \{x, \neg x, y, \neg y\}$, $\neg x \trianglerighteq y$ and $\neg y \trianglerighteq x$. Let $\mathcal{S} = \{x, y\}$. It can be checked that the set $\texttt{Arg}(\mathcal{S})$ is a stable extension of $(\texttt{Arg}(\Sigma), \text{Undercut}, \geq_{GWLP})$. However, $\mathcal{S}$ is not a democratic sub-theory since $\{\neg x, \neg y\} \succeq_d \mathcal{S}$.

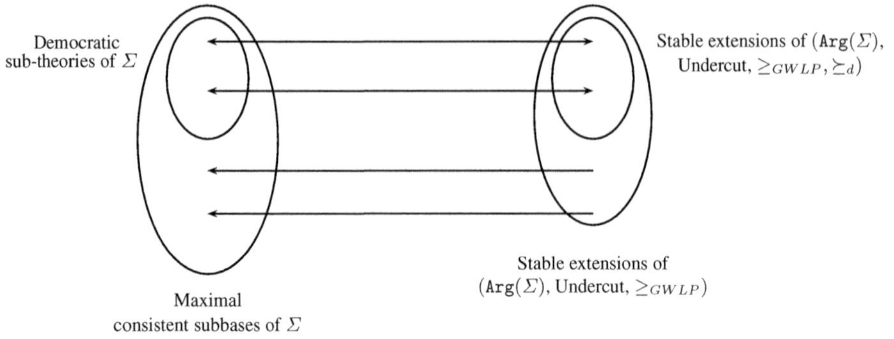

Fig. 2. Summary

It can also be shown that the converse of the above theorem is not true. Indeed, a knowledge base may have a maximal consistent subbase $\mathcal{S}$ and $\mathtt{Arg}(\mathcal{S})$ is not a stable extension of ($\mathtt{Arg}(\Sigma)$, Undercut, $\geq_{GWLP}$). Let us consider the following example.

Example 6. Let $\Sigma = \{x, \neg x\}$ and $x \rhd \neg x$. It is clear that $\{\neg x\}$ is a maximal consistent subbase of Σ while $\mathtt{Arg}(\{\neg x\})$ is not a stable extension of ($\mathtt{Arg}(\Sigma)$, Undercut, $\geq_{GWLP}$).

The following result establishes a link between the 'best' maximal consistent subbases of Σ wrt the democratic relation $\succeq_d$ and the 'best' sets of arguments wrt the same relation $\succeq_d$.

Theorem 6. *Let $\mathcal{S}, \mathcal{S}' \subseteq \Sigma$ be two maximal (for set inclusion) consistent subbases of Σ. It holds that $\mathcal{S} \succeq_d \mathcal{S}'$ iff $\mathtt{Arg}(\mathcal{S}) \succeq_d \mathtt{Arg}(\mathcal{S}')$.*

We also show that from each democratic sub-theory of Σ, one can build a stable extension of the corresponding rich PAF, and each stable extension of the rich PAF is built from a democratic sub-theory.

Theorem 7. *Let Σ be equipped with a partial preorder $\trianglerighteq$.*

- *For all democratic sub-theory $\mathcal{S}$ of Σ, $\mathtt{Arg}(\mathcal{S})$ is a stable extension of the rich PAF ($\mathtt{Arg}(\Sigma)$, Undercut, $\geq_{GWLP}, \succeq_d$).*
- *For each stable extension $\mathcal{E}$ of ($\mathtt{Arg}(\Sigma)$, Undercut, $\geq_{GWLP}, \succeq_d$), $\mathtt{Base}(\mathcal{E})$ is a democratic sub-theory of Σ.*

Finally, we show that there is a one-to-one correspondence between the democratic sub-theories of a base Σ and the stable extensions of its corresponding rich PAF.

Theorem 8. *The stable extensions of ($\mathtt{Arg}(\Sigma)$, Undercut, $\geq_{GWLP}, \succeq_d$) are exactly the $\mathtt{Arg}(\mathcal{S})$ where $\mathcal{S}$ ranges over the democratic subtheories of Σ.*

Figure 2 synthetizes the different links between the democratic sub-theories of a knowledge base Σ and the stable extensions of its corresponding PAF and rich PAF.

5 Conclusion

The paper has proposed a new approach for preference-based argumentation frameworks. This approach allows to encode two roles of preferences between arguments: handling critical attacks and refining the result of the evaluation. It is clearly argued in the paper that the two roles are completely independent and should be modeled in different ways and at different steps of the evaluation process. Then, we have shown that the approach is well-founded since it allows to recover very well known works on handling inconsistency in knowledge bases, namely the ones that restore the consistency of the knowledge base. Indeed, we have shown full correspondences between instances of the new PAF and respectively the preferred sub-theories defined by Brewka in [6] and the democratic sub-theories proposed by Cayrol, Royer and Saurel in [8].

References

1. Amgoud, L., Cayrol, C.: Inferring from inconsistency in preference-based argumentation frameworks. International Journal of Automated Reasoning 29(2), 125–169 (2002)
2. Amgoud, L., Cayrol, C.: A reasoning model based on the production of acceptable arguments. Annals of Mathematics and Artificial Intelligence 34, 197–216 (2002)
3. Amgoud, L., Vesic, S.: On the role of preferences in argumentation frameworks. Technical report, IRIT–Paul Sabatier University (May 2010)
4. Bench-Capon, T.J.M.: Persuasion in practical argument using value-based argumentation frameworks. Journal of Logic and Computation 13(3), 429–448 (2003)
5. Benferhat, S., Dubois, D., Prade, H.: Argumentative inference in uncertain and inconsistent knowledge bases. In: Proceedings of the 9th Conference on Uncertainty in Artificial Intelligence, pp. 411–419 (1993)
6. Brewka, G.: Preferred subtheories: An extended logical framework for default reasoning. In: Proceedings of the Eleventh International Joint Conference on Artificial Intelligence, pp. 1043–1048 (1989)
7. Cayrol, C.: On the relation between argumentation and non-monotonic coherence-based entailment. In: Proceedings of the 14th International Joint Conference on Artificial Intelligence, pp. 1443–1448. Morgan Kaufmann, San Francisco (1995)
8. Cayrol, C., Royer, V., Saurel, C.: Management of preferences in assumption-based reasoning. In: Valverde, L., Bouchon-Meunier, B., Yager, R.R. (eds.) IPMU 1992. LNCS, vol. 682, pp. 13–22. Springer, Heidelberg (1993)
9. Dung, P.M.: On the acceptability of arguments and its fundamental role in nonmonotonic reasoning, logic programming and n-person games. Artificial Intelligence Journal 77, 321–357 (1995)
10. Rescher, N., Manor, R.: On inference from inconsistent premises. Journal of Theory and decision 1, 179–219 (1970)

Appendix

Proof of Property 1. Every set $\mathcal{E} \subseteq \mathcal{A}$ is conflict-free wrt $\mathcal{R}$ iff it is conflict-free wrt $\mathcal{R}_r$. Since extensions are conflict-free wrt $\mathcal{R}_r$, then they are conflict-free wrt $\mathcal{R}$.

Proof of Property 3. Let $\mathcal{S}$ be a democratic sub-theory. From Definition 9, $\mathcal{S}$ is consistent. Assume now that $\mathcal{S}$ is not a maximal (for set inclusion) consistent set. Thus,

$\exists x \in \Sigma \setminus \mathcal{S}$ s.t. $\mathcal{S} \cup \{x\}$ is consistent. It is clear that $\mathcal{S} \cup \{x\} \succ_d \mathcal{S}$. This contradicts the fact that $\mathcal{S}$ is a democratic sub-theory.

Proof of Property 4. Let $\mathcal{E}, \mathcal{E}'$ be two stable extensions of $(\mathtt{Arg}(\Sigma), \text{Undercut}, \geq)$, and let $\mathcal{E} \succeq_d \mathcal{E}'$ with $\mathcal{E} \neq \mathcal{E}'$. It is clear that $\neg(\mathcal{E} \subseteq \mathcal{E}')$ and $\neg(\mathcal{E}' \subseteq \mathcal{E})$. Let $a' \in \mathcal{E}' \setminus \mathcal{E}$ be such that $\forall a'' \in \mathcal{E}' \setminus \mathcal{E}$ it holds that $a' \geq a''$ (this is possible since $\geq$ is a total preorder). From $\mathcal{E} \succeq_d \mathcal{E}'$, we have that $\exists a \in \mathcal{E} \setminus \mathcal{E}'$ s.t. $a > a'$. This means that $\forall b' \in \mathcal{E}' \setminus \mathcal{E}$, $a > b'$. Since $\mathcal{E}'$ is a stable extension, then $\exists a'' \in \mathcal{E}'$ s.t. $a'' \mathcal{R}_r a$, i.e. $(a'' \mathcal{R} a$ and $\neg(a > a''))$ or $(a \mathcal{R} a''$ and $a'' > a)$. Sets $\mathcal{E}$ and $\mathcal{E}'$ are both conflict-free, so $a'' \in \mathcal{E}' \setminus \mathcal{E}$. Contradiction, since $\forall a'' \in \mathcal{E}' \setminus \mathcal{E}$ we have $a > a''$.

Proof of Property 6.

- We show that $x \in \mathcal{S}$ iff $x \in \mathtt{Base}(\mathtt{Arg}(\mathcal{S}))$ where $\mathcal{S}$ is a consistent subbase of Σ.
  $(\Rightarrow)$ Let $x \in \mathcal{S}$. Since $\mathcal{S}$ is consistent, then the set $\{x\}$ is consistent as well. Thus, $(\{x\}, x) \in \mathtt{Arg}(\mathcal{S})$. Consequently, $x \in \mathtt{Base}(\mathtt{Arg}(\mathcal{S}))$.
  $(\Leftarrow)$ Assume that $x \in \mathtt{Base}(\mathtt{Arg}(\mathcal{S}))$. Thus, $\exists a \in \mathtt{Arg}(\mathcal{S})$ s.t. $x \in \mathtt{Supp}(a)$. From the definition of an argument, $\mathtt{Supp}(a) \subseteq \mathcal{S}$. Consequently, $x \in \mathcal{S}$.
- Let us show that the function Base is surjective. Let $\mathcal{S} \subseteq \Sigma$. From the first item of this property, the equality $\mathtt{Base}(\mathtt{Arg}(\mathcal{S})) = \mathcal{S}$ holds. It is clear that $\mathtt{Arg}(\mathcal{S}) \in \mathcal{P}(\mathtt{Arg}(\Sigma))$ ($\mathcal{P}(\mathtt{Arg}(\Sigma))$ being the power set of $\mathtt{Arg}(\Sigma)$).
  The following counter-example shows that the function Base is not injective: Let $\Sigma = \{x, x \to y\}$, $\mathcal{E} = \{(\{x\}, x), (\{x \to y\}, x \to y)\}$ and $\mathcal{E}' = \{(\{x\}, x), (\{x, x \to y\}, y)\}$. Since $\mathtt{Base}(\mathcal{E}) = \mathtt{Base}(\mathcal{E}') = \Sigma$, with $\mathcal{E} \neq \mathcal{E}'$ then Base is not injective.
- If $a \in \mathcal{E}$ where $\mathcal{E} \subseteq \mathtt{Arg}(\Sigma)$, then $\mathtt{Supp}(a) \subseteq \mathtt{Base}(\mathcal{E})$. Consequently, $a \in \mathtt{Arg}(\mathtt{Base}(\mathcal{E}))$.
- Let us prove that Arg is injective. Let $\mathcal{S}, \mathcal{S}' \subseteq \Sigma$ with $\mathcal{S} \neq \mathcal{S}'$. Then, it must be that $\mathcal{S} \setminus \mathcal{S}' \neq \emptyset$ or $\mathcal{S}' \setminus \mathcal{S} \neq \emptyset$ (or both). Without loss of generality, let $\mathcal{S} \setminus \mathcal{S}' \neq \emptyset$ and let $x \in \mathcal{S} \setminus \mathcal{S}'$. If $\{x\}$ is consistent, then, $(\{x\}, x) \in \mathtt{Arg}(\mathcal{S}) \setminus \mathtt{Arg}(\mathcal{S}')$. Thus, $\mathtt{Arg}(\mathcal{S}) \neq \mathtt{Arg}(\mathcal{S}')$.
  We will now present an example that shows that this function is not surjective. Let $\Sigma = \{x, x \to y\}$ and $\mathcal{E} = \{(\{x\}, x), (\{x \to y\}, x \to y)\}$. It is clear that there exists no $\mathcal{S} \subseteq \Sigma$ s.t. $\mathcal{E} = \mathtt{Arg}(\mathcal{S})$, since such a set $\mathcal{S}$ would contain Σ and, consequently, $\mathtt{Arg}(\mathcal{S})$ would contain $(\{x, x \to y\}, y)$, an argument not belonging to $\mathcal{E}$.

Proof of Property 7. Let $\mathcal{S} \subseteq \Sigma$.

- Assume that $\mathcal{S}$ is consistent and $\mathtt{Arg}(\mathcal{S})$ is not conflict-free. This means that there exist $a, a' \in \mathtt{Arg}(\mathcal{S})$ s.t. a undercuts a'. From Definition 2 of undercut, it follows that $\mathtt{Supp}(a) \cup \mathtt{Supp}(a')$ is inconsistent. Besides, from the definition of an argument, $\mathtt{Supp}(a) \subseteq \mathcal{S}$ and $\mathtt{Supp}(a') \subseteq \mathcal{S}$. Thus, $\mathtt{Supp}(a) \cup \mathtt{Supp}(a') \subseteq \mathcal{S}$. Then, $\mathcal{S}$ is inconsistent. Contradiction.
- Assume now that $\mathcal{S}$ is inconsistent. This means that there exists a finite set $\mathcal{S}' = \{h_1, \ldots, h_k\}$ s.t.
  - $\mathcal{S}' \subseteq \mathcal{S}$

- $\mathcal{S}' \vdash \bot$
- $\mathcal{S}'$ is minimal (wrt. set inclusion) s.t. previous two items hold.

Since $\mathcal{S}'$ is a minimal inconsistent set, then $\{h_1, \ldots, h_{k-1}\}$ and $\{h_k\}$ are consistent. Thus, $(\{h_1, \ldots, h_{k-1}\}, \neg h_k), (\{h_k\}, h_k) \in \texttt{Arg}(\mathcal{S})$. Furthermore, those two arguments are conflicting (the former undercuts the latter). This means that $\texttt{Arg}(\mathcal{S})$ is not conflict-free.

Proof of Theorem 1. Let $\mathcal{S}$ be a preferred sub-theory of a knowledge base Σ. Thus, $\mathcal{S}$ is consistent. From Property 7, it follows that $\texttt{Arg}(\mathcal{S})$ is conflict-free. Assume that $\exists a \notin \texttt{Arg}(\mathcal{S})$. Since $a \notin \texttt{Arg}(\mathcal{S})$ and $\mathcal{S}$ is a maximal consistent subbase of Σ (according to Property 2), then $\exists h \in \texttt{Supp}(a)$ s.t. $\mathcal{S} \cup \{h\} \vdash \bot$. Assume that $h \in \Sigma_j$. Thus, $\texttt{Level}(\texttt{Supp}(a)) \geq j$.

Since $\mathcal{S}$ is a preferred sub-theory of Σ, then $\mathcal{S}_1 \cup \ldots \cup \mathcal{S}_j$ is a maximal (for set inclusion) consistent subbase of $\Sigma_1 \cup \ldots \cup \Sigma_j$. Thus, $\mathcal{S}_1 \cup \ldots \cup \mathcal{S}_j \cup \{h\} \vdash \bot$. This means that there exists an argument $(\mathcal{S}', \neg h) \in \texttt{Arg}(\mathcal{S})$ s.t. $\mathcal{S}' \subseteq \mathcal{S}_1 \cup \ldots \cup \mathcal{S}_j$. Thus, $\texttt{Level}(\mathcal{S}') \leq j$. Consequently, $(\mathcal{S}', \neg h) \geq_{WLP} a$. Moreover, $(\mathcal{S}', \neg h)$ undercuts a. Thus, $(\mathcal{S}', \neg h)$ undercuts$_r a$.

The second part of the theorem follows directly from Property 6.

A Possibility Theory-Oriented Discussion of Conceptual Pattern Structures

Zainab Assaghir[1], Mehdi Kaytoue[1], and Henri Prade[2]

[1] Laboratoire Lorrain de Recherche en Informatique et ses Applications (LORIA) Campus Scientique, B.P. 235 54500 Vanduvre-l'es-Nancy France

[2] IRIT, Université Paul Sabatier, 31062 Toulouse Cedex 09, France

Zainab.Assaghir@loria.fr, kaytouem@loria.fr, prade@irit.fr

Abstract. A fruitful analogy between possibility theory and formal concept analysis has recently contributed to show the interest of introducing new operators in this latter setting. In particular, another Galois connection, distinct from the classical one that defines formal concepts, has been laid bare which allows for the decomposition of a formal context into sub-contexts when possible. This paper pursues a similar investigation by considering pattern structures which are known to offer a generalization of formal concept analysis. The new operators as well as the other Galois connection are introduced in this framework, where an object is associated to a structured description rather than just to its set of properties. The description may take many different forms. In this paper, we more particularly focus on two important particular cases, namely ordered lists of intervals, and propositional knowledge bases, which both allow for incomplete descriptions. They are then extended to fuzzy and uncertain descriptions by introducing fuzzy intervals and possibilistic logic bases respectively in these two settings.

1 Introduction

The problem of the description of items is at the basis of any representation. An item (or object) may be associated with a set of properties, more generally with a set of attribute values. Some attributes may be imprecisely known for some objects, or known with uncertainty. Logical descriptions may be also used under the form of propositions that hold true for a considered object. Possibility theory [29,9] provides a framework for modeling imprecise and uncertain information under the form of fuzzy restrictions on ill-known attribute values, or possibilistic logic knowledge bases. Possibility distributions are generally used for describing individual objects, or classes of "similar" objects. They are much less used for associating an attribute value with a set of more or less possible objects, with the noticeable exception of possibilistic classifiers, e.g. [26].

In formal concept analysis [19], objects and properties play symmetric roles in a relation associating them in a formal context. This enables the association of sets of objects with their respective sets of properties, and conversely. This association is based on one operator from which a Galois connection is defined whose

A. Deshpande and A. Hunter (Eds.): SUM 2010, LNAI 6379, pp. 70–83, 2010.

fixed points are pairs of sets of objects, and of sets of properties representing formal concepts. Viewing the set of properties associated with an object (or the set of objects associated with a property) as a possibility distribution has revealed the existence of three other operators [5], and of another Galois connection useful for decomposing a formal context into sub-contexts [12,4]. Formal concept analysis has been enlarged to pattern structures in order to associate objects with more general forms of descriptions [21,18]. A similar objective has motivated independently the introduction of logical information systems [16,15,17]. See also [2]. This enables the use of a Galois connection between objects and descriptions.

In this paper, we continue to investigate the mutual enrichment of possibility theory and formal concept analysis by introducing the new operators existing in possibility theory in the setting of pattern structures. Then we more particularly study the case of possibilistic descriptions, either in terms of fuzzy intervals, or in terms of possibilistic propositional logic bases. The paper is structured as follows. Basic notations (and their meanings) are first introduced in Section 2. In Section 3, a twofold background restates the four basic operators, first in the possibility theory setting, and then in the formal concept analysis setting. Then Section 4 introduces pattern structures, while Section 5 extends their settings with the new operators. Section 6 considers two types of pattern structures, namely ordered lists of intervals (and then of fuzzy intervals), and propositional knowledge bases (and then briefly possibilistic logic knowledge bases). Section 7 concludes by pointing out directions for further research and shows how the setting presented may contribute to a general theory of descriptions.

2 Basic Notations and Their Meanings

First, we consider items, or equivalently, objects. An object will be denoted by x, or x_i in case we consider several ones in the same time. It is interesting to notice that in fact an object may refer to a particular, unique item, or as well to a generic item representative of a class of items *sharing the same description*. The set of all objects is just denoted Obj. A subset of objects will be denoted by a capital letter X, or X_s if necessary, and we shall write $X = \{x_1, \ldots, x_i, \ldots, x_m\}$. The set of all properties is denoted $Prop$. A set of objets associated with their respective sets of properties defines a formal context $R \subseteq Obj \times Prop$ [19].

An object x may be associated with a description denoted $\partial(x)$. In the paper we shall consider many different types of description. The simplest type is a subset Y of properties y_j, namely, $Y = \{y_1, \ldots, y_j, \ldots, y_n\}$. In such a case, we shall write $\partial(x) = Y$.

A useful kind of structured description is in terms of attributes. An attribute, a subset of attributes, the set of all attributes are respectively denoted a, $A = \{a_1, \ldots, a_k, \ldots, a_r\}$, and Att. A property refers to a subset of attribute values. The value of attribute a for x is denoted $a(x) = u$, where u belongs to the attribute domain U_a. In this case, we shall write $\partial(x) = (a_1(x), \ldots, a_k(x), \ldots, a_r(x)) = (u_1, \ldots, u_k, \ldots, u_r)$. This corresponds to a completely informed situation where all the considered attributes values are known for x. When it is not the case, the precise value $a_k(x)$ will be replaced by the possibility distribution $\pi_{a_k(x)}$. Such a possibility distribution [29] is a mapping from U_{a_k} to $[0, 1]$, or more generally any linearly ordered scale. Then $\pi_{a_k(x)}(u) \in [0, 1]$ estimates to what extent it is possible that the value of a_k for x is u. 0 means impossibility; several distinct values may be fully possible (degree 1). The characteristic function of an ordinary subset is a particular case of a possibility distribution. Precise information corresponds to the characteristic function of singletons. An elementary property y can be viewed as a subset of a single attribute domain, i.e. $y \subseteq U$. Note that while Y is a *conjunctive* set of properties (for instance an object possesses all properties in Y), y as a subset of some attribute domain U is a *disjunctive* set of mutually exclusive values for *the value* of a single-valued attribute ill-known for some x).

Another useful kind of description considered in the following is a propositional knowledge base $K(x)$ that describes x. Such a knowledge base may provide a complete as well as an incomplete description (when its set of models induced by the set of literals is not a singleton). The two previous representation may be combined into a description $\partial(x) = (K_1(x), \ldots, K_k(x), \ldots, K_r(x))$, where $K_k(x)$ is the knowledge base that constrains the possible values of attribute a_k for x.

Other specific notations will be introduced when needed.

3 The Four Descriptors

Taking inspiration from the existence of four set functions in possibility theory [11], new operators have been suggested in the setting of formal concept analysis [5]. These set functions are now recalled, emphasizing the symmetrical roles played by the object x and the attribute value u, a point of view unusual in possibility theory.

3.1 Possibility Theory: Mono-Attribute Case

Let $\pi_{a(x)}(u)$ denote the possibility that object x has value $u \in U$ (for attribute a). We assume that π_a is bi-normalized: $\forall x\ \exists u\ \pi_{a(x)}(u) = 1$ and $\forall u\ \exists x\ \pi_{a(x)}(u) = 1$. This means that for any object x, there is some fully possible value for attribute a, and that for any value u there is an object x that takes this value. Let X be a set of objects, $y \subseteq U$ a property. Then, one can define

- i) a possibility measure [29] Π (or "potential possibility"):

$$\Pi(X) = \max_{x \in X} \pi_{a(x)}(u) \text{ and } \Pi(y) = \max_{u \in y} \pi_{a(x)}(u).$$

  $\Pi(X)$ estimates to what extent it is possible that there is an object in X having value u, while $\Pi(y)$ is the possibility that object x has property y. Π is an indicator of non-empty intersection of the fuzzy set induced by the possibility distribution with an ordinary subset.

– ii) a dual measure of necessity N (or "actual necessity") [8]:

$$N(X) = \min_{x \notin X} 1 - \pi_{a(x)}(u) \text{ and } N(y) = \min_{u \notin y} 1 - \pi_{a(x)}(u)$$

$N(X)$ estimates to what extent it is certain (necessarily true) that an object have value u is in X, while $N(y)$ is the certainty that object x has property y. Note that $N(y) = 1 - \Pi(\overline{y})$ where $\overline{y} = U \setminus y$. N may be viewed as a degree of inclusion of the fuzzy set induced by the possibility distribution into an ordinary subset.

– iii) a measure of "actual (or guaranteed) possibility" [10]

$$\Delta(X) = \min_{x \in X} \pi_{a(x)}(u) \text{ and } \Delta(y) = \min_{u \in y} \pi_{a(x)}(u)$$

$\Delta(X)$ estimates to what extent it is possible that *all* objects in X have value u, while $\Delta(y)$ estimates the possibility that object x takes any value in y. Δ may be viewed as a degree of inclusion of an ordinary subset into the fuzzy set induced by the possibility distribution.

– iv) a dual measure of "potential necessity or certainty" [10]

$$\nabla(X) = 1 - \min_{x \notin X} \pi_{a(x)}(u) \text{ and } \nabla(y) = 1 - \min_{u \notin y} \pi_{a(x)}(u)$$

$\nabla(X)$ estimates to what extent there exists at least one object outside X that has a low degree of possibility of having value u, while $\nabla(y)$ is the degree of impossibility that an object x has a value outside y. Note that $\nabla(y) = 1 - \Delta(\overline{y})$. ∇ is an indicator of non-full coverage of the considered universe by the fuzzy set induced by the possibility distribution together with an ordinary subset.

3.2 Formal Context Setting

In [5], the classical setting of formal concept analysis defined from a formal context, and which relies on one operator that associates a subset of objects with the set of properties shared by them (and the dual operator), has been enlarged with the introduction of three other operators. We now recall these four operators, first in the setting of a formal context, before introducing their expressions in the cases of structured attribute, and logical descriptions.

Namely, let R be the formal context, $R(x) = \{y \in Prop | (x, y) \in R\}$ be the set of properties of object x, $R^{-1}(y) = \{x \in Obj | (x, y) \in R\}$ be the set of objects having properties y. Then, four remarkable sets can be associated with a subset X of objects (the notations have been chosen here in relation with the modal semantics underlying these sets, and also in parallel with possibility theory):

– the set $R^{\Diamond}(X)$ of properties that are possessed by *at least one* object in X:

$$R^{\Diamond}(X) = \{y \in Prop | R^{-1}(y) \cap X \neq \emptyset\} = \cup_{x \in X} R(x).$$

Clearly, we have $R^{\Diamond}(X_1 \cup X_2) = R^{\Diamond}(X_1) \cup R^{\Diamond}(X_2)$.

- the set $R^{\Box}(X)$ of properties s. t. any object that satisfies *one* of them is necessarily in X:

$$R^{\Box}(X) = \{y \in Prop | R^{-1}(y) \subseteq X\} = \cap_{x \notin X} \overline{R(x)}.$$

  In other words, having any property in $R^{\Box}(X)$ is a sufficient condition for belonging to X. Moreover, we have $R^{\Box}(X) = \overline{R^{\Diamond}(\overline{X})} = Prop \setminus R^{\Diamond}(\overline{X})$, and $R^{\Box}(X_1 \cap X_2) = R^{\Box}(X_1) \cap R^{\Box}(X_2)$.
- the set $R^{\triangle}(X)$ of properties shared by *all* objects in X:

$$R^{\triangle}(X) = \{y \in Prop | R^{-1}(y) \supseteq X\} = \cap_{x \in X} R(x).$$

  In other words, satisfying all properties in $R^{\triangle}(X)$ is a necessary condition for an object for belonging to X. $R^{\triangle}(X)$ is a partial conceptual characterization of objects in X: objects in X have all the properties of $R^{\triangle}(X)$ and may have some others (that are not shared by all objects in X). It is worth noticing that $\overline{R^{\Diamond}(X)}$ provides a negative conceptual characterization of objects in X since it gathers all the properties that are never satisfied by any object in X. Moreover, we have $R^{\triangle}(X_1 \cup X_2) = R^{\triangle}(X_1) \cap R^{\triangle}(X_2)$. Besides, as can be seen, $R^{\Box}(X) \cap R^{\triangle}(X)$ is the set of properties possessed by all objects in X and only by them.
- the set $R^{\nabla}(X)$ of properties that are not satisfied by at least one object in $\overline{X}$.

$$R^{\nabla}(X) = \{y \in Prop | R^{-1}(y) \cup X \neq Obj\} = \cup_{x \notin X} \overline{R(x)}.$$

  Note that $R^{\nabla}(X) = \overline{R^{\triangle}(\overline{X})} = Prop \setminus R^{\triangle}(\overline{X})$. In other words, in context R, for any property in $R^{\nabla}(X)$, there exists at least one object outside X that misses it. Moreover, we have $R^{\nabla}(X_1 \cap X_2) = R^{\nabla}(X_1) \cup R^{\nabla}(X_2)$.

Note that $R^{\Diamond}(X)$ and $R^{\Box}(X)$ become larger when X increases, while $R^{\triangle}(X)$ and $R^{\nabla}(X)$ get smaller. The four subsets $R^{\Diamond}(X)$, $R^{\Box}(X)$, $R^{\triangle}(X)$, and $R^{\nabla}(X)$ have been considered (with different notations) without any mention of possibility theory by different authors. Düntsch *et al.* [14,13] calls $R^{\triangle}$ a *sufficiency* operator, and its representation capabilities are studied in the theory of Boolean algebras. Taking inspiration as the previous authors from rough sets [23], Yao [28,27] also considers these four subsets. In both cases, the four operators were introduced. See also [24,20].

In such a setting, a formal concept [19] is defined as a pair $(X, Y) \in Obj \times Prop$ such that

$$R^{\triangle}(X) = Y \text{ and } R^{-1\triangle}(Y) = X,$$

where $R^{-1\triangle}(Y) = \{x \in Obj | R(x) \supseteq Y\} = \cap_{y \in Y} R^{-1}(y)$ is the set X of objects having all properties in Y, and in this case Y is also the maximal set of properties shared by all objects in X. A formal concept (X, Y) is a sub-rectangle in the formal context, i.e. is such that $X \times Y \subseteq R$.

Another Galois connection can be defined from $R^{\Box}$ [4,12,3] in a similar formal way which focuses on pairs (X, Y) such that:

$$R^{\Box}(X) = Y \text{ and } R^{-1\Box}(Y) = X,$$

where $R^{-1\Box}(Y) = \{x \in Obj | Y \supseteq R(x)\} = \cap_{x \notin Y} \overline{R(x)}$ is the set X of objects that are the only ones to have a property in Y. Conversely, Y is a set of properties that cannot be found outside X. Then (X, Y) constitutes an independent sub-context in R, in the sense that (X, Y) is a pair such that $(X \times Y) + (\overline{X} \times \overline{Y}) \supseteq R$, where $+$ is the dual of $\times$ [3], i.e. there is no object / property pair (x, y) of the context R either in $X \times \overline{Y}$ or in $\overline{X} \times Y$.

Note that $R^{\Diamond}$ induces the same Galois connection as $R^{\Box}$, while R^{∇} gives back the one defined from R^{Δ}.

4 Pattern Structures

Taking inspiration from previous work in concept learning [21], Ganter and Kuznetsov [18] have proposed an extension of formal concept analysis, called *pattern structures*. The idea is to associate to any object x its description $\partial(x)$, where X is supposed to belong to a complete semilattice structure $\mathcal{D}$ equipped with a meet $\sqcap$. Then two derivation operators are defined as follows:

$$\forall X \subseteq Obj, X^{\Delta} = \sqcap_{x \in X} \partial(x) \text{ and } \forall d \in \mathcal{D}, d^{\Delta} = \{x \in Obj | d \sqsubseteq \partial(x)\}.$$

where a subsumption relation $\sqsubseteq$ is defined between descriptions in the classical way with respect to $\sqcap$, namely, $c \sqsubseteq d \iff c \sqcap d = c$.

This can be clearly paralleled with the construction recalled in Section 3.2

$$\forall X \subseteq Obj, R(X)^{\Delta} = \cap_{x \in X} R(x) \text{ and } \forall Y \subseteq Prop, R^{-1}(Y)^{\Delta} = \{x \in Obj | Y \subseteq R(x)\},$$

with $\partial(x) = R(x)$, $d = Y$, $\mathcal{D} = 2^{Prop}$, and $\sqcap = \cap$.

As in standard formal concept analysis, the operator $()^{\Delta}$ makes a Galois connection between 2^{Obj} and $\mathcal{D}$ (equipped with $\sqsubseteq$). A pair (X, d) such that

$$X \subseteq Obj, d \in \mathcal{D}, X^{\Delta} = d \text{ and } X = d^{\Delta}$$

is called a *pattern concept* [18].

Interestingly enough, starting from a different viewpoint motivated by logical representation concerns, Ferré and Ridoux [16] (see also [17]) have proposed a similar construction where $\partial(x) = K(x)$ is a propositional knowledge base, d is a set of propositional formulas, $\mathcal{D}$ is the set of knowledge bases induced by a propositional language, and $\sqcap$ is the logical conjunction $\wedge$, while $\sqsubseteq$ is the logical subsumption. Indeed, Ganter and Kuznetsov [18] have called *pattern implication* the relation $c \rightarrow d$ defined by $a^{\triangle} \subseteq b^{\triangle}$ which means that the set of objects associated with description a is included in the one corresponding to description b, while *object implication* corresponds to the relation $X \rightarrow Y$ defined by $X^{\triangle} \subseteq Y^{\triangle}$ which states that the description of the objects in Y subsumes the description of the objects in X.

5 The Extended Setting

Let us consider a complete lattice structure $\mathcal{D}$ equipped with a meet $\sqcap$ and a dual joint $\sqcup$, top $\top$ and bottom elements $\bot$. As in Section 3.2 one may define the following derivation operators defined as follows:

$$\forall X \subseteq Obj, X^{\Diamond} = \sqcup_{x \in X} \partial(x) \text{ and } \forall d \in \mathcal{D}, d^{\Diamond} = \{x \in Obj | d \sqcap \partial(x) \neq \bot\}.$$

$X^{\Diamond}$ provides the description of the set of objects in X, while $d^{\Diamond}$ is the set of objects whose description has something in common with the one of object x. This can be clearly paralleled with the situation in the formal concept analysis setting

$$\forall X \subseteq Obj, R(X)^{\Diamond} = \cup_{x \in X} R(x) \text{ and } \forall Y \subseteq Prop, R^{-1}(Y)^{\Diamond} = \{x \in Obj | Y \cap R(x) \neq \emptyset\},$$

with $\partial(x) = R(x)$, $d = Y$, $\mathcal{D} = 2^{Prop}$, and $\sqcup = \cup$ If the lattice is also complemented, i.e. for each $d \in \mathcal{D}$, there is a unique $\overline{d}$ such that $\overline{d} \sqcap d = \bot$ and $\overline{d} \sqcup d = \top$, one may also define

$$\forall X \subseteq Obj, X^{\Box} = \sqcap_{x \notin X} \overline{\partial(x)} \text{ and } \forall d \in \mathcal{D}, d^{\Box} = \{x \in Obj | d \sqsupseteq \partial(x)\}.$$

where $\overline{\partial(x)}$ is the complement of $\partial(x)$. $\overline{\partial(x)}$ is the opposite description to the one of x; it subsumes all the descriptions that are not consistent with the one of x. Thus, $X^{\Box} = \sqcap_{x \in \overline{X}} \overline{\partial(x)}$ is the common part of the descriptions that are the opposite to the ones of the objects outside X, which means that having a description subsumed by $X^{\Box}$ is enough for belonging to X (since then this is a description having nothing in common with any of the descriptions of the objects outside X). Clearly, $X^{\Box} = \overline{\overline{X}^{\Diamond}}$. Besides, $d^{\Box}$ is the of objects having a description subsumed by d. We have $d^{\Box} = \overline{\overline{d}^{\Diamond}}$.

The duals $X^{\triangledown} = \overline{\overline{X}^{\triangle}}$ and $d^{\triangledown} = \overline{\overline{d}^{\triangle}}$ of $X^{\triangle}$ and $d^{\triangle}$ can be defined as well:

$$\forall X \subseteq Obj, X^{\triangledown} = \sqcup_{x \notin X} \overline{\partial(x)} \text{ and } \forall d \in \mathcal{D}, d^{\triangledown} = \{x \in Obj | d \sqcup \partial(x) \neq \top\}.$$

As in the case of standard formal concept analysis, the operator $()^{\Box}$ (or equivalently $()^{\Diamond}$) defines another Galois connection between 2^{Obj} and $\mathcal{D}$ (equipped with $\sqsubseteq$). A pair (X, d) such that

$$X \subseteq Obj, d \in \mathcal{D}, X^{\Box} = d \text{ and } X = d^{\Box}$$

or equivalently

$$X \subseteq Obj, d \in \mathcal{D}, X^{\Diamond} = d \text{ and } X = d^{\Diamond}$$

will be called a *pattern subcontext.*

6 Two Noticeable Particular Cases

In this section, we consider two important particular types of representation setting, respectively in attribute / object / value format, and in logical format. Although apparently different, these two formats agree with a representation in terms of possibility distribution, which can encompass the handling of uncertainty.

6.1 Structured Attribute Setting

We assume here that each object x is described in terms of attributes a_k, by means of possibility distributions $\pi_{a_k(x)}$ for $k = 1, \ldots, r$, which restrict the possible values of each attribute for the object. Then, $\partial(x) = (\partial_1(x), \ldots, \partial_r(x))$ where $\partial_k(x)$ denotes the fuzzy set whose membership function is $\pi_{a_k(x)}$, i.e. $\partial(x) = (\pi_{a_1(x)}, \ldots, \pi_{a_r(x)})$. Subsumption relation will defined in terms of fuzzy set inclusion, namely $\partial(x) \sqsupseteq \partial'(x)$ iff $\forall k \ \ \forall u,\ \pi_{a_k(x)} \supset \pi'_{a_k(x)}$.

An important particular case that we consider first is when the possibility distributions are the membership functions of ordinary sets, i.e. are $\{0,1\}$-valued. Then $\pi_{a_k(x)}(u) = 1$ if $u \in \partial_k(x)$ and $\pi_{a_k(x)}(u) = 0$ if $u \notin \partial_k(x)$. When several distinct values of u are in $\partial_k(x)$, the information about a_k for x is imprecise.

The above situation includes the case of numerical values known to belong to intervals. An example of such data is given in Table.1 for objects x_i described by means of two attributes a_1 and a_2 whose values are restricted by intervals, e.g. $\partial(x_1) = ([2,5],[3,6])$. Then, for a subset X of objects, a description $d = (y_1, y_2)$

Table 1. Interval-valued attributes

	a_1	a_2
x_1	$[2,5]$	$[3,6]$
x_2	$[1,6]$	$[1,7]$
x_3	$[4,7]$	$[5,9]$
x_4	$[8,10]$	$[13,18]$
x_5	$[9,12]$	$[11,14]$

Table 2. Fuzzy interval-valued attributes

	a_1		a_2	
	Conservative	best-estimate	Conservative	best-estimate
x_1	$[2,5]$	$[3,4]$	$[3,6]$	$[4,5]$
x_2	$[1,6]$	$[2,6]$	$[1,7]$	$[2,6]$
x_3	$[4,7]$	$[4,6]$	$[5,9]$	$[5,6]$
x_4	$[8,13]$	$[9,12]$	$[13,18]$	$[14,17]$
x_5	$[9,12]$	$[10,11]$	$[14,17]$	$[12,13]$

represents a pair of intrvals y_1 and y_2 respectively for a_1 and a_2 and $(k = 1, 2)$, we apply the four descriptors detailed in section 5 where $\sqcup = \cup$ and $\sqcap = \cap$. Subsomption are defined in terms of interval inclusion i.e. $\sqsubseteq = \subseteq$ and $\sqsupseteq = \supseteq$.

$$X^{\triangle} = \{\cap_{x \in X} \partial_k(x)\} \quad \text{and} \quad d^{\triangle} = \{x \in Obj | \ y_k \subseteq \partial_k(x)\}$$

$X^{\triangle}$ returns the maximal set of values shared by all objects in X. For example, in the Table 1, considering $X = \{x_1, x_2, x_3\}$ then, $X^{\triangle} = (\cap_{x \in X} \partial_1(x), \cap_{x \in X} \partial_2(x)) = ([4,5],[5,6])$. However, $d^{\triangle}$ returns the maximal set of objects having the description (y_1, y_2). Going back to Table 1 and considering the description $d = ([4,5],[5,6])$ then $d^{\triangle} = \{x \in Obj | ([4,5],[5,6]) \subseteq \partial_k(x) | k = 1,2\} = \{x_1, x_2, x_3\}$.

$$X^{\diamond} = \{\cup_{x \in X} \partial_k(x)\} \quad \text{and} \quad d^{\diamond} = \{x \in Obj | y_k \cap \partial_k(x) \neq \bot\}$$

$X^{\diamond}$ is the set of all values of objects in X corresponding to each attribute. For example, in Table 1, $X = \{x_1, x_2, x_3\}$ and $X^{\diamond} = \cup_{x \in \{x_1, x_2, x_3\}} \partial_k(x) = ([1,7],[1,9])$ for $k = 1,2$. It means that $[1,7]$ (resp. $[1,9]$) are all possible values taken by a_1 (resp. a_2) for the subset X. However, for any description $d = (y_1, y_2)$, $d^{\diamond}$ returns the set of objects whose attribute values are possibly in d, i.e. objects whose description has at least a value in common with one object in X e.g. $([1,7],[1,9])^{\diamond} = \{x \in Obj | ([1,7],[1,9]) \cap \partial_k(x) \neq \emptyset\} = \{x_1, x_2, x_3\}$. The descriptor $X^{\diamond}$ is similar to the meet operator considered in [22], but the author there rather uses the convex hull of the union of intervals. Now, we will consider operators based on the necessity measure the dual operator of the possibility measure. Then, for our example we have:

$$X^{\square} = \{\cap_{x \notin X} \overline{\partial_k(x)}\} \quad \text{and} \quad d^{\square} = \{x \in Obj |\ y_k \supseteq \partial_k(x)\}$$

For $X = \{x_1, x_2, x_3\}$ in Table 1, $X^{\square} = \cap_{x \notin \{x_1, x_2, x_3\}} \overline{\partial_k(x)} = \cap_{x \in \{x_4, x_5\}} \overline{\partial_k(x)} = \overline{\cup_{x \in \{x_4, x_5\}} \partial_k(x)} = \overline{([8,12],[11,18])} = ([1,7],[1,9])$, that means only objects in X satisfy the value $[1,7]$ for the attribute a_1 (resp. $[1,9]$ for a_2). Considering a description $d = (y_1, y_2)$ and applying the dual descriptor, we obtain $([1,7],[1,9])^{\square} = \{x \in Obj | ([1,7],[1,9]) \supseteq \partial_k(x)\} = \{x_1, x_2, x_3\}$ meaning that $\{x_1, x_2, x_3\}$ have description in d. As for last descriptors detailed in section 5, we have:

$$X^{\nabla} = \{\cup_{x \notin X} \overline{\partial_k(x)}\} \quad \text{and} \quad d^{\nabla} = \{x \in Obj |\ y_k \cup \partial_k(x) \neq \top\}$$

Going back to the Table 1, considering $X = \{x_1, x_2, x_3\}$ and $d = ([1,7],[1,9])$. Then, $X^{\nabla} = \cup_{x \notin \{x_1, x_2, x_3\}} \overline{\partial_k(x)} = \cup_{x \in \{x_4, x_5\}} \overline{\partial_k(x)} = \overline{\cap_{x \in \{x_4, x_5\}} \partial_k(x)} = \overline{([8,9],[13,14])} = ([1,7],[1,9])$ and $d^{\nabla} = \{x_1, x_2, x_3\}$. It means that attributes values of a_1 in $[1,7]$ are not satisfied by any object in X (resp. $[1,9]$) for a_2).

It becomes natural to consider the pairs (X, d), where X is a subset of objects and d is a description, w.r.t. these four descriptors. Then the pair (X, d) is a concept such that $X^{\triangle} = d$ and $d^{\triangle} = X$, e.g. $(\{x_1, x_2, x_3\}, ([4,5],[5,6]))$ is a patern concept. Considering the operators $\diamond$ or equivalently $\square$, the pairs (X, d) where $X^{\diamond} = d$ and $d^{\diamond} = X$ are concepts if the pairs $\overline{X}^{\square} = \overline{d}$ and $\overline{d}^{\square} = \overline{X}$ hold. For example, $(\{x_1, x_2, x_3\}, ([1,7],[1,9]))$ and $(\{x_4, x_5\}, ([8,12],[11,18]))$ are pairs such that $X^{\square} = d$ and $d^{\square} = X$. Moreover, they are disjoint w.r.t. both Obj and the set of values of $Prop$. Finding such pairs, aims at decomposing the context into independant blocks without object and values of propreties in common, as in the context given in Table 1. When such a decomposition no longer holds, for example, if we consider the object x_6 where $\partial(x_6) = ([6,10],[8,12])$ with the existing objects in Table 1, then pairs (X,d) where $X^{\Box} = d$ and $d^{\Box} = X$ no longer exist, except for the trivial pair $(Obj, \top)$.

We can also generalise the case of intervals i.e. we consider that the object gives the description by mean of possibility distribution in $[0,1]$. The Table 2 represents an example where objects describe the attributes in terms of two intervals, a best-estimate and a conservative intervals. The best-estimate and the conservative intervals correspond respectively to the support and the core of the possibility distribution. Then, the possibility distributions are supposed to have trapezoidal shapes. For example, the possibility distribution that represents values given by the object x_1 for the attribute a_1 is defined by its core $[3,4]$ with possibility 1, and its support $[2,5]$ outside of which the possibility is 0, and we write $\langle[2,5];[3,4]\rangle$. Similarly to the case of intervals, and considering $\partial(x) = (\pi_{a_1(x)}, \ldots, \pi_{a_r(x)})$, we can apply the four descriptors defined in Section 5. Then, we will consider here after a subset X of objects, a description $d = (y_1, y_2)$ representing a pair of possibility distributions y_1 and y_2 respectively for a_1 and a_2 and $(k = 1,2)$, and $\sqcup = \max$ and $\sqcap = \min$. Subsomption are defined in terms of fuzzy set inclusion, e.g. $\langle[4,5];[5,5]\rangle \sqsubseteq \langle[3,4];[2,5]\rangle$ since $[4,5] \subseteq [3,4]$ and $[5,5] \subseteq [2,5]$.

$$X^{\triangle} = \{min_{x\in X}\pi_{a_k(x)}\} \quad \text{and} \quad d^{\triangle} = \{x \in Obj|\; y_k \sqsubseteq \pi_{a_k(x)}\}$$

$X^{\triangle}$ represents the maximal fuzy set of values shared by objects in X for each attribute, e.g. , for $X = \{x_1, x_2, x_3\}$ in Table 2, $X^{\triangle} = \min_{x\in X} \pi_{a_k(x)} = (\langle[4,5];[5,5]\rangle, \langle[5,6];[5,5]\rangle)$. However, $d^{\triangle}$ represents the fuzzy set of objects having in common the description d, e.g. $(\langle[4,5];[5,5]\rangle, \langle[5,6];[5,5]\rangle)^{\triangle} = \{x \in Obj|\langle[4,5];[5,5]\rangle \sqsubseteq \pi_{a_1(x)}$ and $\langle[5,6];[5,5]\rangle \sqsubseteq \pi_{a_2(x)}\} = \{x_1, x_2, x_3\}$.

$$X^{\Diamond} = \{\max_{x\subset X} \pi_{a_k(x)}\} \quad \text{and} \quad d^{\Diamond} = \{x \in Obj|\min(y_k, \pi_{a_k(x)}) \neq \bot\}$$

$X^{\Diamond}$ is the the fuzzy sets of possible values taken by each attribute for an object in X. Considering $X = \{x_1, x_2, x_3\}$ in Table 2, $X^{\Diamond} = \max_{x\in X} \pi_{a_k(x)} = (\langle[1,7];[2,6]\rangle, \langle[1,9];[2,6]\rangle)$. As for $d = (y_1, y_2)$, $d^{\Diamond}$ returns the fuzzy set of objects whose attribute values are possibly in (y_1, y_2), e.g. $(\langle[1,7];[2,6]\rangle, \langle[1,9];[2,6]\rangle)^{\Diamond} = \{x \in Obj|\min(\langle[1,7];[2,6]\rangle, \pi_{a_k(x)}) \neq \bot$ and $\min(\langle[1,9];[2,6]\rangle, \pi_{a_k(x)}) \neq \bot\} = \{x_1, x_2, x_3\}$.

$$X^{\Box} = \{\min_{x\notin X} \overline{\pi_{a_k(x)}}\} \quad \text{and} \quad d^{\Box} = \{x \in Obj|\; \pi_{a_k(x)} \sqsubseteq y_k\}$$

where the first operator returns the fuzzy set of values that only objects in X satisfy them. Having any value in $X^{\Box}$ is a sufficient condition for belonging to X. In Table 2 and $X = \{x_1, x_2, x_3\}$, $X^{\Box} = \min_{x\notin X} \overline{\pi_{a_k(x)}} = \min_{x\in\{x_4,x_5\}} \overline{\pi_{a_k(x)}} = \overline{\max_{x\in\{x_4,x_5\}} \pi_{a_k(x)}} = \overline{\{x_4, x_5\}^{\Diamond}} = (\langle[1,7];[2,6]\rangle, \langle[1,9];[2,6]\rangle)$.

Then, the second operator returns the fuzzy set of objects that necessary give a value (or a subset of values) in $(y_1, \ldots, y_k, \ldots, y_r)$, e.g.

$(\langle[1,7];[2,6]\rangle, \langle[1,9];[2,6]\rangle)^{\Box} = \{x \in Obj | \pi_{a_k(x)} \sqsubseteq \langle[1,7];[2,6]\rangle$ and $\pi_{a_k(x)} \sqsubseteq \langle[1,9];[2,6]\rangle\}$.

$$X^{\triangledown} = \{\max_{x \notin X} \overline{\pi_{a_k(x)}}\} \quad \text{and} \quad d^{\triangledown} = \{x \in Obj| \ \max(y_k, \pi_{a_k(x)}) \neq \top\}$$

where the first operator returns the set of values that are not satisfied by at least one object in $\overline{X}$. For example, in Table 2, $\{x_1, x_2, x_3\}^{\triangledown} = \max_{x \notin \{x_1,x_2,x_3\}} \overline{\pi_{a_k(x)}} = \max_{x \in \{x_4,x_5\}} \overline{\pi_{a_k(x)}} = \overline{\{x_4, x_5\}^{\triangle}} = \overline{(\langle[9,12];[10,11]\rangle, \langle[14,17];[15,16]\rangle)} = (\langle[1,7];[2,6]\rangle, \langle[1,9];[2,6]\rangle)$

Therefore, the pair $(\{x_1, x_2, x_3\}, \langle[4,5];[5,5]\rangle, \langle[5,6];[5,5]\rangle)$ is a pattern concept since $\{x_1, x_2, x_3\}^{\triangle\triangle} = \{x_1, x_2, x_3\}$.

The pairs $(\{x_1, x_2, x_3\}, \langle[1,7];[2,6]\rangle, \langle[1,9];[2,6]\rangle)$ and $(\{x_4, x_5\}, \langle[8,13];[9,12]\rangle, \langle[13,18];[14,17]\rangle)$ are such that $X^{\Diamond} = d$ and $d^{\Diamond} = X$, then these pairs are pattern subcontext. These pairs aims to decomopose the context into independent subcontexts.

If we consider the object x_6 where $\partial(x_6) = (\langle[6,12];[7,11]\rangle, \langle[4,8];[5,6]\rangle)$. Then, pairs (X, d) s.t. $X^{\Box} = d$ and $d^{\Box} = X$ no longer exist, except for the trivial pair $(Obj, \top)$, and hence the table is no longer decomposable.

6.2 Logical Setting

Logic is a natural setting for the qualitative representation of information. Thus, it is convenient in some situations to use logic for the description of properties associated to objects in a formal concept analysis setting. Although it may be still organized in an attribute format (i.e., there is one knowledge base per attribute), it is simpler to use a unique knowledge base for each object (which allows for the modeling of possible dependencies between attributes). Pattern structures allow for such descriptions. In fact it was the departure point of [16,17] when they proposed logical information systems. At the semantic level, the construction will be quite similar to the one of the previous subsection, since the semantics of a propositional knowledge base, and more generally of a possibilistic knowledge base (where propositional formulas are associated with lower bounds of necessity measures that encode levels of certainty), can be expressed in terms of a possibility distribution [6]. Due to space limitation this is only outlined.

Let $\mathcal{K}$ denote the set of all propositional knowledge bases (induced by some language). Let $K(x)$ denote the propositional knowledge base associated with object x. Then the four basic operators now write

- $\forall X \subseteq Obj, X^{\Diamond} = \bigvee_{x \in X} K(x)$ and $\forall K \in \mathcal{K}, K^{\Diamond} = \{x \in Obj | K \wedge K(x) \neq \bot\}$, where $X^{\Diamond}$ is the disjunction of all the knowledge bases describing objects in X just providing the least common subsumer of all these descriptions, while $K^{\Diamond}$ is the set all objects whose description is consistent with K;
- $\forall X \subseteq Obj, X^{\Box} = \bigwedge_{x \notin X} \neg K(x)$ and $\forall K \in \mathcal{K}, K^{\Box} = \{x \in Obj | K(x) \vDash K\}$, where $X^{\Box}$ is the knowledge base representing the information which characterizes any object of X, while $K^{\Box}$ is the set of objects whose description is more specific than K;

- $\forall X \subseteq Obj, X^{\triangle} = \bigwedge_{x \in X} K(x)$ and $\forall K \in \mathcal{K}, K^{\triangle} = \{x \in Obj | K \vDash K(x)\}$, where $X^{\triangle}$ is the description of what the objects in X have in common, while $K^{\triangle}$ is the set of objects whose description is more general than K;
- $\forall X \subseteq Obj, X^{\triangledown} = \bigvee_{x \notin X} \neg K(x)$ and $\forall K \in \mathcal{K}, K^{\triangledown} = \{x \in Obj | K \vee K(x) \neq \top\} = \{x \in Obj | \neg K \wedge \neg K(x) \neq \bot\}$, where $X^{\triangledown}$ is the least common subsumer of all the negation of the descriptions of the object outside X, while $K^{\triangledown}$ is the set of objects "negatively" consistent with K.

Then a pair (X, K) such that $X \subseteq Obj, K \in \mathcal{K}$,
$X^{\triangle} = \bigwedge_{x \in X} K(x) = K$ and $X = K^{\triangle} = \{x \in Obj | K \vDash K(x)\}$
is the logical counterpart of a pattern concept [17].
The extended pattern structure setting leads us also to consider the pairs (X, K) such that $X \subseteq Obj, K \in \mathcal{K}$,
$X^{\square} = K$ and $X = K^{\square}$
or equivalently,
$X^{\Diamond} = \bigvee_{x \in X} K(x) = K$ and $X = K^{\Diamond} = \{x \in Obj | K \wedge K(x) \neq \bot\}$ which are the logical counterpart of a pattern sub-context. Note that any object x outside X is such that its description $K(x)$ is inconsistent with K, the least common subsumer of all the descriptions of objects in X. In other words, the objects outside X have nothing in common with those in X, they belong to independent "worlds". See [4] for the particular case of this Galois connection in standard formal concept analysis.

All this can be formally extended to possibilistic logic [6], where subsumption (or entailment) is semantically equivalent to a fuzzy set inclusion between possibility distributions. But what makes the extension to possibilistic logic more interesting is its graded handling of inconsistency, and the existence of a graded entailment. Namely, let $\widehat{K}$ be a possibilistic logic base, then we have $\widehat{K} \vDash (K(x), \alpha)$ if and only if we have the classical entailment $\widehat{K}_\alpha \vDash K(x)$, where $\widehat{K}_\alpha$ is the set of formulas in $\widehat{K}$ whose certainty level is greater or equal to α. Thus, one can, for instance, define what formal concepts holds at a given certainty level. This would straightforwardly apply to the particular case of standard formal concept analysis where objets and properties are associated with various degrees of certainty.

7 Concluding Remarks

The main contribution of this paper is to show the full compatibility of possibility theory as a setting for representing imprecise and uncertain information with formal concept analysis generalized under the form of pattern structures. As illustrated in the paper, this applies to quantitative information, for instance represented by (fuzzy) intervals, as well as to qualitative information expressed in logical terms, just providing a very general setting for a theory of object descriptions. It is also worth noticing that the parallel between possibility theory and formal concept analysis leads to introduce more operators, as well as another Galois connection, in the pattern structures setting. Moreover, possibility provides a natural way for handling uncertain pieces of information in this setting. This generalized framework could be in turn related to description logic and to version space learning. Indeed, relations between possibility theory and these two areas have been laid bare, see ([7] and [25]), while pattern structures have been motivated to some extent by concerns coming from these two points of view [1] [21]. Bridging all these areas through possibility theory is clearly a topic for further research, and is a source for mutual enrichment.

Acknowledgements

The authors are grateful to Didier Dubois for useful discussions in relation with the topic of this paper.

References

1. Baader, F., Molitor, R.: Building and structuring description logic knowledge bases using least common subsumers and concept analysis. In: Ganter, B., Mineau, G.W. (eds.) ICCS 2000. LNCS, vol. 1867, pp. 292–305. Springer, Heidelberg (2000)
2. Chaudron, L., Maille, N.: Generalized formal concept analysis. In: Ganter, B., Mineau, G.W. (eds.) ICCS 2000. LNCS, vol. 1867, pp. 357–370. Springer, Heidelberg (2000)
3. Djouadi, Y., Dubois, D., Prade, H.: Différentes extensions floues de l'analyse formelle de concepts. In: Rencontres Francophones sur la Logique Floue et ses Applications (LFA), Annecy, France, November 5-6, pp. 141–148. Cépadues Editions (2009)
4. Djouadi, Y., Dubois, D., Prade, H.: Possibility theory and formal concept analysis: Context decomposition and uncertainty handling. In: Hüllermeier, E., Kruse, R., Hoffmann, F. (eds.) Computational Intelligence for Knowledge-Based Systems Design. LNCS, vol. 6178, pp. 260–269. Springer, Heidelberg (2010)
5. Dubois, D., Dupin de Saint Cyr, F., Prade, H.: A possibilty-theoretic view of formal concept analysis. Fundamenta Informaticae (1-4), 195–213 (2007)
6. Dubois, D., Lang, J., Prade, H.: Possibilistic logic. In: Gabbay, D.M., Hogger, C.J., Robinson, J.A., Nute, D. (eds.) Handbook of Logic in Artificial Intelligence and Logic Programming, vol. 3, pp. 439–513. Oxford University Press, Oxford (1994)
7. Dubois, D., Mengin, J., Prade, H.: Possibilistic uncertainty and fuzzy features in description logic. A preliminary discussion. In: Sanchez, E. (ed.) Fuzzy Logic and the Semantic Web, pp. 101–113. Elsevier, Amsterdam (2006)
8. Dubois, D., Prade, H.: Fuzzy Sets and Systems: Theory and Applications (1980)
9. Dubois, D., Prade, H.: Possibility Theory. Plenum Press, New York (1988)
10. Dubois, D., Prade, H.: Possibility theory as a basis for preference propagation in automated reasoning. In: Proc. 1st IEEE Inter. Conf. on Fuzzy Systems 1992 (FUZZ-IEEE 1992), San Diego, Ca., March 8-12, pp. 821–832 (1992)
11. Dubois, D., Prade, H.: Possibility theory: qualitative and quantitative aspects. In: Gabbay, D., Smets, P. (eds.) Quantified Representation of Uncertainty and Imprecision. Handbook of Defeasible Reasoning and Uncertainty Management Systems, vol. 1, pp. 169–226. Kluwer Acad. Publ., Dordrecht (1998)
12. Dubois, D., Prade, H.: Possibility theory and formal concept analysis in information systems. In: Proc. Inter. Fuzzy Systems Assoc. World Congress and Conf. of the Europ. Soc. for Fuzzy Logic and Technology (IFSA-EUSFLAT 2009), Lisbon, July 20-24, pp. 1021–1026 (2009)

13. Düntsch, I., Gediga, G.: Approximation operators in qualitative data analysis. In: Theory and Application of Relational Structures as Knowledge Instruments, pp. 216–233 (2003)
14. Düntsch, I., Orlowska, E.: Mixing modal and sufficiency operators. Bulletin of the Section of Logic, Polish Academy of Sciences 28(2), 99–106 (1999)
15. Ferré, S.: Complete and incomplete knowledge in logical information systems. In: Benferhat, S., Besnard, P. (eds.) ECSQARU 2001. LNCS (LNAI), vol. 2143, pp. 782–791. Springer, Heidelberg (2001)
16. Ferré, S., Ridoux, O.: A logical generalization of formal concept analysis. In: Ganter, B., Mineau, G.W. (eds.) ICCS 2000. LNCS, vol. 1867, pp. 371–384. Springer, Heidelberg (2000)
17. Ferré, S., Ridoux, O.: Introduction to logical information systems. Inf. Process. Management 40(3), 383–419 (2004)
18. Ganter, B., Kuznetsov, S.O.: Pattern structures and their projections. In: Alexandrov, V.N., Dongarra, J., Juliano, B.A., Renner, R.S., Tan, C.J.K. (eds.) ICCS 2001. LNCS, vol. 2074, pp. 129–142. Springer, Heidelberg (2001)
19. Ganter, B., Wille, R.: Formal Concept Analysis. Springer, Heidelberg (1999)
20. Georgescu, G., Popescu, A.: Non-dual fuzzy connections. Arch. Math. Log. 43(8), 1009–1039 (2004)
21. Kuznetsov, S.O.: Learning of simple conceptual graphs from positive and negative examples. In: Żytkow, J.M., Rauch, J. (eds.) PKDD 1999. LNCS (LNAI), vol. 1704, pp. 384–391. Springer, Heidelberg (1999)
22. Kuznetsov, S.O.: Pattern structures for analyzing complex data. In: Sakai, H., Chakraborty, M.K., Hassanien, A.E., Ślezak, D., Zhu, W. (eds.) RSFDGrC 2009. LNCS, vol. 5908, pp. 33–44. Springer, Heidelberg (2009)
23. Pawlak, Z.: Rough Sets. Theoretical Aspects of. Reasoning about Data. Kluwer Acad. Publ., Dordrecht (1991)
24. Popescu, A.: A general approach to fuzzy concepts. Mathematical Logic Quarterly 50, 265–280 (2004)
25. Prade, H., Serrurier, M.: Bipolar version space learning. Inter. J. of Intelligent Systems 23(10), 1135–1152 (2008)
26. Benferhat, S., Tabia, K.: An efficient algorithm for naive possibilistic classifiers with uncertain inputs. In: Greco, S., Lukasiewicz, T. (eds.) SUM 2008. LNCS (LNAI), vol. 5291, pp. 63–77. Springer, Heidelberg (2008)
27. Yao, Y.Y., Chen, Y.: Rough set approximations in formal concept analysis. In: Peters, J.F., Skowron, A. (eds.) Transactions on Rough Sets V. LNCS, vol. 4100, pp. 285–305. Springer, Heidelberg (2006)
28. Yao, Y.Y.: A comparative study of formal concept analysis and rough set theory in data analysis. In: Tsumoto, S., Słowiński, R., Komorowski, J., Grzymała-Busse, J.W. (eds.) RSCTC 2004. LNCS (LNAI), vol. 3066, pp. 59–68. Springer, Heidelberg (2004)
29. Zadeh, L.A.: Fuzzy sets as a basis for a theory of possibility. Fuzzy Sets and Systems 1, 3–28 (1978)

DK-BKM: Decremental K Belief K-Modes Method

Sarra Ben Hariz and Zied Elouedi

LARODEC, Institut Supérieur de Gestion de Tunis, 41 Avenue de la Liberté, 2000 Le Bardo, Tunisie
sarra.benhariz@gmail.com, zied.elouedi@gmx.fr

Abstract. This paper deals with the dynamic clustering under uncertainty by developing a decremental K Belief K-modes method (DK-BKM). Our clustering DK-BKM method tackles the problem of decreasing the number of clusters in an uncertain context using the Transferable Belief Model (TBM).

The proposed approach generalizes belief K-modes method (BKM) to a dynamic environment. Thus, this so-called DK-BKM method provides a new clustering technique handling uncertain categorical attribute's values of dataset objects where dynamic clusters' number is considered. Using the dissimilarity measure concept makes us to update the partition without performing complete reclustering. Experimental results of this dynamic approach show good performance on well-known benchmark datasets.

Keywords: Clustering, Transferable belief model (TBM), Belief K-modes method (BKM), Number of clusters.

1 Introduction

Clustering is the process of organizing data into groups such that the objects in the same cluster have a high degree of similarity. This data clustering, also called cluster analysis or unsupervised classification, has been addressed in many fields like marketing, medicine, banking, finance, security, etc.

These techniques are very much associated with data types. They have typically focused on numerical datasets using methods such as K-means [11]. More recently, considerable efforts have been on clustering data with categorical or qualitative attributes [10]. The K-modes method [9] is considered as one of the most popular of such techniques since it is inspired from the well-known K-means method, and its efficiency in dealing with large categorical databases. This method is based on a simple matching dissimilarity measure to compute distances, modes instead of means for clusters' representatives, and a frequency-based method to update modes using the K-means paradigm.

However, in practical applications, there is a large amount of data with imperfect characteristics. So, the necessity of handling imprecision and uncertainty of data lead to use several mathematical theories, such as the belief function

A. Deshpande and A. Hunter (Eds.): SUM 2010, LNAI 6379, pp. 84–97, 2010.

theory [17]. Therefore, the idea is to combine clustering methods with this theory for representing and managing uncertainty. In this work, we will focus on the already developed belief K-modes method (BKM) [2], using the belief function theory as interpreted in the Transferable Belief model [20] to overcome this limitation. Several works have been developed within this uncertain framework such as EVCLUS [8] to cluster proxitmity data, ECM [12] which is a direct extension of fuzzy clustering algorithms for vectorial data, while RECM [13] was developed as an evidential counterpart of relational fuzzy clustering algorithm. All these belief clustering approaches generate a credal partition. Moreover, a belief clustering method [15] was proposed for clustering different types of belief functions. Note that, contrary to all these latter clustering methods, the BKM method deals with objects characterized by uncertain attributes' values.

On the other hand, data mining fields confront, generally, the problem of pattern maintenance because the update of data is a fundamental task in data management process. Most existing approaches suppose that databases are static, and their updates require, thus, scanning the entire old and new information. Such algorithms are known as off-line techniques, while the proposition of dynamic mining techniques for addition or deletion of any information will be more efficient. This on-line category takes into account model evolution over time. Recently, various dynamic approaches were introduced. Works in [4,16] propose incremental clustering methods dealing with the increasing of the attributes' set by one. In [1,21], the clusters are created by adding incrementally the dataset objects. Furthermore, in [22], incremental as well as decremental datasets procedures are provided to cluster dynamic objects sets. However, most of these non-standard methods, except [4], assume that the datasets are certain and then could not deal with the challenge of uncertainty management. Moreover, a preliminary dynamic uncertain clustering approach is proposed in [5], handling only incremental clusters' number.

As a real example, we mention the problem of target markets by using cluster analysis. We get, for instance, as a segmentation of the market four meaningful groups respectively customers of less than 20 years old, between 21 and 35, between 36 and 55, and finally customers over 55 years old. Based on these 'natural' groups resulting from clustering process, and when we will conduct the clustering interpretation in order to develop potential new products' opportunities, the following case may occur namely the elimination of one group among the above structure. Indeed by observing the customers, their detected behaviors must belong to only three groups to correctly define the classes' model from dataset. So, each customer must be assigned to one group among these persistent ones. Note that, the customers' characteristics may be pervaded with uncertainty.

To solve such previous exposed problems namely uncertainty and dynamic ones, the objective of this paper is to propose a new dynamic clustering method in an uncertain context that uses the belief K-modes paradigm and based on the dissimilarity measure concept called decremental K belief K-modes (DK-BKM). Developed to discover an optimal partition when an old cluster is eliminated reducing the clusters' number. Thus, by applying this new approach, the clusters are efficiently maintained without reclustering.

The remainder of this paper is organized as follows. Section 2 recalls the basics of belief function theory within TBM framework. Section 3 introduces the extension of the K-modes method to this uncertain environment. In Section 4, we present our decremental belief K-modes method, denoted DK-BKM, where the algorithm will be detailed. Following that, Section 5 reports and analyzes experimental results carried out on belief versions of the UCI machine learning repository datasets [14]. Finally, Section 6 concludes this work.

2 Belief Function Theory

In this Section, we briefly review the main concepts underlying the theory of belief functions as interpreted in the Transferable Belief Model (TBM). More details can be found in [17,18,20].

2.1 Basic Concepts

Let Θ be a finite non empty set of elementary events to a given problem, called *the frame of discernment.* The set of all the subsets of Θ is referred to *the power set* of Θ, denoted by 2^Θ.

The basic belief assignment (bba) is a function $m : 2^\Theta \rightarrow [0,1]$ such that $\sum_{A \subseteq \Theta} m(A) = 1$. The value m(A), named a basic belief mass (bbm), represents the part of belief supporting exactly that the actual event belongs to A and nothing more specific. The subsets A in Θ such that $m(A) > 0$ are called focal elements.

The belief function *bel* expresses the total belief fully committed to the subset A of Θ without being also committed to $\bar{A}$. This function is defined as follows: $bel : 2^\Theta \mapsto [0,1]$, $bel(A) = \sum_{\phi \neq B \subseteq \Theta} m(B)$, where ϕ is the empty set.

A vacuous bba [17] is such that $m(\Theta) = 1$ and $m(A) = 0$, $\forall A \subseteq \Theta$, $A \neq \Theta$. It represents a state of total ignorance. However, a certain bba expresses the total certainty. It is defined [17] as follows: $m(A) = 1$ and $m(B) = 0$ for all $B \neq A$ and $B \subseteq \Theta$, where A is a singleton event of Θ. Besides, a bayesian belief function is a belief function where the focal elements are all singletons, it is defined by [17]: $bel(\emptyset) = 0$, $bel(\Theta) = 1$, and $bel(A \cup B) = bel(A) + bel(B)$, whenever $A, B \subset \Theta$, $A \cap B = \emptyset$

2.2 Combination Rules

Let m_1 and m_2 be two basic belief assignments induced from two distinct pieces of evidence, defined on the same frame Θ. These bba's can be combined either conjunctively or disjunctively [19]. These TBM combination rules are defined as follows:

The Conjunctive Rule
If both sources of information are fully reliable, then the bba representing the combined evidence satisfies [19]:

$$(m_1 \textcircled{\cap} m_2)(A) = \sum_{B,C \subseteq \Theta; B \cap C = A} m_1(B) m_2(C) \tag{1}$$

The Disjunctive Rule
If at least one of these sources of information is to be accepted, but we do not know which one. So, the disjunctive rule is proposed and is defined as follows [19]:

$$(m_1 \textcircled{\cup} m_2)(A) = \sum_{B,C \subseteq \Theta; B \cup C - A} m_1(B) m_2(C) \tag{2}$$

2.3 Pignistic Transformation

When a decision should be made, beliefs are transformed into probability measures denoted *BetP* [18]. The link between these two functions is achieved by the pignistic transformation defined by:

$$BetP(A) = \sum_{B \subseteq \Theta} \frac{|A \cap B|}{|B|} \frac{m(B)}{(1 - m(\phi))}, \text{ for all } A \in \Theta \tag{3}$$

3 From K-Modes to Belief K-Modes

In this Section, the standard K-modes method will be briefly exposed, before detailing its extension into an uncertain belief context. The general notations of the two versions certain as well as uncertain are exposed in the following.

3.1 Notations

$T = \{X_1, X_2, ..., X_n\}$: The set of n objects defined by s categorical attributes whith certain and precise values.
UT: a given uncertain counterpart of the dataset T.
A=$\{A_1$, A_2,...,$A_s\}$: the set of s attributes.
$\Theta_j = \{a_{j,1}, a_{j,2}, ..., a_{j,p_j}\}$ describes a domain of p_j categories or values of the attribute A_j, where $1 \leq j \leq s$. Its power set is denoted by 2^{Θ_j}.
X_i: an object or instance, for $1 \leq i \leq n$. It can be represented via a conjunction of attribute values as follows: $(x_{i,1}, x_{i,2}, ..., x_{i,s})$.
$x_{i,j}$: the value of the attribute A_j for the object X_i.
$m_j = \{m(c_j) \mid c_j \in 2^{\Theta_j}\}$ represents the bba of the attribute A_j.
$m_{i,j}$: expresses the bba of the attribute A_j corresponding to the object X_i.
$m_i(c_j)$: is the bbm given to $c_j \in 2^{\Theta_j}$ relative to object X_i.

3.2 Standard K-Modes Approach

To tackle the problem of clustering large categorical datasets in data mining, the K-modes algorithm [9] was proposed as an extension of the K-means one [11], since the standard version of this last one deals only with numerical datasets. By using the K-means paradigm, this categorical clustering technique considers a simple matching dissimilarity measure, modes instead of means for cluster representatives, and a frequency-based approach to update modes during the clustering process. These K-means modifications will be discussed later.

Let T be the set to cluster, the clusters' modes are represented by $Q_l = \{q_{l,1}, q_{l,2}, ..., q_{l,s}\}$, for $1 \leq l \leq K$, where K is the number of clusters to build. Each mode is defined by assigning to $q_{l,j}$, the category that is most frequently encountered in $\{x_{1,j}, ..., x_{np_l,j}\}$ where $np_l = |C_l|$, represents the objects' number in this considered cluster C_l.

The dissimilarity measure between an object X_i and the cluster mode Q_l, which is the representative vector of the cluster C_l, is defined by the total mismatches of the corresponding attributes' categories. It is defined as follows:

$$d(X_i, Q_l) = \sum_{j=1}^{s} \delta(x_{i,j}, q_{l,j}) \tag{4}$$

where $\delta(x_{i,j}, q_{l,j}) = \begin{cases} 0 & if\ x_{i,j} = q_{l,j} \\ 1 & if\ x_{i,j} \neq q_{l,j} \end{cases}$

3.3 Belief K-Modes Extension into an Uncertain Belief Framework

The standard version of K-modes method presents good results for precise and certain data. However, this algorithm shows serious limitations when dealing with uncertainty. Indeed, this uncertainty may appear in the attributes' values of instances belonging to the training set used to ensure the cluster construction phase. A belief K-modes method (BKM) [2] was developed, as a new clustering technique based on the K-modes technique and using the belief function theory in order to deal with uncertainty that may characterize datasets. For this belief version of K-modes method, building clusters needs the definition of its fundamental parameters, namely, cluster modes computation and the dissimilarity distance measure as for the standard procedure, but these parameters must take into account the uncertainty pervading the attributes' values of the dataset instances.

In the following part of this paper, we present the two major parameters of this K-modes extension needed to ensure the clustering task in an uncertain context within the belief function framework. We start by presenting the belief structure of the considered training sets.

Structure of Training Sets under Belief Function Framework

The structure of the training set UT is different from the standard one T. Contrary to the certain training set which includes only precise information, here, we deal with n objects where each of their s attributes is represented by a bba expressing beliefs on its values.

The corresponding bba of an attribute A_j, where $1 \leq j \leq s$, is then given by m_j. Each attribute is represented via one conjunction of all possible values and their corresponding masses. Each $x_{i,j} = \{(c_j, m_i(c_j)) \mid c_j \in 2^{\Theta_j}\}$. This training set offers a more generalized framework than the standard one.

Cluster Mode

Given a cluster $C_l = \{X_1, ..., X_p\}$ of p objects, for $1 \leq l \leq K$. An intuitive definition of one strategy of computing the clusters' modes within the belief function theory context could be the conjunctive rule of combination [19] generally used as an aggregation operator in such framework combining between two or several bba's. This operator is particularly suitable when distinct sources provide pieces of evidence respectively to the same object. However, in our case, the induced pieces of evidence are related to different objects' attributes. So, the conjunctive (even disjunctive) rule is not appropriate. On the other hand, the mean operator permits combining bba's respectively to each attribute provided by all p objects of one cluster as well as satisfying the commutativity and the idempotency properties [7]. Then, the idea was to apply this operator to this uncertain context [2].

Thus, the mode of the cluster C_l can be defined by the following belief mode $Q_l = (q_{l,1}, .., q_{l,j}, .., q_{l,s})$, such that:

$$q_{l,j} = \{(c_j, m_l(c_j)) | c_j \in 2^{\Theta_j}\} \tag{5}$$

Where $m_l(c_j)$ is the relative bbm of attribute value c_j within C_l which is defined as follows:

$$m_l(c_j) = \frac{\sum_{i=1}^{p} m_i(c_j)}{|C_l|} \tag{6}$$

Where $|C_l|$ is the number of objects in C_l while, $m_l(c_j)$ expresses the part of belief about the value c_j of the attribute A_j corresponding to this cluster mode. It is the average of all masses provided by the objects in this cluster C_l corresponding to this attribute category c_j. Note that the associativity property is ensured via the arithmetic addition of all masses.

Dissimilarity Measure

The dissimilarity measure has to take into account the new dataset structure, where a bba is defined for each attribute per object. It should be able to compute the distance between any object and each cluster mode, respectively to all considered attributes, within such context (bba's representation). In BKM framework, the idea was to adapt the belief distance defined by [6] to this uncertain clustering context. It can be expressed as follows:

$$D(X_i, Q_l) = \sum_{j=1}^{m} d(m_{i,j}, m_{l,j}) \tag{7}$$

where $m_{i,j}$ and $m_{l,j}$ are the relative bba's of the attribute A_j provided by respectively the object X_i and the mode Q_l. For d component is more detailed in [2].

Based on these two major parameters, the BKM algorithm can be applied and it has the same skeleton as the standard K-modes method [2].

4 Decremental Belief K-Modes Method

The most clustering techniques are static meaning that all clustering parameters and data are available at a time when starting the clustering process and operate in a certain environment. However, new information, which may be uncertain, can be added to or an old one can be deleted from existing clustering results without having to recalculate the complete analysis. Such problems make the standard clustering approaches not able to ensure their task. Then, the development of appropriate dynamic techniques dealing with these limitations is unavoidable.

Several incremental algorithms were proposed as examples let us mention [1,4,5,16,21,22], while for decremental concept, one clustering approach was developed in [22]. All these non-standard techniques are proposed to learn evolving clustering in a non-stationary environment to deal with dynamic data objects (addition or deletion aspects) or dynamic attribute set (new attribute) as well as the dynamic cluster's number (increasing number). Thus, these dynamic classifiers can provide optimal update models of the dynamic data. Furthermore, [4,5], which are based on BKM approach handling in an uncertain framework, whereas the other methods operate within standard context where clustering parameters are known with total certainty.

In this present study, we propose our dynamic clustering method in an uncertain framework for updating the clusters' partition, when the number of clusters decreases by one, instead of complete reclustering all uncertain data.

4.1 DK-BKM Principal

Like in the *BKM* static framework, within *DK-BKM* context objects may be characterized by uncertain attributes represented through bba's . The key idea is as follows: having $K+1$ clusters, we have to return only K groups. To decrease this clusters' number, we have, initially, to run the BKM algorithm to obtain $K+1$ clusters and their belief modes respectively to our uncertain dataset. This final partition P will be considered as the initial one for the decremental process in order to produce P' smaller partitions. Then, how to process to reduce this clustering parameter without reclustering all the dataset?

So, we have, at first, to find which two groups among the initial ones will be merged initially. It is the *initialization phase*. After that, many rearrangements between objects on the remaining K clusters will be possible during the second main phase namely the *updating phase*.

We recall that the aim of all unsupervised clustering algorithms is to group objects according to their similarity computed using a suitable distance metric.

An intuitive idea will be based on this. However, the proposed approach will define different distance measures namely the inter-clusters, the intra-cluster, and the inter-objects ones, needed during the two already mentioned phrases. Thus, these measures can serve as basics of our proposal technique. A brief description of the two *DK-BKM* phases is presented as follows.

Initialization Phase

Let us assume that the available partition of $K+1$ clusters, resulting by applying BKM procedure, constitutes the starting point of our decremental method. Note that this initialization input respects this condition: each object is correctly assigned which means that its distance respectively to its cluster is the smallest one. For the initialization step, in order to improve the partition quality, we subtract one cluster. In consists in choosing among the input clusters, the two appropriate ones to merge based on their inter-dissimilarity measure. To this end, the inter-clusters dissimilarities matrix $K+1 * K+1$ must be defined as follows:

$$D_{Inter}(C_l, C_r) = \sum_{i=1}^{n_{pl}} D(X_i, Q_r) + \sum_{j=1}^{n_{pr}} D(X_j, Q_l) \tag{8}$$

With $l, r \in \{1, .., K+1\}$ and $l \neq r$, while n_{pl} and n_{pr} are respectively the number of objects in clusters C_l and C_r and D component is as defined by Equation 7. It expresses the dissimilarity between any two clusters C_l and C_r. Thus, the most similar ones, providing the lowest D_{Inter}, will be grouped together and the partitions modes must be updated consequently. Seen that the sum of the intra-cluster dissimilarities expresses the quality of this initial partition, it should be also computed in order to minimize it within the next phase. It is defined as follows:

$$SD_{Intra} = \sum_{l=1}^{K} D_{Intra}(C_l) \tag{9}$$

Where the intra-cluster dissimilarities matrix $1 * K$, representing the respective quality of the cluster C_l, where $l \in \{1, .., K\}$ and n_l is the cardinal of the cluster C_l, is computed by:

$$D_{Intra}(C_l) = \sum_{i=1}^{n_l} D(X_i, Q_l) \tag{10}$$

Updating Phase

This procedure aims at updating the cluster's organization after the fusion of the two clusters made in the previous step. It consists in the rearrangement of all the dataset objects into the appropriate clusters using similarity/dissimilarity measures. Indeed, the inter-objects distance is used to decide of the objects' movements. Moreover, the intra-cluster dissimilarity measure is considered to judge the clustering quality.

At the beginning, based on the all already computed intra-cluster distances (Equation 9), we decide which one among all groups must be rearranged. It is done according to the highest D_{Intra} which represents the worst cluster. So, this distance should be minimized by the rearrangement of data objects which is guaranteed by checking all inter-objects distances sum and assigning the most far one(s), respectively to all other objects, into this cluster providing the lowest D_{Intra}. The definition of the used inter-objects distance sum $DS(X_i)$ of an object X_i, belonging to C_l, respectively to all other objects of this same cluster is as follows:

$$DS(X_i) = \sum_{j=1, j \neq i}^{np_l} D(X_i, X_j) \tag{11}$$

Where the inter-objects dissimilarity $n \times n$ matrix $D(X_i, X_j)$, for i_1 and $i_2 \in \{1, .., n\}$, is defined by an adaptation of the proposed one in BKM framework see Equation 7. It is defined by:

$$D(X_{i_1}, X_{i_2}) = \sum_{j=1}^{s} d(m_{i_1,j}, m_{i_2,j}) \tag{12}$$

After the objects movements, we have to recompute the sum of the intra-cluster dissimilarities (cost function) in order to minimize it comparing to the previous step. If this improvement is not reached (stable partition) or we have iterated a maximal number of times, this updating process is stopped.

Several iterations can be generated but let us note, for our decremental process and during each iteration, we have only to compute the intra-cluster dissimilarity measures and check for the worst one its inter-objects distances based on the inter-objects matrix calculated initially. It makes our dynamic approach faster than the static approach since in each iteration,for this latter method, all distance measures between each object respectively to each cluster mode has to be computed.

4.2 DK-BKM Algorithm

The final partition P of $K+1$ clusters resulting from BKM approach, the uncertain dataset UT represented via bba's concept and the maximum number of iterations, which must be allowed $noMaxIter$, are the inputs of our *DK-BKM* method. The above phases are formalized in the DK-BKM algorithm to reduce the clusters' number. Thus, the DK-BKM algorithm is summarized in the follows.

Note that, when the decreasing of the partition cardinality will be by more than one cluster, we have to re-run our proposed method many times. Every time, we have to decrease by one cluster our partition until obtain the wished groups' number. Besides, this proposed belief dynamic clustering approach can be considered as probabilistic technique when the attribute uncertainty is expressed via the bayesian belief function. Moreover, the standard certain database can be handled via certain bba's within DK-BKM framework.

DK-BKM Algorithm (P,UT,$noMaxIter$)

Data: P, UT, $noMaxIter$
Result: P'
begin

Initialization phase
1. Compute $K+1$ inter-clusters dissimilarities D_{Inter} by using Equation 8.
2. The two clusters providing $min(D_{Inter})$ (the most similar ones) must be merged to decrease the clusters' number by one.
3. Compute the K intra-cluster dissimilarities D_{Intra} (Equation 10) and their sum SD_{Intra} (Equation 9).

Updating phase
4. Compute the inter-objects dissimilarities matrix $n \times n$ by applying Equation 12.
5. Set $t = 1$. The intra-dissimilarity of the cluster C_l providing the $max(D_{Intra})$ (the worst arranged) may be improved by any object movement thus:
- Compute for any X_i belonging to C_l its distances' sum (DS_i) (Equation 11).
- The object(s) with $max(DS)$, will be assigned to the cluster with $min(D_{Intra})$.
- Update the partition of two altered clusters and their respective modes as well as their intra-dissimilarity measures and the corresponding sum (Equation 9 and Equation 10).
- Compare the new NSD_{Intra} to previous one SD_{Intra}.

  if SD_{Intra} - $NSD_{Intra} > \epsilon$ **then**
    $SD_{Intra} \leftarrow NDS_{Intra}$.
    Set $t \leftarrow t + 1$.
    if $t <= noMaxIter$ **then**
      Reiterate (go to step 5)
    else
      Stop
  else
    Stop

  Return the clusters' partition obtained before this last assignment.

end

5 Experimentation

In our experiments, we have performed several tests on real databases obtained from the U.C.I. repository [14] to evaluate our method in such dynamic uncertain environment. A brief description of these databases is given in Table 1. #instances, #attributes, #classes denote respectively the total number of instances, the number of attributes and the number of classes.

These databases are artificially modified in order to include uncertainty in attributes' values. For this purpose, we consider different parameters namely the certain attributes' values of the instances, the degree of uncertainty p per attribute (varying in [0, 1]), and the uncertain objects percent for the dataset, i.e. the percent of generated uncertain objects for the given dataset to cluster.

To create the belief datasets, the basic idea is to assign to each attribute a bba on the set of its possible values using the already mentioned parameters. This is done for all dataset objects. The resulting bba's describe our belief about the value of the attributes which any object has.

A set of experiments has been conducted for these belief versions of UCI datasets, for testing our proposed approach. The different results carried out from tests will be presented and analyzed in order to evaluate our clustering method in such uncertain and dynamic framework.

One criterion used to judge the performance of our proposal is the *PCC* expressing the percent of the correctly classified instances. It is equivalent to the proposed measure of clustering K-modes results in [9], called the clustering accuracy. Besides to PCC, we take the convergence speed, expressed by the iterations' number needed to obtain the final stable partition, as another criterion to compare the dynamic and static algorithms of BKM method. Indeed, we assume as within static framework the choice of clusters' number to form is the same as the number of classes of actual datasets seen the presence of true labels (see Table 1).

Table 1. Description of UCI databases

Databases	#instances	#attributes	#classes
Solar Flare	1389	10	3
Zoo	101	17	7
Congressional voting records	435	16	2
Breast Cancer Wisconsin	699	9	2
Hayes-Roth	160	5	3
Mushroom	5644	22	2
Car evaluation	1728	6	4
Lymphography	148	18	4
Spect heart	267	22	2
Tic-Tac-Toe Endgame	958	9	2

Two cases must be considered, namely decremental and static contexts in order to compare them. To this end, we have to proceed by applying the BKM approach initially, we build $K+1$ clusters for each uncertain dataset, where K is the actual class's number. We consider then the resulting output partition as DK-BKM input. Thus, by running this dynamic clustering technique, the final stable partition of K clusters is obtained after NB iterations with PCC evaluation criterion. To judge its quality, the proposed approach must be compared to the static version. We have to retain the corresponding PCC as well as the iterations number NB resulting by using BKM method to cluster dataset in only K groups. Table 2 reports comparatively, the resulting values of the two considered evaluation criteria. For each run, only K initial same clusters' modes were considered within two algorithms using the already proposed method for its selection [3]. We run the algorithm ten times. Thus, the accuracy of our results is measured according to the mean values corresponding to the two considered criteria.

Table 2. Experimental results

Databases	DK-BKM $PCC(\%)$	NB	BKM $PCC(\%)$	NB
Solar Flare	90.45	9	88.25	11
Zoo	74.5	5	71.8	8
Congressional voting records	87.9	5	88.15	7
Breast Cancer Wisconsin	79.9	7	79.7	8
Hayes-Roth	78.4	9	76.7	12
Mushroom	83.7	9	81.5	11
Car evaluation	82.9	11	83.2	14
Lymphography	79.4	7	80.3	11
Spect heart	75.8	8	71.9	10
Tic-Tac-Toe Endgame	78.9	7	78.5	11

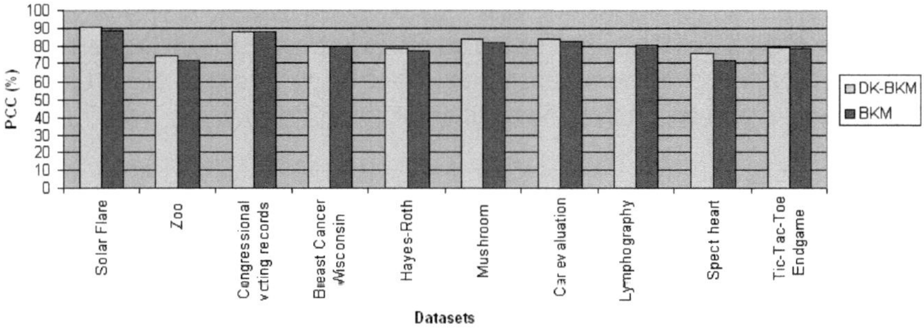

Fig. 1. PCC of DK-BKM and BKM methods

From Figure 1 and Table 2, we can conclude that by applying the dynamic method, an improvement of the $PCC's$ values is mentioned, for the most datasets, comparing to those obtained by the BKM method. In fact, only for two datasets, the PCC of BKM is a little bit better than the one given by DK-BKM. However, this difference is very weak (respectively 0.25% and 0.9%).

We observe, by the exposed iterations' number, that reclustering the whole data objects after the decreasing of the clusters number cannot be considered as a good practice comparing to the decremental developed process due to the great number of iterations needed to find the final stable partitions compared to the ones obtained via DK-BKM approach. The provided NB by applying DK-BKM are smaller than BKM's ones for all the datasets, which improves the efficiency of the developed dynamic method besides the already effectiveness quality. Indeed, the advantage of such dynamic approach is the speed of convergence expressed by the iterations' number which is significantly decreases compared to static method, moreover the clustering quality. Figure 2 allows us to show the comparative resulting iterations' numbers of two versions static and dynamic BKM approaches.

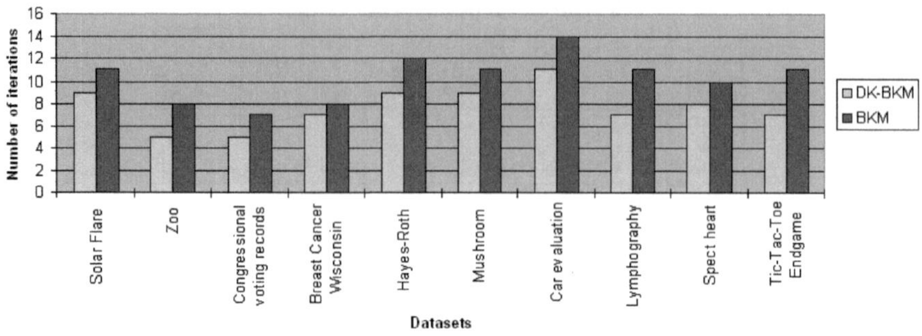

Fig. 2. Number of iterations of DK-BKM & BKM methods

Based on PCC and number of iterations, we draw the conclusion that our developed DK-BKM is more appropriate than BKM to use in order to update the cluster's partition when their number decreases by one.

6 Conclusion

In this paper, a new dynamic clustering approach named DK-BKM has been developed. This approach is based on K-modes paradigm within the belief function framework to handle uncertain attribute values. It consists in incrementally deletion of one cluster from the resulting BKM ones, it is done by considering the inter-dissimilarity measure concept. Our proposed method was tested on real datasets versions soiled by uncertainty and the corresponding experiments have shown its efficiency, when considering the above advanced evaluation criterion as well as the convergence speed. The obtained DK-BKM results are competitive to the static BKM ones.

We plan to use another dissimilarity measure in order to compare the obtained results to the current ones. As a future work, we suggest to improve our developed approach by handling other dynamic aspects and uncertainty forms of the considered environment. Thus, the proposed method will be more flexible.

References

1. Aranganayagi, S., Thangavel, K.: Incremental Algorithm to Cluster the Categorical Data with Frequency Based Similarity Measure. International Journal of Computational Intelligence 6, 24–32 (2010)
2. Ben Hariz, S., Elouedi, Z., Mellouli, K.: Clustering Approach using Belief Function Theory. In: Euzenat, J., Domingue, J. (eds.) AIMSA 2006. LNCS (LNAI), vol. 4183, pp. 162–171. Springer, Heidelberg (2006)
3. Ben Hariz, S., Elouedi, Z., Mellouli, K.: Selection Initial Modes for Belief K-Modes Method. International Journal of Applied Science, Engineering and Technology (IJASET) 4, 233–242 (2008)

4. Ben Hariz, S., Elouedi, Z.: An Incremental Clustering Approach within Belief Function Framework. In: The Twelfth IASTED International Conference on Artificial Intelligence and Soft Computing (ASC), pp. 98–103 (2008)
5. Ben Hariz, S., Elouedi, Z.: IK-BKM: An Incremental Clustering Approach Based on Intra-Cluster Distance. To appear in The eighth ACS/IEEE International Conference on Computer Systems and Applications, AICCSA 2010 (2010)
6. Bosse, E., Grenier, D., Jousselme, A.L.: A new distance between two bodies of evidence. Information Fusion 2, 91–101 (2001)
7. Catherine Murphy, K.: Combining belief functions when evidence conflicts. Decision Support Systems 29, 1–9 (2000)
8. Denoeux, T., Masson, M.: EVCLUS: Evidential Clustering of Proximity Data. IEEE Transactions on Systems, Man and Cybernetics B 34(1), 95–109 (2004)
9. Huang, Z.: Extensions to the k-means algorithm for clustering large data sets with categorical values. Data Mining Knowledge Discovery 2(2), 283–304 (1998)
10. Ganti, V., Gehrke, J.E., Ramakrishnan, R.: CACTUS–Clustering Categorical Data Using Summaries. In: Proceedings of the 1999 SIGKDD Conference, San Diego, California, pp. 73–83 (1999)
11. MacQueen, J.: Some methods for classification and analysis of multivariate observations. In: Proceeding of the Fifth Berkeley Symposium on Math., Stat. and Prob., pp. 281–296 (1967)
12. Masson, M.-H., Denoeux, T.: ECM: An evidential version of the fuzzy c-means algorithm. Pattern Recognition 41, 1384–1397 (2008)
13. Masson, M.-H., Denoeux, T.: RECM: Relational Evidential c-means algorithm. Pattern Recognition Letters 30, 1015–1026 (2009)
14. Murphy, M., Aha, D.W.: Uci repository databases (1996), `http://www.ics.uci.edu/mlearn`
15. Schubert, J.: Clustering decomposed belief functions using generalized weights of conflict. International journal of approximate reasoning 48(2), 466–480 (2008)
16. Serban, G., Campan, A.: Incremental Clustering Using a Core-Based Approach. Informatica L(1), 854–863 (2005)
17. Shafer, G.: A mathematical theory of evidence, p. 30. Princeton Univ. Press, Princeton (1976)
18. Smets, P.: The Transferable Belief Model and Other Interpretations of Dempster-Shafer's Model. In: Bonissone, P.P., Henrion, M., Kanal, L.N., Lemmer, J.F. (eds.) Uncertainty in Artificial Intelligence 6, pp. 375–384. North Holland, Amsteram (1991)
19. Smets, P.: Belief functions: The disjunctive rule of combination and the generalized bayesian theorem. International Journal of Approximate Reasoning 9, 1–35 (1993)
20. Smets, P.: The transferable belief model for quantified belief representation. In: Gabbay, D.M., Smets, P. (eds.) Handbook of Defeasible Reasoning and Uncertainty Management Systems, vol. 1, pp. 267–301 (1998)
21. Su, X., et al.: A Fast Incremental Clustering Algorithm. In: Proceedings of the 2009 International Symposium on Information Processing (ISIP 2009), pp. 175–178 (2009)
22. Truta, T.M., Campan, A.: K-Anonymization Incremental Maintenance and Optimization techniques. In: Adams, C., Miri, A., Wiener, M. (eds.) SAC 2007. LNCS, vol. 4876, pp. 380–387. Springer, Heidelberg (2007)

On the Use of Fuzzy Cardinalities for Reducing Plethoric Answers to Fuzzy Queries

Patrick Bosc[1], Allel Hadjali[1], Olivier Pivert[1], and Grégory Smits[2]

[1] Irisa ENSSAT
Univ. Rennes 1 Lannion France
{bosc,hadjali,pivert}@enssat.fr
[2] Irisa IUT Lannion
Univ. Rennes 1 Lannion France
gregory.smits@univ-rennes1.fr

Abstract. Retrieving data from large-scale databases sometimes leads to plethoric answers especially when queries are under-specified. To overcome this problem, we propose to strengthen the initial query with additional predicates. These predicates are selected among predefined ones according principally to their degrees of semantic correlation with the initial query in order to avoid an excessive modification of its initial scope. According to the size of the initial answer set and the number of expected results specified by the user, fuzzy cardinalities are used to assess the reduction capability of these correlated predefined predicates.

Keywords: Plethoric answers, query strengthening, correlation, fuzzy cardinalities, fuzzy association rules.

1 Introduction

The practical need for endowing intelligent information systems with the ability to exhibit cooperative behavior has been recognized since the early nineties. As pointed out in [8], the main intent of cooperative systems is to provide correct, non-misleading and useful answers, rather than literal answers to user queries.

Two dual problems are addressed in this field. The first one is known as the "Empty Answer" (EA) problem, that is, the problem of providing the user with alternative data when there is no item fitting his/her query. The second one is the "Plethoric Answers" (PA) problem which occurs when the amount of returned data is too large to be manageable. This paper focuses on this latter issue in the context of fuzzy queries.

The PA problem has been intensively addressed by the information retrieval community and two main approaches have been proposed for Boolean queries. The first one, that may be called data-oriented, aims at ranking the answers in order to return the best k ones to the user. However, this strategy is often faced with the difficulty of comparing and distinguishing between tuples that satisfy the initial query. In this data-oriented approach, we can also mention works which aim at summarizing the answer set to a query [14].

A. Deshpande and A. Hunter (Eds.): SUM 2010, LNAI 6379, pp. 98–111, 2010.

The second type of approach may be called query-oriented as it performs a modification of the initial query in order to make it more selective. For instance, a strategy consists in strengthening the specified predicates (as an example, a predicate $A \in [a_1, a_2]$ becomes $A \in [a_1 + \gamma, a_2 - \gamma]$) [3]. However, for some predicates, this strengthening leads to a deep modification of the meaning of the initial predicate. For example, if we consider a query looking for fast-food restaurants located in a certain district delimited by geographical coordinates, a strengthening of the condition related to the location could lead to the selection of restaurants in a very small area, and the final answers would not necessarily fit the user's need. Another type of approach advocates the use of user-defined preferences on attributes which are not involved in the initial query [10,7,2]. Such a subjective knowledge can then be used to select the most preferred items among the initial answer set. Still another category of query-oriented approaches aims at automatically completing the initial query with additional predicates to make it more demanding. Our work belongs to this last family of approaches but its specificity concerns the way additional predicates are selected.

Indeed, we consider that the predicates added to the query must respect two properties: i) they must reduce the size of the initial answer set, ii) they must modify the scope of the initial query as little as possible. Based on a predefined vocabulary materialized by fuzzy partitions that linguistically describes the attribute domains, we propose to identify the predicates which are the most correlated to the initial query. Such correlation properties are inferred from the data and express semantic links between possible additional predicates and those present in the initial query. Moreover, we consider that user queries involve a user-specified quantitative threshold K corresponding to the approximate number of expected results. A useful strengthening method is a method that assists the user through the reduction of a plethoric answer set to a subset containing approximately K results. To reach this latter objective, we propose to precompute fuzzy cardinalities as a predefined knowledge about the data distributions.

The remainder of the paper is structured as follows. Section 2 introduces the basic necessary notions and Section 3 describes the two main steps of our query strengthening approach: i) retrieving the predicates the most correlated to a query; ii) identifying the best ones taking into account the selectivity of the augmented query. Section 4 describes the query strengthening process and the way it can be made efficient. An experimentation illustrates this interactive approach in Section 5. Before concluding and drawing perspectives in Section 7, our approach is put into perspective with related works in Section 6.

2 Preliminaries

2.1 Plethoric Answers and Under-Specified Fuzzy Queries

We consider a database fuzzy querying framework such as the SQLf language introduced in [4] that is used to formulate queries involving fuzzy predicates. On top of the predicates used to express the user requirements, we also consider

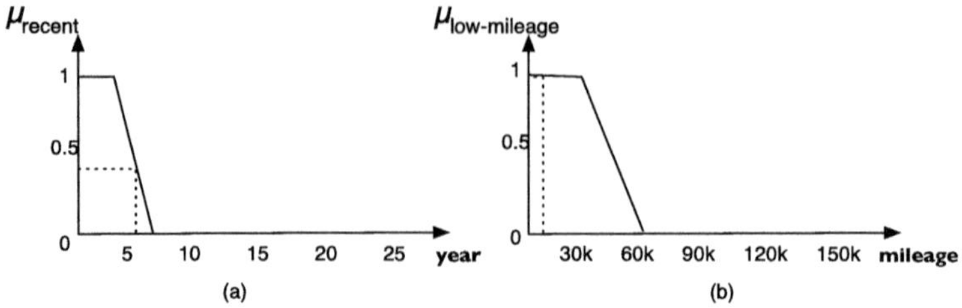

Fig. 1. Fuzzy predicates (a) *recent* and (b) *low-mileage* (where 30K means 30.000 km)

that he/she specifies a quantitative threshold defining the approximate number of expected results, denoted by K.

A typical example of a fuzzy query is: `"retrieve the recent and low-mileage cars"`, where `recent` and `low-mileage` are gradual predicates represented by means of fuzzy sets as illustrated in Figure 1. Let Q be a fuzzy query. We denote by Σ_Q the answer set to Q when addressed to a regular relational database D. Σ_Q contains the items of the database that *somewhat* satisfy the fuzzy requirements involved in D. Formally,

$$\Sigma_Q = \{t \in D/\mu_Q(t) > 0\},$$

where t stands for a database tuple. Let h, $h \in]0,1]$ be the height of Σ_Q, *i.e.* the highest membership degree assigned to an item of Σ_Q. Let now $\Sigma_Q^* \ (\subseteq \Sigma_Q)$ denote the set of answers that satisfy Q with the degree h.

$$\Sigma_Q^* = \{t \in D/\mu_Q(t) = h\}$$

Definition 1. Let Q be a fuzzy query, we say that Q leads to a PA problem if the set Σ_Q^* is too large, *i.e.*, is significantly greater than K.

To reduce Σ_Q^*, we propose an approach that integrates additional predicates as new conjuncts to Q. By doing so, we obtain a more restrictive query Q' which may lead to a reduced set of answers $\Sigma_Q'^* \subset \Sigma_Q^*$ with a cardinality as close as possible to K. This strengthening strategy based on predicate correlation is mainly dedicated to what we call under-specified queries.

Definition 2. An under-specified query Q typically involves few predicates (between 1 and 3) to describe an expected set of answers that can be more precisely described by properties not specified in Q.

For example, consider a user looking for *almost new* cars through a query like `"select secondHandCars which are very recent"`. The answer set to this query can be reduced through the integration of additional properties like low mileage, high security and high comfort level, *i.e.*, properties which are usually possessed by very recent cars.

This strengthening approach aims at identifying correlation links between additional properties and an initial query. These additional correlated properties are suggested to the user as candidates for the strengthening of his/her initial query. This interactive process is iterated until the result is of an acceptable size for the user and corresponds to what he/she was really looking for.

2.2 Fuzzy Cardinalities and Association Rules

Fuzzy Cardinalities. In the context of flexible querying, fuzzy cardinalities appear to be a convenient formalism to represent how many tuples from a relation satisfy a fuzzy predicate to various degrees. It is considered that these various membership degrees are defined by a finite scale $1 = \sigma_1 > \sigma_2 > ... > \sigma_f > 0$. Such fuzzy cardinalities can be incrementally computed and maintained for each linguistic label and for the diverse conjunctive combinations of these labels. Fuzzy cardinalities are represented by means of a possibilistic distribution [9] like $F_{P^a} = 1/0 + ...1/(n-1) + 1/n + \lambda_1/(n+1) + ... + \lambda_k/(n+k) + 0/(n+k+1) + ...$, where $1 > \lambda_1 \geq ... \geq \lambda_k > \lambda_{k+1} = 0$ for a predicate P^a. In this paper, without loss of information, we use a more compact representation $F_{P^a} = 1/c_1 + \sigma_2/c_2 + ... + \sigma_f/c_f$, where $c_i, i = 1..f$ is the number of tuples in the concerned relation that are P^a with a degree at least equal to σ_i. For the computation of cardinalities concerning a conjunction of q fuzzy predicates, like $F_{P^a \wedge P^b \wedge ... \wedge P^q}$, one takes into account the minimal satisfaction degree obtained by each tuple t for the concerned predicates, $min(\mu_{P^a}(t), \mu_{P^b}(t), ..., \mu_{P^q}(t))$.

Association Rules and Correlation. Given two predicates P^a and P^b, an association rule denoted by $P^a \Rightarrow P^b$ aims at quantifying the fact that tuples that are P^a are also P^b (P^a and P^b can be replaced by any conjunction of predicates). As suggested in [5], fuzzy cardinalities can be used to quantify the confidence of such an association by means of a scalar or by means of a fuzzy set detailing the subsequent confidence for each σ-cut. The first representation of the confidence as a scalar is used in our approach as it is more convenient to compare fuzzy sets using scalar measures. Thus, the confidence of an association rule $P^a \Rightarrow P^b$, denoted by $confidence(P^a \Rightarrow P^b)$, is computed as follows:

$$confidence(P^a \Rightarrow P^b) = \frac{\Gamma_{P^a \wedge P^b}}{\Gamma_{P^a}}$$

Here, $\Gamma_{P^a \wedge P^b}$ and Γ_{P^a} correspond to scalar cardinalities, which are computed as the weighted sum of the elements belonging to the concerned fuzzy cardinalities. On the example above, the scalar version of $\Gamma_{recent} = 1/6 + 0.6/7 + 0.2/8$ is $\Gamma_{recent} = 1 \times 6 + 0.6 \times (7-6) + 0.2 \times (8-7) = 6.8$.

Example 1. Let us consider a relation named `secondHandCars` containing ads for second hand cars with {`brand`, `model`, `type`, `year`, `mileage`, `optionLevel`, `securityLevel`, `horsePower`}. A sample of its extension is given in Table 1.

We assume that the finite scale of degrees used for the computation of the fuzzy cardinalities is $\sigma_1 = 1 > \sigma_2 = 0.8 > \sigma_3 = 0.6 > \sigma_4 = 0.4 > \sigma_5 = 0.2 > 0$. From Table 1, we can compute the cardinalities of the predicates *recent* and *low-mileage* (Fig. 1) and of their conjunction:

Table 1. Extension of relation `secondHandCars`

t_i	brand	model	type	year	mileage	optionLvl.	securityLvl.	horsePw.
t_1	vw	golf	sedan	06	95K	6	8	90
t_2	seat	ibiza	sport	01	150K	3	6	80
t_3	audi	A3	sport	08	22K	8	7	120
t_4	seat	cordoba	sedan	04	220K	4	4	100
t_5	ford	focus	estate	05	80K	7	5	70
t_6	vw	polo	city	98	120K	2	3	50
t_7	vw	golf	estate	07	40K	5	5	80
t_8	ford	ka	city	03	240K	4	3	50
t_9	kia	rio	city	09	10K	2	3	60
t_{10}	seat	leon	sport	09	25K	8	8	115
t_{11}	ford	focus	sedan	07	53K	6	7	90
t_{12}	rover	223	sedan	97	100K	9	8	120

- $F_{recent} = 1/6 + 0.8/6 + 0.6/7 + 0.4/7 + 0.2/8$
- $F_{low\text{-}mileage} = 1/3 + 0.8/4 + 0.6/4 + 0.4/4 + 0.2/5$
- $F_{low\text{-}mileage \wedge recent} = 1/3 + 0.8/4 + 0.6/4 + 0.4/4 + 0.2/5$.

From this dataset, one can deduce that the association rule $P_{recent} \Rightarrow P_{low\text{-}mileage}$ expressing that recent cars have a low mileage has a confidence of $\frac{1\times3+0.8\times(4-3)+0.2\times(5-4)}{1\times6+0.6\times1+0.2\times1} \approx 0.59$ and that the association rule $P_{low\text{-}mileage} \Rightarrow P_{recent}$ has a maximal confidence of 1 as all low-mileage cars are also recent in Table 1.

2.3 Shared Vocabulary

Fuzzy sets constitute an interesting framework for extracting knowledge on data that can be easily comprehensible by humans. Indeed, associated with a membership function and a linguistic label, a fuzzy set is a convenient way to formalize a gradual property. As mentioned before, especially in [12], such prior knowledge can be used to represent what the authors call a "macro expression of the database". Contrary to the approach presented in [12] where this knowledge is computed by means of a fuzzy classification process, it is, in our approach, defined *a priori* by means of a `Ruspini` partition of each attribute domain. Let us recall that a Ruspini partition is composed of fuzzy sets, where a set, say P_i, can only overlap with its predecessor P_{i-1} or/and its successor P_{i+1} (when they exist) and for each tuple t, $\sum_{i=1}^{n_i} \mu_{P_i}(t) = 1$, where n_i is the number of partition elements of the concerned attribute. These partitions are specified by an expert during the database design step and represent "common sense partitions" of the domains instead of the result of an automatic process which may be difficult to interpret. Indeed, we consider that the predefined fuzzy sets involved in a partition constitute a shared vocabulary and that these fuzzy sets are used by the users when formulating their queries. Moreover, in our approach, the predicates that are added to the initial query also belong to this vocabulary.

Let us consider a relation R containing w tuples $\{t_1, t_2, \ldots, t_w\}$ defined on a set Z of q categorical or numerical attributes $\{Z_1, Z_2, \ldots, Z_q\}$. A shared predefined vocabulary on R is defined by means of partitions of the q domains. A partition $\mathscr{P}_i$ associated with the domain of attribute Z_i is composed of m_i fuzzy predicates $\{P^p_{i,1}, P^p_{i,2}, ..., P^p_{i,m_i}\}$, such that $\forall z_i \in \mathscr{D}(Z_i)$, $\sum_{j=1}^{m_i} \mu_{P^p_{ij}}(z_i) = 1$. A predefined predicate is denoted by $P^p_{i,j}$, which corresponds to the jth element of the partition defined on attribute Z_i. Each $\mathscr{P}_i$ is associated with a set of linguistic labels $\{L^p_{i,1}, L^p_{i,2}, \ldots, L^p_{i,m_i}\}$, each of them corresponding to an adjective which gives the meaning of the fuzzy predicate.

A query Q to this relation R is composed of fuzzy predicates chosen among the predefined ones which form the partitions. A predicate involved in a user query is said to be *specified* and is denoted by $P^s_{k,l}$ and corresponds to the lth element of the partition associated with the domain of Z_k.

If Q leads to a plethoric answer set, we propose to strengthen Q in order to obtain a more restrictive query Q' such that $\Sigma^*_{Q'} \subset \Sigma^*_Q$. Query Q' is obtained through the integration of additional predefined fuzzy predicates chosen from the shared vocabulary on R.

As an example, let us consider again the relation `secondHandCars` introduced in Section 2.2. A common sense partition and labelling of the domain of attribute *year* is illustrated in Fig. 2.

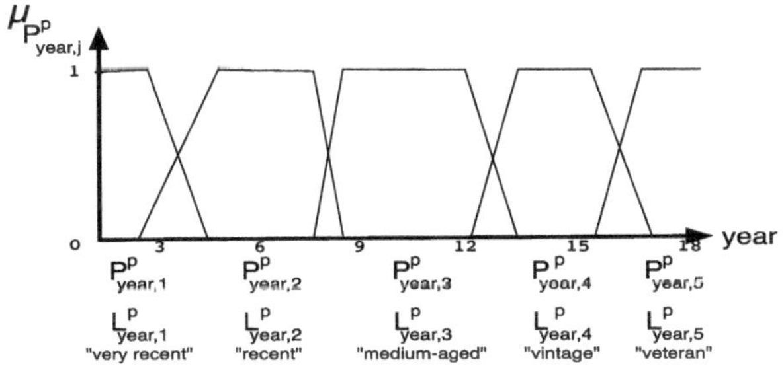

Fig. 2. A partition of the domain of attribute `year`

3 Strengthening Steps

3.1 Correlation-Based Ranking of the Candidates

In the approach we propose, the new conjuncts to be added to the initial query are chosen among a set of possible predicates pertaining to the attributes of the schema of the database queried (see Section 2.3). This choice is mainly made according to their correlation with the initial query. A user query Q is composed of $n, n \geq 1$, specified fuzzy predicates, denoted by $P^{s_1}_{k_1,l_1}, P^{s_2}_{k_2,l_2}, ...,$ $P^{s_n}_{k_n,l_n}$, which come from the shared vocabulary associated with the database (Section 2.3). The first step of the strengthening approach is to identify the predefined predicates the most correlated to the initial query Q.

The notion of correlation is used to qualify and quantify the extent to which two fuzzy sets (one associated with a predefined predicate $P^p_{i,j}$, the other associated with the initial query Q) are somewhat "semantically" linked. This degree of correlation is denoted by $\mu_{cor}(P^p_{i,j}, Q)$. Roughly speaking, we consider that a predicate $P^p_{i,j}$ is somewhat correlated with a query Q if the group of items characterized by $P^p_{i,j}$ is somewhat similar to Σ_Q. For instance, one may notice that a fuzzy predicate "highly powerful engine" is more correlated to a query aimed at retrieving "fast cars" than cars having a "low consumption". Adding predicates that are correlated to the user-specified ones makes it possible to preserve the scope of the query (i.e., the user's intent) while making it more demanding. It is worth mentioning that the term *correlation* used in this approach means a mutual semantic relationship between two concepts, and does not have the meaning it has in statistics where it represents similarities between series variations.

As it has been illustrated in Section 2.2, fuzzy association rules express such a semantic link that can be quantified by their confidence. Thus, if we consider a query Q and a predefined predicate $P^p_{i,j}$, we compute the confidence of the association rules $Q \Rightarrow P^p_{i,j}$ and $P^p_{i,j} \Rightarrow Q$ according to the fuzzy cardinalities F_Q, $F_{P^p_{i,j}}$ and $F_{Q \wedge P^p_{i,j}}$. We then quantify the correlation degree between Q and $P^p_{i,j}$, denoted by $\mu_{cor}(P^p_{i,j}, Q)$, as:

$$\mu_{cor}(P^p_{i,j}, Q) = \top(confidence(Q \Rightarrow P^p_{i,j}), confidence(P^p_{i,j} \Rightarrow Q))$$

where $\top$ stands for a t-norm and the minimum is used in our experimentation (Section 5). Thus, this correlation degree is both reflexive $\mu_{cor}(Q, Q) = 1$ and symmetrical $\mu_{cor}(P^p_{i,j}, Q) = \mu_{cor}(Q, P^p_{i,j})$.

Based on this measure we can identify the predefined predicates the most correlated to an under-specified query Q. In practice, we only consider the η most correlated predicates to a query, where η is a technical parameter which has been set to 5. This limitation is motivated by the fact that a strengthening process involving more than η iterations, i.e., the addition of more than η predicates, could lead to important modifications of the scope of the initial query. Those η predicates the the most correlated to Q are denoted by $P^{c^1}_Q, P^{c^2}_Q, ..., P^{c^\eta}_Q$.

3.2 Reduction-Based Reranking of the Candidates

In Section 3.1 we have shown how to retrieve the η most correlated predicates to an initial query Q. The second step of this strengthening process aims at reranking those η predicates according to their reduction capability. As it has been said in Section 1, we consider that the user specifies a value for the parameter K which defines his/her expected number of answers. Let $F_{Q \wedge P^{c^r}_Q}, r = 1..\eta$, be the fuzzy cardinality of the result set when Q is augmented with $P^{c^r}_Q$. $P^{c^r}_Q$ is all the more interesting for strengthening Q as $Q \wedge P^{c^r}_Q$ contains a σ_i-cut ($\sigma_i \in]0, 1]$) with a cardinality c_i close to K and a σ_i close to 1. To quantify how interesting $P^{c^r}_Q$ is, we compute for each σ_i-cut of $F_{Q \wedge P^{c^r}_Q}$ a strengthening degree which represents a compromise between its membership degree σ_i and its associated cardinality c_i. The global strengthening degree assigned to $F_{Q \wedge P^{c^r}_Q}$, denoted by $\mu_{stren}(F_{Q \wedge P^{c^r}_Q})$, is the maximal strengthening degree of its σ_i-cuts:

$$\mu_{stren}(F_{Q \wedge P_Q^{c^r}}) = sup_{1 \leq i \leq f} \top(1 - \frac{|c_i - K|}{max(K, |\Sigma_Q^*| - K)}, \sigma_i)$$

where $\top$ stands for a t-norm and the minimum is used in our experimentation. This reranking of the predicates the most correlated to Q can be carried out using the fuzzy cardinalities associated with each conjunction $Q \wedge P_Q^{c^r}$, $r = 1..\eta$.

Example 2. To illustrate this reranking strategy, let us consider a user query Q resulting in a PA problem ($|\Sigma_Q^*| = 123$), where K has been set to 50. As an example, let us consider the following candidates $P_Q^{c^1}, P_Q^{c^2}, P_Q^{c^3}, P_Q^{c^4}, P_Q^{c^5}$ and the respective fuzzy cardinalities:

- $F_{Q \wedge P_Q^{c^1}} = \{1/72 + 0.8/74 + 0.6/91 + 0.4/92 + 0.2/121\}$, $\mu_{stren}(F_{Q \wedge P_Q^{c^1}}) \simeq 0,7$
- $F_{Q \wedge P_Q^{c^2}} = \{1/89 + 0.8/101 + 0.6/135 + 0.4/165 + 0.2/169\}$, $\mu_{stren}(F_{Q \wedge P_Q^{c^2}}) \simeq 0,47$
- $F_{Q \wedge P_Q^{c^3}} = \{1/24 + 0.8/32 + 0.6/39 + 0.4/50 + 0.2/101\}$, $\mu_{stren}(F_{Q \wedge P_Q^{c^3}}) \simeq 0,75$
- $F_{Q \wedge P_Q^{c^4}} = \{1/37 + 0.8/51 + 0.6/80 + 0.4/94 + 0.2/221\}$, $\mu_{stren}(F_{Q \wedge P_Q^{c^4}}) \simeq 0,8$
- $F_{Q \wedge P_Q^{c^5}} = \{1/54 + 0.8/61 + 0.6/88 + 0.4/129 + 0.2/137\}$, $\mu_{stren}(F_{Q \wedge P_Q^{c^5}}) \simeq 0,95$.

According to the problem definition ($K = 50$) and the fuzzy cardinalities above, the following ranking is suggested to the user: 1) $P_Q^{c^5}$, 2) $P_Q^{c^4}$, 3) $P_Q^{c^3}$, 4) $P_Q^{c^1}$, 5) $P_Q^{c^2}$. Of course, to make this ranking more intelligible to the user, the candidates are proposed with their associated linguistic labels (cf. the concrete example about used cars given in Section 5).

4 Strengthening Process

4.1 Pre-computed Knowledge

As the predicates specified by the user and those that we propose to add to the initial query are chosen among the predicates that form the domain partitions (Section 2.3), one can precompute some useful knowledge that will make the strengthening process faster (Section 4.2). We propose to compute and maintain pre-computed knowledge which is stored in two tables.

The first one contains pre-computed fuzzy cardinalities. It is of course impossible to pre-compute the fuzzy cardinalities for all possible conjunctions of predefined predicates as the number of conjunctions growths exponentially with the size of the shared vocabulary. It is even useless to store the fuzzy cardinalities of each possible conjunction as our approach mainly focuses on under-specified queries and produces strengthened queries that contain correlated predicates only. Indeed, queries involving noncorrelated predicates are less likely to return a plethoric answer set. Thus, this table only contains the fuzzy cardinalities of conjunctions involving sufficiently correlated predicates. A threshold γ is used to prune some branches of the exploration tree of all possible conjunctions.

This table can be easily maintained as the fuzzy cardinalities can be updated incrementally. Moreover, as we will see in Section 4.2, this precomputed knowledge is used to give interesting information — in constant time — about any under-specified query, without computing the query. In particular, it provides the size of the answer set associated with a query and the predicates that can be used to strengthen it. An index computed on the string representation of each conjunction makes this fast access possible.

The second table stores the correlation degrees between each pair of predefined predicates (Section 3.1). According to these correlation degrees, one can also store for each predefined predicate its most correlated predefined predicates ranked in decreasing order of their correlation degree. This table is checked to retrieve the η predicates the most correlated to each specified predicate and thus to generate the candidates.

Both tables have to be updated after each (batch of) modification(s) performed on the data but these updates imply a simple incremental calculus.

4.2 Interactive Strengthening Process

Let us recall that a query Q is assumed to be composed of fuzzy predicates chosen among predefined ones that form the shared vocabulary. One first checks the table of fuzzy cardinalities (Sec. 4.1) in order to test if its fuzzy cardinality is available. If it is the case, one then determines if the user is faced with a PA problem according to the value he/she has assigned to K. If so, one retrieves – still in constant time – up to η candidates that are then reranked according to K and presented to the user. Finally, as it is illustrated in Section 5, the user can decide to process the initial query, to process one of the suggested strengthened queries, or to ask for another strengthening iteration of a strengthened query.

Using the aforementioned tables, this approach guides the users in an interactive way from an under-specified query to a more demanding but semantically close query that returns an acceptable number of results.

5 A First Experimentation

5.1 Context

We consider a relation `secondHandCars` containing 10.604 ads about used cars. This relation has the schema: `{idAd, model, year, mileage, optionLevel, securityLevel, ComfortLevel, horsePower, engineSize, price}`. The following shared vocabulary composed of 41 fuzzy predicates has been predefined to query this relation and to strengthen under-specified queries. The quadruplet (a, b, c, d) associated with each linguistic label defines its trapezoidal membership function (where $[a, d]$ represents the support and $[b, c]$ the core).

year: 'vintage' ($-\infty$, $-\infty$, 1981, 1983), 'very_old' (1981, 1983, 1991, 1992), 'old' (1991, 1992, 1997, 1999), 'medium' (1997, 1999, 2004, 2005), 'recent' (2004, 2005, 2007, 2008), 'very recent' (2007, 2008, 2009, 2010), 'last model' (2009, 2010, $+\infty$, $+\infty$)

mileage: 'very_low' ($-\infty$, $-\infty$, 20.000, 30.000), 'low' (20.000, 30.000, 60.000, 70.000), medium (60.000, 70.000, 130.000, 150.000) , 'high' (130.000, 150.000, 260.000, 300.000), 'very_high' (260.000, 300.000, $+\infty$, $+\infty$)

optionLevel: 'very_low'($-\infty$, $-\infty$,1, 2), 'low' (1, 2, 4, 5), 'medium' (4, 5, 6, 7), 'high' (6, 7, 10, 12), 'very_high' (10, 12, $+\infty$, $+\infty$)

securityLevel: 'very_low' ($-\infty$,$-\infty$, 1, 2), 'low' (1, 2, 4, 5), 'medium' (4, 5, 6, 7), 'high' (6, 7, 8, 9), 'very_high' (8, 9, $+\infty$, $+\infty$)

comfortLevel: 'very_low' ($-\infty$, $-\infty$,1, 2), 'low' (1, 2, 4, 5), 'medium' (4, 5, 6, 7), 'high' (6, 7, 8, 9), 'very_high' (8, 9, $+\infty$, $+\infty$)

horsePower: 'very_low' ($-\infty$, $-\infty$, 30, 50), 'low' (30, 50, 70, 80), 'high' (70, 80, 120, 140), 'very_high' (120, 140, $+\infty$, $+\infty$)

engineSize: 'very_small' ($-\infty$, $-\infty$, 0.8, 1.0), 'small' (0.8, 1.0, 1.4, 1.6) , 'big' (1.4, 1.6, 2.0, 2.4), 'very_big' (2, 2.4, $+\infty$, $+\infty$)

price: 'very_low' ($-\infty$, $-\infty$, 2.000, 2.500), 'low' (2.000, 2.500, 6.000, 6.500), 'medium' (6.000, 6.500, 12.000, 13.000), 'high' (12.000,13.000, 20.000, 22.000), 'very_high' (20.000, 22.000, 35.000, 40.000), 'excessively_high' (35.000, 40.000, $+\infty$, $+\infty$)

According to these partitions, fuzzy cardinalities of conjunctions involving correlated predicates (Sect. 4.1) have been computed in order to populate the table of fuzzy cardinalities and then to identify the $\eta = 5$ predicates the most correlated to each partition element. On a laptop with a classical configuration (Intel Core 2 Duo 2.53GHz with 4Go 1067 MHz of DDR3 ram), it took less than 13 minutes to compute the useful fuzzy cardinalities and to store them in a dedicated table. This process can even be made faster using indexes that could be defined on the concerned relation (i.e., `secondHandCars`). Concretely, using a low correlation threshold of 0.05 (see Sect. 4.1), only 7.436 conjunctions involve predicates that are sufficiently correlated to be considered as possible under-specified queries among the 604.567 possible conjunctions that can be constructing using the shared vocabulary. Among the 7.436 conjunctions stored in the fuzzy cardinalities table, 41 concern one predicate (one for each partition element), 170 fuzzy cardinalities for conjunctions of two predicates, 382 for three predicates, 695 for four predicates, 1040 for five predicates, 1422 for six predicates, 1808 for seven predicates and 1878 for eight predicates.

5.2 Example of an Under-Specified Query

To illustrate the strengthening method, let us take an example of a query Q composed of fuzzy predicates chosen among the shared vocabulary (Sec. 5.1): $Q = \ select * from\ secondHandCars\ where\ year\ is\ old\ with\ K = 50$.

From a string representation of Q and the table of fuzzy cardinalities, we directly check if Q corresponds to an under-specified query. If it is the case, its fuzzy cardinality is presented to the user (obviously in a more linguistic and user-friendly way): $F_Q = \{1/179 + 0.8/179 + 0.6/179 + 0.4/323 + 0.2/323\}$.

At the same time, predicates corresponding to properties correlated to the initial query are suggested, if they exist, and ranked in decreasing order of their reduction capability w.r.t. K. On this example, the following candidates are suggested with the fuzzy cardinality of the corresponding strengthened queries:

1. mileage IS 'medium' ($\mu_{cor}(Q, P^p_{mileage,\text{'}medium\text{'}}) = 0.11$)

$$F_{Q \wedge P^p_{mileage,\ \text{'}medium\text{'}}} = \{1/24 + 0.8/27 + 0.6/28 + 0.4/72 + 0.2/77\}$$

2. mileage IS 'very high' ($\mu_{cor}(Q, P^p_{mileage,\text{'}veryhigh\text{'}}) = 0.19$)

$$F_{Q \wedge P^p_{mileage,\ \text{'}veryhigh\text{'}}} = \{1/7 + 0.8/7 + 0.6/8 + 0.4/18 + 0.2/19\}$$

3. mileage IS 'high' ($\mu_{cor}(Q, P^p_{mileage,\text{'}high\text{'}}) = 0.37$)

$$F_{Q \wedge P^p_{mileage,\ \text{'}high\text{'}}} = \{1/101 + 0.8/106 + 0.6/110 + 0.4/215 + 0.2/223\}.$$

For each candidate query Q', the user can decide to process Q' (i.e. retrieve the results) or to repeat the strengthening process on Q'. If this latter option is chosen, the table of fuzzy cardinalities is checked in order to retrieve strengthening candidates for Q' (i.e. properties correlated to Q') and their associated fuzzy cardinalities that are ranked according to K. In this example, even if K (50) items can be ranked and returned to the user according to this strengthened query, let us consider that the user selects $Q' = year\ IS\ \text{'}old\text{'}\ AND\ mileage\ IS\ \text{'}medium\text{'}$ for a second step of strengthening. The following candidates are suggested with their fuzzy cardinalities:

1. optionLevel IS 'low' ($\mu_{cor}(Q', P^p_{optionLevel,\text{'}low\text{'}}) = 0.34$)

$$F_{Q' \wedge P^p_{optionLevel,\ \text{'}low\text{'}}} = \{1/18 + 0.8/20 + 0.6/21 + 0.4/46 + 0.2/51\}$$

2. optionLevel IS 'very low' ($\mu_{cor}(Q', P^p_{optionLevel,\text{'}verylow\text{'}}) = 0.15$)

$$F_{Q' \wedge P^p_{optionLevel,\ \text{'}verylow\text{'}}} = \{1/6 + 0.8/7 + 0.6/7 + 0.4/22 + 0.2/22\}.$$

For this latter strengthening step, the augmented queries Q'' return less than K items. This is why they are ranked in a decreasing order of their relative correlation degree with Q'.

5.3 Remarks about This Experimentation

From this first experimentation on a concrete database, we can observe that this query strengthening approach based on correlation links gives the users interesting information about data distributions and the possible queries that can be formulated in order to retrieve coherent answer sets. By coherent answer set, we mean a group of items that share correlated properties and that may correspond to what the user was looking for without knowing initially how to retrieve them. Moreover, thanks to the precomputed knowledge tables, it is not necessary to process correlated queries (i.e., queries containing correlated predicates) to provide the user with interesting information about the size of their answer sets and the predicates that can be used to strengthen them.

This experimentation showed that the predicates suggested to strengthen the queries are meaningful and coherent according to the initial under-specified queries. One can find below some examples of suggested augmented queries Q' starting from under-specified queries Q:

- $Q = year\ is\ `old'\ AND\ mileage\ is\ `high'\ AND\ optionLevel\ is\ `low'$ intensified after two iterations into $Q' = year\ is\ `old'\ AND\ mileage\ IS\ `high'\ AND\ optionLevel\ is\ `low'\ AND\ comfortLevel is\ `very\ low'\ AND\ securityLevel\ is\ `low'$, with $|\Sigma^*_Q| = 63$ and $|\Sigma^*_{Q'}| = 26$.
- $Q = year\ is\ `recent'$ intensified after two iterations into $Q' = year\ is\ `recent'\ AND\ mileage\ is\ `low'\ AND\ optionLevel\ is\ `high'$, with $|\Sigma^*_Q| = 4.060$ and $|\Sigma^*_{Q'}| = 199$.
- $Q = comfortLevel\ is\ `high'$ intensified after one iteration into $Q' = comfortLevel\ is\ `high'\ AND\ optionLevel\ is\ `high'$, with $|\Sigma^*_Q| = 180$ and $|\Sigma^*_{Q'}| = 45$.
- $Q = year\ is\ `very\ old'$ intensified after two iterations into $Q' = year\ is\ `very\ old'\ AND\ optionLevel\ is\ `low'\ AND\ comfortLevel\ is\ `very\ low'$, with $|\Sigma^*_Q| = 35$ and $|\Sigma^*_{Q'}| = 6$.

Finally, we noticed that the parameter η was useless in this applicative context as the correlation threshold used during the computation of fuzzy cardinalities already restricts the size of the correlation table to the most correlated conjunction of predicates. However, for some applicative contexts involving much more predefined predicates and a low correlation threshold, η can be useful.

6 Related Work

In their probabilistic ranking model, Chaudhuri *et al.* [6] also propose to use a correlation property between attributes and to take it into account when computing ranking scores. However, correlation links are identified between attributes and not predicates, and the identification of these correlations relies on a workload of past submitted queries.

Su *et al.* [13] have emphasized the difficulty to manage such a workload of previously submitted queries or users feedbacks. This is why they have proposed to learn attribute importances regarding to a *price* attribute and to rank retrieved items according to their commercial interest. Nevertheless, this method is domain-dependent and can only be applied for e-commerce databases.

The approach advocated by Ozawa *et al.* [11,12] is also based on the analysis of the database itself, and aims at providing the user with information about the data distributions and the most efficient constraints to add to the initial query in order to reduce the initial set of answers. The approach we propose in this paper is somewhat close to that introduced in [11], but instead of suggesting an attribute on which the user should specify a new constraint, our method directly suggests a set of fuzzy predicates along with some information about their relative interest with respect to the user needs. The main limitation of the approach advocated in [11] is that the attribute chosen is the one which maximizes the dispersion of the initial set of answers, whereas most of the time, it does not have any semantic link with the predicates that the user specified in his/her initial query. To illustrate this, let us consider again the relation `secondHandCars` introduced in Section 2.2. Let Q be a fuzzy query on `secondHandCars`: *"select estate cars which are recent"* resulting in a PA problem. In such a situation, Ozawa *et al.* [11] first apply a fuzzy c-means algorithm [1] to classify the data, and each fuzzy cluster is associated with a predefined linguistic label. After having attributed a weight to each cluster according to its representativity of the initial set of answers, a global dispersion degree is computed for each attribute. The user is then asked to add new predicates on the attribute for which the dispersion of the initial set of answers is maximal. In this example, this approach may have suggested that the user should add a condition on the attributes *mileage* or *brand*, on which the recent estate cars are probably the most dispersed. We claim that it is more relevant to reduce the initial set of answers with additional conditions which are in the semantic scope of the initial query. Here for instance, it would be more judicious to focus on cars with a high level of security and comfort as well as a low mileage, which are features usually related to recent estate cars. This issue has been concretely illustrated in Section 5.

The problem of plethoric answers to fuzzy queries has been addressed in [3] where a query strengthening mechanism is proposed. Let us consider a fuzzy set $F = (A,\ B,\ a,\ b)$ representing a fuzzy query Q. Bosc *et al.* [3] define a fuzzy tolerance relation E which can be parameterized by a tolerance indicator Z, where Z is a fuzzy interval centered in 0 that can be represented in terms of a trapezoidal membership function by the quadruplet $Z = (-z,\ z,\ \delta,\ \delta)$. From a fuzzy set $F = (A,\ B,\ a,\ b)$ and a tolerance relation $E(Z)$, the erosion operator builds a set F_Z such that $F_Z \subseteq F$ and $F_Z = F \ominus Z = (A + z,\ B - z,\ a - \delta,\ b - \delta)$. However, as mentioned in Section 1, such an erosion-based approach can lead to a deep modification of the meaning of the user query.

7 Conclusion and Perspectives

The approach presented in this paper deals with the plethoric answer problem by determining predicates that can be used to strengthen the initial query. These predicates are selected among a set of predefined fuzzy terms according to their degree of semantic correlation with the initial query. From these correlated predicates, the approach uses fuzzy cardinalities to identify which predicate(s) should be integrated to the initial query in order to obtain a new answer set containing a number of items close to K, where K is specified by the user in his/her query as the number of expected results. What makes the approach tractable is the fact that it uses a table which stores the correlation degrees between the predefined predicates, as well as their fuzzy cardinalities.

This work opens many perspectives for future research. While preserving the main principles of this approach, it could be interesting to let the user specify his/her own fuzzy predicates when querying a database instead of forcing him/her to use a predefined vocabulary. To keep the strengthening process efficient, it is necessary to identify the predefined predicate the closest to each user-defined predicate in order to use the tables that store the correlation degrees and fuzzy cardinalities. We are thus currently working on a measure that could be used to precisely quantify the bias introduced by this approximation.

Another perspective of improvement concerns the dimension of the needed pre-computed knowledge. Indeed, even if we have shown that in practice it is not necessary to compute the fuzzy cardinalities for all possible conjunctions of predicates, the size of the stored knowledge is all the same significant. A perspective of improvement is to use a workload of submitted queries or a classification method so as to focus on the most frequent or typical associations of predicates.

Another important aspect concerns the qualitative assessment of the approach. To this end, an evaluation protocol over our used cars database to collect qualitative evaluations from users feedbacks is underway.

References

1. Bezdek, J.: Pattern recognition with fuzzy objective function algorithm. Plenum Press, New York (1981)
2. Bodenhofer, U., Küng, J.: Fuzzy ordering in flexible query answering systems. Soft Computing 8, 512–522 (2003)
3. Bosc, P., Hadjali, A., Pivert, O.: Empty versus overabundant answers to flexible relational queries. Fuzzy sets and systems 159(12), 1450–1467 (2008)
4. Bosc, P., Pivert, O.: SQLf: a relational database language for fuzzy querying. IEEE Transactions on Fuzzy Systems 3(1), 1–17 (1995)
5. Bosc, P., Pivert, O., Dubois, D., Prade, H.: On fuzzy association rules based on fuzzy cardinalities. In: FUZZ-IEEE, pp. 461–464 (2001)
6. Chaudhuri, S., Das, G., Hristidis, V., Weikum, G.: Probabilistic ranking of database query results. In: Proc. of VLDB 2004, pp. 888–899 (2004)
7. Chomicki, J.: Querying with intrinsic preferences. In: Jensen, C.S., Jeffery, K., Pokorný, J., Šaltenis, S., Bertino, E., Böhm, K., Jarke, M. (eds.) EDBT 2002. LNCS, vol. 2287, pp. 34–51. Springer, Heidelberg (2002)
8. Corella, F., Lewison, K.: A brief overview of cooperative answering. Technical report (2009), http://www.pomcor.com/whitepapers/cooperative_responses.pdf
9. Dubois, D., Prade, H.: Fuzzy cardinalities and the modeling of imprecise quantification. Fuzzy sets and systems 16, 199–230 (1985)
10. Kiessling, W.: Foundations of preferences in database systems. In: Proc. of VLDB 2002 (2002)
11. Ozawa, J., Yamada, K.: Cooperative answering with macro expression of a database. In: Bouchon-Meunier, B., Yager, R.R., Zadeh, L.A. (eds.) IPMU 1994. LNCS, vol. 945, pp. 17–22. Springer, Heidelberg (1995)
12. Ozawa, J., Yamada, K.: Discovery of global knowledge in database for cooperative answering. In: Proc. of Fuzz-IEEE 1995, pp. 849–852 (1995)
13. Su, W., Wang, J., Huang, Q., Lochovsky, F.: Query result ranking over e-commerce web databases. In: Proc. of CIKM 2006 (2006)
14. Ughetto, L., Voglozin, W.A., Mouaddib, N.: Database querying with personalized vocabulary using data summaries. Fuzzy Sets and Systems 159(15), 2030–2046 (2008)

From Bayesian Classifiers to Possibilistic Classifiers for Numerical Data

Myriam Bounhas[1], Khaled Mellouli[1], Henri Prade[2], and Mathieu Serrurier[2]

[1] Laboratoire LARODEC, ISG de Tunis, 41 rue de la liberté, 2000 Le Bardo, Tunisie
[2] Institut de Recherche en Informatique de Toulouse (IRIT), UPS-CNRS.118 route de Narbonne, 31062 Toulouse Cedex-France
Myriam_Bounhas@yahoo.fr, Khaled.Mellouli@topnet.tn,
{Prade,Serrurier}@irit.fr

Abstract. Naïve Bayesian classifiers are well-known for their simplicity and efficiency. They rely on independence hypotheses, together with a normality assumption, which may be too demanding, when dealing with numerical data. Possibility distributions are more compatible with the representation of poor data. This paper investigates two kinds of possibilistic elicitation methods that will be embedded into possibilistic naïve classifiers. The first one is derived from a probability-possibility transformation of Gaussian distributions (or mixtures of them), which introduces some further tolerance. The second kind is based on a direct interpretation of data in fuzzy histogram or possibilistic formats that exploit an idea of proximity between attribute values in different ways. Besides, possibilistic classifiers may be allowed to leave the classification open between several classes in case of insufficient information for choosing one (which may be of interest when the number of classes is large). The experiments reported show the interest of possibilistic classifiers.

Keywords: Naive Possibilistic Classifier, Possibility Theory, Naive Bayesian Classifier, Gaussian Distribution, Kernel Density, Numerical Data.

1 Introduction

Inductive reasoning moves from a set of specific facts to general statements, and machine learning consists in designing algorithms that produce a general theory from externally supplied examples, which can be used for making predictions about the classification of new data.

Classification tasks can be handled by mainly three classes of approaches: those based on empirical risk minimization (decision trees [28], artificial neural networks [19]), approaches based on maximum likelihood estimation (such as Bayesian networks [17], k-nearest neighbors [20]), and the ones based on Kolmogorov complexity [29]. See for instance Kotsiantis [4] for a comparative study between these methods.

In this paper we are mainly interested in the second class of methods and we intend to deal with the classification of numerical data. Given a new piece of data

A. Deshpande and A. Hunter (Eds.): SUM 2010, LNAI 6379, pp. 112–125, 2010.

to classify, this family of approaches seeks to estimate the plausibility of each class with respect to its description (built from the training set of examples), and assigns the class having the highest plausibility value. There are principally two methods: the k-Nearest Neighbors which are local methods and the Naive Bayesian Classifiers. The latter assumes independence of variables (attributes) in the context of classes in order to estimate the probability distribution on the classes for a given observed data.

The objective of this paper is to use Naive Bayesian Classifiers as a reference and to test the feasibility of using other kinds of representation for the distributions associated with attributes. We choose to investigate the use of Possibilistic Classifiers using a counterpart of Bayes rule in the settings of Possibility Theory [11], which leads to estimate possibility distributions. In spite of the fact that these distributions are useful for representing imperfect knowledge, there have been only few works that use naive possibilistic classifiers [30]. Moreover, we also investigate the idea of allowing for multiple-classification when several classes have very close plausibility estimates.

The paper is structured as follows: in the next section, we briefly recall Bayesian classification. Section 3 restates possibilistic classification. In Section 4, we study the two kinds of elicitation methods for building possibility distributions: i) the first one is based on a transformation method from probability to possibility, whereas ii) the second one makes a direct, fuzzy histogram-based, or possibilistic, interpretation of data, taking advantage of the idea of proximity. The experimentation results of the proposed approaches are in Section 5. The experiments reported show the interest of possibilistic classifiers. In particular, flexible possibilistic classifiers perform well for data agreeing with the normality assumption, while proximity-based possibilistic classifiers outperform other classifiers in the other cases. Related works are discussed in Section 6. Finally, Section 7 concludes and suggests some directions for future research.

2 Background: Bayesian Classification of Continuous Data

The Naive Bayesian Classifier (NBC) is based on Bayes rule. It assumes the independence of the input variables. Despite its simplicity, NBC can often outperform more sophisticated classification methods [27]. NBC can be seen as a Bayesian network in which predictive attributes are assumed to be conditionally independent given the class attribute. A variety of NBC have been proposed to handle an arbitrary number of independent attributes [27,26,13,24]. A semi-naive Bayesian classifier (SNBC) [1] takes into account correlation between dependent attributes as it allows joining highly correlated attributes and treating them as one.

Given a new vector $X = \{x_1, x_2, ..., x_M\}$ to classify, a NBC calculates the posterior probability for each possible class $c_j (j = 1, ..., C)$ and labels the vector X with the class c_j that achieves the highest posterior probability, that is:

$$c^* = arg \max_{c_j} p(c_j|X) = arg \max_{c_j} p(c_j) \prod_{i=1}^{M} p(x_i|c_j) \quad (1)$$

A supplementary common assumption made by NBCs in the continuous case, is that within each class the values of numerical attributes are normally distributed. The NBCs represent such distribution in terms of its mean and standard deviation and compute the probability of an observed value from such estimates as follows:

$$p(x_i|c_j) = g(x_i, \mu_j, \sigma_j) = \frac{1}{\sqrt{2\Pi}\sigma_j} e^{-\frac{(x_i-\mu_j)^2}{2\sigma_j^2}} \tag{2}$$

If the normality assumption is violated, classification results of NBC may deteriorate. John and Langley [14] proposed a Flexible Naive Bayesian Classifier (FNBC) that no longer assumes the normality of the distribution, and instead uses nonparametric kernel density estimation for each conditional distribution. The FNBC has the same properties as those of NBC; the only difference is that the density for each continuous attribute x_i is estimated as the average of a set of Gaussian distributions:

$$p(x_i|c_j) = \frac{1}{N_j} \sum_{k=1}^{N_j} g(x_i, \mu_{ik}, \sigma_j) \tag{3}$$

Where N_j is the number of instances belonging to the class c_j. The mean μ_{ik} is equal to the real value of attribute i of the instance k belonging to the class j, e.g.$\mu_{ik} = x_{ik}$. The FNBC estimates the standard deviation by:

$$\sigma_j = \frac{1}{\sqrt{N_j}} \tag{4}$$

Pérez et al. [2] have recently proposed a new approach for Flexible Bayesian classifiers based on kernel density estimation that extends the FNBC proposed by [14] in order to handle dependent attributes and abandon the independence assumption.

3 Possibilistic Classification

There are only few works that study possibilistic classifiers in spite of the fact that they have similar architecture to Bayesian ones, known for their capability to handle a variety of datasets. In this paper, we investigate possibilistic classifiers, viewed as natural counterpart of Bayesian ones, for several reasons. First, possibility distributions can be considered as representing, or more generally approximating, a family of probability measures. Given a probability distribution on a finite set of elementary events one can define a possibility distribution defining a possibility measure which is an upper bound of the corresponding probability measure for any event [9]. The transformation from probability to possibility distributions [8], which has been extended to continuous universes, accounts for an epistemic uncertainty. It yields the most restrictive possibility distribution which is co-monotone with the probability distribution and which obeys to the above upper bound condition on events. Second, we know that

NBCs may be improved (in the continuous case) by a mixture of Gaussians using a flexible version in order to obtain a data representation more suitable and closer to data. Indeed, the probabilistic model may be too rich in terms of information compared to available data. Besides, possibilistic classification may be viewed as an intermediate between the Bayesian and a purely set-based classifier (such classifiers use as distributions the convex hull for each attribute of the data values to identify classes, usually leading to too many multiple classifications).

As in the case of Bayesian classification, possibilistic classification is based on the possibilistic version of the Bayes theorem. Given a vector $X = \{x_1, x_2, ..., x_M\}$ of n observed variables and the set of classes $C = \{c_1, c_2, ..., c_C\}$, the classification problem consists in estimating a possibility distribution on classes and in choosing the class with the highest possibility for the vector X, i.e.:

$$\pi(c_j|x_1, x_2, ..., x_M) = \frac{\pi(x_1, x_2, ..., x_M|c_j) * \pi(c_j)}{\pi(x_1, x_2, ..., x_M)} \tag{5}$$

where * stands for product in quantitative possibility settings. Assuming the *-independence between variables x_i in the context of classes [22], this possibility distribution can easily be specified by the *product* (or the *minimum*) of the conditional possibilities $\pi(x_i|c_j)$ for all variables x_i. Each conditional possibility represents the possibility of x_i knowing c_j.

In this paper, considering an unknown test instance I_t with attribute values $(a_1, ..., a_M)$, the classification task amounts to calculating values of possibilities for each class: $\Pi(c_j|I_t)$. Assuming attribute independence, the plausibility of each class for a given instance is calculated as:

$$\Pi(c_j|I_t) = \prod_{i=1}^{M} \Pi(a_i|c_j) = \Pi(a_1|c_j) * ... * \Pi(a_M|c_j) \tag{6}$$

In a product-based setting, a given instance is assigned to the most plausible class c*:

$$c^* = arg \max_{c_j} \prod_{i=1}^{M} \Pi(x_i|c_j) \tag{7}$$

Using the min-based setting, the classification is based on selecting the class having the highest minimum:

$$c^* = arg \max_{c_j} \min_{i=1}^{M} \Pi(x_i|c_j) \tag{8}$$

The main problem of Bayesian or possibilistic classification is that classification accuracy can significantly deteriorate if the classifier fails to distinguish between classes. In some particular cases, classes may have too close plausibility estimates. It is why we also propose a $multiple - classification$ *approach* in such cases in the following. Instead of considering only the most plausible class, the idea is to consider more than one class at a time for classifying new instances when the plausibility difference between the most relevant classes is negligible.

4 Possibilistic Distributions for Describing Numerical Data Sets

In this paper, we try to build possibilistic distributions that describe data sets. We investigate two kinds of approaches, either based on probability-possibility transformation [12,8,18], or on a direct interpretation of data taking advantage of the idea of proximity.

4.1 Probability to Possibility Transformation Method Applied to Normal Distributions

Dubois et al. [8] have justified a probability-possibility transformation method in the continuous case based on confidence intervals (with level ranging from 0 to 1) built around a nominal value which is the mode. It generalizes a previously proposed method for the discrete case [12]. For symmetric densities, the mode is equal to the mean and to the median. In this case, Dubois et al. [8] defines a possibility distribution in the continuous case as follows:

$$\pi^*(x) = sup\{1 - P(I^*_\alpha), x \in I^*_\alpha\}. \tag{9}$$

where I_α is the $\alpha\%$ confidence interval. This possibility distribution satisfies the following properties:

a) Possibility - Probability consistency: For any probability density p, the possibility distribution π^* is consistent with p, that is: $\forall A, \Pi^*(A) \geq P(A)$, with Π^* and P being the possibility and probability measures associated respectively to π^* and p.

b) Preserving consistency on events: $\Pi(A) \geq P(A)$ for all A, and order-equivalence on distributions, i.e. $\pi(x) > \pi(x')$ if and only if $p(x) > p(x')$.

The rationale behind this transformation is that given a probability p, one tries to preserve as much information as possible. This leads to select the most specific element in the set $PI(P) = \{\Pi : \Pi \geq P\}$ of possibility measures dominating P such that $\pi(x) > \pi(x')$ iff $p(x) > p(x')$.

To satisfy all the previously cited properties, the most specific possibility distribution consistent with p, and ordinally equivalent to it is obtained such that (see [12,8] for details):

$$\forall L > 0, \pi(a_L) = \pi(a_L + L) = 1 - P(I_L). \tag{10}$$

where I_L is the smallest confidence interval, of length L, that contains a_L.

We apply this transformation to the case of the NBC where the distribution is assumed to be normal and then to its flexible extension.

Let us consider a Gaussian distribution $g_{ij} = g(a_i, \mu_{ij}, \sigma_{ij})$ that corresponds to the conditional probability of a_i knowing c_j, where μ_{ij} is the mean of the attribute a_i for the class c_j and σ_{ij} is its standard deviation for the same class. If I_{a_i} is the confidence interval centered at μ_{ij}, its probability $P(I_{a_i}|c_j)$ can be estimated by:

$$P(I_{a_i}|c_j) \simeq 2 * G(a_i, \mu_{ij}, \sigma_{ij}) - 1. \tag{11}$$

where G is a Gaussian cumulative distribution easily evaluated using the table of the Standard Normal Distribution. We propose to estimate $\pi(a_i|c_j)$ by $1 - P(I_{a_i}|c_j)$ using the following formula:

$$\pi(a_i|c_j) = 1 - (2 * G(a_i, \mu_{ij}, \sigma_{ij}) - 1) = 2 * (1 - G(a_i, \mu_{ij}, \sigma_{ij})). \tag{12}$$

Hence, in the training phase we should simply calculate the mean μ_{ij} and the standard deviation σ_{ij} for each attribute a_i of instances belonging the class c_j. This step enables us to estimate the possibility distribution: $\pi(a_i|c_j)$ using (12).

In this approach and for all the rest of this work, all attribute values a_i's are normalized as follows:

$$a_{i_n} = \frac{a_i - min(a_i)}{max(a_i) - min(a_i)}. \tag{13}$$

The FNPC is mainly based on the FNBC as previously introduced. For this classifier, the building procedure is reduced to the calculation of the standard deviation: σ. The FNPC is exactly the same as the NPC in all respects, the only difference between the two classifiers is the method used for density estimation on continuous attributes. Although using a single Gaussian to estimate each continuous attribute, we choose to investigate kernel density estimation as in the FNBC. Kernel estimation with Gaussian kernels looks much the same except that the estimated density is averaged over a large set of kernels.

$$\pi(a_i|c_j) = \frac{1}{N_j} \sum_{k=1}^{N_j} \pi(a_i, c_{jk}). \tag{14}$$

with:

$$\pi(a_i, c_{jk}) = 2 * (1 - G(a_i, \mu_{ik}, \sigma)) \tag{15}$$

where k ranges over the N_j instances of the training set in class c_j and $\mu_{ik} = a_{ik}$. For all distributions, the standard deviation is estimated by:

$$\sigma = \frac{1}{\sqrt{N}} \tag{16}$$

Estimating possibility distribution using N Gaussian kernels for the FNPC leads to an increase in the computational complexity of the classification algorithm if compared to the NPC. In fact, when classifying a new instance, the NPC estimates $\pi(a_i|c_j)$ by evaluating the Gaussian G once whereas FNPC has to evaluate this G, N times per observed value of attribute a_{ik} in each training instance I_k.

4.2 Approximate Equality-Based Interpretations of Data

In this section we make use of two other methods for building a distribution from a set of data without computing a Gaussian probability distribution first. The two methods make use of the idea that a value is all the more plausible for an attribute as this value is close to other values that have been observed in the examples. They were first suggested in [10]. Both use an *approximate equality relation* between numerical values. Let d be the distance between the two values, this fuzzy relation, namely $\mu_E(d(x,y))$ estimates to what extent x is close to y as follows (in other words E is a fuzzy set with decreasing membership function on $[0,+\infty)$ with a bounded support and such that $\mu_E(0)=1$) :

$$\mu_E(d) = max(0, min(1, \frac{\alpha+\beta-d}{\beta})), \alpha \geq 0; \beta > 0. \tag{17}$$

This relation is parameterized by α and β. In the first method, we use the approximate equality function to build a fuzzy histogram [25] for attribute a_i given a class c_j.

$$\pi(a_i|c_j) = \frac{1}{N_j}\sum_{k=1}^{N_j}\mu_E(d(a_i, a_{ik})). \tag{18}$$

where N_j is the number of instances belonging to the class c_j. The idea is here to be more faithful to the way the data are distributed (rather than assuming a normal distribution), and to take advantage of the approximate equality for obtaining a smooth distribution on the numerical domain, and may be supplying the scarcity of data. In that respect, the parameters of the approximate equality relation, depending on their values, not only reflect the expression of a tolerance on values that are not significantly different for a given attribute, but may also express a form of extrapolation from the observed data. The distribution (18) can then be directly used in the classification procedure. The algorithm based on this method will be named Fuzzy Histogram Classifier (FuHC) in the following.

We propose a second approach, named Nearest Neighbor-based Possibilistic Classifier (NNPC), which is based only on the analysis of the proximities between the attribute values a_{ik} belonging to each class c_j without counting them. The main idea of this classifier is to search for the nearest neighbor attribute value a_{ik} for the attribute value a_i of the item to be classified, in the training set of each class. The approximate equality function calculated between a_i and its nearest neighbor a_{ik} is then used to estimate the possibility distribution of the attribute value a_i given a class c_j as follows:

$$\pi(a_i|c_j) = \max_{k=1}^{N_j}\mu_E(d(a_i, a_{ik})). \tag{19}$$

In this approach, the closer an attribute value a_i to other attribute values of instances belonging to a class c_j, the greater the possibility to belong to the class (w.r.t. the considered attribute). The expression (19) may be considered as a genuine possibility distribution [31]. An attribute value having a possibility 0 means that this value is not compatible with the associated class (it is the case when the value is not close to any other observed value of the attribute for the class). If the possibility is equal or close to 1, then the value is relevant for describing the class (a value having a small distance to instances of a class is considered as a possible candidate value in the representation of the class for a considered attribute).

5 Experiments and Discussion

This section provides experimental results for the NPC, NBC, FNPC, FNBC, FuHC and NNPC methods. The experimental study is based on several datasets taken from the U.C.I repository of machine learning databases [16]. A brief description of these datasets is given in Table 1. Since we have chosen to deal only with numerical attributes in this study, all these datasets have numerical attribute values.

The experimental study is divided in two parts. First, we evaluate the NPC, FNPC, FuHC and NNPC methods and compare our results to those of a classical NBC [14] and FNBC [14]. Second, we test the case of multiple-classification for the NPC, NBC and NNPC for datasets having high number of classes (more than two classes). For each dataset, we used a ten-fold cross validation to evaluate the generalization accuracy of classifiers.

All possibilistic classifiers exploit a product-based setting in the classification step except for the NNPC where we tested the three options (product, minimum, and a leximin-based refinement of minimum enabling us to solve ties between evaluations). For this experimental report, we keep only the min-based NNPC version because the three versions have a competitive efficiency. Note that we only considered normalized attribute values in this paper. For the FuHC and NNPC, in order to guarantee in (17) a significant value of the approximate equality function ($0 < \mu_E(d(x, y)) < 1$), α and β are respectively fixed to 0 and 1, once d is normalized in [0, 1], for all attributes.

Table 1. Description of datasets

Database	Data	Attributes	Classes
Iris	150	4	3
W. B. Cancer	699	8	2
Wine	178	13	3
Diabetes	768	7	2
Magic gamma telescope	1074	10	2
Transfusion	748	4	2
Satellite Image	1090	37	6
Segment	1500	20	7
Yeast	1484	9	10
Ecoli	336	8	8
Glass	214	10	7

Table 2 shows the classification performance obtained with NPC, NBC, FNPC, FNBC, FuHC and NNPC for the eleven mentioned datasets. By comparing the classification performance of the six classifiers we note that:

• For the two classifiers assuming normality distribution (NPC and NBC), we remark that NPC is more accurate than NBC in three databases (Yeast, Ecoli and Glass) and less accurate in the remaining databases except Iris where the

Table 2. Experimental results given as the mean and the standard deviation of 10 cross-validations

	NPC	NBC	FNPC	FNBC	FuHC	NNPC
Iris	95.33±6.0	95.33±6.0	**96.0±5.33**	95.33±5.21	94.66±4.0	90.66±4.42
Cancer	95.03±2.26	96.34±0.97	97.37±1.82	**97.65±1.76**	96.05±1.96	93.41±2.49
Wine	94.37±5.56	**97.15±2.86**	96.6±3.73	96.67±5.67	93.26±4.14	92.64±5.12
Diabetes	69.01±3.99	74.34±4.44	**74.36±4.57**	74.35±3.38	73.44±5.31	67.96±6.05
Magic	59.24±7.09	66.02±5.37	**73.37±2.96**	72.8±3.29	68.34±6.69	64.80±2.41
Transfusion	61.67±6.6	72.6±4.56	67.43±7.43	70.09±7.68	72.76±7.19	**76.50±5.94**
Sat. Image	88.26±2.62	90.55±2.46	92.02±2.81	90.0±4.39	86.88±3.67	**93.58±1.88**
Segment	71.47±4.15	80.73±2.16	**90.73±1.8**	88.27±3.19	81.07±3.51	90.73±2.15
Yeast	49.67±4.87	48.65±4.42	52.02±5.05	**55.93±3.36**	53.36±4.57	43.06±2.53
Ecoli	83.37±4.46	82.53±5.32	**83.55±9.4**	79.02±10.0	77.7±13.31	80.65±6.98
Glass	49.18±11.8	33.74±9.0	58.46±9.59	53.42±16.0	39.26±13.9	**65.84±9.70**

two classifiers have the same accuracy. A normality test (test of Shapiro-Wilk) done on the three databases (Yeast, Ecoli and Glass) show that they contain attributes that are not normally distributed. We can conclude that applying a Probability-Possibility transformation on the NBC (which leads to NPC) enables it to be less sensitive to normality violation.

• As previously published in [14], overall FNBC is better than a classical NBC. In fact FNBC is more accurate than the NBC in five of 11 datasets and less accurate in three datasets and not significantly different in three cases (Iris, Diabetes and Satellite Image).

• For the four classifiers using Gaussian distribution (NPC, NBC, FNPC and FNBC), classification results of our FNPC are significantly better than other classifiers for all datasets expect in the case of "Transfusion" and "Yeast" databases where FNPC performs worse than others.

• If we compare results for the two Flexible classifiers (FNPC and FNBC), we note that the FNPC performs better with the highest accuracy for the majority of datasets. For this classifier, the greatest increase in accuracy compared to the FNBC has occurred for databases "Glass", "Ecoli", "Satellite image" and "Segment" (Table 2). In Table 1, we note that the attributes for these databases range from 8 to 37, and the number of classes from 6 to 8. So the FNPC is significantly more efficient than FNBC (and also than NPC and NBC) for datasets with high number of attributes and classes.

• Experiments of the second family of approximate equality-based classifiers having a direct interpretation of data (FuHC and NNPC), either in terms of a fuzzy histogram, or in terms of a possibility distribution, show that they have a competitive efficiency with respect to other possibilistic classifiers for the majority of databases. Besides, we note that the Min-based NNPC, not only outperforms the FuHC but also all other classifiers for 4 datasets (Transfusion, Satellite Image, Segment and Glass) (Fig. 1.). By testing the normality assumption for the

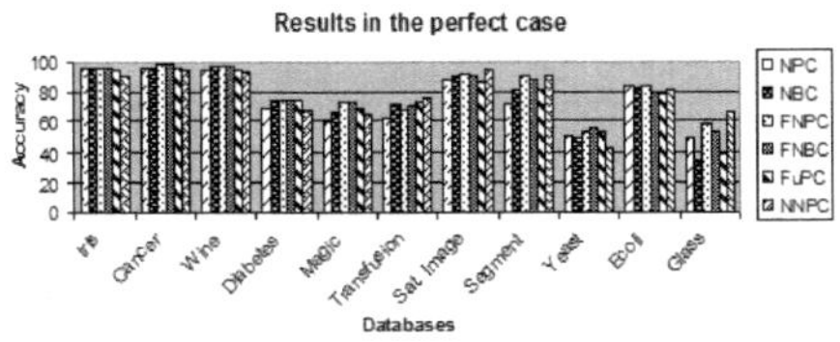

Fig. 1. Classification Accuracy of classifiers

Table 3. Experimental results for the Multiple Classification

Database	NPC	NBC	NNPC
Iris	99.33 ± 2.0, CP: 0.05	95.33 ± 5.21, CP: 0.0	94.0 ± 5.54, CP: 0.04
Wine	99.38 ± 1.88, CP: 0.08	97.22 ± 2.78, CP: 0.0	95.55 ± 4.16, CP: 0.03
Sat. Image	98.07 ± 1.33, CP: 0.10	91.10 ± 1.93, CP: 0.01	96.88 ± 1.65, CP: 0.04
Segment	87.53 ± 3.5, CP: 0.16	80.74 ± 2.88, CP: 0.004	97.80 ± 1.2, CP: 0.07
Yeast	66.3 ± 4.43, CP: 0.18	51.42 ± 2.45, CP: 0.003	72.71 ± 2.26, CP : 0.3
Ecoli	91.28 ± 5.59, CP: 0.09	82.74 ± 6.1, CP: 0.003	88.08 ± 4.39, CP: 0.09
Glass	69.2 ± 10.77, CP: 0.22	37.37 ± 11.26, CP: 0.015	78.97 ± 7.95, CP: 0.15

attributes of theses datasets, we note that the frequency of attributes violating normality is high if compared to other datasets. Thus, the Min-based NNPC seems to be the most efficient classifier for non normal datasets.

These results show that the proposed FNPC is the most efficient classifier even in the presence of a significant mass of attributes and classes if the normality assumption holds. However the NNPC may be preferred for datasets violating normality.

Table 3 includes experimental results for NPC, NBC and NNPC in the case of Multiple-Classification. In this case, we consider the two most relevant classes to classify a new instance (instead of considering only one class), if classes have very close plausibility evaluations, i.e., if the difference between their plausibility or probability is less than a fixed level. In our experimental study, this level is fixed to 0.02. CP in Table 3 denotes the *Confusion Probability* that is the frequency of instances where the classifier succeeds to correctly classify a current instance using the second most relevant class.

Table 3 shows that, for all datasets, classification accuracy of NPC and NNPC is significantly increased in the case of Multiple-Classification if compared to the classical classification in Table 2. Besides, the accuracy of NBC seems to be more stable with a CP$\leq$ 1.5 for all datasets. This can be explained by the use of the Gaussian function g in NBC that has an exponential nature enabling class probabilities to be obviously distinct.

By comparing results for the NPC and NNPC, we remark that overall the NPC confuses more between classes than the NNPC except for "Yeast" databases (Fig. 2.). In fact, using the probability-possibility transformation method in NPC for a non normal dataset, leads to too weak class plausibility's (because in that

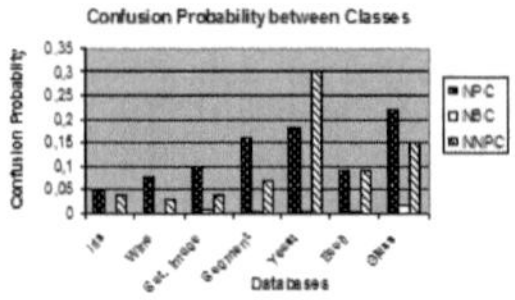

Fig. 2. Results for the Confusion Probability between classes

case probability of confidence intervals is high) which makes the NPC unable to distinguish between near classes (mainly for datasets "Glass", "Segment" and "Yeast"). Results in Table 2 prove that NNPC is less sensible to normality violation than NPC.

6 Related Works

Some approaches have already proposed the use of a possibilistic data representation in classification methods. Let us cite possibilistic decision trees [15] induced from instances with vaguely defined linguistic attributes and classes. A qualitative approach for classification of objects having possibilistic uncertain attribute values within the decision tree technique is proposed by Ben Amor et al. [23]. This last work aims to search the most plausible class labeling a vector, knowing its possibility distribution on attribute values given by an expert.

A Naive Bayes Style Possibilistic Classifier (NBSPC) developed by Borgelt et al. [6] is induced from imprecise training sets. Imprecision is localized in attribute values of instances excepting the class attribute, whereas the testing set is perfect. The possibility distribution of an attribute given the class is inferred from the computation of the maximum-based projection [7] over the set S of precise instances (S is included in the extended dataset) that contains both the target value of the considered attribute with the class.

A naive possibilistic network classifier, proposed by Haouari et al. [3], presents a building procedure that deals with imperfect dataset attributes and classes, and a classification procedure used to classify new instances that may be characterized by imperfect attributes. This imperfection is represented by means of a possibility distribution given by an expert who expresses its partial ignorance, due to a lack of a priori knowledge. Benferhat and Tabia proposes in [30] an efficient algorithm for revising possibilistic knowledge encoded by a naive product-based possibilistic network classifier given uncertain inputs using Jeffrey's rule which can not be directly used since it is exponential in the number of attributes and attribute domains. The main advantage of the proposed algorithm is its capability to ensure classification task in polynomial time in the number of attributes. All the previous cited works [23,3,15,6,30] deal only with discrete attribute values and are not appropriate for continuous attribute values. These approaches require a preliminary discretization phase for the continuous attribute values.

An attempt to treat uncertainty in continuous data is proposed in [5], where authors developed a classification algorithm able to generate rules from uncertain continuous data. In this work, uncertainty is represented through intervals with probability distribution function which models only imprecision and not genuine uncertainty.

7 Conclusion and Summary Discussion

In this paper we have proposed and tested the performance of two families of possibilistic classifiers for numerical attributes: the first family, assuming normality distribution, is based on a probability-possibility transformation method transforming a classical NBC to a NPC, which introduces some further tolerance in the description of classes. We have also tested the feasibility of a Flexible Naive Possibilistic Classifier which is the possibilistic counterpart of the Flexible Naive Bayesian Classifier. The second family of possibilistic classifiers abandons the normality assumption and has a direct representation on data. We have proposed two classifiers named Fuzzy Histogram Classifier and Nearest Neighbor-based Possibilistic Classifier in this context. The first one exploits an idea of proximity between attribute values in an additive manner whereas the second one is based only on the analysis of proximities between attributes without counting them. Moreover, possibilistic classifiers may be allowed to use multiple-classification instead of classifying new instances using only one class. This strategy should be fruitful when classes have close plausibility estimates (which may be of interest when the number of classes is large). Indeed, it is completely in the spirit of possibilistic classifiers to allow for multiple class classification in case data information is insufficient for a more precise classification, rather than choosing too arbitrarily between classes having very close plausibility estimates.

We have tested the proposed possibilistic classifiers on several datasets. Experimental results show the performance of these classifiers to deal with numerical input data. However, while the NPC is less sensible than NBC to normality violation, the FNPC shows high classification accuracy and good ability to deal with any type of databases when compared to other classifiers in the same family. On the other hand, possibilistic classifiers exploiting proximity between attributes are competitive with others. Besides, the NNPC seems to be the most efficient classifier in particular for databases where normality assumption is strongly violated.

As future works, we first intend to develop a refinement approach used to improve classifier performance when classes have very close plausibility evaluations. We propose to exploit a nearest neighbor heuristic to separate indistinguishable classes. Second, we will orient our research to extend our possibilistic classifiers to handle uncertainty in data representation and we aim to deal with imprecise/uncertain attributes and classes. Lastly, since quantitative possibility measures are a special type of imprecise probabilities, it would be interesting to compare the approach with naive credal classifiers [21].

References

1. Denton, A., Perrizo, W.: A kernel-based semi-naive bayesian classifier using p-trees. In: Proc. of the 4th SIAM Inter. Conf. on Data Mining (2004)
2. Pérez, A., Larraoaga, P., Inza, I.: Bayesian classifiers based on kernel density estimation:flexible classifiers. Inter. J. of Approximate Reasoning 50, 341–362 (2009)
3. Haouari, B., Ben Amor, N., Elouadi, Z., Mellouli, K.: Naïve possibilistic network classifiers. Fuzzy Set and Systems 160(22), 3224–3238 (2009)
4. Kotsiantis, S.B.: Supervised machine learning: A review of classification techniques. Informatica 31, 249–268 (2007)
5. Qin, B., Xia, Y.: and Prabhakar S., and Tu Y. A rule-based classification algorithm for uncertain data. In: IEEE International Conference on Data Engineering (2009)
6. Borgelt, C., Gebhardt, J.: A naïve bayes style possibilistic classifier. In: Proc. 7th European Congress on Intelligent Techniques and Soft Computing, pp. 556–565 (1999)
7. Borgelt, C., Kruse, R.: Efficient maximum projection of database-induced multivariate possibility distributions. In: Proc. 7th IEEE Int. Conf. on Fuzzy Systems, pp. 663–668 (1998)
8. Dubois, D., Laurent, F., Gilles, M., Prade, H.: Probability-possibility transformations, triangular fuzzy sets, and probabilistic inequalities. Reliable Computing 10, 273–297 (2004)
9. Dubois, D., Prade, H.: When upper probabilities are possibility measures. Fuzzy sets and Systems 49, 65–74 (1992)
10. Dubois, D., Prade, H.: On data summarization with fuzzy sets. In: Proc. of the 5th Inter. Fuzzy Systems Assoc. World Congress, IFSA 1993 (1993)
11. Dubois D. and Prade H. Possibility theory: Qualitative and quantitative aspects. *D. Gabbay and P. Smets. editors. Handbook on Defeasible Reasoning and Uncertainty Management Systems*, 1:169–226, 1998.
12. Dubois, D., Prade, H., Sandri, S.: On possibility/probability transformations. In: Fuzzy Logic, pp. 103–112 (1993)
13. Grossman, D., Dominigos, P.: Learning bayesian maximizing conditional likelihood. In: Proc. on Machine Learning, pp. 46–57 (2004)
14. John, G.H., Langley, P.: Estimating continuous distributions in bayesian classifiers. In: Proceedings of the 11th Conference on Uncertainty in Artificial Intelligence (1995)
15. Jenhani, I., Ben Amor, N., Elouedi, Z.: Decision trees as possibilistic classifiers. Inter. J. of Approximate Reasoning 48(3), 784–807 (2008)
16. Mertz, J., Murphy, P.M.: Uci repository of machine learning databases, `ftp://ftp.ics.uci.edu/pub/machine-learning-databases`
17. Pearl, J.: Probabilistic Reasoning in Intelligent Systems: Networks of Plausible Inference. Morgan Kaufmman, San Francisco (1988)
18. Yamada, K.: Probability-possibility transformation based on evidence theory. In: IFSA World Congress, vol. 10, pp. 70–75 (2001)
19. Bishop, C.M.: Neural Networks for Pattern Recognition. Oxford University Press, New York (1996)
20. Cover, T.M., Hart, P.E.: Nearest neighbour pattern classification. IEEE Transactions on Information Theory 13, 21–27 (1967)
21. Zaffalon, M.: The naive credal classifier. Journal of statistical planning and inference 105, 5–21 (2002)

22. Ben Amor, N., Mellouli, K., Benferhat, S., Dubois, D., Prade, H.: A theoretical framework for possibilistic independence in a weakly ordered setting. International Journal of Uncertainty, Fuzziness and Knowledge-Based Systems 10, 117–155 (2002)
23. Ben Amor, N., Benferhat, S., Elouedi, Z.: Qualitative classification and evaluation in possibilistic decision trees. In: FUZZ-IEEE 2004, vol. 1, pp. 653–657 (2004)
24. Friedman, N., Geiger, D., Goldszmidt, M.: Bayesian network classifiers. Machine Learning 29, 131–161 (1997)
25. Strauss, O., Comby, F., Aldon, M.J.: Rough histograms for robust statistics. In: Proc. Inter. Conf. on Pattern Recognition (ICPR 2000), Barcelona, pp. II:2684–2687. IEEE Computer Society, Los Alamitos (2000)
26. Langley, P., Sage, S.: Induction of selective bayesian classifiers. In: Proceedings of 10th Conference on Uncertainty in Artificial Intelligence UAI 1994, pp. 399–406 (1994)
27. Langley, P., Iba, W., Thompson, K.: An analysis of bayesian classifiers. In: Proceedings of AAAI 1992, vol. 7, pp. 223–228 (1992)
28. Quinlan, J.R.: Induction of decision trees. Machine Learning 1, 81–106 (1986)
29. Solomonoff, R.: A formal theory of inductive inference. Information and Control 7, 224–254 (1964)
30. Benferhat, S., Tabia, K.: An efficient algorithm for naive possibilistic classifiers with uncertain inputs. In: Greco, S., Lukasiewicz, T. (eds.) SUM 2008. LNCS (LNAI), vol. 5291, pp. 63–77. Springer, Heidelberg (2008)
31. Sudkamp, T.: Similarity as a foundation for possibility. In: Proc. 9th IEEE Inter. Conf. on Fuzzy Systems, San Antonio, pp. 735–740 (2000)

Plausibility of Information Reported by Successive Sources

Laurence Cholvy

ONERA, 2 avenue Edouard Belin,
31055 Toulouse, France
cholvy@onera.fr

Abstract. This paper deals with the evaluation of a piece of information when successively reported by several sources. It describes a model based on Dempster-Shafer's theory in which the evaluation of a reported piece of information is defined by a plausibility degree. This value depends on the degrees at which sources are correct and the degrees at which they are wrong.

Keywords: Information reported, Dempster-Shafer's Theory.

1 Introduction

Before making a decision, a rational agent tries to know what is the current state of the world. Thus, the agent has to acquire information about the current state of the world and different ways exist for doing so. First, the agent can itself acquire information if it has got the capacity. For instance, in order to know if I take my umbrella before going out, I can glance at the sky through the window. The agent can also get information it needs via another agent which can provide it. For instance, in order to know if I take my umbrella before going out, I can look at the web site of Météo-France. Sometimes, the process is more complex and the agent gets information it needs via a long chain of agents. This is the case when, in order to know if I take my umbrella before going out, I read the forecast provided by Météo-France in my newspaper. Or when I ask my neighbour to read the forecast provided by Météo-France in his newspaper. Here, between the agent which provides the report (Météo-France) and I who need it, there is a sequence of agents: the newspaper, the neighbour, each one getting the report provided by the previous one.

Once a piece of information is acquired, the problem of evaluating its plausibility is raised. If it is not easy to answer this question when the report is acquired via another agent, it is even less when it is reported by several successive agents. Indeed, how can I estimate how plausible is the report "rainy weather today" given by my neighbour after he reads the forecast provided by Météo France in his newspaper?

The question of computing the plausibility of a piece of information reported by several successive agents is the object of our current research.

A. Deshpande and A. Hunter (Eds.): SUM 2010, LNAI 6379, pp. 126–136, 2010.

In a recent paper [5], we have defined a model based on Dempster-Shafer's theory, in which the information evaluation depends on *the degree of validity* of the successive agents. This degree intended to model how valid, i.e how correct, we think an information source is. In this model, any information source was associated with a validity degree $d \in [0, 1]$ so that: d was assigned to our belief in the fact "if the source provides a report then this report is true", and $1 - d$ was assigned to our belief in the fact "if the source provides a report, then this report is false".

Here, we present an extension of this model and we will associate any information source with two degrees: the degree at which it is correct (i.e delivers true reports) and the degree at which it is wrong (i.e delivers false reports).

This paper is organized as follows. Section 2 presents a state of the art of the question. Section 3 presents our model. Finally, section 4 points outs the limitations of this model and its possible extensions.

2 State of the Art

A domain in which the question of information evaluation is very important is military Intelligence. For this reason, NATO has defined a standard in order to guide intelligence officers to associate information with its evaluation [11]. According to this standard, this evaluation is a pair of two values. The first one corresponds to the reliability degree of the source which provides the piece of information and the second value refers to the information credibility. Informal comments define such values. More precisely, the reliability degree of a source depends on how, from its past use, we can trust the source for delivering true information. The credibility degree of a piece of information depends on the fact that it is confirmed or not by several sources and also depends on the fact that it is more or less in conflict with other information.

This way of defining information evaluation has been criticized in [3] where it is shown that the two values are not independent. Several proposals have been proposed in order to circumvent the main limitations of the NATO guidelines [3], [9]. However none of these proposals deal with reported information as informally defined in introduction and which is precisely the main point of interest of this paper.

As far as we know, the work described in [8] was the first one which studied the case of reported information. This work considers French Press Agency (AFP) dispatches, each press dispatch being represented by a sentence, the most representative one, which can mention several successive sources like, for instance: "President X said that country Y is probably developing nuclear power". Here, AFP reports "country Y is probably developing nuclear power" reported by X. In this work, parameters which are considered as having an influence on the evaluation of such information are: *the quality of the source*, which can be defined from its past use if the source is already known, or be defined a-priori if we know which type the source is. Notice that this notion of quality is close to the notion of reliability already mentioned; *the opinion of the source on its own report* which is drawn from text analysis and in particular, from the analysis of the subjective modalities mentioned in the report such "I am certain that ..." or "this is highly probable", "this is impossible", "I am fearing that ..."...; *relations which may exist between the agent which makes the report and agents mentioned in this report*; *relations which may exist between successive sources.* Such relations can be neutral, hostile or friendly. For instance, in case of non neutral relation, the source may make a non sincere report, thus propagating a piece of information it knows to be false.

We have also found an interesting contribution in the domain of logic. More precisely, in his work about the notion of trust, [7], R. Demolombe studies the relations which exist between a piece of information, its truth and the mental attitudes of the agent which produces this piece of information. The formalism he uses is a modal logic, [2], some operators of which are: B_i ($B_i p$ means "agent i believes that p"), I_i^j ($I_i^j p$ means "agent i informs agent j that p"). Operator B_i obeys KD system which is quite usual for beliefs and operator I_i^j only obeys rule of equivalence substitutivity. Before focusing on the notion of trust which is his main subject, the author defines several properties agents can have, called epistemic properties, among which the following are interesting for our problem:

- *Sincerity*: Agent i is sincere with regard to j for information p iff, if i informs j that p, then i believes p. I.e. a sincere agent believes what he says. Thus $sincere(i, j, p) \equiv I_i^j p \rightarrow B_i p$.
- *Competence*: Agent i is competent about p iff, if i believes p, then p is true. I.e. the beliefs of a competent agent are true. Thus $competent(i, p) \equiv B_i p \rightarrow p$.
- *Validity*: Agent i is valid with regard to j for p iff, if i informs j about p, then p is true. I.e. a valid agent tells the truth. Thus $valid(i, j, p) \equiv I_i^j p \rightarrow p$.

Thus we have: $sincere(i, j, p) \wedge competent(i, p) \rightarrow valid(i, j, p)$.

In a recent work, [6], D. Dubois and T. Denoeux address very close questions by using Dempster-Shafer's Theory [10]. In this theory, the concept which corresponds to the notion of evaluation is the concept of plausibility. In their work, the authors propose a mechanism for computing the plausibility of a piece of information which is emitted by an agent i given our uncertain belief about i's reliability. In this work, the reliability of an agent is defined by its relevance[1] and its sincerity. For Dubois and Denoeux, an agent is relevant if it is competent in the topic of the piece of information it provides; an agent is sincere if it does not lie (a non-sincere agent says the opposite of what it believes). Thus Dubois and Denoeux's notion of "reliability" and Demolombe's notion of "validity" are very close, despite being modelled in different formalisms. Consider that an agent i provides information ϕ. The belief one has about i's reliability is used in Dubois and Denoeux's model as follows. If i is not competent in the topic of ϕ, then ϕ is replaced by the tautology $\phi \vee \neg\phi$; If i is competent in the topic of ϕ, then, if it is sincere then we keep ϕ, else ϕ is replaced by $\neg\phi$. Competence and sincerity can be considered as two independent notions. Thus, if p is the probability of i's

[1] The term "pertinent" used by the authors is questionnable. We think that "competence" would be more appropriate.

being competent and q is the probability of i's being sincere, then the plausibility of ϕ can be shown to be equal to $p.q + 1 - p$.

In their work, Dubois and Denoeux assume that any piece of information is provided by a single agent. They do not assume that information is reported by several successive agents. However, like Dubois and Denoeux, we think that Demspter-Shafer's Theory is an interesting formalism when one has to deal with uncertainty. Indeed, this formalism offers two interesting concepts which are the concept of mass assignment, which allows us to express degrees of beliefs on information and the concept of plausibility function which will allow us to quantify the evaluation of a piece of information.

This is why, in [5] we have used this theory to express a graded validity[2] and use it to get a graded plausibility of reported information. More precisely, in this work, an information source i is *valid for ϕ at the degree d_i*, $d_i \in [0, 1]$, if and only if our beliefs can be modelled by a mass assignment which assigns d_i to our belief in the fact "if source i provides information ϕ then ϕ is true"; and $1 - d_i$ to our belief in the fact "if source i provides informationϕ then ϕ is false".

According to this model, when provided by successive source $i_1, i_2, ... i_n$, information ϕ is strictly more plausible than $\neg\phi$ if and only if the degrees of validity of i_2 ... i_n are not equal to 0 and i_1 is valid at a degree strictly greater than 0.5. In any other case, we cannot conclude.

For example, consider that the reports about the weather I need are given to me by my neighbour who reads them in his newspaper. According to this model:

- If I consider that the validity degree of my neighbour is 1 (i.e. he really tells me what he reads in his newspaper) and if I consider the validity degree of the newspaper is 1 also (the forecast is always true in this newspaper) then, I can conclude that the forecast my neighbour gives me is true.
- If I consider that the validity degree of my neighbour is 1 (i.e. he really tells me what he reads in his newspaper) and if I consider that the validity degree of his newspaper is 0 (the forecast is always false in this newspaper), then, I can conclude that the forecast my neighbour gives me is false.
- If I consider that the validity degree of my neighbour is 0 (i.e he does not tell me what he reads in his newspaper because for instance the newspaper were not distributed today) I cannot conclude: the forecast he reports may be true but it may be false as well.

This model can be criticized because the assignment it defines is dogmatic i.e., no mass is assigned to the total ignorance. This is why we have refined this model.

3 The Proposed Model

We assume that the reader is familiar with Demspter-Shafer's Theory and also with the propositional logic.

[2] We prefer to use the term "validity" or equivalently "correctness" instead of the term "reliability".

This section is divided in two. First, we assume we face only one reported information and we define its plausibility. Several subcases are studied depending on the length of the chain of sources which report it. Then we assume we face several reported information and address the classical problem of merging information.

3.1 Only One Reported Information

First case: one agent. In this first case, we consider that an agent i reports a piece of information ϕ. This is denoted $R_i\phi$. The question we deal with is: how plausible ϕ is ? In order to answer this question, we take as a starting point our previous work [5] and we generalize it by taking into account the degrees at which sources are correct and the degrees at which they are wrong.

We consider a propositional language the two letters of which are: ϕ and $R_i\phi$, representing respectively the facts "information ϕ is true" and "agent i reported information ϕ". The four interpretations of this language are $\{w_1, w_2, w_3, w_4\}$. w_1 represents the situation in which i has reported information ϕ and ϕ is true. It is denoted $w_1 = \{R_i\phi, \phi\}$; w_2 represents the situation in which i has reported information ϕ and ϕ is false. It is denoted $w_2 = \{R_i\phi, \neg\phi\}$; w_3 represents the situation in which i did not report information ϕ and ϕ is true. It is denoted $w_3 = \{\neg R_i\phi, \phi\}$.; w_4 represents the situation in which i did not report information ϕ and ϕ is false. It is denoted $w_4 = \{\neg R_i\phi, \neg\phi\}$. We consider as discernment frame, the set $\Theta = \{w_1, w_2, w_3, w_4\}$.

Definition 1. *Consider a source i and a piece of information ϕ. Let $d_i \in [0,1]$ and $d'_i \in [0,1]$ so that $0 \leq d_i + d'_i \leq 1$. We say that i is correct for ϕ at degree d_i and wrong for ϕ at degree d'_i written $CW(i, \phi, d_i, d'_i)$ iff our beliefs can be modelled by the mass assignment $m^{(i,\phi,d_i,d'_i)}$ defined by:*

$$m^{(i,\phi,d_i,d'_i)}(w_1 \vee w_3 \vee w_4) = d_i$$
$$m^{(i,\phi,d_i,d'_i)}(w_2 \vee w_3 \vee w_4) = d'_i$$
$$m^{(i,\phi,d_i,d'_i)}(w_1 \vee w_2 \vee w_3 \vee w_4) = 1 - (d_i + d'_i)$$

Let us recall that assigning a mass on a disjunction of w_i is equivalent to assigning this mass on any propositional formula satisfied by all the w_i in the disjunction. The equivalence is proved in [4]. Consequently, the mass assignment defined in the previous definition can be reformulated by:

$$m^{(i,\phi,d_i,d'_i)}(R_i\phi \rightarrow \phi) = d_i$$
$$m^{(i,\phi,d_i,d'_i)}(R_i\phi \rightarrow \neg\phi) = d'_i$$
$$m^{(i,\phi,d_i,d'_i)}(True) = 1 - (d_i + d'_i)$$

Thus, according to this definition, we consider that i is correct for ϕ at degree d_i and wrong for ϕ at degree d'_i iff our belief degree in the fact "if i reports ϕ then ϕ is true" is d_i, our belief degree in the fact "if i reports ϕ then ϕ is false" is d'_i, and our total ignorance degree is $1 - (d_i + d'_i)$.

It must be noticed that that for any source i and any information ϕ, degrees d_i and d'_i are considered as unique. In particular, it is assumed that these degrees do not depend on the current environment in which the correctness and the wrongness of the source are evaluated. This is questionnable and obviously simplistic.

The following particular cases are worth detailing:

- $(d_i = 1)$ and $(d'_i = 0)$ In this case, we say that i is *correct* for ϕ. We have $m^{(i,\phi,1,0)}(R_i\phi \rightarrow \phi) = 1$. I.e. we are certain that if i reports ϕ then ϕ is true.
- $(d_i = 0)$ and $(d'_i = 1)$ In this case, we say that i is *wrong* for ϕ. We have $m^{(i,\phi,0,1)}(R_i\phi \rightarrow \neg\phi) = 1$. I.e. we are certain that if i reports ϕ then ϕ is false.

Definition 2. *$m^{R_i\phi}$ is the mass assignment defined by: $m^{R_i\phi}(R_i\phi) = 1$ or equivalently, $m^{R_i\phi}(w_1 \vee w_2) = 1$.*

The mass function defined by this definition represents the fact that, for sure, agent i has reported information ϕ.

Definition 3. *Let us consider a source i such that $CW(i, \phi, d_i, d'_i)$. After i reports ϕ, our beliefs are modelled by the mass assignment m obtained by Dempster's combination of $m^{(i,\phi,d_i)}$ and $m^{R_i\phi}$. I.e.,*

$$m = m^{(i,\phi,d_i)} \oplus m^{R_i\phi}$$

Proposition 1

$$m(R_i\phi \wedge \phi) = d_i$$
$$m(R_i\phi \wedge \neg\phi) = d'_i$$
$$m(R_i\phi) = 1 - (d_i + d'_i)$$

Let pl be the plausibility function associated with assignment m.

Proposition 2. *When $R_i\phi$ and $CW(i, \phi, d_i, d'_i)$ we have:*

$$pl(\phi) = 1 - d'_i$$
$$pl(\neg\phi) = 1 - d_i$$

Proposition 3. *When $R_i\phi$ and $CW(i, \phi, d_i, d'_i)$ we can conclude that ϕ is more plausible than $\neg\phi$ iff $d_i > d'_i$.*

This result is quite obvious.

Second case: two agents. Here, we consider that agent j reports that agent i has reported ϕ. This is denoted: $R_jR_i\phi$. The question is: how plausible ϕ is ? or saying it differently, what is the influence of correctness degrees and wrongness degrees of i and j on the plausibility of ϕ ?

We consider a propositional language the letters of which are: ϕ, $R_i\phi$, and $R_jR_i\phi$. This language has got 8 interpretations $w_1 = \{R_jR_i\phi, R_i\phi, \phi\}$; $w_2 = \{R_jR_i\phi, R_i\phi, \neg\phi\}$; $w_3 = \{R_jR_i\phi, \neg R_i\phi, \phi\}$; $w_4 = \{R_jR_i\phi, \neg R_i\phi, \neg\phi\}$; $w_5 = \{\neg R_jR_i\phi, R_i\phi, \phi\}$; $w_6 = \{\neg R_jR_i\phi, R_i\phi, \neg\phi\}$; $w_7 = \{\neg R_jR_i\phi, \neg R_i\phi, \phi\}$; $w_8 = \{\neg R_jR_i\phi, \neg R_i\phi, \neg\phi\}$; The frame of discernment is the set $\Theta = \{w_1, ...w_8\}$.

As before, we will assign mass on formulas and not on disjunctions of w_i.

Definition 4. *Consider that $R_jR_i\phi$ such that $CW(i, \phi, d_i, d'_i)$ and $CW(j, R_i\phi, d_j, d'_j)$. Then, our beliefs are defined by the mass assignment denoted m defined by:*

$$m = m^{(i,\phi,d_i,d'_i)} \oplus m^{(j,R_j\phi,d_j,d'_j)} \oplus m^{R_jR_i\phi}$$

Thus, when $R_jR_i\phi$, our beliefs are defined by combining by Dempster's rule our beliefs on the fact that i is correct at degree d_i and wrong at degree d'_i for ϕ, and our beliefs on the fact that j, assumed to be correct at degree d_j and wrong at degree d'_j for $R_i\phi$, has reported information $R_i\phi$.

Proposition 4

$$m(R_jR_i\phi \wedge R_i\phi \wedge \phi) = d_i.d_j$$

$$m(R_jR_i\phi \wedge R_i\phi \wedge \neg\phi) = d'_i d_j$$

$$m(R_jR_i\phi \wedge R_i\phi) = (1 - (d_i + d'_i)).d_j$$

$$m(R_jR_i\phi \wedge \neg R_i\phi \wedge (R_i\phi \rightarrow \phi)) = d_i.d'_j$$

$$m(R_jR_i\phi \wedge \neg R_i\phi \wedge (R_i\phi \rightarrow \neg\phi)) = d'_i.d'_j$$

$$m(R_jR_i\phi \wedge \neg R_i\phi) = (1 - (d_i + d'_i)).d'_j$$

$$m(R_jR_i\phi \wedge (R_i\phi \rightarrow \phi)) = d_i.(1 - (d_j + d_j))$$

$$m(R_jR_i\phi \wedge (R_i\phi \rightarrow \neg\phi)) = d'_i.(1 - (d_j + d'_j))$$

$$m(R_jR_i\phi) = (1 - (d_i + d'_i)).(1 - (d_j + d'_j))$$

Proposition 5. *When $RjRi\phi$ such that $CW(i, \phi, d_i, d'_i)$ and $CW(j, R_i\phi, d_j, d'_j)$ we have:*

$$pl(\phi) = 1 - d'_i.d_j$$

$$pl(\neg\phi) = 1 - d_i.d_j$$

The two following cases are worth examining:

- $(d_j = 1)$ and $(d'_j = 0)$. This means that j is correct for information $R_i\phi$ i.e, it is true that i has reported ϕ. Then we get $pl(\phi) = 1 - d'_i$ and $pl(\neg\phi) = 1 - d_i$. We come to the "one agent" case.
- $(d_j = 0)$ and $(d'_j = 1)$. This means that j is wrong for information $R_i\phi$ thus $R_i\phi$ is false. In this case we have: $pl(\phi) = 1$ and $pl(\neg\phi) = 1$. Thus we cannot decide among ϕ and $\neg\phi$ which is the most plausible.

Proposition 6. *When $RjRi\phi$ such that $CW(i, \phi, d_i, d'_i)$ and $CW(j, R_i\phi, d_j, d'_j)$ we have:*

$$pl(\phi) > pl(\neg\phi) \iff d_j \neq 0 \text{ and } d_i > d'_i$$

I.e. we can conclude that ϕ is strictly more plausible than $\neg\phi$ iff j is not wrong, and i's correctness degree is strictly greater that its wrongness degree.

Example 1. Let us illustrate this on the example given in introduction. Consider that the reports about the weather are given to me by my neighbour who read them in his newspaper.

- If I consider that my neighbour is correct (i.e. he really tells me what he reads in his newspaper) and not wrong at all and if I consider his newspaper entirely correct (the forecast is always true in this newspaper) and not wrong at all then, I can conclude that the forecast my neighbour gives me is true.
- If I consider that my neighbour is correct and not wrong at all and if I consider his newspaper wrong (the forecast is always false in this newspaper) then I can conclude that the forecast my neighbour gives me is false.
- But If I consider that my neighbour is wrong (i.e he does not tell me what he reads in his newspaper because for instance the newspaper were not distributed today) I cannot decide: the forecast he reports may be true but it may be false as well.

General case. We consider here the case when a source i_n reports that a source i_{n-1} has reported that ..... i_1 has reported ϕ. This is denoted $R_{i_n}...R_{i_1}\phi$. The question is again the influence of the degrees of correctness and wrongness of sources on the plausibility of information.

We consider a propositional language the $n+1$ letters of which are ϕ, $R_{i_1}\phi$, $R_{i_2}R_{i_1}\phi$, ..., $R_{i_n}...R_{i_1}\phi$. This language has got 2^{n+1} interpretations which form the frame of discernment we consider but we do not detail them because, as before, we assign masses to formulas.

Definition 5. *Assume* $R_{i_n}...R_{i_1}\phi$, $CW(i_1, \phi, d_1, d'_1)$, $CW(i_2, R_{i_1}\phi, d_2, d'_2)$,.., $CW(i_n, R_{i_{n-1}}...R_{i_1}\phi, d_n, d'_n)$. *Then, our beliefs are defined by the following mass assignment:*

$$m = m^{(i_1, \phi, d_1, d'_1)} \oplus ... \oplus m^{(i_n, R_{i_{n-1}}...R_{i_1}\phi, d_n, d'_n)} \oplus m^{R_{i_n}...R_{i_1}\phi}$$

By this definition, when $R_{i_n}...R_{i_2}R_{i_1}\phi$, our beliefs are defined by combining: our beliefs in the fact that i_1 is correct at degree d_1 and wrong at degree d'_1 for ϕ and .... our beliefs in the fact that i_n, assumed to be correct at degree d_{i_n} and wrong at degree d'_{i_n} for $R_{i_{n-1}}...R_{i_1}\phi$, has reported $R_{i_{n-1}}...R_{i_2}R_{i_1}\phi$.

Proposition 7. *With the hypothesis of this general case we have:*

$$pl(\phi) = 1 - d_n...d_2.d'_1$$
$$pl(\neg\phi) = 1 - d_n...d_2.d_1$$

Proposition 8. *With the hypothesis of this general case we have:*

$$pl(\phi) > pl(\neg\phi) \iff \forall i = 2...n \ d_i \neq 0 \text{ and } d_1 > d'_1$$

I.e. we can conclude that ϕ is strictly more plausible than $\neg\phi$ iff i_2 ... i_n are not totally wrong and i_1's correctness degree for ϕ is strictly greater that its wrongness degree for ϕ.

3.2 Merging Several Reported Information

Here, we examine the case of several reported information, each one of them being reported by one or several successive sources.

In order to be as general as possible, we suppose r reports ϕ_1 ... ϕ_r and we suppose that each report ϕ_k is reported by successive sources $k_1, ...k_{n_k}$. Thus we have, for $k = 1...r$: $R_{k_{n_k}} ... R_{k_1}\phi_k$. Furthermore, we suppose that we have, for $k = 1...r$: $CW(k_1, \phi_k, d_{k_1}, d'_{k_1})$ ...$CW(k_{n_k}, R_{k_{n_k-1}} ... R_{k_1}\phi_k, d_{k_{n_k}}, d'_{k_{n_k}})$

Thus, for any $k = 1...r$, we consider a propositional language L_k the $n_k + 1$ letters of which are ϕ_k, $R_{k_1}\phi_k$, $R_{k_2}R_{k_1}\phi_k$, ..., $R_{k_n}...R_{k_1}\phi_k$. This language has got 2^{n_k+1} interpretations which form the frame of discernment Θ_k we consider. Thus by definition 5, we get the following mass assigment:

$m_k = m_k^{(k_1,\phi_k,d_{k_1},d'_{k_1})} \oplus ... \oplus m_k^{(k_n,R_{k_{n_k-1}}...R_{k_1}\phi_k,d_{n_k},d'_{n_k})} \oplus m^{R_{k_n}...R_{k_1}\phi_k}$

Each m_k represents our beliefs if we only assume that $R_{k_{n_k}}...R_{k_1}\phi_k$.

Since there are several reported information, we have to merge these beliefs. This implies that we have to combine assigments $m_1, ..., m_r$.

But these assigments are respectively defined on $\Theta_1, ...\Theta_r$ which are different sets. Thus we face a problem of combining assigments defined on different frames. which can be solved by using one of the methods described in [1]. Assuming that one of these methods has been used, we get a new assigment m which expresses our merged beliefs. Finally, the plausibility of any piece of information ψ is defined by $pl(\psi)$ where pl is the plausibility function associated with m.

Example 2. Consider four sources named a, b, c, d.

Assume that b reports that a reported ϕ. I.e. $R_bR_a\phi$. We suppose that $CW(a, \phi, 1, 0)$ and $CW(b, R_a\phi, 0.7, 0.1)$.

Assume that c reports that d reported $\neg\phi$. I.e. $R_cR_d\neg\phi$. We suppose that $CW(d, \neg\phi, 1, 0)$ and $VD(c, R_d\neg\phi, 0.2, 0.1)$.

Then, according to proposition 3, we get two mass assigments m_1 and m_2:

$$m_1(R_bR_a\phi \wedge R_a\phi \wedge \phi) = 0.7$$

$$m_1(R_bR_a\phi \wedge \neg R_a\phi \wedge (R_a\phi \rightarrow \phi)) = 0.1$$

$$m_1(R_bR_a\phi \wedge (R_a\phi \rightarrow \phi)) = 0.2$$

and

$$m_2(R_cR_d\neg\phi \wedge R_d\neg\phi \wedge \neg\phi) = 0.2$$

$$m_2(R_cR_d\neg\phi \wedge \neg R_d\neg\phi \wedge (R_d\neg\phi \rightarrow \neg\phi)) = 0.1$$

$$m_2(R_cR_d\neg\phi \wedge (R_d\neg\phi \rightarrow \neg\phi)) = 0.7$$

For combining these assigments, we can use the following method:

1. Consider the propositional language: $L = \{\phi, R_a\phi, R_d\neg\phi, R_bR_a\phi, R_cR_d\neg\phi\}$

2. We reformulate m_1 and m_2 in this language.
We still get: $m_1(R_bR_a\phi \wedge R_a\phi \wedge \phi) = 0.7$, $m_1(R_bR_a\phi \wedge \neg R_a\phi \wedge (R_a\phi \rightarrow \phi)) = 0.1$, $m_1(R_bR_a\phi \wedge (R_a\phi \rightarrow \phi)) = 0.2$ and $m_2(R_cR_d\neg\phi \wedge R_d\neg\phi \wedge \neg\phi) = 0.2$, $m_2(R_cR_d\neg\phi \wedge \neg R_d\neg\phi \wedge (R_d\neg\phi \rightarrow \neg\phi)) = 0.1$, $m_2(R_cR_d\neg\phi \wedge (R_d\neg\phi \rightarrow \neg\phi)) = 0.7$.
3. The two assigments being on the same frame, we then can use any combination rule we want.
Assume that we use the classical Dempster's rule. Then we get an assignment m so that: $pl(\phi) = \frac{0.4}{0.43}$ and $pl(\neg\phi) = \frac{0.15}{0.43}$
Thus finally, we can conclude $pl(\phi) > pl(\neg\phi)$.

Intuitively, this can be explained as follows: the degree of correctness of b for reporting $R_a\phi$ is higher that the degree of correctness of c for reporting $R_d\neg\phi$, then we trust b's report more than c's report. And, since, a is correct for reporting ϕ, we will believe ϕ more than its negation.

4 Discussion

The main contribution of this paper is a model for characterizing the plausibility of information when reported by several successive sources of information. This model assumes that this value mainly depends on the ability of the reporting sources to report something true or to report something false. More precisely, this present model extends a previous one and assumes that the plausibility of information depends on the degree at which the reporting sources are correct and the degree at which they are wrong.

This work takes credit of the fact that the problem of information reported by successive sources is rather original and many open issues exist.

First, we have to study the parameters which influence the degrees at which sources are correct and the degrees at which they are wrong. Indeed, in real applications, we will have to provide guides to define them. For doing so, we could take into account more knowledge about the reporting sources(for instance, knowldege about their past behaviour, their competence, their sincerity but also their current aim) and also more knowledge about relations between sources (for instance relations of hostility, neutrality or friendship) that may influence their attitude towards delivering truth.

Concerning the different choices we make (choices of a combination method for combining assignments defined on the same frame, choice of a method for combining assignments defined on different frames...), each of them can be discussed. Analysing the plausibility function we would get by making other choices defines an interesting research direction.

Another interesting open issue is to extend the type of information reported by the agents. In particular, we are currently investigating means of handling information of the type "agent i reports that he believes that ϕ is highly probable".

Acknowledgements

This work has been granted by ANR (Agence Nationale de Recherche) under project CAHORS.

References

1. Appriou, A., Janez, F.: Theory of Evidence and non-exhaustive frames of discernment: Plausibilities correction methods. International Journal of Approximate Reasoning 18(1-2), 1–19 (1998)
2. Chellas, B.F.: Modal logic: An introduction. Cambridge University Press, Cambridge (1980)
3. Cholvy, L.: Information Evaluation in fusion: a case stud. In: Proceedings of 10th International Conference on Information Processing and Management of Uncertainty in Knowledge-Based Systems (IPMU 2004), Perugia (July 2004)
4. Cholvy, L.: Using Logic to Understand relations between DSmT and Dempster-Shafer Theory. In: Sossai, C., Chemello, G. (eds.) ECSQARU 2009. LNCS, vol. 5590. Springer, Heidelberg (2009)
5. Cholvy, L.: Evaluation of Information Reported: A model in the Theory of Evidence. In: Hüllermeier, E., Kruse, R., Hoffmann, F. (eds.) Computational Intelligence for Knowledge-Based Systems Design. LNCS, vol. 6178. Springer, Heidelberg (2010)
6. Dubois, D., Denoeux, T.: Pertinence et Sincerité en Fusion d'Informations Rencontres Françaises sur la Logique Floue et ses Applications. In: LFA 2009, Annecy (Novembre 2009)
7. Demolombe, R.: Reasoning about trust: a formal logical framewor. In: Jensen, C., Poslad, S., Dimitrakos, T. (eds.) iTrust 2004. LNCS, vol. 2995, pp. 291–303. Springer, Heidelberg (2004)
8. Jacquelinet, J.: Pertinence et cotation du renseignement Mémoire de stage de l'ENSIIE, effectué sous la direction de Ph. Capet, Thalès (2007)
9. Revault d'Allonnes, A., Besombes, J.: Critères d'évaluation contextuelle pour le traitement automatique. In: Atelier Qualité des Données et des Connaissances (QDC), Strasbourg, France, Janvier (2009)
10. Shafer, G.: A mathematical Theory of Evidence. Princeton University Press, Princeton (1976)
11. NATO: Intelligence Reports. STANAG 2511 (January 2003)

Combining Semantic Web Search with the Power of Inductive Reasoning

Claudia d'Amato[1], Nicola Fanizzi[1], Bettina Fazzinga[2], Georg Gottlob[3,4], and Thomas Lukasiewicz[3,5]

[1] Dipartimento di Informatica, Università degli Studi di Bari, Italy
{claudia.damato,fanizzi}@di.uniba.it
[2] Dipartimento di Elettronica, Informatica e Sistemistica, Università della Calabria, Italy
bfazzinga@deis.unical.it
[3] Computing Laboratory, University of Oxford, UK
{georg.gottlob,thomas.lukasiewicz}@comlab.ox.ac.uk
[4] Oxford-Man Institute of Quantitative Finance, University of Oxford, UK
[5] Institut für Informationssysteme, TU Wien, Austria

Abstract. With the introduction of the Semantic Web as a future substitute of the Web, the key task for the Web, namely, Web Search, is evolving towards some novel form of *Semantic Web search*. A very promising recent approach to Semantic Web search is based on combining standard Web pages and search queries with ontological background knowledge, and using standard Web search engines as the main inference motor of Semantic Web search. In this paper, we continue this line of research. We propose to further enhance this approach by the use of inductive reasoning. This increases the robustness of Semantic Web search, as it adds the important ability to handle inconsistencies, noise, and incompleteness, which are all very likely to occur in distributed and heterogeneous environments such as the Web. In particular, inductive reasoning allows to infer (from training individuals) new knowledge, which is not logically deducible. We also report on a prototype implementation of the new approach and its experimental evaluations.

1 Introduction

Web search is a key technology of the Web, since it is the primary way to access content in the ocean of Web data. Current Web search technologies are essentially based on a combination of textual keyword search with an importance ranking of documents via the link structure of the Web [4]. For this reason, however, current standard Web search does not allow for a *semantic processing of Web search queries*, which analyzes both Web search queries and Web pages with respect to their meaning, and returns exactly the semantically relevant pages to a query. For the same reason, current standard Web search also does not allow for *evaluating complex Web search queries that involve reasoning over the Web*.

Many experts predict that the next huge step forward in Web information technology will be achieved by adding such structure and/or semantics to Web contents and exploiting them when processing Web search queries. Indeed, the *Semantic Web* (*SW*) [3] as a vision of a more powerful future Web goes in this direction. It is a common framework

A. Deshpande and A. Hunter (Eds.): SUM 2010, LNAI 6379, pp. 137–150, 2010.

that allows data to be shared and reused in different applications, enterprises, and communities. The SW is an extension of the current Web by standards and technologies that help machines to understand the information on the Web so that they can support richer discovery, data integration, navigation, and automation of tasks. It consists of several hierarchical layers, where the *Ontology layer*, in form of the *OWL Web Ontology Language* [16,19,2], is the highest layer that has currently reached a sufficient maturity. Some important layers below the Ontology layer are the *RDF* and *RDF Schema layers* along with the *SPARQL* query language. For the higher *Rules*, *Logic*, and *Proof layers* of the SW, one has especially developed languages integrating rules and ontologies, and languages supporting more sophisticated forms of knowledge.

The development of a new search technology for the SW, called *SW search*, is currently an extremely hot topic, both in Web-related companies and in academic research (see [13] for a recent survey). In particular, there is a fast growing number of commercial and academic SW search engines. There are essentially two main research directions. The first (and most common) one is to develop a new form of search for searching the pieces of data and knowledge that are encoded in the new representation formalisms of the SW (e.g., [8]), while the second (and nearly unexplored) direction is to use the formalisms of the SW in order to add some semantics to Web search (e.g., [14,17]).

A very promising recent representative of the second direction is given in [12], where an ontologically enriched Web along with complex ontology-based search on the Web are achieved on top of the existing Web and using existing Web search engines. Intuitively, standard Web pages are first connected to (and via) an ontological knowledge base, which then allows for formulating and processing complex ontology-based (unions of conjunctive) search queries (eventually containing negated subqueries) that involve reasoning over the data of the Web. The query processing step is based on new techniques (i) for pre-compiling the ontological knowledge using standard ontology reasoning techniques and (ii) for translating complex ontology-based Web queries into (sequences of) standard Web queries that are answered by standard Web search. That is, essential parts of ontological search on the Web are actually reduced to state-of-the-art search engines such as *Google search*. As important advantages, this approach can immediately be applied to the whole existing Web, and it can be done with existing Web search technology (and so does not require completely new technologies).

In this paper, we continue this line of research. We propose to further enhance our SW search in [12] by the use of inductive reasoning for the offline ontology compilation step. This allows for an increased robustness of our approach to SW search, as it adds the ability to handle inconsistencies, noise, and incompleteness, which are all very likely to occur in distributed and heterogeneous environments such as the Web. In particular, this also allows to infer (from existing training individuals) new knowledge, which is not logically deducible. To our knowledge, this is the first combination of SW search with inductive reasoning. The main contributions of this paper are summarized as follows:

- We develop a combination of our approach to SW search as presented in [12] with inductive reasoning (based on similarity search [18] for retrieving the resources that likely have a query property [7]). Here, inductive reasoning serves in an offline ontology compilation step to compute completed semantic annotations.

- Importantly, the inductive approach to SW search is more robust, as it can handle inconsistencies, noise, and incompleteness in SW knowledge bases, which are all very likely to occur in distributed and heterogeneous environments such as the Web. We provide several examples illustrating these important advantages.
- We report on a prototype implementation of our approach to SW search in the framework of desktop search. We also provide experimental results on (1) the running time of the online query processing step and (2) the precision and the recall of our inductive approach to SW search compared to the deductive one.

The rest of this paper is organized as follows. In Sections 2 and 3, we give a brief overview of our SW search system and the underlying theoretical model, respectively. Sections 4 and 5 combine our SW search with inductive reasoning and describe its main advantages, respectively. In Sections 6 and 7, we report on a prototype implementation along with its experimental evaluations and provide a conclusion, respectively.

2 Conceptual Overview

Our SW search system consists of an *Interface*, a *Query Evaluator* (on top of standard Web *Search Engines*), and an *Inference Engine*. Standard *Web* pages and their objects are enriched by *Annotation* pages, based on an *Ontology*. We now briefly describe these components and ingredients as well as their interplay in some more detail.

Ontology. Our approach to SW search is done relative to a fixed underlying *ontology*, which defines an alphabet of elementary ontological ingredients, as well as terminological relationships between these ingredients. The ontology may either describe fully general knowledge (such as the knowledge encoded in Wikipedia) for general ontology-based search on the Web, or it may describe some specific knowledge (such as biomedical knowledge) for vertical ontology-based search on the Web. The former results into a general ontology-based interface to the Web similar to Google, while the latter produces different vertical ontology-based interfaces to the Web. There are many existing ontologies that can be used, which have especially been developed for the SW, but also in biomedical and technical areas. Such ontologies are generally created and updated by human experts in a knowledge engineering process. Recent research attempts are also directed towards an automatic generation of ontologies from text documents, eventually coming along with existing ontological knowledge [5,11].

For example, an ontology may contain the knowledge that (i) conference and journal papers are articles, (ii) conference papers are not journal papers, (iii) *isAuthorOf* relates scientists and articles, (iv) *isAuthorOf* is the inverse of *hasAuthor*, and (v) *hasFirstAuthor* is a functional binary relationship, which is formalized as follows:

$$\begin{array}{l} \mathit{ConferencePaper} \sqsubseteq \mathit{Article},\ \mathit{JournalPaper} \sqsubseteq \mathit{Article},\ \mathit{ConferencePaper} \sqsubseteq \neg \mathit{JournalPaper}, \\ \exists \mathit{isAuthorOf} \sqsubseteq \mathit{Scientist},\ \exists \mathit{isAuthorOf}^{-} \sqsubseteq \mathit{Article},\ \mathit{isAuthorOf}^{-} \sqsubseteq \mathit{hasAuthor}, \\ \mathit{hasAuthor}^{-} \sqsubseteq \mathit{isAuthorOf},\ (\mathsf{funct}\ \mathit{hasFirstAuthor})\,. \end{array} \quad (1)$$

Annotations. As a second ingredient of our SW search, we assume the existence of assertional pieces of knowledge about Web pages and their objects, also called *(semantic) annotations*, which are defined relative to the terminological relationships of the underlying ontology. Such annotations are starting to be widely available for a large class of Web resources, especially user-defined annotations with the Web 2.0. They may also be automatically learned from Web pages and their objects (see, e.g., [6]). As a midway between such fully user-defined and fully automatically generated annotations, one can also automatically extract annotations from Web pages using user-defined rules [12].

For example, in a very simple scenario relative to the ontology in Eq. 1, a Web page i_1 may contain information about a Ph.D. student i_2, called Mary, and two of her papers: a conference paper i_3 with title "*Semantic Web search*" and a journal paper i_4 entitled "*Semantic Web search engines*" and published in 2008. There may now exist one semantic annotation each for the Web page, the Ph.D. student Mary, the journal paper, and the conference paper. The annotation for the Web page may simply encode that it mentions Mary and the two papers, while the one for Mary may encode that she is a Ph.D. student with the name Mary and the author of the papers i_3 and i_4. The annotation for i_3 may encode that i_3 is a conference paper and has the title "*Semantic Web search*", while the one for i_4 may encode that i_4 is a journal paper, authored by Mary, has the title "*Semantic Web search engines*", was published in 2008, and has the keyword "RDF". The semantic annotations of i_1, i_2, i_3, and i_4 are then formally expressed as the following sets of ontological axioms $\mathcal{A}_{i_1}$, $\mathcal{A}_{i_2}$, $\mathcal{A}_{i_3}$, and $\mathcal{A}_{i_4}$, respectively:

$$\begin{aligned}
\mathcal{A}_{i_1} &= \{\mathit{contains}(i_1, i_2),\ \mathit{contains}(i_1, i_3), \mathit{contains}(i_1, i_4)\},\\
\mathcal{A}_{i_2} &= \{\mathit{PhDStudent}(i_2),\ \mathit{name}(i_2, \text{“}\mathit{mary}\text{”}),\ \mathit{isAuthorOf}(i_2, i_3),\ \mathit{isAuthorOf}(i_2, i_4)\},\\
\mathcal{A}_{i_3} &= \{\mathit{ConferencePaper}(i_3),\ \mathit{title}(i_3, \text{“}\mathit{Semantic\ Web\ search}\text{”})\},\\
\mathcal{A}_{i_4} &= \{\mathit{JournalPaper}(i_4),\ \mathit{hasAuthor}(i_4, i_2),\ \mathit{title}(i_4, \text{“}\mathit{Semantic\ Web\ search\ engines}\text{”}),\\
&\qquad \mathit{yearOfPublication}(i_4, 2008),\ \mathit{keyword}(i_4, \text{“RDF”})\}.
\end{aligned} \tag{2}$$

Inference Engine. Differently from the ontology, the semantic annotations can be directly published on the Web and searched via standard Web search engines. To also make the ontology visible to standard search engines, it is compiled into the semantic annotations: all semantic annotations are completed in an offline ontology compilation step, where the *Inference Engine* adds all properties (that is, ground atoms) that can be derived (deductively in [12] and inductively here) from the ontology and the semantic annotations. The resulting (*completed*) semantic annotations are then published as Web pages, so that they can be searched by standard search engines.

For example, considering again the running scenario, using the ontology in Eq. 1, in particular, we can derive from the semantic annotations in Eq. 2 that the two papers i_3 and i_4 are also articles, and both authored by Mary.

HTML Encoding of Annotations. The above searchable (completed) semantic annotations of (objects on) standard Web pages are published as HTML Web pages with pointers to the respective object pages, so that they (in addition to the standard Web pages) can be searched by standard search engines. For example, the HTML pages for the completed semantic annotations of the above $\mathcal{A}_{i_1}$, $\mathcal{A}_{i_2}$, $\mathcal{A}_{i_3}$, and $\mathcal{A}_{i_4}$ are shown in Fig. 1. We here use the HTML address of the Web page/object's annotation page as an identifier for that Web page/object. The plain textual representation of the completed semantic annotations allows their processing by existing standard search engines for the Web. It is important to point out that this textual representation is simply a list of

i_1 : www.xyuni.edu/mary/an1.html

```
<html>
<body>
www.xyuni.edu/mary <br>
WebPage <br>
contains i2 <br>
contains i3 <br>
contains i4 <br>
</body>
</html>
```

i_3 : www.xyuni.edu/mary/an3.html

```
<html>
<body>
www.xyuni.edu/mary <br>
Article <br>
ConferencePaper <br>
hasAuthor i2 <br>
title Semantic Web search <br>
</body>
</html>
```

i_2 : www.xyuni.edu/mary/an2.html

```
<html>
<body>
www.xyuni.edu/mary <br>
PhDStudent <br>
name mary <br>
isAuthorOf i3 <br>
isAuthorOf i4 <br>
</body>
</html>
```

i_4 : www.xyuni.edu/mary/an4.html

```
<html>
<body>
www.xyuni.edu/mary <br>
Article <br>
JournalPaper <br>
hasAuthor i2 <br>
title Semantic Web search engines <br>
yearOfPublication 2008 <br>
keyword RDF <br>
</body>
</html>
```

Fig. 1. Four HTML pages encoding (completed) semantic annotations

properties, each eventually along with an identifier or a data value as attribute value, and it can thus immediately be encoded as a list of RDF triples. Similarly, the completed semantic annotations can be easily encoded in RDFa or microformats.

Query Evaluator. The *Query Evaluator* reduces each SW search query of the user in an online query processing step to a sequence of standard Web search queries on standard Web and annotation pages, which are then processed by a standard Web *Search Engine*. The Query Evaluator also collects the results and re-transforms them into a single answer which is returned to the user. As an example of a SW search query, one may ask for all Ph.D. students who have published an article in 2008 with RDF as a keyword, which is formally expressed as follows:

$$Q(x) = \exists y\, (\mathit{PhDStudent}(x) \wedge \mathit{isAuthorOf}(x, y) \wedge \mathit{Article}(y) \wedge \mathit{yearOfPublication}(y, 2008) \wedge \mathit{keyword}(y, \text{“}RDF\text{”}))\,. \tag{3}$$

This query Q is transformed into the two queries $Q_1 = \mathit{PhDStudent}$ AND $\mathit{isAuthorOf}$ and $Q_2 = \mathit{Article}$ AND "*yearOfPublication* 2008" AND "*keyword* RDF", which can both be submitted to a standard Web search engine. The result of the original Q is then built from the results of Q_1 and Q_2. A graphical user interface, such as the one of Google's advanced search, and ultimately a natural language interface (for queries in written or spoken natural language) can help to hide the conceptual complexity of ontological queries to the user.

3 Semantic Web Search

In this section, we briefly recall from [12] SW knowledge bases and the syntax and the semantics of SW search queries to such knowledge bases, as well as how our approach to SW search is realized. We assume that the reader is familiar with Description Logics (DLs) [1], which we use as underlying ontology languages.

Semantic Web Knowledge Bases. Intuitively, a SW knowledge base consists of a background TBox and a collection of ABoxes, one for every concrete Web page and for every object on a Web page. For example, the homepage of a scientist may be such a concrete Web page and be associated with an ABox, while the publications on the homepage may be such objects, which are also associated with one ABox each.

We assume pairwise disjoint sets $\mathbf{D}$, $\mathbf{A}$, $\mathbf{R}_A$, $\mathbf{R}_D$, $\mathbf{I}$, and $\mathbf{V}$ of *atomic datatypes*, *atomic concepts*, *atomic roles*, *atomic attributes*, *individuals*, and *data values*, respectively. Let $\mathbf{I}$ be the disjoint union of two sets $\mathbf{P}$ and $\mathbf{O}$ of *Web pages* and *Web objects*, respectively. Informally, every $p \in \mathbf{P}$ is an identifier for a concrete Web page, while every $o \in \mathbf{O}$ is an identifier for a concrete object on a concrete Web page. A *semantic annotation* $\mathcal{A}_a$ for a Web page or object $a \in \mathbf{P} \cup \mathbf{O}$ is a finite set of concept membership axioms $A(a)$, role membership axioms $P(a,b)$, and attribute membership axioms $U(a,v)$ (which all have a as first argument), where $A \in \mathbf{A}$, $P \in \mathbf{R}_A$, $U \in \mathbf{R}_D$, $b \in \mathbf{I}$, and $v \in \mathbf{V}$. A *SW knowledge base* $KB = (\mathcal{T}, (\mathcal{A}_a)_{a \in \mathbf{P} \cup \mathbf{O}})$ consists of a TBox $\mathcal{T}$ and one semantic annotation $\mathcal{A}_a$ for every Web page and object $a \in \mathbf{P} \cup \mathbf{O}$. For example, let $\mathbf{I} = \mathbf{P} \cup \mathbf{O}$, where $\mathbf{P} = \{i_1\}$ is the set of Web pages, and $\mathbf{O} = \{i_2, i_3, i_4\}$ is the set of Web objects on i_1. Then, a SW knowledge base is defined by $KB = (\mathcal{T}, (\mathcal{A}_a)_{a \in \mathbf{P} \cup \mathbf{O}})$, where the TBox $\mathcal{T}$ contains the axioms in Eq. 1, and the semantic annotations of the individuals in $\mathbf{P} \cup \mathbf{O}$ are the ones in Eq. 2.

Semantic Web Search Queries. We use unions of conjunctive queries with conjunctive and negated conjunctive subqueries as SW search queries to SW knowledge bases. We now first define the syntax of SW search queries and then their semantics.

Syntax. Let $\mathbf{X}$ be a finite set of variables. A *term* is either a Web page $p \in \mathbf{P}$, a Web object $o \in \mathbf{O}$, a data value $v \in \mathbf{V}$, or a variable $x \in \mathbf{X}$. An *atomic formula* (or *atom*) α is of one of the following forms: (i) $d(t)$, where d is an atomic datatype, and t is a term; (ii) $A(t)$, where A is an atomic concept, and t is a term; (iii) $P(t,t')$, where P is an atomic role, and t, t' are terms; and (iv) $U(t,t')$, where U is an atomic attribute, and t, t' are terms. An *equality* has the form $=(t,t')$, where t and t' are terms. A *conjunctive formula* $\exists \mathbf{y}\, \phi(\mathbf{x}, \mathbf{y})$ is an existentially quantified conjunction of atoms α and equalities $=(t,t')$, which have free variables among $\mathbf{x}$ and $\mathbf{y}$. A *SW search query* $Q(\mathbf{x})$ is an expression $\bigvee_{i=1}^{n} \exists \mathbf{y}_i\, \phi_i(\mathbf{x}, \mathbf{y}_i)$, where each ϕ_i with $i \in \{1, \ldots, n\}$ is a conjunction of atoms α (also called *positive atoms*), conjunctive formulas ψ, negated conjunctive formulas $not\ \psi$, and equalities $=(t,t')$, which have free variables among $\mathbf{x}$ and $\mathbf{y}_i$. For example, $Q(x)$ of Eq. 3 is a SW search query.

Semantics of Positive Search Queries. The semantics of positive search queries is defined in terms of ground substitutions via the notion of logical consequence.

A search query $Q(\mathbf{x})$ is *positive* iff it contains no negated conjunctive subqueries. A *(variable) substitution* θ maps variables from $\mathbf{X}$ to terms. A substitution θ is *ground* iff it maps to Web pages $p \in \mathbf{P}$, Web objects $o \in \mathbf{O}$, and data values $v \in \mathbf{V}$. A closed first-order formula ϕ is a *logical consequence* of a knowledge base $KB = (\mathcal{T}, (\mathcal{A}_a)_{a \in \mathbf{P} \cup \mathbf{O}})$, denoted $KB \models \phi$, iff every first-order model $\mathcal{I}$ of $\mathcal{T} \cup \bigcup_{a \in \mathbf{P} \cup \mathbf{O}} \mathcal{A}_a$ also satisfies ϕ. Given a SW knowledge base KB and a positive SW search query $Q(\mathbf{x})$, an *answer* for $Q(\mathbf{x})$ to KB is a ground substitution θ for the variables $\mathbf{x}$ with $KB \models Q(\mathbf{x}\theta)$. For example, an answer for $Q(x)$ of Eq. 3 to the running KB is $\theta = \{x/i_2\}$ (recall that i_2 represents the scientist Mary).

Semantics of General Search Queries. We next define the semantics of general search queries by reduction to the semantics of positive ones, interpreting negated conjunctive subqueries $not\ \psi$ as the lack of evidence about the truth of ψ. That is, negations are interpreted by a closed-world semantics on top of the open-world semantics of DLs.

Given a SW knowledge base KB and search query

$$Q(\mathbf{x}) = \bigvee_{i=1}^{n} \exists \mathbf{y}_i (\phi_{i,1}(\mathbf{x}, \mathbf{y}_i) \wedge \cdots \wedge \phi_{i,l_i}(\mathbf{x}, \mathbf{y}_i) \wedge not\ \phi_{i,l_i+1}(\mathbf{x}, \mathbf{y}_i) \wedge \cdots \wedge not\ \phi_{i,m_i}(\mathbf{x}, \mathbf{y}_i)),$$

an *answer* for $Q(\mathbf{x})$ to KB is a ground substitution θ for the variables $\mathbf{x}$ such that $KB \models Q^+(\mathbf{x}\theta)$ and $KB \not\models Q^-(\mathbf{x}\theta)$, where $Q^+(\mathbf{x})$ and $Q^-(\mathbf{x})$ are defined as follows:

$$Q^+(\mathbf{x}) = \bigvee_{i=1}^{n} \exists \mathbf{y}_i\, (\phi_{i,1}(\mathbf{x},\mathbf{y}_i) \wedge \cdots \wedge \phi_{i,l_i}(\mathbf{x},\mathbf{y}_i)) \text{ and}$$
$$Q^-(\mathbf{x}) = \bigvee_{i=1}^{n} \exists \mathbf{y}_i\, (\phi_{i,1}(\mathbf{x},\mathbf{y}_i) \wedge \cdots \wedge \phi_{i,l_i}(\mathbf{x},\mathbf{y}_i) \wedge (\phi_{i,l_i+1}(\mathbf{x},\mathbf{y}_i) \vee \cdots \vee \phi_{i,m_i}(\mathbf{x},\mathbf{y}_i)))\,.$$

Roughly, a ground substitution θ is an answer for $Q(\mathbf{x})$ to KB iff (i) θ is an answer for $Q^+(\mathbf{x})$ to KB, and (ii) θ is not an answer for $Q^-(\mathbf{x})$ to KB, where $Q^+(\mathbf{x})$ is the positive part of $Q(\mathbf{x})$, while $Q^-(\mathbf{x})$ is the positive part of $Q(\mathbf{x})$ combined with the complement of the negative one. Notice that both $Q^+(\mathbf{x})$ and $Q^-(\mathbf{x})$ are positive queries.

Realizing Semantic Web Search. Processing SW search queries Q is divided into

- an offline ontology reasoning step, where the TBox $\mathcal{T}$ of a SW knowledge base KB is compiled into the ABox $\mathcal{A}$ of KB via completing all semantic annotations of Web pages and objects by membership axioms entailed from KB, and
- an online reduction to standard Web search, where Q is transformed into standard Web search queries whose answers are used to construct the answer for Q.

In the offline ontology reasoning step, we check whether the SW knowledge base is satisfiable, and we compute the completion of all semantic annotations, that is, we augment the semantic annotations with all concept, role, and attribute membership axioms that can be derived (deductively in [12] and inductively here) from the semantic annotations and the ontology. In the online reduction to standard Web search, we decompose a given SW search query Q into a collection of standard Web search queries, of which the answers are then used to construct the answer for Q. These standard Web search queries are processed with existing search engines on the Web.

Note that our approach to SW search comes along with a ranking on Web pages and objects, called *ObjectRank*, which generalizes the standard PageRank ranking, and which can be computed by reduction to the computation of the PageRank ranking.

4 Inductive Offline Ontology Compilation

In this section, we propose to use inductive inference based on a notion of similarity as an alternative to deductive inference for offline ontology compilation in our SW search. Rather than obtaining the simple completion of a semantic annotation by adding all logically entailed membership axioms, we now obtain it by adding all inductively entailed membership axioms. Section 5 then summarizes the central advantages of this proposal, namely, an increased robustness due to the additional ability to handle inconsistencies, noise, and incompleteness.

Inductive Inference Based on Similarity Search. The *inductive inference* (or *classification*) problem here can be briefly described as follows. Given a SW knowledge base $KB = (\mathcal{T}, (\mathcal{A}_a)_{a \in \mathbf{P} \cup \mathbf{O}})$, a set of training individuals *TrExs* $\subseteq \mathbf{P} \cup \mathbf{O}$, a Web page or object a, and a property $Q(x)$, decide whether KB and *TrExs* inductively entail $Q(a)$. Here, (i) a *property* $Q(x)$ is either a concept membership $A(x)$, a role membership $P(x, b)$, or an attribute membership $U(x, v)$, where $A \in \mathbf{A}$, $P \in \mathbf{R}_A$, $U \in \mathbf{R}_D$, $b \in \mathbf{I}$, and $v \in \mathbf{V}$, and (ii) the notion of *inductive entailment* is defined using a notion of similarity between individuals as follows.

We now review the basics of the k-*nearest-neighbor* (k-NN) method applied to the SW context [7]. Informally, exploiting a notion of nearness, i.e., a similarity (or dissimilarity) measure [18], the most similar individual(s) to a given individual a to be classified can be selected and the classification of a can be decided based on their properties and proximity.

Formally, the method aims at inducing an approximation for a discrete-valued target hypothesis function $h\colon \mathit{IS} \to V$ from a space of individuals *IS* to a set of values $V = \{v_1, \ldots, v_s\}$, standing for the properties that have to be predicted. The approximation moves from the availability of training individuals $\mathit{TrExs} \subseteq \{x \in \mathit{IS} \mid \exists v \in V\colon\ h(x) = v\} \subseteq \mathit{IS}$, which is a subset of all prototypical individuals whose correct classification $h(\cdot)$ is known.

Let x_q be the query individual whose property is to be determined. Using a dissimilarity measure $d\colon\ \mathit{IS} \times \mathit{IS} \mapsto \mathbb{R}$, we select the set of the k-nearest training individuals (neighbors) of *TrExs* relative to x_q, denoted $\mathit{NN}(x_q) = \{x_1, \ldots, x_k\}$. Hence, the k-NN procedure approximates h for classifying x_q on the grounds of the values that h assumes for the neighbor training individuals in $\mathit{NN}(x_q)$. Precisely, the value is decided by means of a weighted majority voting procedure: it is the most *voted* value by the neighbor individuals in $\mathit{NN}(x_q)$ weighted by their similarity. The estimate of the hypothesis function for the query individual is as follows:

$$\hat{h}(x_q) = \operatorname{argmax}_{v \in V} \sum_{i=1}^{k} w_i \cdot \delta(v, h(x_i))\,, \tag{4}$$

where the indicator function δ returns 1 in case of matching arguments and 0 otherwise, and the weights w_i are determined by $w_i = 1 \,/\, d(x_i, x_q)$.

But this approximation determines a value that stands for one in a set of disjoint properties. Indeed, this is intended for simple settings with attribute-value representations [15]. In a multi-relational context, like with typical representations of the SW, this is no longer valid, since one deals with multiple properties, which are generally not implicitly disjoint. A further problem is related to the open-world assumption (OWA) generally adopted with SW representations; the absence of information of an individual relative to some query property should not be interpreted negatively, as in knowledge discovery from databases, where the closed-world assumption (CWA) is adopted; rather, this case should count as neutral (uncertain) information. Therefore, under the OWA, the multi-classification problem is transformed into a number of ternary problems (one per property), adopting $V = \{-1, 0, +1\}$ as the set of classification values relative to each query property Q, where the values denote explicitly membership ($+1$), non-membership (-1), and uncertainty (0) relative to Q.

So, inductive inference can be re-stated as follows: given $\mathit{KB} = (\mathcal{T}, (\mathcal{A}_a)_{a \in \mathbf{P} \cup \mathbf{O}})$, $\mathit{TrExs} \subseteq \mathit{IS} = \mathbf{P} \cup \mathbf{O}$, and a query property Q, find an approximation $\hat{h}_Q$ (on *IS*) of the hypothesis function h_Q, whose value $h_Q(x)$ for every training individual $x \in \mathit{TrExs}$ is as follows:

$$h_Q(x) = \begin{cases} +1 & \mathit{KB} \models Q(x) \\ -1 & \mathit{KB} \not\models Q(x),\ \mathit{KB} \models \neg Q(x) \\ 0 & \textit{otherwise.} \end{cases}$$

That is, the value of h_Q for the training individuals is determined by logical entailment. Alternatively, a mere look-up for the assertions $(\neg)Q(x)$ in $(\mathcal{A}_a)_{a \in \mathbf{P} \cup \mathbf{O}}$ could be considered, to simplify the inductive process, but also adding a further approximation.

Once the set of training individuals *TrExs* has been constructed, the inductive classification $\hat{h}_Q(x_q)$ of an individual x_q through the k-NN procedure is done via Eq. 4.

To assess the similarity between individuals, a totally semantic and language-independent family of dissimilarity measures is used [7]. They are based on the idea of comparing the semantics of the input individuals along a number of dimensions represented by a committee of concepts $\mathsf{F} = \{F_1, F_2, \ldots, F_m\}$, which stands as a context of discriminating *features* expressed in the considered DL; they are defined as follows [7]:

Definition 1 (family of measures). Let $KB = (\mathcal{T}, \mathcal{A} = (\mathcal{A}_a)_{a \in \mathbf{P} \cup \mathbf{O}})$ be a SW knowledge base. Given a set of concepts $\mathsf{F} = \{F_1, F_2, \ldots, F_m\}$, $m \geq 1$, weights $w_1, \ldots, w_m$, and $p > 0$, a family of dissimilarity functions $d_p^{\mathsf{F}} : \mathbf{P} \cup \mathbf{O} \times \mathbf{P} \cup \mathbf{O} \mapsto [0, 1]$ is defined by:

$$\forall a, b \in \mathbf{P} \cup \mathbf{O}: \quad d_p^{\mathsf{F}}(a, b) = \frac{1}{m} \left[\sum_{i=1}^{m} w_i \mid \delta_i(a, b) \mid^p \right]^{1/p},$$

where the dissimilarity function δ_i $(i \in \{1, \ldots, m\})$ is defined as follows:

$$\delta_i(a, b) = \begin{cases} 0 & (F_i(a) \in \mathcal{A} \wedge F_i(b) \in \mathcal{A}) \vee (\neg F_i(a) \in \mathcal{A} \wedge \neg F_i(b) \in \mathcal{A}) \\ 1 & (F_i(a) \in \mathcal{A} \wedge \neg F_i(b) \in \mathcal{A}) \vee (\neg F_i(a) \in \mathcal{A} \wedge F_i(b) \in \mathcal{A}) \\ \frac{1}{2} & \textit{otherwise}. \end{cases}$$

An alternative definition for the functions δ_i requires the logical entailment of the assertions $(\neg)F_i(x)$, rather than their simple ABox look-up; this makes the measure more accurate, but also more complex to compute. Moreover, using logical entailment, induction is done on top of deduction, thus making it a kind of completion of deduction.

The weights w_i in the family of measures should reflect the impact of the single feature F_i relative to the overall dissimilarity. This is determined by the quantity of information conveyed by the feature, which is measured in terms of its entropy. Namely, the probability of belonging to F_i may be quantified in terms of a measure of the extension of F_i relative to the whole domain of objects (relative to the canonical interpretation $\mathcal{I}$): $P_{F_i} = \mu(F_i^{\mathcal{I}})/\mu(\Delta^{\mathcal{I}})$. This can be roughly approximated by $|\{x \in \mathbf{P} \cup \mathbf{O} \mid F_i(x) \in \mathcal{A}\}| \,/\, |\mathbf{P} \cup \mathbf{O}|$. Hence, considering also the probability $P_{\neg F_i}$ related to its negation and the one related to the unclassified individuals (relative to F_i), denoted P_{U_i}, one obtains an entropic measure for the feature as follows:

$$H(F_i) = -\left[P_{F_i} \log(P_{F_i}) + P_{\neg F_i} \log(P_{\neg F_i}) + P_{U_i} \log(P_{U_i}) \right].$$

Alternatively, these weights may be based on the variance related to each feature [11]. Note that the measures strongly depend on F. Here, we make the assumption that the feature set F represents a sufficient number of (possibly redundant) features that are able to discriminate really different individuals. However, finding an optimal discriminating feature set may represent a preliminary learning task [10]. Experimentally, we obtained good results by using the set of all both primitive and defined concepts that occur in the knowledge base [7].

Measuring the Likelihood of an Answer. The inductive inference made by the procedure presented above is not guaranteed to be deductively valid. Indeed, it naturally yields a certain degree of uncertainty. So, from a more general perspective, the main idea behind the above inductive inference for SW search is closely related to the idea of using probabilistic ontologies to increase the precision and the recall of querying databases and of information retrieval in general. However, rather than learning probabilistic ontologies from data, representing them, and reasoning with them, we directly use the data in the inductive inference step.

To measure the likelihood of the inductive decision (x_q has the query property Q denoted by the value v, maximizing the argmax argument in Eq. 4), given $NN(x_q) = \{x_1, \ldots, x_k\}$, the quantity that determined the decision should be normalized:

$$l(Q(x_q) = v|NN(x_q)) = \frac{\sum_{i=1}^{k} w_i \cdot \delta(v, h_Q(x_i))}{\sum_{v' \in V} \sum_{i=1}^{k} w_i \cdot \delta(v', h_Q(x_i))} \,. \tag{5}$$

Hence, the likelihood of $Q(x_q)$ corresponds to the case when $v = +1$. The computed likelihood can be used for building a probabilistic ABox, which is a collection of pairs, each consisting of a classical ABox axiom and a probability value $(Q(x_q), \ell)$.

5 Towards Robustness of Semantic Web Search

In this section, we illustrate the main advantages of using inductive rather than deductive inference in SW search. In detail, inductive inference can better handle cases of inconsistency, noise, and incompleteness in SW knowledge bases than deductive inference. These cases are all very likely to occur when knowledge bases are fed by multiple heterogeneous sources and maintained on distributed peers on the Web.

Inconsistency and Noise. Since our inductive inference is triggered by factual knowledge (assertions concerning prototypical neighboring individuals in the presented algorithm), it can provide a correct classification even in the case of knowledge bases that are inconsistent due to wrong assertions. This is illustrated by the following example.

Example 1. Consider the following DL knowledge base $KB = (\mathcal{T}, \mathcal{A})$:

$$\begin{aligned}
\mathcal{T} = \{\, & \mathit{Professor} \equiv \mathit{Graduate} \sqcap \exists \mathit{worksAt.University} \sqcap \exists \mathit{teaches.Course};\\
& \mathit{Researcher} \equiv \mathit{Graduate} \sqcap \exists \mathit{worksAt.Institution} \sqcap \neg\exists \mathit{teaches.Course}; \ldots\};\\
\mathcal{A} = \{\, & \mathit{Professor}(\mathit{FRANZ});\ \mathit{teaches}(\mathit{FRANZ}, \mathit{COURSE00});\ \mathit{Professor}(\mathit{JIM});\\
& \mathit{teaches}(\mathit{JIM}, \mathit{COURSE01});\ \mathit{Professor}(\mathit{FLO});\ \mathit{teaches}(\mathit{FLO}, \mathit{COURSE02});\\
& \mathit{Researcher}(\mathit{NICK});\ \mathit{Researcher}(\mathit{ANN});\ \mathit{teaches}(\mathit{NICK}, \mathit{COURSE03}); \ldots\}.
\end{aligned}$$

Suppose that *NICK* is actually a *Professor*, and he is indeed asserted to be a lecturer of some course. However, by mistake, he is also asserted to be a *Researcher*, and because of the definition of *Researcher*, he cannot teach any course. Hence, this KB is inconsistent, and thus logically entails anything under deductive inference. Under inductive inference as described above, on the contrary, *Professor*(*NICK*) is inductively entailed, because of the similarity of *NICK* to other individuals (*FRANZ*, *JIM*, and *FLO*) known to be instances of *Professor*.

In the former case, noisy assertions may be pinpointed as the very source of inconsistency. An even trickier case is when noisy assertions do not produce any inconsistency, but are indeed wrong relative to the intended true models. Inductive reasoning can also provide a correct classification in such a presence of incorrect assertions on concepts, roles, and/or attributes relative to the intended true models.

Example 2. Consider the knowledge base $KB = (\mathcal{T}', \mathcal{A})$, where the ABox $\mathcal{A}$ does not change relative to Example 1 and the TBox $\mathcal{T}'$ is obtained from $\mathcal{T}$ of Example 1 by simplifying the definition of *Researcher* dropping the negative restriction:

$$Researcher \equiv Graduate \sqcap \exists worksAt.Institution\,.$$

Again, suppose *NICK* is actually a *Professor*, but by mistake asserted to be a *Researcher*. Due to the new definition of *Researcher*, there is no inconsistency. But by deductive reasoning, *NICK* turns out to be a *Researcher*, while by inductive reasoning, the returned classification result is that *NICK* is an instance of *Professor*, as above, because the most similar individuals (*FRANZ*, *JIM*, and *FLO*) are all instances of *Professor*.

Incompleteness. Clearly, inductive reasoning may also be able to give a correct classification in the presence of incompleteness in a knowledge base. That is, inductive reasoning is not necessarily deductively valid, and can suggest new knowledge.

Example 3. Consider yet another slightly different knowledge base $KB = (\mathcal{T}', \mathcal{A}')$, where the TBox $\mathcal{T}'$ is as in Example 2 and the ABox $\mathcal{A}'$ is obtained from the ABox $\mathcal{A}$ of Example 1 by removing the axiom *Researcher*(*NICK*). Then, the resulting knowledge base is neither inconsistent nor noisy, but we know less about *NICK*. Nonetheless, by the same line of argumentation as in the previous examples, *NICK* is inductively entailed to be an instance of *Professor*.

6 Implementation and Experiments

We now describe our prototype implementation for a semantic desktop search engine. We also report on the running time of the online query processing step, and the precision and the recall under deductively and inductively completed semantic annotations. For further experimental results (especially for deductive SW search), we refer to [12].

Implementation. We have implemented a prototype for a semantic desktop search engine, which is based on an offline inference step for generating the completed semantic annotation for every considered resource. We have implemented both a deductive and an inductive version of the offline inference step. The deductive one uses PELLET (http://www.mindswap.org), while the inductive one is based on the k-NN technique, integrated with an entropic measure, as proposed in Section 4. Specifically, each individual i of a SW knowledge base is classified relative to all atomic concepts and all restrictions $\exists R^{-}.\{i\}$ with roles R. The parameter k was set to $\log(|\mathbf{P} \cup \mathbf{O}|)$. The simpler distances d_1^{F} were employed, using all the atomic concepts in the knowledge base for determining the set F. The implementation also realizes an online query processing step, which reduces semantic desktop search queries to a collection of standard desktop search queries over resources and their completed semantic annotations.

Table 1. Overall time (ms) used for online query processing

	Onto-logy	**Query**	No. Results	Overall Time (ms)
1	FSM	$\exists y\,(Transition(y) \wedge target(y, finalState) \wedge source(y,x) \wedge State(x))$	2	6
2	FSM	$\exists y,z\,(State(z) \wedge State(y) \wedge Transition(x) \wedge source(x,y) \wedge target(x,z))$	11	11
3	FSM	$\exists y\,(State(y) \wedge entry(y, accountNameIndexEntryAction) \wedge Transition(x) \wedge target(x,y))$	1	6
4	FSM	$\exists y\,((Initial(y) \wedge Transition(x) \wedge source(x,y)) \vee (Transition(x) \wedge target(x,y) \wedge Final(y)))$	3	13
5	FSM	$\exists y\,(State(y) \wedge not\ entry(y, accountNameIndexEntryAction) \wedge Transition(x) \wedge target(x,y))$	10	7
6	SWM	$\exists y\,(Developer(y) \wedge hasCountry(y, usa) \wedge Model(x) \wedge hasDeveloper(x,y))$	36	15
7	SWM	$Model(x) \wedge not\ hasdomain(x, river) \wedge not\ hasdomain(x, lake) \wedge not\ \exists y\,(Developer(y) \wedge hasDeveloper(x,y) \wedge hasCountry(y, usa))$	5	16
8	SWM	$\exists x\,(Model(y) \wedge not\ hasModelDimension(y, two\_Dimensional) \wedge hasDeveloper(y,x) \wedge University(x))$	1	10
9	SWM	$(Numerical(x) \wedge hasModelDimension(x, two\_Dimensional) \wedge hasAvailability(x, public)) \vee (Numerical(x) \wedge hasModelDimension(x, three\_Dimensional) \wedge hasAvailability(x, commercial))$	9	12
10	SWM	$(Model(x) \wedge not\ hasModelDimension(x, three\_Dimensional) \wedge hasDomain(x, estuary) \wedge not\ hasDomain(x, lake) \wedge not\ \exists y\,(Developer(y) \wedge hasDeveloper(x,y) \wedge hasCountry(y, italy))) \vee (Model(x) \wedge not\ hasModelDimension(x, two\_Dimensional) \wedge hasDomain(x, estuary) \wedge not\ hasDomain(x, channelNetworks) \wedge not\ \exists y\,(Developer(y) \wedge hasDeveloper(x,y) \wedge hasCountry(y, uk)))$	12	14

Efficiency of Online Query Processing. We provide experimental results on the running time of online query processing relative to the FINITE-STATE-MACHINE (FSM) and the SURFACE-WATER-MODEL (SWM) ontologies from the Protégé Ontology Library[1]. They are given in Table 1, which shows the overall time used by our system for processing 10 different search queries on completed annotations relative to the two ontologies FSM and SWM. For example, Query (1) asks for all states that are directly leading to a final state, while Query (6) asks for all models developed by an American. Observe that this overall time (for the decomposition of the query and the composition of the query result) is very small (at most 16 ms in the worst case). Table 1 also shows the different numbers of returned resources.

Precision and Recall of Inductive SW Search. We finally give an experimental comparison between SW search under inductive and under deductive inference. We do this by providing the precision and the recall of the latter vs. the former (where the precision and the recall under deductive inference are both 1). Our experimental results with queries relative to the two ontologies FSM and SWM are summarized in Table 2. For example, Query (8) asks for all transitions having no target state, while Query (16) asks for all numerical models having either the domain "lake" and public availability, or the domain "coastalArea" and commercial availability. The experimental results in Table 2 essentially show that the answer sets under inductive reasoning are very close to the ones under deductive reasoning.

[1] http://protegewiki.stanford.edu/index.php/Protege_Ontology_Library

Table 2. Precision and recall of inductive vs. deductive SW search

	Ontology	Query	No. Results Deduction	No. Results Induction	No. Correct Results Induction	**Precision Induction**	**Recall Induction**
1	FSM	$State(x)$	11	11	11	1	1
2	FSM	$StateMachineElement(x)$	37	37	37	1	1
3	FSM	$Composite(x) \wedge hasStateMachineElement(x, accountDetails)$	1	1	1	1	1
4	FSM	$State(y) \wedge StateMachineElement(x) \wedge hasStateMachineElement(x, y)$	3	3	3	1	1
5	FSM	$Action(x) \vee Guard(x)$	12	12	12	1	1
6	FSM	$\exists y, z\ (State(y) \wedge State(z) \wedge Transition(x) \wedge source(x, y) \wedge target(x, z))$	11	2	2	1	0.18
7	FSM	$StateMachineElement(x) \wedge not\ \exists y\ (StateMachineElement(y) \wedge hasStateMachineElement(x, y))$	34	34	34	1	1
8	FSM	$Transition(x) \wedge not\ \exists y\ (State(y) \wedge target(x, y))$	0	5	0	0	1
9	FSM	$\exists y\ (StateMachineElement(x) \wedge not\ hasStateMachineElement(x, accountDetails) \wedge hasStateMachineElement(x, y) \wedge State(y))$	2	2	2	1	1
10	SWM	$Model(x)$	56	56	56	1	1
11	SWM	$Mathematical(x)$	64	64	64	1	1
12	SWM	$Model(x) \wedge hasDomain(x, lake) \wedge hasDomain(x, river)$	9	9	9	1	1
13	SWM	$Model(x) \wedge not\ \exists y\ (Availability(y) \wedge hasAvailability(x, y))$	11	11	11	1	1
14	SWM	$Model(x) \wedge hasDomain(x, river) \wedge not\ hasAvailability(x, public)$	2	8	0	0	0
15	SWM	$\exists y\ (Model(x) \wedge hasDeveloper(x, y) \wedge University(y))$	1	1	1	1	1
16	SWM	$Numerical(x) \wedge hasDomain(x, lake) \wedge hasAvailability(x, public) \vee Numerical(x) \wedge hasDomain(x, coastalArea) \wedge hasAvailability(x, commercial)$	12	9	9	1	0.75

7 Summary and Outlook

In this paper, we have presented a combination of our approach to SW search in [12] with inductive reasoning based on similarity search [18] for retrieving the resources that likely have a query property [7]. As crucial advantages, the new approach to SW search has an increased robustness, as it allows for handling inconsistencies, noise, and incompleteness in SW knowledge bases, which are all very likely in distributed and heterogeneous environments, such as the Web. In particular, inductive reasoning allows to infer (from training individuals) new knowledge, which is not logically deducible. We have also reported on a prototype implementation and positive experimental results on (1) the running time of the online query processing step, and (2) the precision and the recall of the new (inductive) approach compared to the previous (deductive) one.

From a more general perspective, the main idea behind this paper is closely related to the idea of using probabilistic ontologies to increase the precision and the recall of querying databases and of information retrieval in general. But, rather than learning probabilistic ontologies from data, representing them, and reasoning with them during query processing, we directly use the data in the inductive inference step.

In the future, we aim especially at extending the desktop implementation to a real Web implementation, using existing search engines, such as Google. Another interesting topic is to explore how search expressions formulated in plain natural language can be translated into the ontological (unions of conjunctive) queries of our approach.

Acknowledgments. This work was supported by the European Research Council under the EU's 7th Framework Programme (FP7/2007-2013)/ERC grant 246858 – DIADEM, the EPSRC grant EP/E010865/1 "Schema Mappings and Automated Services for Data Integration", by a Yahoo! Research Fellowship, and by the DFG under the Heisenberg Programme. Georg Gottlob, whose work was partially carried out at the Oxford-Man Institute of Quantitative Finance, also gratefully acknowledges support from the Royal Society as the holder of a Royal Society-Wolfson Research Merit Award.

References

1. Baader, F., Calvanese, D., McGuinness, D.L., Nardi, D., Patel-Schneider, P.F. (eds.): The Description Logic Handbook. Cambridge University Press, Cambridge (2003)
2. Bao, J., Kendall, E.F., McGuinness, D.L., Wallace, E.K.: OWL2 Web ontology language: Quick reference guide (2008), http://www.w3.org/TR/owl2-quick-reference/
3. Berners-Lee, T., Hendler, J., Lassila, O.: The Semantic Web. Sci. Am. 284, 34–43 (2001)
4. Brin, S., Page, L.: The anatomy of a large-scale hypertextual web search engine. Comput. Netw. 30(1-7), 107–117 (1998)
5. Buitelaar, P., Cimiano, P.: Ontology Learning and Population: Bridging the Gap Between Text and Knowledge. IOS Press, Amsterdam (2008)
6. Chirita, P.-A., Costache, S., Nejdl, W., Handschuh, S.: P-TAG: Large scale automatic generation of personalized annotation TAGs for the Web. In: Proc. WWW 2007, pp. 845–854. ACM Press, New York (2007)
7. d'Amato, C., Fanizzi, N., Esposito, F.: Query answering and ontology population: An inductive approach. In: Bechhofer, S., Hauswirth, M., Hoffmann, J., Koubarakis, M. (eds.) ESWC 2008. LNCS, vol. 5021, pp. 288–302. Springer, Heidelberg (2008)
8. Ding, L., Finin, T.W., Joshi, A., Peng, Y., Pan, R., Reddivari, P.: Search on the Semantic Web. IEEE Computer 38(10), 62–69 (2005)
9. Ding, L., Pan, R., Finin, T.W., Joshi, A., Peng, Y., Kolari, P.: Finding and ranking knowledge on the Semantic Web. In: Gil, Y., Motta, E., Benjamins, V.R., Musen, M.A. (eds.) ISWC 2005. LNCS, vol. 3729, pp. 156–170. Springer, Heidelberg (2005)
10. Fanizzi, N., d'Amato, C., Esposito, F.: Induction of classifiers through non-parametric methods for approximate classification and retrieval with ontologies. International Journal of Semantic Computing 2(3), 403–423 (2008)
11. Fanizzi, N., d'Amato, C., Esposito, F.: Metric-based stochastic conceptual clustering for ontologies. Inform. Syst. 34(8), 725–739 (2009)
12. Fazzinga, B., Gianforme, G., Gottlob, G., Lukasiewicz, T.: Semantic Web search based on ontological conjunctive queries. In: Link, S., Prade, H. (eds.) FoIKS 2010. LNCS, vol. 5956, pp. 153–172. Springer, Heidelberg (2010)
13. Fazzinga, B., Lukasiewicz, T.: Semantic search on the Web. Semantic Web — Interoperability, Usability, Applicability (forthcoming)
14. Guha, R.V., McCool, R., Miller, E.: Semantic search. In: Proc. WWW 2003, pp. 700–709. ACM Press, New York (2003)
15. Hastie, T., Tibshirani, R., Friedman, J.: The Elements of Statistical Learning – Data Mining, Inference, and Prediction. Springer, Heidelberg (2001)
16. Horrocks, I., Patel-Schneider, P.F., van Harmelen, F.: From $\mathcal{SHIQ}$ and RDF to OWL: The making of a Web ontology language. J. Web. Sem. 1(1), 7–26 (2003)
17. Lei, Y., Uren, V.S., Motta, E.: SemSearch: A search engine for the Semantic Web. In: Staab, S., Svátek, V. (eds.) EKAW 2006. LNCS (LNAI), vol. 4248, pp. 238–245. Springer, Heidelberg (2006)
18. Zezula, P., Amato, G., Dohnal, V., Batko, M.: Similarity Search — The Metric Space Approach. In: Advances in Database Systems, vol. 32. Springer, Heidelberg (2006)
19. W3C. OWL Web ontology language overview, 2004. W3C Recommendation (February 10, 2004), http://www.w3.org/TR/2004/REC-owl-features-20040210/

Evaluating Trust from Past Assessments with Imprecise Probabilities: Comparing Two Approaches

Sebastien Destercke

INRA/CIRAD, UMR1208, 2 place P. Viala, F-34060 Montpellier cedex 1, France
sebastien.destercke@supagro.inra.fr

Abstract. In this paper, we consider a trust system where the trust in an agent is evaluated from past assessments made by other agents. We consider that trust is evaluated by values given on a finite scale. To model the agent trustworthiness, we propose to build imprecise probabilistic models from these assessments. More precisely, we propose to derive probability intervals (i.e., bounds on singletons) using two different approaches: Goodman's multinomial confidence regions and the imprecise Dirichlet model (IDM). We then use these models for two purposes: (1) evaluating the chances that a future assessments will take particular values, and (2) computing an interval summarizing the agent trustworthiness, eventually fuzzyfying this interval by letting the confidence value vary over the unit interval. We also give some elements of comparison between the two approaches.

Keywords: trustworthiness, probability intervals, expectations bounds.

1 Introduction

The notion of trust and how to evaluate it has taken more and more importance in computer science with the emergence of the semantic web (particularly in the field of e-commerce or security) and multi-agent systems. Once done, trust evaluation can be used to compare agents or to make an absolute judgement whether an agent can be trusted. To perform such an evaluation, many trust systems have been developed in the past years (see Sabater and Sierra [1] for a review).

Note that the notion of trust as well as the information used to evaluate it can take many forms [2]. One can differentiate between individual-level and system-level trusts, the former concerning the trust one has in a particular agent, while the latter concerns the overall system and the way it ensures that no one will be able to use the system in a selfish way (i.e., to its own profit). The collected information about the trustworthiness of an agent may be direct (coming from past transactions one has done with this agent) or indirect (provided by third-party agents), and when it is indirect, it may be a direct evaluation of the agent reputation or information concerning some of its characteristics.

In this paper, we consider that the information whether an agent (called here the *trustee*) can be trusted or not is given in the form of past evaluations provided

A. Deshpande and A. Hunter (Eds.): SUM 2010, LNAI 6379, pp. 151–162, 2010.

by other agents on a numerical scale $\mathcal{X} = \{-n, \ldots, -1, 0, 1, \ldots, n\}$ ranging from $-n$ to n. In this bipolar scale, a rate of $-n$ means that the trustee is totally untrusted, while n means that it is totally trusted, 0 standing for neutral. For sake of clarity, we will also refer to the elements of $\mathcal{X}$ as $\mathcal{X} = \{x_1, \ldots, x_{2n+1}\}$. Using the classification proposed in Ramchurn *et al.* [2], we are working here with indirect information concerning the individual-level trust of the system and the reputation of an agent.

In [3], Ben Naim and Prade discuss the interest of summarising past evaluations by intervals, as they are more informative than mere mean values and precise points (as interval imprecision is a valuable information reflecting our quantity of knowledge), and are far easier to read than the whole set of evaluations. Eventually, the summarising interval can be reduced to a single evaluation, should such precision be needed.

Consider a counting vector $\Theta = \{\theta_1, \ldots, \theta_{2n+1}\}$ where θ_i is the number of times an agent has given x_i as its evaluation of the *trustee* truthfulness, and $\hat{\theta} = \sum_i \theta_i$ the total number of evaluations. The problem we consider here is how to summarise the information provided by this counting vector in an interval representation describing the past behaviour of the trustee. To do so, we propose to use an imprecise probabilistic model well fit to represent uncertainty on multinomial data (here, the ratings), namely probability intervals [4], and to use the notion of lower and upper expectations to compute the summarising interval. As we shall see, this simple model allows for efficient computations of a summarising interval.

Section 2 recalls some basics of probability intervals and presents the two uncertainty models built from the counting vector Θ. Section 3 then details how a summarising interval can be built from these models. It also provides some elements of comparison by exploring the properties of these intervals with respect to Θ and the possibilities of building fuzzy interval as summary rather than a single one.

2 The Model

Let us first recall some elements about probability intervals, before studying how they can be derived from Θ by using confidence regions.

2.1 Probability Intervals

Probability intervals as uncertainty models have been studied extensively by De Campos et al. [4]. Probability intervals on a space $\mathcal{X} = \{x_1, \ldots, x_{2n+1}\}$ are defined as a set $L = \{[l_i, u_i] | i = 1, \ldots, 2n+1\}$ of intervals such that $l_i \leq p(x_i) \leq u_i$, where $p(x_i)$ is the unknown probability of element x_i. In this paper, built probability intervals satisfy a number of reasonable conditions usually required to work with this uncertainty representation, namely

$$\sum_{i=1}^{2n+1} l_i \leq 1 \leq \sum_{i=1}^{2n+1} u_i, \tag{1}$$

and for $i = 1, \ldots, 2n+1$

$$u_i + \sum_{\substack{j \in \{1,\ldots,2n+1\} \\ j \neq i}} l_j \leq 1 \quad ; \quad l_i + \sum_{\substack{j \in \{1,\ldots,2n+1\} \\ j \neq i}} u_j \geq 1. \tag{2}$$

If probability intervals satisfy these conditions, then they induce a set of probability measures $\mathcal{P}_L$ such that

$$\mathcal{P}_L = \{p \in \mathbb{P}_{\mathcal{X}} | i = 1, \ldots, 2n+1, \; l_i \leq p(x_i) \leq u_i\},$$

with $\mathbb{P}_{\mathcal{X}}$ the set of all probability measures over $\mathcal{X}$. From $\mathcal{P}_L$can be computed lower and upper probabilities on any event A, respectively as $\underline{P}(A) = \inf_{p \in \mathcal{P}_L} P(A)$ and $\overline{P}(A) = \sup_{p \in \mathcal{P}_L} P(A)$. In the case of probability intervals, their computations are facilitated, since we have [4]

$$\underline{P}(A) = \max(\sum_{x_i \in A} l_i, 1 - \sum_{x_i \notin A} u_i) \quad ; \quad \overline{P}(A) = \min(\sum_{x_i \in A} u_i, 1 - \sum_{x_i \notin A} l_i)$$

The question is now how probability intervals can be derived from the counting vector Θ of past evaluations. Had we an infinite number of evaluations at our disposal, it would be reasonable to adopt as a model of the trustee trustworthiness the probability distribution p_∞ corresponding to limiting frequencies. Therefore, we should ask probability intervals to tend towards such frequencies, i.e.,

$$l_i \xrightarrow[\hat{\theta} \to \infty]{} p_\infty(x_i) \quad ; \quad u_i \xrightarrow[\hat{\theta} \to \infty]{} p_\infty(x_i).$$

In practice there may only be a few evaluations available, and in any case a finite quantity of them. Therefore, the chosen uncertainty representation should both tend towards the limiting frequencies and reflect our potential lack of information.

We propose two approaches to build such representations. The first use Goodman's multinomial confidence intervals [5], while the second use the popular Imprecise Dirichlet Model (IDM for short) [6]. The two approaches as a basis of trustee evaluation are then compared in Section 3.

2.2 Building Intervals from Θ: First Approach

In this first approach, we propose to use Goodman's multinomial confidence intervals [5] as our representation. Given a space $\mathcal{X}$ and a counting vector Θ, Goodmans intervals $[l_i^{G,\alpha}, u_i^{G,\alpha}]$ with confidence level α read, for $i = 1, \ldots, 2n+1$,

$$l_i^{G,\alpha} = \frac{b + 2\theta_i - \sqrt{\Delta_i^\alpha}}{2(\hat{\theta} + b)}, \qquad u_i^{G,\alpha} = \frac{b + 2\theta_i + \sqrt{\Delta_i^\alpha}}{2(\hat{\theta} + b)}, \tag{3}$$

where b is the quantile of order $1 - {}^{(1-\alpha)}/_{(2n+1)}$ of the chi-square distribution with one degree of freedom and where

$$\Delta_i^\alpha = b \left(b + \frac{4\theta_i(\hat{\theta} - \theta_i)}{\hat{\theta}} \right).$$

Note that b is an increasing function of α and n, meaning that confidence interval imprecision increases as α increases and as n (the number of possibilities) increases. These probability intervals satisfy Conditions (1) and (2). They tend towards limiting frequencies and the distance between l_i and u_i decreases as more information is collected (i.e. $u_i - l_i$ is a decreasing function of θ_i and $\hat{\theta}$). Also, they are very simple to compute, since only Θ is needed to estimate them. We will denote by $L^{G,\alpha}$ the obtained probability intervals and by $\mathcal{P}_L^{G,\alpha}$ the induced probability set.

Example 1. Consider a space $\mathcal{X} = \{-2, -1, 0, 1, 2\}$ containing 5 possible values. The following counting vector $\Theta = (0, 9, 13, 11, 17)$ summarises the various evaluations given by different agents. The probability intervals obtained with a confidence level $\alpha = 0.95$ are summarised in Table 1

Table 1. Example 1 probability intervals

	x_1	x_2	x_3	x_4	x_5
$u_i^{G,0.95}$	0.117	0.354	0.441	0.398	0.522
$l_i^{G,0.95}$	0	0.081	0.135	0.107	0.196

2.3 Building Intervals from Θ: Second Approach

The second approach we propose consists in using the IDM [6] to build the confidence intervals. The IDM basically extends the classical multinomial Dirichlet model by considering all Dirichlet distributions as the initial set of prior distributions. Intervals extracted from the IDM depend on a hyperparameter $s \geq 0$ that determines the influence of prior information on the posterior information. In the IDM, the value s can be seen as a way to settle the speed of convergence of probability intervals to limiting frequencies p_∞, this speed decreasing when s value increases. An often suggested interpretation for s is that it represents the number of "unseen" observations, and on most applications, $s \in \{1, 2\}$.

Given a space $\mathcal{X}$, a counting vector Θ and a positive value s, intervals $[l_i^{I,s}, u_i^{I,s}]$ resulting from the use of the IDM read, for $i = 1, \ldots, 2n+1$,

$$l_i^{I,s} = \frac{\theta_i}{\hat{\theta} + s}, \qquad u_i^{I,s} = \frac{\theta_i + s}{\hat{\theta} + s}. \tag{4}$$

As for Goodman's interval, their computation only requires to know Θ, and the distance $u_i - l_i$ decreases as more information is collected. We will denote by $L^{I,s}$ the obtained probability intervals and by $\mathcal{P}_L^{I,s}$ the induced probability set.

Example 2. Consider the space and counting vector of Example 1. The probability intervals obtained with the IDM and a value $s = 2$ are summarised in Table 2.

Table 2. Example 2 probability intervals

	x_1	x_2	x_3	x_4	x_5
$u_i^{I,2}$	0.038	0.212	0288	0.25	0.365
$l_i^{I,2}$	0	0.173	0.25	0.212	0.327

As we can see, these intervals are much narrower than the ones obtained in Example 1. This indicates that small values of s may be unwarranted in the current application (as an agent would be most of the time unwilling to make precise inference from a small number of evaluations). Note that it may be difficult to obtain a general result relating the interval imprecision obtained by the two approaches, since the difference $u_i^{I,s} - l_i^{I,s}$ does not depend on the θ_i, while the difference $u_i^{G,\alpha} - l_i^{G,\alpha}$ does.

Note that in both approaches, one can interpret the built intervals L, and the associated probability set $\mathcal{P}_L$ as a predictive model providing information about the next possible evaluations. In particular, lower and upper probabilities of an event A gives an interval $[\underline{P}(A), \overline{P}(A)]$ characterising our uncertainty about wether the next evaluation will fall in the set A.

3 Summarising Interval

In the first part of this section, we consider that we work with a fixed confidence level α (in the first approach) or with a fixed hyper-parameter s (in the second approach), for sake of clarity. These assumptions will be relaxed in the last subsection.

3.1 Lower and Upper Expectations

Let us first recall some elements about the notions of lower and upper expectations. Given a probability set $\mathcal{P}$ defined over a domain $\mathcal{X}$ and a real-valued bounded function $f : \mathcal{X} \to \mathbb{R}$, one can compute the lower and upper expectations of f, $\underline{E}_{\mathcal{P}}(f)$ and $\overline{E}_{\mathcal{P}}(f)$ as

$$\underline{E}_{\mathcal{P}}(f) = \inf_{p \in \mathcal{P}} E_p(f), \qquad \overline{E}_{\mathcal{P}}(f) = \sup_{p \in \mathcal{P}} E_p(f),$$

with $E_p(f)$ the expected value of f with respect to probability distribution p. Lower and upper expectations are dual, in the sense that $\underline{E}(f) = -\overline{E}(-f)$, and have the property that if a constant value μ is added to f, $\underline{E}(f+\mu) = \underline{E}(f) + \mu$ and $\overline{E}(f+\mu) = \overline{E}(f) + \mu$.

When the lower (resp. upper) probabilities of a credal set $\mathcal{P}$ satisfies the property of 2-monotonicity (resp. 2-alternance), that is when, for any two events $A, B \subseteq \mathcal{X}$, we have $\underline{P}(A) + \underline{P}(B) \leq \underline{P}(A \cup B) + \underline{P}(A \cap B)$ (resp. $\overline{P}(A) + \overline{P}(B) \geq \overline{P}(A \cup B) + \overline{P}(A \cap B)$), one can use the Choquet integral [7] to evaluate the lower and upper expectations. Consider a positive bounded function[1] f. If we denote by $()$ a reordering of elements of $\mathcal{X}$ such that $f(x_{(1)}) \leq \ldots \leq f(x_{(2n+1)})$, Choquet integrals giving lower and upper expectations are given by

$$\underline{E}(f) = \sum_{i=1}^{N} (f(x_{(i)}) - f(x_{(i-1)})\underline{P}(A_{(i)}), \tag{5}$$

$$\overline{E}(f) = \sum_{i=1}^{N} (f(x_{(i)}) - f(x_{(i-1)})\overline{P}(A_{(i)}), \tag{6}$$

with $f(x_{(0)}) = 0$ and $A_{(i)} = \{x_{(i)}, \ldots, x_{(N)}\}$. In Walley's [8] behavioural interpretation of lower and upper expectations, $\underline{E}(f)$ represents the maximum buying price an agent would pay for a gamble whose gains are represented by f, and $\overline{E}(f)$ the minimum selling price an agent would be ready to accept for the gamble f.

3.2 Expectation Bounds as a Summarising Interval

Let us now come back to our trust evaluation problem, and consider the first approach. Information about the trustee are given by probability intervals $L^{G,\alpha}$ resulting from the counting vector Θ and inducing a probability set $\mathcal{P}_L^{G,\alpha}$. It is known [4] that probability intervals induce 2-monotone and 2-alternating lower and upper probabilities.

Given this information $L^{G,\alpha}$, we propose to summarise the trustworthiness of the trustee as the interval given by lower and upper expectations of a function f such that $f(x_1) = -n, f(x_2) = -n+1, \ldots, f(x_{n+1}) = 0, \ldots, f(x_{2n+1}) = n$ with respect to the probability set $\mathcal{P}_L^{G,\alpha}$. $\underline{E}^{G,\alpha}(f)$ can then be interpreted as the maximal price an agent would be ready to pay to be in interaction with the trustee, while $\overline{E}^{G,\alpha}(f)$ can be interpreted as the minimal price an agent would be ready to accept for being forbidden to interact with the trustee. Algorithm 1 provides an easy way to compute lower and upper expectations.

Algorithm 1 uses the facts that function f values are always rank-ordered in the same way and that the difference of two consecutive values of f is 1. Therefore, Equations (5) and (6) reduce to sums of lower and upper probabilities in this particular case. The adaptation of Algorithm 1 to the second approach is straightforward, since it consists of replacing $l_i^{G,\alpha}$, $u_i^{G,\alpha}$ with $l_i^{I,s}, u_i^{I,s}$. In this latter case, the resulting interval will be denoted by $I^{I,s} := [\underline{E}^{I,s}(f), \overline{E}^{I,s}(f)]$.

Example 3. The summarising intervals corresponding to interval probabilities of Examples 1 and 2 are

$$I^{G,0.95} = [\underline{E}^{G,0.95}, \overline{E}^{G,0.95}] = [-0.09, 1.225]$$

$$I^{I,2} = [\underline{E}^{I,2}, \overline{E}^{I,2}] = [0.615, 0.769]$$

[1] Note that any bounded function f can be made positive by adding a suitable constant to it.

Algorithm 1. Algorithm giving summarising interval

Input: Θ,α
Output: $I^{G,\alpha} = [\underline{E}^{G,\alpha}(f), \overline{E}^{G,\alpha}(f)]$
$\underline{E}^{G,\alpha}(f) = 0, \overline{E}^{G,\alpha}(f) = 0$;
Evaluate $\hat{\theta} = \sum_{i=1}^{2n+1} \theta_i$;
for $i = 1, \ldots, 2n+1$ **do**
  Evaluate $l_i^{G,\alpha}$ (Eq.(3));
  Evaluate $u_i^{G,\alpha}$ (Eq.(3)) ;
for $i = 1, \ldots, 2n+1$ **do**
  if $i == 1$ **then**
    $\underline{E}^{G,\alpha}(f) = \underline{E}^{G,\alpha}(f) + 1$;
    $\overline{E}^{G,\alpha}(f) = \overline{E}^{G,\alpha}(f) + 1$;
  else
    $\underline{E}^{G,\alpha}(f) = \underline{E}^{G,\alpha}(f) + \max(\sum_{k=i}^{2n+1} l_k^{G,\alpha}, 1 - \sum_{k=1}^{i-1} u_k^{G,\alpha})$;
    $\overline{E}^{G,\alpha}(f) = \overline{E}^{G,\alpha}(f) + \min(\sum_{k=i}^{2n+1} u_k^{G,\alpha}, 1 - \sum_{k=1}^{i-1} l_k^{G,\alpha})$;
$\underline{E}^{G,\alpha}(f) = \underline{E}^{G,\alpha}(f) - (n+1)$;
$\overline{E}^{G,\alpha}(f) = \overline{E}^{G,\alpha}(f) - (n+1)$;

3.3 Some Properties

Let us now study some of the properties of each summarising intervals. The first property, satisfied by the two approaches, show that two similar evaluation profiles (in the sense that empirical frequencies are equal) with different amount of information (quantity of evaluations) give coherent summarising intervals, in the sense that the interval obtained with a greater amount of evaluations is included in the one obtained with less evaluations.

Proposition 1. *Let Θ and Θ' be two counting vectors with $\Theta' = \beta\Theta$, $\beta > 1$. Then, given a confidence value α or a hyper-parameter s, we have*

$$I^{G',\alpha} \subset I^{G,\alpha} \quad \textit{and} \quad I^{I',s} \subset I^{I,s},$$

with $I^{G',\alpha}$, $I^{I',s}$ the summarising intervals obtained from Θ', and $I^{I,s}$, $I^{G,\alpha}$ the summarising intervals obtained from Θ.

Proof. We will only prove the inclusion for the first approach, the proof for the second being similar. $\Theta' = \beta\Theta$ implies that for $i = 1, \ldots, 2n+1$, $\theta'_i = \beta\theta_i$. By Eq. (3), we have that $l_i^{G,\alpha} < l_i^{G',\alpha}$ and $u_i^{G',\alpha} < u_i^{G,\alpha}$, hence $[l_i^{G',\alpha}, u_i^{G',\alpha}] \subset [l_i^{G,\alpha}, u_i^{G,\alpha}]$. This means that $\mathcal{P}_L^{G',\alpha} \subset \mathcal{P}_L^{G,\alpha}$, and that infinimum and supremum of expectations over these two sets are such that $I^{G',\alpha} \subset I^{G,\alpha}$.

Let us now demonstrate a proposition that only holds for the IDM approach, and that basically says that better evaluations should provide a better global score (both higher lower and upper expectations) for the trustee.

Proposition 2. *Let Θ and Θ' be two counting vectors, with $\hat{\theta} = \hat{\theta}'$ and for which there is an index i such that $\forall j \geq i$, $\theta'_j \geq \theta_j$ and $\forall j < i$, $\theta'_j \leq \theta_j$. Then, given a hyper-paramater s, we have*

$$\underline{E}^{I,s} \leq \underline{E}^{I',s} \quad \textit{and} \quad \overline{E}^{I,s} \leq \overline{E}^{I',s}, \tag{7}$$

with $\underline{E}^{I,s}$ and $\underline{E}^{I',s}$ the lower expectation resp. obtained from Θ and Θ', and likewise for the upper expectations.

Proof. Let us consider the initial counting vector Θ. As $\hat{\theta} = \hat{\theta}'$, going from Θ to Θ' can be done by transferring some evaluations, e.g. of index $k < i$ to better ones e.g. of index $i \geq m$ one at a time. Therefore, all we have to do is to consider the counting vector Θ'' such that $\theta''_k = \theta_k - 1$, $\theta''_m = \theta_m + 1$ and $\theta''_i = \theta_i$ for all other indices, and to prove that Eq. (7) holds in this case.

By Eq (4), we have that $l_i^{I'',s} = l_i^{I,s}$ and $u_i^{I'',s} = u_i^{I,s}$ for any i different of k, m. We also have that $l_k^{I'',s} \leq l_k^{I,s}$, $u_k^{I'',s} \leq u_k^{I,s}$, $l_m^{I'',s} \geq l_m^{I,s}$ and $u_m^{I'',s} \geq u_m^{I,s}$. Now, concentrating on the lower expectation and using Eq (5), to prove Eq. (7), we need to prove $\sum_{i=1}^{N} \underline{P}(A_{(i)}) \leq \sum_{i=1}^{N} \underline{P}''(A_{(i)})$, with $\underline{P}$ and $\underline{P}''$ the lower probabilities induced by the probability intervals obtained from Θ and Θ''. The two sums read:

$$\sum_{i=1}^{N} \underline{P}(A_{(i)}) = \sum_{i=1}^{N} \max\{\sum_{j=i}^{2n+1} l_j, 1 - \sum_{j=1}^{i-1} u_j\}$$

$$\sum_{i=1}^{N} \underline{P}''(A_{(i)}) = \sum_{i=1}^{N} \max\{\sum_{j=i}^{2n+1} l''_j, 1 - \sum_{j=1}^{i-1} u''_j\}$$

For $i \leq k$ or $i > m$, we have $\max\{\sum_{j=i}^{2n+1} l_j, 1 - \sum_{j=1}^{i-1} u_j\} = \max\{\sum_{j=i}^{2n+1} l''_j, 1 - \sum_{j=1}^{i-1} u''_j\}$, because $l_k^{I,s} - l_k^{I'',s} = l_m^{I'',s} - l_m^{I,s}$ $(i \leq k)$ and $u_m^{I'',s} - u_m^{I,s} = u_k^{I,s} - u_k^{I'',s}$ $(i < m)$. Now, consider the case where $k < i \leq m$, we do have $l_m^{I'',s} \geq l_m^{I,s}$ and $u_k^{I,s} \geq u_k^{I'',s}$, therefore $\max\{\sum_{j=i}^{2n+1} l_j, 1 - \sum_{j=1}^{i-1} u_j\} \leq \max\{\sum_{j=i}^{2n+1} l''_j, 1 - \sum_{j=1}^{i-1} u''_j\}$. Hence, we have $\underline{E}^{I,s} \leq \underline{E}^{I',s}$. The proof concerning the upper expectation is similar.

As shows the next example, the approach using Goodman's confidence intervals does not satisfy this property, that may seem intuitive at first sight. This is mainly due to the fact that differences between upper and lower probability bounds u_i, l_i derived from Goodman's confidence intervals depend on the number of evaluations θ_i, i.e. more evaluations θ_i will provide a narrower interval $[l_i, u_i]$. This means that the model precision depends on how evaluations are distributed, while it can be argued that it is not the case for the IDM (where differences $u_i - l_i$ depend solely on parameter s and $\hat{\theta}$).

Example 4. Consider a space $\mathcal{X} = \{-2, -1, 0, 1, 2\}$ containing 5 possible values and the two following counting vectors $\Theta = (0, 0, 10, 0, 0)$ and $\Theta' = (0, 0, 8, 2, 0)$. With a confidence degree $\alpha = 0.95$, we have

$$I^{G,0.95} = [-0.8, 0.8] \qquad I^{G',0.95} = [-0.92, 1]$$

From the example, it can be seen that Goodman's intervals somewhat reflect the dispersion of evaluations, i.e. the model imprecision depends on how concentrated evaluations are. Indeed, more dispersed evaluations may improve the upper score, while providing a more imprecise interval $[\underline{E}, \overline{E}]$ (as in Example 4).

It would be interesting to relate this kind of behaviour (interval imprecision increase) with some dispersion measures of the empirical frequencies distributions (e.g., entropy, Gini index, ...). Also, it could be checked whether Goodman's intervals approach satisfy a weaker condition than Proposition 2, namely that for two counting vectors Θ and Θ' satisfying condition of Proposition 2 and a given confidence value α, we have $\overline{E}^{G,\alpha} \leq \overline{E}^{G',\alpha}$.

These two properties may be seen as monotonic properties w.r.t. evaluation quantity and evaluation score, respectively. Other properties, such as adaptation of the ones proposed by Ben-Naim and Prade [3], should be investigated in further studies.

3.4 Towards Fuzzy Evaluations

In this subsection, we relax some of the previous assumptions (i.e. fixed confidence level α and parameter s) and propose some methods to obtain a fuzzy interval as the evaluation summary rather than a crisp interval.

Recall that a fuzzy set μ is a mapping $\mu : \mathcal{X} \to [0, 1]$ from $\mathcal{X}$ (here, the interval $[-n, n]$) to the unit interval, where $\mu(x)$ is called the membership value of x. The β-cut of a fuzzy set μ is the set $A_\beta := \{x \in \mathcal{X} | \mu(x) \geq \beta\}$.

First approach: Goodman's intervals. Extending the first approach to obtain a fuzzy representation is straightforward, since the formalism of fuzzy sets is particularly well suited to the representation of confidence intervals [9]. Indeed, a β-cut can be interpreted as a confidence set or interval with a confidence level $1 - \beta$.

An interval $I^{G,\alpha} = [\underline{E}^{G,\alpha}, \overline{E}^{G,\alpha}]$ for a given α can therefore be directly associated to the $(1 - \alpha)$-cut of a fuzzy set giving a global evaluation of the trustee trustworthiness. The resulting fuzzy set μ^G is such that, for any $\alpha \in (0, 1]$

$$\mu^G(\underline{E}^{G,\alpha}) = 1 - \alpha \qquad \mu^G(\overline{E}^{G,\alpha}) = 1 - \alpha.$$

Example 5. Consider the counting vector $\Theta = (0, 9, 13, 11, 17)$ provided in Example 1. Figure 1 illustrates the obtained summarising fuzzy interval. The representation shows that the trustee has a positive score, centred around 0.7. Only

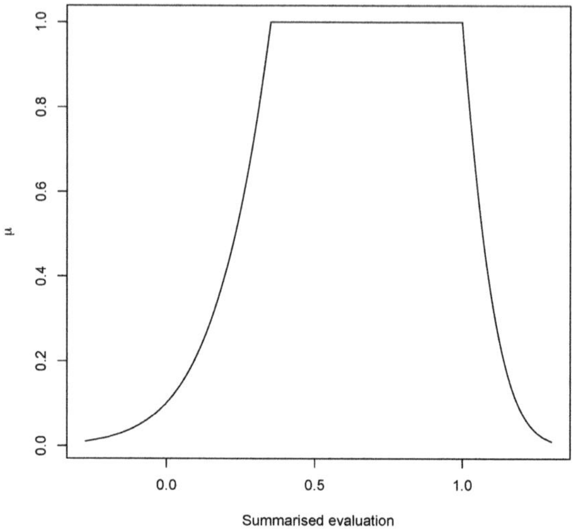

Fig. 1. Fuzzy evaluation with Goodman's intervals (Example 5)

intervals given by conservative confidence values (above 0.9) provide summarising intervals that include negative values.

Second approach: IDM intervals. How to build a fuzzy evaluation by using the IDM approach is less straightforward. An idea (the one we take here) is to let the parameter s vary within some bounds $[0, \overline{s}]$, and to build a fuzzy set μ^I such that, for any $s \in [0, \overline{s}]$,

$$\mu^I(\underline{E}^{I,s}) = \frac{\overline{s} - s}{\overline{s}} \qquad \mu^I(\overline{E}^{I,s}) = \frac{\overline{s} - s}{\overline{s}}.$$

This is indeed a fuzzy set, since for two $s, s' \in [0, \overline{s}]$ such that $s \leq s'$, we do have $[\underline{E}^{I,s}, \overline{E}^{I,s}] \subset [\underline{E}^{I,s'}, \overline{E}^{I,s'}]$. However, the interpretation in terms of confidence intervals is in this case less clear, and the final fuzzy global evaluation is highly dependent of the value $\overline{s}$ (the lower $\overline{s}$, the more precise will be the fuzzy set). Hence this extension is more ad hoc, as well as questionable.

Example 6. Consider the counting vector $\Theta = (0, 9, 13, 11, 17)$ provided in Example 1 and a value $\overline{s} = 50$. Figure 2 illustrates the obtained summarising fuzzy interval. Although the fuzzy set is centred around the same values as in Figure 1, its shape is quite different. Indeed, in this case the imprecision growth decreases as α-value decreases, while in the case of Goodman's intervals the imprecision growth increases as α-value increases.

Note that, since Propositions 1 and 2 are valid for any confidence level or hyperparameter values, their conclusions can directly be extended to the proposed fuzzy extensions.

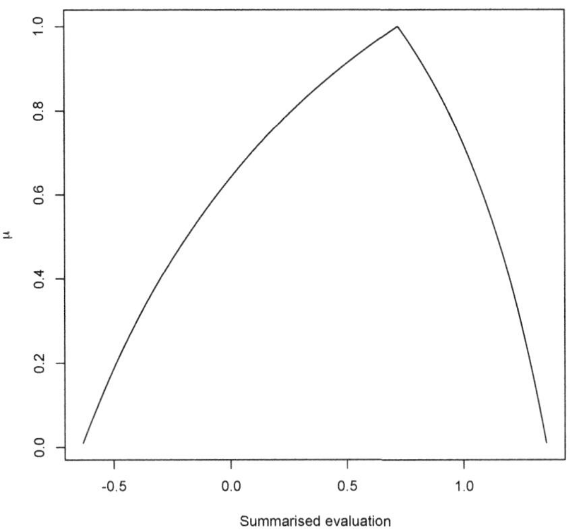

Fig. 2. Fuzzy evaluation with IDM intervals (Example 6)

4 Conclusion

In this paper, we have proposed and compared two imprecise probabilistic models to evaluate the trustworthiness of an agent (the trustee) from previous evaluations made by other agents. The two models are based on the estimation of lower and upper expectations induced by probability intervals, themselves induced by the counting vector of evaluations. In the first model, these probability intervals are given by Goodman's statistical confidence intervals, while in the second, probability intervals are provided by the (popular) Imprecise Dirichlet model. Lower and upper expectations summarise the counting vector of evaluations in a richer way than single point values, since the interval they provide also reflects the dispersion of evaluations and their quantity. Both methods are computationally efficient, and we have proposed for both of them extensions to fuzzy evaluations.

From our study, it appears that the two approaches are at odds. Indeed, Goodman's intervals approach does not satisfy some monotonic properties (Property 2) that intuitively one may wish to satisfy, while the IDM approach does. However, it could be argued that the IDM probability intervals and the induced summarising interval only takes account of the quantity of evaluations, as the imprecision in both of them only depends on the number of evaluations (once s is fixed). On the contrary, Goodman's probability intervals imprecision and the induced summarising interval are also influenced by the evaluations distribution across space $\mathcal{X}$. Goodman's confidence intervals also have a clear statistical interpretation, allowing for a very natural extension of the summarising process to fuzzy intervals. Such an extension, although possible, is more tricky to interpret in the case of the IDM, for which the choice of s and its meaning are still discussed among researchers.

In conclusion, our preference would be to let go of Property 2 and to use Goodman's intervals, since they account for evaluations dispersion in $\mathcal{X}$ and have a clear statistical interpretation. However, if one considers that Property 2 have to be satisfied, then the IDM approach should be used.

A possible improvement of the current approach would be to integrate additional features to the evaluations or the way they are taken into account. For instance, it could be desirable to allow for imprecise evaluations or to consider the time at which evaluations were given (recent evaluations being more reliable than old ones). However, such additional information would also mean that the counting vector Θ would no longer be sufficient "statistic" to provide a summary.

Another interesting topic to explore is how trust information coming from past evaluations can be combined with other trust information sources (e.g., direct interactions).

References

1. Sabater, J., Sierra, S.: Review on computational trust and reputation models. Artificial Intelligence Review 24(33-60) (2005)
2. Ramchurn, S., Huynh, D., Jennings, N.: Trust in multi-agent systems. The Knowledge Engineering Review 19, 1–25 (2004)
3. Ben-Naim, J., Prade, H.: Evaluating trustworthiness from past performances: Interval-based approaches. In: Greco, S., Lukasiewicz, T. (eds.) SUM 2008. LNCS (LNAI), vol. 5291, pp. 33–46. Springer, Heidelberg (2008)
4. de Campos, L., Huete, J., Moral, S.: Probability intervals: a tool for uncertain reasoning. I. J. of Uncertainty, Fuzziness and Knowledge-Based Systems 2, 167–196 (1994)
5. Goodman, L.: On simultaneous confidence intervals for multinomial proportions. Technometrics 7, 247–254 (1964)
6. Bernard, J.M.: An introduction to the imprecise dirichlet model for multinomial data. I. J. of Approximate Reasoning 39, 123–150 (2004)
7. Choquet, G.: Theory of capacities. Annales de l'institut Fourier 5, 131–295 (1954)
8. Walley, P.: Statistical reasoning with imprecise Probabilities. Chapman and Hall, New York (1991)
9. Dubois, D., Foulloy, L., Mauris, G., Prade, H.: Probability-possibility transformations, triangular fuzzy sets, and probabilistic inequalities. Reliable Computing 10, 273–297 (2004)

Range-Consistent Answers of Aggregate Queries under Aggregate Constraints

Sergio Flesca, Filippo Furfaro, and Francesco Parisi

DEIS - Università della Calabria
Via Bucci - 87036 Rende (CS) Italy
{flesca,furfaro,fparisi}@deis.unical.it

Abstract. A framework for computing range-consistent answers of aggregate queries in the presence of aggregate constraints is introduced. The range-consistent answer of an aggregate query is the narrowest interval containing all the answers of the query evaluated on every possible repaired database. A wide form of aggregate constraints is considered, consisting of linear inequalities on aggregate-sum functions. In this setting, three types of aggregate queries are investigated, namely SUM, MIN, MAX queries. Our approach computes consistent answers by solving Integer Linear Programming (ILP) problem instances, thus enabling well-established techniques for ILP resolution to be exploited.

1 Introduction

A great deal of attention has been recently devoted to the problem of extracting reliable information from data inconsistent w.r.t. integrity constraints. Most of the work dealing with this problem is based on the notions of *repair* and *consistent query answer* (CQA) introduced in [1]. A repair of an inconsistent database is a new database instance, on the same scheme as the original database, satisfying the given integrity constraints and which is "minimally" different from the original database instance (the minimality criterion aims at preserving the information in the original database as much as possible). Thus, an answer of a given query posed on an inconsistent database is said to be consistent if the same answer is obtained from every possible repair of the database. Based on this notion of CQA, several works investigated the problem of querying inconsistent data considering different classes of queries and constraints. Most of these works deal with "classical" integrity constraints (such as keys, foreign keys, functional dependencies). Indeed, these kinds of constraint often do not suffice to manage data consistency, as they cannot be used to define algebraic relations between stored values. In fact, this issue frequently occurs in several scenarios, such as scientific databases, statistical databases, and data warehouses, where numerical values in some tuples result from aggregating values in other tuples. In our previous work [11], we introduced a new form of integrity constraint, namely *aggregate constraint*, which enables conditions to be expressed on aggregate values extracted from the database. In that work, we characterized the computational complexity of the CQA problem for atomic ground queries in the presence of aggregate constraints.

In this paper, we consider a more expressive form of queries (namely, *aggregate queries*), consisting of the evaluation of an aggregate operator (SUM, MIN, MAX) over the

A. Deshpande and A. Hunter (Eds.): SUM 2010, LNAI 6379, pp. 163–176, 2010.

tuples of a relation satisfying the desired condition. We consider a more specific notion of consistency of answers, namely *range-consistent query answer* (range-CQA), which was introduced and shown to be more suitable for queries evaluating aggregates in [3]. Basically, the range-consistent answer of an aggregate query is the narrowest interval containing all the answers of the query evaluated on every possible repair of the original database. In this setting, we devise a strategy for computing range-CQAs of aggregate queries. Before presenting our contribution in detail, we provide an example describing an application scenario of our work, and make the reader acquainted with the notions of aggregate constraint and aggregate query.

Example 1. The balance sheet of a company is a financial statement providing information on what the company owns (its assets), what it owes (its liabilities), and the value of the business to its stockholders. A thorough analysis of balance sheets is extremely important for both stock and bond investors, since it allows potential liquidity problems of a company to be detected, thus determining the company financial reliability as well as its ability to satisfy financial obligations.

Generally balance sheets are available as paper documents, thus they cannot be automatically processed by balance analysis tools, since these work on electronic data only. Hence, the automatic acquisition of balance-sheet data from paper documents is often performed as the preliminary phase of the decision making process, as it yields data prone to be analyzed by suitable tools for discovering information of interest.

Table 1 represents a relation *BalanceSheets* obtained from the balance sheets of two consecutive years of a company. These data were acquired by means of an OCR (*Optical Character Recognition*) tool from paper documents. Values '*det*', '*aggr*' and '*drv*' in column *Type* stand for *detail*, *aggregate* and *derived*, respectively. Specifically, an item is *aggregate* if it is obtained by aggregating items of type *detail* of the same section, whereas a *derived* item is an item whose value can be computed using the values of other items of any type and belonging to any section.

Relation *BalanceSheets* must satisfy the following integrity constraints:

κ_1 : for each section and year, the sum of the values of all *detail* items must be equal to the value of the *aggregate* item of the same section and year;

Table 1. Relation *BalanceSheets*

	Year	*Section*	*Subsection*	*Type*	*Value*
t_1	2008	Receipts	beginning cash	drv	50
t_2	2008	Receipts	cash sales	det	100
t_3	2008	Receipts	receivables	det	120
t_4	2008	Receipts	total cash receipts	aggr	250
t_5	2008	Disbursements	payment of accounts	det	120
t_6	2008	Disbursements	capital expenditure	det	20
t_7	2008	Disbursements	long-term financing	det	80
t_8	2008	Disbursements	total disbursements	aggr	220
t_9	2008	Balance	net cash inflow	drv	30
t_{10}	2008	Balance	ending cash balance	drv	80
t_{11}	2009	Receipts	beginning cash	drv	80
t_{12}	2009	Receipts	cash sales	det	110
t_{13}	2009	Receipts	receivables	det	90
t_{14}	2009	Receipts	total cash receipts	aggr	200
t_{15}	2009	Disbursements	payment of accounts	det	130
t_{16}	2009	Disbursements	capital expenditure	det	40
t_{17}	2009	Disbursements	long-term financing	det	20
t_{18}	2009	Disbursements	total disbursements	aggr	120
t_{19}	2009	Balance	net cash inflow	drv	10
t_{20}	2009	Balance	ending cash balance	drv	90

κ_2 : for each year, the *net cash inflow* must be equal to the difference between *total cash receipts* and *total disbursements*;

κ_3 : for each year, the *ending cash balance* must be equal to the sum of the *beginning cash* and the *net cash inflow*.

Although the original balance sheet (in paper format) was consistent, its digital version is not, as some symbol recognition errors occurred during the digitizing phase. In fact, constraints κ_1, κ_2 and κ_3 are not satisfied on the acquired data shown in Table 1. For instance, for year 2008, in section *Receipts*, the aggregate value of *total cash receipts* is not equal to the sum of detail values of the same section: $100 + 120 \neq 250$.

The analysis of the financial conditions of a company can be supported by evaluating aggregate queries on its balance sheets. For instance, in our example, it may be useful to know:

q_1 : the maximum value of *cash sales* over years 2008 and 2009;
q_2 : the minimum value of *cash sales* over years 2008 and 2009;
q_3 : the sum of *cash sales* for both years 2008 and 2009.

Clearly, since the available data are inconsistent, the mere evaluation of these queries on them may yield a wrong picture of the real world. However, the range-consistent answers of these queries can still support several analysis tasks. For instance, knowing that, in every "reasonable" repair of the data, the maximum and the minimum of *cash sales* are in the intervals $[110, 130]$ and $[100, 110]$, respectively, and that the sum of *cash sales* for the considered years is in $[210, 240]$, can give a sufficiently accurate picture of the trend of cash sales. □

Besides the typical scenario of numerical inconsistencies due to OCR recognition errors, the problem of extracting reliable aggregate information from data inconsistent w.r.t. the same kind of constraints used in Example 1 arises in several scenarios, such as sensor networks, where errors in the collected data can be due to wrong sensor readings.

In this context, our contribution is a technique for computing range-CQAs of *aggregate queries* (such as queries q_1, q_2, q_3 of Example 1) on data which are not consistent w.r.t. *aggregate constraints* (such as κ_1, κ_2, κ_3 in the same example). Our work builds on the strategy proposed in [11] for repairing data inconsistent w.r.t. a given set of aggregate constraints. According to this approach, reasonable repairs (namely, *card*-minimal repairs) are sets of updates making the database consistent and having minimum cardinality. Correspondingly, range-CQAs are the narrowest intervals containing all the answers of the aggregate queries that can be obtained from every possible *card*-minimal repair. Specifically, our contribution consists in showing that range-CQAs of aggregate queries can be evaluated without computing every possible *card*-minimal repair, but only solving three *Integer Linear Programming* (ILP) problems. Thus, our approach enables the computation of range-CQAs by means of well-known techniques for solving ILP.

Related Work. The most widely-used notion of CQA for non-aggregate queries (which is that presented at the beginning of the introduction of this work) was introduced in [1].

The query rewriting technique proposed in [1] was extended in [12,13] and further generalized in [17]. The computational complexity of the CQA problem was studied in [6] and in [8]. Several works [2,14] exploited logic-based frameworks for investigating the problem of computing repairs and evaluating consistent query answers. A framework for computing CQAs was presented in [7].

The notion of CQA was adapted for queries involving aggregates in [3], where the notion of range-CQA was introduced and the problem of computing range-CQAs was studied in the presence of functional dependencies. This problem was further investigated in [12] for aggregate queries with grouping under key constraints.

All the above-cited approaches assume that tuple insertions and deletions are the basic primitives for repairing inconsistent data. In [4,5,11,16], repairs consisting of also value-update operations were considered.

However, none of these works investigated the problem of computing (range-) consistent answers to aggregate queries in the presence of aggregate constraints. The form of aggregate constraints considered in this paper was introduced in [11], where the complexity was characterized of several problems regarding the extraction of reliable information from inconsistent numerical data (i.e. repair existence, minimal repair checking, as well as consistent query answer for atomic ground queries). In [9], the architecture of a tool for acquiring and repairing numerical data inconsistent w.r.t. a restricted form of aggregate constraints was presented, along with a strategy for computing reasonable repairs, whereas in [10] the problem of computing reasonable repairs w.r.t. a set of both strong and weak aggregate constraints was addressed.

2 Preliminaries

We assume classical notions of database scheme, relation scheme, and relation instances. Relation schemes will be represented by means of sorted predicates of the form $R(A_1 : \Delta_1, \ldots, A_n : \Delta_n)$, where R is the name of the relation scheme, $A_1, \ldots, A_n$ are attribute names (composing the set denoted as $\mathcal{A}_R$), and $\Delta_1, \ldots, \Delta_n$ are the corresponding domains. Each Δ_i can be either the domain of strings or the domain of signed integers bounded in absolute value by a constant M. Attributes [resp. constants] defined over (M-bounded) integers will be said to be *numerical attributes* [resp. *constants*]. Observe that the assumption that the numerical domain is integer yields no loss of generality, as our framework can be easily extended to the case of rationals.

A tuple over a relation scheme $R(A_1 : \Delta_1, \ldots, A_n : \Delta_n)$ is a member of $\Delta_1 \times \cdots \times \Delta_n$. A relation instance of R is a set r of tuples over R. A database scheme $\mathcal{D}$ is a set of relation schemes, and a database instance D of $\mathcal{D}$ is a set of instances of the relation schemes of $\mathcal{D}$. Given a tuple t, the value of attribute A of t will be denoted as $t[A]$.

On each relation scheme R, a key constraint is assumed. Specifically, we denote as $\mathcal{K}_R$ the subset of $\mathcal{A}_R$ consisting of the names of the attributes which are a key for R. For instance, in "Balance Sheets" example, $\mathcal{K}_R = \{$*Year, Subsection*$\}$. Given a relation scheme R, we will denote the set of its numerical attributes representing measure data as $\mathcal{M}_R$ (namely, *Measure attributes*). That is, $\mathcal{M}_R$ specifies the set of attributes representing measure values, such as weights, lengths, prices, etc. For instance, in "Balance Sheets" example, $\mathcal{M}_R$ consists of attribute *Value* only.

Given a boolean formula β consisting of comparison atoms of the form $X \diamond Y$, where X, Y are either attributes of a relation scheme R or constants, and $\diamond$ is a comparison operator in $\{=, \neq, \leq, \geq, <, >\}$, we say that a tuple t over R satisfies β (denoted as $t \models \beta$) if replacing the occurrences of each attribute A in β with $t[A]$ makes β true.

Assumptions. Our framework is based on the following two assumptions: (i) for any relation scheme R, $\mathcal{K}_R \cap \mathcal{M}_R = \emptyset$, i.e., measure attributes are not used to identify tuples; (ii) the absolute values of measure attributes are bounded by a constant M. Although these assumptions lead to a loss of generality, they are acceptable from a practical point of view, since the situations excluded by them are unlikely to occur in real-life scenarios. The former assumption, as a matter of fact, was used in [4,11], and it is easy to see that it holds in our "Balance Sheets" example. As regards the latter, it is often possible to pre-determine a specific range for numerical attributes. For instance, in our example, it can be reasonably assumed that the items in balance sheets are bounded by \$ 10^9. A brief discussion on possible extensions beyond this limitation (which could be interesting from a theoretical perspective) is provided in Section 4.

2.1 Aggregate Constraints

Given a relation scheme R, an *attribute expression* e on R is either a constant or a numerical attribute of R. Given an attribute expression e on R and a tuple t over R, we denote as $e(t)$ the value e, if e is a constant, or the value $t[e]$, if e is an attribute.

Given a relation scheme R and a sequence $\boldsymbol{y}$ of variables, an *aggregation function* $\chi(\boldsymbol{y})$ on R is a triplet $\langle R, e, \alpha(\boldsymbol{y})\rangle$, where e is an *attribute expression* on R and $\alpha(\boldsymbol{y})$ is a boolean combination of atomic comparisons of the form $X \diamond Y$, where X and Y are constants, attributes of R, or variables in $\boldsymbol{y}$, and $\diamond \in \{=, \neq, \leq, \geq, <, >\}$.

Given an aggregation function $\chi(\boldsymbol{y}) = \langle R, e, \alpha(\boldsymbol{y})\rangle$ and a sequence $\boldsymbol{a}$ of constants with $|\boldsymbol{a}| = |\boldsymbol{y}|$, $\chi(\boldsymbol{a})$ maps every instance r of R to $\sum_{t \in r \wedge t \models \alpha(\boldsymbol{a})} e(t)$, where $\alpha(\boldsymbol{a})$ is the (ground) boolean combination of atomic comparisons obtained from $\alpha(\boldsymbol{y})$ by replacing each variable in $\boldsymbol{y}$ with the corresponding value in $\boldsymbol{a}$. We assume that, in the case that the set of tuples selected by the evaluation of an aggregation function χ is empty, χ evaluates to 0.

Example 2. The following aggregation functions are defined on the relational scheme *BalanceSheets(Year, Section, Subsection, Type, Value)* of Example 1:

$$\chi_1(x, y, z) = \langle \textit{BalanceSheets, Value, } (\textit{Section}{=}x \wedge \textit{Year}{=}y \wedge \textit{Type}{=}z)\,\rangle$$
$$\chi_2(x, y) = \langle \textit{BalanceSheets, Value, } (\textit{Subsection}{=}x \wedge \textit{Year}{=}y)\,\rangle$$

Function χ_1 returns the sum of *Value* of all the tuples having *Subection* x, *Year* y, and *Type* z. For instance, χ_1('Disbursements', 2008, 'det') returns $120 + 20 + 80 = 220$, and χ_1('Receipts', 2009, 'aggr') returns 200. In our running example, as *Year, Subsection* is a key for *BalanceSheets*, the sum returned by χ_2 is an attribute value of a single tuple. For instance, χ_2(2008, 'cash sales') = 100, and χ_2(2008, 'receivables') = 120. □

Definition 1 (Aggregate constraint). *Given a database scheme $\mathcal{D}$, an aggregate constraint on $\mathcal{D}$ is of the form:* $\forall \boldsymbol{x}\ (\phi(\boldsymbol{x}) \implies \sum_{i=1}^{n} c_i \cdot \chi_i(\boldsymbol{y}_i) \leq K)$, *where:*

1. *n is a positive integer, and $c_1, \dots, c_n, K$ are rational constants;*
2. *$\phi(\boldsymbol{x})$ is a (possibly empty) conjunction of atoms constructed from relation names, constants, and all the variables in $\boldsymbol{x}$;*
3. *each $\chi_i(\boldsymbol{y}_i)$ is an aggregation function, where $\boldsymbol{y}_i$ is a list of variables and constants, and every variable that occurs in $\boldsymbol{y}_i$ also occurs in $\boldsymbol{x}$.*

A database D satisfies an aggregate constraint ac, denoted $D \models ac$, if, for all the substitutions θ of the variables in $\boldsymbol{x}$ with constants making $\phi(\theta(\boldsymbol{x}))$ *true* on D, the inequality $\sum_{i=1}^{n} c_i \cdot \chi_i(\theta(\boldsymbol{y}_i)) \leq K$ holds on D. For a set of aggregate constraint $\mathcal{AC}$, D satisfies $\mathcal{AC}$ (denoted as $D \models \mathcal{AC}$) if $D \models ac$ for each $ac \in \mathcal{AC}$. Correspondingly, we say that D is consistent [resp. inconsistent] w.r.t. $\mathcal{AC}$ if $D \models \mathcal{AC}$ [resp. $D \not\models \mathcal{AC}$].

Observe that aggregate constraints enable equalities to be expressed as well, since an equality can be viewed as a pair of inequalities. In the following, for the sake of brevity, equalities will be written explicitly and universal quantification will be omitted.

Example 3. Constraints κ_1, κ_2 and κ_3 of Example 1 can be expressed as follows:

κ_1 : $BalanceSheets(x_1, x_2, x_3, x_4, x_5) \implies \chi_1(x_2, x_1, \text{'det'}) - \chi_1(x_2, x_1, \text{'aggr'}) = 0$

κ_2: $BalanceSheets(x_1, x_2, x_3, x_4, x_5) \implies \chi_2(\text{'net cash inflow'}, x_1) - (\chi_2(\text{'total cash receipts'}, x_1) - \chi_2(\text{'total disbursements'}, x_1)) = 0$

κ_3: $BalanceSheets(x_1, x_2, x_3, x_4, x_5) \implies \chi_2(\text{'ending cash balance'}, x_1) - (\chi_2(\text{'beginning cash'}, x_1) + \chi_2(\text{'net cash inflow'}, x_1)) = 0$ □

Let $R(A_1, \dots, A_n)$ be a relation scheme and $R(x_1, \dots, x_n)$ an atom, where each x_j is either a variable or a constant. For each $j \in [1..n]$, we say that the term x_j is *associated* with the attribute A_j. Moreover, we say that a variable x_i is a *measure variable* if it is associated with a measure attribute.

We now provide the definition of a restricted form of aggregate constraints, namely *steady aggregate constraints*, which were introduced in [11]. As observed in [11] and as will be clearer in the following, the steadiness restriction limits the expressiveness of aggregate constraints, but not dramatically, since they suffice to model algebraic conditions ensuring data consistency in several scenarios (such as that of our running example). For this reason, along with the fact that considering steady aggregate constraints will allow us to devise a technique for computing consistent query answers, most of the discussions and the results in the remainder of this paper will deal with this restricted form of constraints.

Definition 2 (Steady aggregate constraint). *An aggregate constraint ac is* steady *if:*

1. *for every aggregation function $\langle R, e, \alpha \rangle$ on the right-hand side of ac, no measure attribute occurs in α;*
2. *measure variables occur at most once in ac;*
3. *no constant occurring in the conjunction of atoms ϕ on the left-hand side of ac is associated with a measure attribute.*

It is easy to see that the aggregate constraints κ_1, κ_2, κ_3 of Example 3 are steady.

2.2 Repairing Inconsistent Databases

Updates at attribute-level will be used as the basic primitives for repairing data.

Definition 3 (Atomic update). *Let* $t = R(v_1, \ldots, v_n)$ *be a tuple on the relation scheme* $R(A_1 : \Delta_1, \ldots, A_n : \Delta_n)$. *An* atomic update *on* t *is a triplet* $< t, A_i, v'_i >$, *where* $A_i \in \mathcal{M}_R$ *and* v'_i *is a value in* Δ_i *different from* v_i.

Definition 4 (Consistent database update). *Let* D *be a database and* $U = \{u_1, \ldots, u_n\}$ *be a set of atomic updates on tuples of* D. *The set* U *is said to be a* consistent database update *iff, for each pair of distinct updates* $u_1 =< t_1, A_1, v_1 >$, $u_2 =< t_2, A_2, v_2 >$ *in* U, *either* $t_1 \neq t_2$ *or* $A_1 \neq A_2$.

Given a tuple $t = R(v_1, \ldots, v_n)$, and the (atomic) update $u =< t, A_i, v'_i >$, we denote the tuple resulting from applying u on t as $u(t)$, i.e., $u(t) = R(v_1, \ldots, v_{i-1}, v'_i, v_{i+1}, \ldots, v_n)$. Moreover, given a consistent database update U, we denote the database resulting from performing all the atomic updates in U on a given database D as $U(D)$.

Definition 5 (Repair). *Let* $\mathcal{D}$ *be a database scheme,* $\mathcal{AC}$ *a set of aggregate constraints on* $\mathcal{D}$, *and* D *an instance of* $\mathcal{D}$ *such that* $D \not\models \mathcal{AC}$. *A* repair ρ *for* D *is a consistent database update such that* $\rho(D) \models \mathcal{AC}$.

Example 4. A repair ρ_1 for *BalanceSheets* w.r.t. $\mathcal{AC} = \{\kappa_1, \kappa_2, \kappa_3\}$ consists of increasing attribute *Value* in the tuples t_2 and t_{18} up to 130 and 190 respectively, that is, $\rho_1 = \{< t_2, Value, 130 >, < t_{18}, Value, 190 >\}$. Another repair for *BalanceSheets* is: $\rho' = \{< t_2,\ Value, 130 >, < t_{15},\ Value, 120 >, < t_{16},\ Value, 50 >, < t_{18},\ Value, 190 >\}$. □

Given a database D inconsistent w.r.t. a set of aggregate constraints $\mathcal{AC}$, different repairs can be performed on D yielding a new consistent database. To evaluate whether a repair should be considered "relevant" or not, we use the ordering criterion stating that a repair ρ_1 precedes a repair ρ_2 if the number of changes issued by ρ_1 is less than ρ_2.

Definition 6 (*Card*-minimal repair). *Let* $\mathcal{D}$ *be a database scheme,* $\mathcal{AC}$ *a set of aggregate constraints on* $\mathcal{D}$, *and* D *an instance of* $\mathcal{D}$. *A repair* ρ *for* D *w.r.t.* $\mathcal{AC}$ *is a* card-*minimal repair iff there is no repair* ρ' *for* D *w.r.t.* $\mathcal{AC}$ *such that* $|\rho'| < |\rho|$.

Example 5. In our running example, the set of *card*-minimal repairs is $\{\rho_1, \rho_2\}$, where ρ_1 is the repair defined in Example 4 and $\rho_2 = \{ < t_3, Value, 150 >, < t_{18}, Value, 190 >\}$. □

2.3 Aggregate Queries

We consider aggregate queries involving scalar functions `MIN`, `MAX` and `SUM` returning a single value for each relation.

Definition 7 (Aggregate Query). *An aggregate query* q *on a database scheme* $\mathcal{D}$ *is an expression of the form* `SELECT` f `FROM` R `WHERE` α, *where:*

i) R *is a relation scheme in* $\mathcal{D}$;
ii) f *is one of* `MIN`(A), `MAX`(A) *or* `SUM`(A), *where* A *in an attribute of* R; *and*

iii) *α is boolean combination of atomic comparisons of the form $X \diamond Y$, where X and Y are constants or non-measure attributes of R, and $\diamond \in \{=, \neq, \leq, \geq, <, >\}$.*

Basically, the restriction that no measure attribute occurs in the WHERE clause of an aggregate query means considering queries which satisfy a "steadiness" condition analogous to that imposed on steady aggregate constraints. Given an instance D of $\mathcal{D}$, the evaluation of an aggregate query q on D will be denoted as $q(D)$.

Example 6. Queries q_1, q_2 and q_3 defined in Example 1 can be expressed as follows:

$q_1 =$ SELECT MAX($Value$) FROM $BalanceSheets$ WHERE $Subsection =$ 'cash sales'
$q_2 =$ SELECT MIN($Value$) FROM $BalanceSheets$ WHERE $Subsection =$ 'cash sales'
$q_3 =$ SELECT SUM($Value$) FROM $BalanceSheets$ WHERE $Subsection =$ 'cash sales' □

We now introduce the fundamental notion of range-consistent answer of an aggregate query. Basically, it consists in the narrowest range *[greatest-lower bound (glb), least-upper bound (lub)]* containing all the answers resulting from evaluating the query on every database resulting from the application of a *card*-minimal repair.

Definition 8 (Range-consistent query answer). *Let $\mathcal{D}$ be a database scheme, $\mathcal{AC}$ a set of aggregate constraints on $\mathcal{D}$, q an aggregate query on $\mathcal{D}$, and D an instance of $\mathcal{D}$. The* range-consistent query answer *of q on D, denoted as $CQA^q_{\mathcal{D},\mathcal{AC}}(D)$ is the empty interval $\emptyset$, if D admits no repair w.r.t. $\mathcal{AC}$, or the interval $[glb, lub]$, otherwise, where:*

i) *for each* card-*minimal repair ρ for D w.r.t. $\mathcal{AC}$, it holds that $glb \leq q(\rho(D)) \leq lub$;*
ii) *there is a pair ρ', ρ'' of* card-*minimal repairs for D w.r.t. $\mathcal{AC}$ such that $q(\rho'(D)) = glb$ and $q(\rho''(D)) = lub$.*

Example 7. In our running example, the narrowest range including the evaluations of query q_1 on every database resulting from the application of a *card*-minimal repair is $[110, 130]$ (as shown in Example 5, the *card*-minimal repairs are ρ_1 and ρ_2; q_1 evaluates to 130 and 110 on the databases repaired by ρ_1 and ρ_2, respectively). Hence, the range-CQA of query q_1 is $[110, 130]$. Similarly, it is easy to see that the range-CQAs of q_2 and q_3 are $[100, 110]$ and $[210, 240]$, respectively.

3 Query Answering

The data-complexity of the problem of computing range-consistent answers of aggregate queries in the presence of general and steady aggregate constraints is characterized in the following theorem.

Theorem 1. *Let $\mathcal{D}$ be a fixed database scheme, $\mathcal{AC}$ a fixed set of aggregate constraints on $\mathcal{D}$, q a fixed aggregate query on $\mathcal{D}$, D an instance of $\mathcal{D}$, and $[\ell, u]$ a fixed interval. Then:*

i) deciding whether $CQA^q_{\mathcal{D},\mathcal{AC}}(D) \neq \emptyset$ is NP-complete;
ii) deciding whether $CQA^q_{\mathcal{D},\mathcal{AC}}(D) \subseteq [\ell, u]$ is $\Delta^p_2[\log n]$-complete;
iii) the lower complexity bounds still hold in the case that $\mathcal{AC}$ is steady.

An important result stated in Theorem 1 is that the range-CQA problem is hard also when the aggregate constraints are steady. This means that the loss in expressiveness yielded by the steadiness restriction not only has no dramatic impact on the practical usefulness of aggregate constraints (as explained in the previous section), but also on the computational complexity of the range-CQA problem. However, the steadiness restriction can be exploited to devise a technique for computing range-CQAs which does not work for general aggregate constraints.

Our technique is based on a translation of the range-CQA problem into the Integer Linear Programming (ILP) problem [15], thus enabling the computation of consistent answers by means of any of the well-established techniques for solving ILP. Observe that the result stated in Theorem 1 on the high complexity of range-CQA backs the use of our approach in the following sense: we solve a hard problem by translating it into another hard problem (in fact, ILP is NP-hard), for which a resolution method is known.

Our technique is progressively introduced in the rest of this section, which is organized as follows. First, we explain how aggregate constraints can be translated into sets of inequalities (Section 3.1). Then, we show how this translation can be used to define an ILP instance which computes the cardinality of the *card*-minimal repairs (Section 3.2). Finally, in Section 3.3, we explain how, starting from the knowledge of this cardinality and the translation of constraints into inequalities, the problem of computing range-CQAs for the three types of aggregate queries (SUM, MAX, MIN) can be solved through a pair of further ILP instances.

3.1 Expressing Steady Aggregate Constraints as a Set of Inequalities

Given a database scheme $\mathcal{D}$, a set of steady aggregate constraints $\mathcal{AC}$ on $\mathcal{D}$, and an instance D of $\mathcal{D}$, we show how the triplet $\langle \mathcal{D}, \mathcal{AC}, D \rangle$ can be translated into a set of linear inequalities $\mathcal{S}(\mathcal{D}, \mathcal{AC}, D)$ such that every solution of $\mathcal{S}(\mathcal{D}, \mathcal{AC}, D)$ corresponds to a (possibly not-minimal) repair for D w.r.t. $\mathcal{AC}$.

We first describe the translation for a single steady aggregate constraint ac (which has the form: $\forall \boldsymbol{x}\ \phi(\boldsymbol{x}) \implies \sum_{i=1}^{n} c_i \cdot \chi_i(\boldsymbol{y}_i) \leq K$, where $\forall i \in [1..n]$, $\chi_i(\boldsymbol{y}) = \langle R_i, e_i, \alpha_i(\boldsymbol{y}_i) \rangle$). The translation results from the following three steps (for every relation scheme R_ℓ in $\mathcal{D}$, we will denote its instance in D as r_ℓ):

1) *Associating variables with pairs* $\langle$tuple, measure attribute$\rangle$*:*
   For each tuple t of a relation instance r_ℓ in D and measure attribute $A_j \in \mathcal{M}_{R_\ell}$, we create the integer variable z_{t,A_j};
2) *Translating each* χ_i *into sums of variables and constants:*
   Let $\Theta(ac)$ be the set of the ground substitutions of variables in $\boldsymbol{x}$ with constants such that $\forall \theta \in \Theta(ac)\ \phi(\theta \boldsymbol{x})$ is $true$ on D. For every ground substitution $\theta \in \Theta(ac)$ and every χ_i, we denote as $T_{\chi_i}(\theta)$ the set of tuples involved in the evaluation of χ_i w.r.t. θ, that is $T_{\chi_i}(\theta) = \{t : t \in r_i \wedge t \models \alpha_i(\theta \boldsymbol{y_i})\}$, where r_i is the instance in D of the relation scheme R_i in χ_i. Then, for every ground substitution $\theta \in \Theta(ac)$, we define the translation of χ_i w.r.t. θ as:

$$\mathcal{P}(\chi_i, \theta) = \begin{cases} \sum_{t \in T_{\chi_i}(\theta)} z_{t,A_j} & \text{if } e_i \text{ is the measure attribute } A_j; \\ \sum_{t \in T_{\chi_i}(\theta)} e_i(t) & \text{otherwise.} \end{cases}$$

3) *Translating ac into a set of linear inequalities:*
The constraint ac is translated into the set $\mathcal{S}(\mathcal{D}, ac, D)$ of linear inequalities containing, for every ground substitution $\theta \in \Theta(ac)$, the inequality $\sum_{i=1}^{n} c_i \cdot \mathcal{P}(\chi_i, \theta) \leq K$.

The system of linear inequalities $\mathcal{S}(\mathcal{D}, \mathcal{AC}, D)$ (which takes into account all the aggregate constraints in $\mathcal{AC}$) is then defined as $\mathcal{S}(\mathcal{D}, \mathcal{AC}, D) = \cup_{ac \in \mathcal{AC}} \mathcal{S}(\mathcal{D}, ac, D)$.

For the sake of simplicity, in the following we assume that the pairs $\langle t, A_j \rangle$, where A_j is the name of a measure attribute of tuple t, are associated with distinct integer indexes (the set of these indexes will be denoted as $\mathcal{I}$). Therefore, being i the integer associated with the pair $\langle t, A_j \rangle$, the variable z_{t,A_j} will be denoted as z_i.

Example 8. In "Balance Sheets" example, we associate each pair $\langle t_i, Value \rangle$ with the integer i, thus $\mathcal{I} = \{1, \ldots, 20\}$. The translation of the aggregate constraints of Example 3 is the following (we explicitly write equalities instead of inequalities):

$$\begin{cases} z_2 + z_3 = z_4; \quad z_5 + z_6 + z_7 = z_8; \quad z_{12} + z_{13} = z_{14}; \quad z_{15} + z_{16} + z_{17} = z_{18}; \\ z_4 - z_8 = z_9; \quad z_{14} - z_{18} = z_{19}; \quad z_1 + z_9 = z_{10}; \quad z_{11} + z_{19} = z_{20}. \end{cases}$$

A solution of this system assigns to each z_i the value $t_i[Value]$, except for z_2, z_{15}, z_{16}, z_{18}, which are assigned 130, 120, 50, 190, respectively. This solution corresponds to the non-minimal repair ρ' of Example 4. □

Looking at the steps of the translation, it is easy to see that there is a biunique correspondence between the (non-minimal) repairs for D w.r.t. $\mathcal{AC}$ and $\mathcal{D}$ and the solutions of $\mathcal{S}(\mathcal{D}, \mathcal{AC}, D)$ whose absolute values are bounded by M. In particular, the solution corresponding to a repair ρ assigns to each z_i the value taken by $t[A_j]$ in the database $\rho(D)$, where $\langle t, A_j \rangle$ is the pair *tuple, attribute* associated with z_i.

3.2 Computing the Cardinality of *Card*-Minimal Repairs by Solving an ILP Instance

In the previous section, we have shown how aggregate constraints can be translated into a set of linear inequalities. We now exploit this translation to encode the problem of computing the cardinality of *card*-minimal repairs as an ILP instance. This translation will be used as the core of a strategy for computing range-CQAs by solving three ILP instances.

We start by introducing a fundamental system of inequalities.

Definition 9 **($\mathcal{ILP}(\mathcal{D}, \mathcal{AC}, D)$).** *Given a database scheme $\mathcal{D}$, a set $\mathcal{AC}$ of steady aggregate constraints on $\mathcal{D}$, and an instance D of $\mathcal{D}$, $\mathcal{ILP}(\mathcal{D}, \mathcal{AC}, D)$ is:*

$$\begin{cases} \boldsymbol{A} \times \boldsymbol{z} \leq \boldsymbol{B}; & & \\ z_i - M \leq 0; & -z_i - M \leq 0; & \forall i \in \mathcal{I} \\ z_i - v_i - (M + |v_i| + 1) \cdot \delta_i \leq 0; & -z_i + v_i - (M + |v_i| + 1) \cdot \delta_i \leq 0; & \forall i \in \mathcal{I}; \\ z_i \in \mathbb{Z}; & \delta_i \in \{0, 1\}; & \forall i \in \mathcal{I}; \end{cases}$$

where:

(i) $\mathbf{A} \times \boldsymbol{z} \leq \boldsymbol{B}$ *is the set of inequalities* $\mathcal{S}(\mathcal{D}, \mathcal{AC}, D)$ *(*$\boldsymbol{z}$ *is the vector of variables* z_i*);*
(ii) for each $i \in \mathcal{I}$*,* v_i *is the database value corresponding to the variable* z_i*, that is, if* z_i *is associated with the pair* $\langle t, A_j \rangle$*, then* $v_i = t[A_j]$*;*
(iii) M *is the constant bounding the absolute value of measure attributes.*

Let $s[z]$ be the value taken by variable z in a solution s of an ILP problem.

Basically, for every solution of $\mathcal{ILP}(\mathcal{D}, \mathcal{AC}, D)$, the variables z_i are assigned values which satisfy both $\mathbf{A} \times \boldsymbol{z} \leq \mathbf{B}$ and $-M \leq z_i \leq M$. Hence, given a solution s of $\mathcal{ILP}(\mathcal{D}, \mathcal{AC}, D)$, the consistent set of updates assigning each value $s[z_i]$ to the pair $\langle t, A_j \rangle$ associated with z_i is a (possibly not minimal) repair $\rho(s)$ for D w.r.t $\mathcal{AC}$.

The inequalities in $\mathcal{ILP}(\mathcal{D}, \mathcal{AC}, D)$ other than $\mathbf{A} \times \boldsymbol{z} \leq \mathbf{B}$ and $-M \leq z_i \leq M$ define a mechanism for counting the number of variables z_i which are assigned a value different from the corresponding pair *tuple*, *attribute* in the original data. In fact, for every solution s:

- if $s[z_i] > v_i$ (i.e., z_i is assigned a value greater than the "original" value v_i), then $s[\delta_i] = 1$ (this is entailed by the inequality $z_i - v_i - (M + |v_i| + 1) \cdot \delta_i \leq 0$).
- if $s[z_i] < v_i$, then $s[\delta_i] = 1$ too (this is entailed by $-z_i + v_i - (M + |v_i| + 1) \cdot \delta_i \leq 0$);
- if $s[z_i] = v_i$, then $s[\delta_i]$ is either 0 or 1.

Hence, for every solution, the sum of the values taken by variables δ_i is an upper bound on the number of variables z_i taking a value different from the corresponding v_i.

Theorem 2. *There is a biunique correspondence between the solutions of* $\mathcal{ILP}(\mathcal{D}, \mathcal{AC}, D)$ *and the repairs for* D *w.r.t* $\mathcal{AC}$*. In particular, every solution* s *of* $\mathcal{ILP}(\mathcal{D}, \mathcal{AC}, D)$ *corresponds to a repair* $\rho(s)$ *such that:*

(i) for each z_i *associated with the pair* $\langle t, A_j \rangle$ *and such that* $s[z_i] \neq t[A_j]$*,* $\rho(s)$ *contains the atomic update* $\langle t, A_i, s[z_i] \rangle$*;*
(ii) $|\rho(s)| \leq \sum_{i \in \mathcal{I}} s[\delta_i]$*.*

Thus, the solutions of $\mathcal{ILP}(\mathcal{D}, \mathcal{AC}, D)$ correspond to the repairs for D w.r.t. $\mathcal{AC}$, and vice versa. Moreover, looking at any solution s of $\mathcal{ILP}(\mathcal{D}, \mathcal{AC}, D)$, we can get an upper bound on the cardinality of the corresponding repair $\rho(s)$: this upper bound is given by the sum of the values taken by variables δ_i in s.

We point out that removing the steadiness restriction from aggregate constraints may result in breaking the biunique relation between the solutions of $\mathcal{ILP}(\mathcal{D}, \mathcal{AC}, D)$ and the repairs for D w.r.t. $\mathcal{AC}$. Intuitively, this derives from the fact that, in the presence of non-steady aggregate constraints, applying to D the set of updates corresponding to a solution of $\mathcal{ILP}(\mathcal{D}, \mathcal{AC}, D)$ may trigger violations of some aggregate constraints in $\mathcal{AC}$ which were not encoded in $\mathcal{ILP}(\mathcal{D}, \mathcal{AC}, D)$.

The following corollary strengthens the result of Theorem 2, and provides a method for computing the cardinality of any *card*-minimal repair.

Corollary 1. *A repair for* D *w.r.t.* $\mathcal{AC}$ *exists iff* $\mathcal{ILP}(\mathcal{D}, \mathcal{AC}, D)$ *has at least one solution, and the optimal value of the optimization problem:*

$$\mathcal{OPT}(\mathcal{D}, \mathcal{AC}, D) := \text{minimize} \sum_{i \in \mathcal{I}} \delta_i \text{ subject to } \mathcal{ILP}(\mathcal{D}, \mathcal{AC}, D)$$

coincides with the cardinality of any card*-minimal repair for* D *w.r.t.* $\mathcal{AC}$*.*

3.3 Computing Range-Consistent Query Answers

In this section we show how range-CQAs can be computed by solving ILP instances. The following corollary, which straightforward follows from Theorem 2, addresses the the case that the range-CQA is the empty interval (i.e., there is no repair for the given database w.r.t. the given set of aggregate constraints).

Corollary 2. *$CQA^q_{\mathcal{D},\mathcal{AC}}(D) = \emptyset$ iff $\mathcal{ILP}(\mathcal{D}, \mathcal{AC}, D)$ has no solution.*

Let $q =$ SELECT f FROM R WHERE α be an aggregate query over the relation scheme R, where f is one of MIN(A_j), MAX(A_j) or SUM(A_j) and A_j is an attribute of R. Given an instance r of R, we define the translation of q as $\mathcal{T}(q) = \sum_{t:\, t \in r \wedge t \models \alpha} z_{t,A_j}$.

SUM-queries. Consider the following optimization ILP problems:

$$\mathcal{OPT}^{SUM}_{glb}(\mathcal{D}, \mathcal{AC}, q, D) \quad \begin{cases} \textit{minimize } \mathcal{T}(q) \\ \textit{subject to } \mathcal{ILP}(\mathcal{D}, \mathcal{AC}, D) \cup \{\lambda = \sum_{i \in \mathcal{I}} \delta_i\} \end{cases}$$

$$\mathcal{OPT}^{SUM}_{lub}(\mathcal{D}, \mathcal{AC}, q, D) \quad \begin{cases} \textit{maximize } \mathcal{T}(q) \\ \textit{subject to } \mathcal{ILP}(\mathcal{D}, \mathcal{AC}, D) \cup \{\lambda = \sum_{i \in \mathcal{I}} \delta_i\} \end{cases}$$

where λ is the value returned by $\mathcal{OPT}(\mathcal{D}, \mathcal{AC}, D)$. Intuitively enough, since the solutions of $\mathcal{ILP}(\mathcal{D}, \mathcal{AC}, D)$ correspond to the repairs for D w.r.t. $\mathcal{AC}$, the solutions of $\mathcal{ILP}(\mathcal{D}, \mathcal{AC}, D) \cup \{\lambda = \sum_{i \in \mathcal{I}} \delta_i\}$ correspond to the repairs whose cardinality is equal to λ, that is, *card*-minimal repairs. Hence, the above-introduced $\mathcal{OPT}^{SUM}_{glb}$ and $\mathcal{OPT}^{SUM}_{lub}$ return the minimum and the maximum value of the query q on all the consistent databases resulting from applying *card*-minimal repairs. These values are the boundaries of the range-CQA of q, as stated in the following theorem.

Theorem 3. *For a SUM-query q, either $CQA^q_{\mathcal{D},\mathcal{AC}}(D) = \emptyset$, or $CQA^q_{\mathcal{D},\mathcal{AC}}(D) = [\ell, u]$, where ℓ is the value returned by $\mathcal{OPT}^{SUM}_{glb}(\mathcal{D}, \mathcal{AC}, q, D)$ and u the value returned by $\mathcal{OPT}^{SUM}_{lub}(\mathcal{D}, \mathcal{AC}, q, D)$.*

MAX- and MIN-queries. We consider MAX-queries, since MIN-queries can be handled symmetrically. Given a MAX-query q, we denote the set of indices in $\mathcal{I}$ of the variables z_i occurring in $\mathcal{T}(q)$ as $\mathcal{I}(q)$. Let $In(q)$ be the following set of inequalities:

$$\begin{cases} z_j - z_i - 2M \cdot \mu_i \leq 0 & & \forall j,i \in \mathcal{I}(q), j \neq i \\ \sum_{i \in \mathcal{I}(q)} \mu_i = |\mathcal{I}(q)| - 1 & & \\ x_i - M \cdot \mu_i \leq 0; & -x_i - M \cdot \mu_i \leq 0; & \forall i \in \mathcal{I}(q) \\ z_i - x_i - 2M \cdot (1 - \mu_i) \leq 0; & -z_i + x_i - 2M \cdot (1 - \mu_i) \leq 0; & \forall i \in \mathcal{I}(q); \\ x_i - M \leq 0; & -x_i - M \leq 0; & \forall i \in \mathcal{I}(q); \\ x_i \in \mathbb{Z}; & \mu_i \in \{0, 1\}; & \forall i \in \mathcal{I}(q); \end{cases}$$

and let $\mathcal{ILP}^*(\mathcal{D}, \mathcal{AC}, D, q) = \mathcal{ILP}(\mathcal{D}, \mathcal{AC}, D) \cup \{\lambda = \sum_{i \in \mathcal{I}} \delta_i\} \cup In(q)$, where λ is the value returned by $\mathcal{OPT}(\mathcal{D}, \mathcal{AC}, D)$. It is easy to see that, for every solution s of $\mathcal{ILP}^*(\mathcal{D}, \mathcal{AC}, D, q)$:

1) s can be obtained from a solution of $\mathcal{ILP}(\mathcal{D}, \mathcal{AC}, D) \cup \{\lambda = \sum_{i \in \mathcal{I}} \delta_i\}$ by appropriately setting the new variables x_i and μ_i;

2) for each $i \in \mathcal{I}(q)$, the inequalities $z_j - z_i - 2M \cdot \mu_i \leq 0$ occurring in $In(q)$ (where $j \in \mathcal{I}(q) \setminus \{i\}$) imply that μ_i can take the value 0 only if z_i is not less than every other z_j (that is, if z_i has the maximum value among all z_j);
3) the equality $\sum_{i \in \mathcal{I}(q)} \mu_i = |\mathcal{I}(q)| - 1$ occurring in $In(q)$ imposes that there is exactly one i such that $s[\mu_i] = 0$, while for every $j \neq i$ it is the case that $s[\mu_j] = 1$;
4) considering both the inequalities discussed in 2) and 3) imposes that, if $s[\mu_i] = 0$, then z_i takes the maximum value among variables z_j;
5) the inequalities $x_i - M \cdot \mu_i \leq 0$ and $-x_i - M \cdot \mu_i \leq 0$ impose that $s[x_i] = 0$ if $s[\mu_i] = 0$. Hence, there is exactly one i such that x_i is assigned 0 in s, and this i is such that z_i has the maximum value among the variables z_j. Observe that these inequalities do not impose any restriction on a variable x_i if $s[\mu_i] = 1$.
6) the inequalities $z_i - x_i - 2M \cdot (1 - \mu_i) \leq 0$ and $-z_i + x_i - 2M \cdot (1 - \mu_i) \leq 0$ impose that $s[z_i] - s[x_i] = 0$ if $s[\mu_i] = 1$.

On the whole, for any solution s of $\mathcal{ILP}^*(\mathcal{D}, \mathcal{AC}, D, q)$, there is exactly one x_i which is assigned 0, while every other x_j is assigned the same value as z_j. In particular, the index i such that $s[x_i] = 0$ corresponds to a variable z_i having the maximum value among all the variables z_j. Hence, $\sum_{i \in \mathcal{I}(q)} (s[z_i] - s[x_i])$ results in the maximum value assigned to variables z_i in s.

Now, consider the following optimization ILP problems:

$$\mathcal{OPT}^{MAX}_{glb}(\mathcal{D}, \mathcal{AC}, q, D) \qquad \begin{cases} \textit{minimize } \sum_{i \in \mathcal{I}(q)} (z_i - x_i) \\ \textit{subject to } \mathcal{ILP}^*(\mathcal{D}, \mathcal{AC}, D, q) \end{cases}$$

$$\mathcal{OPT}^{MAX}_{lub}(\mathcal{D}, \mathcal{AC}, q, D) \qquad \begin{cases} \textit{maximize } \sum_{i \in \mathcal{I}(q)} (z_i - x_i) \\ \textit{subject to } \mathcal{ILP}^*(\mathcal{D}, \mathcal{AC}, D, q) \end{cases}$$

Basically, the problems above return the maximum and the minimum values taken by $\sum_{i \in \mathcal{I}(q)} (z_i - x_i)$ among all the solutions of $\mathcal{ILP}^*(\mathcal{D}, \mathcal{AC}, D, q)$. Since the solutions of $\mathcal{ILP}^*(\mathcal{D}, \mathcal{AC}, D, q)$ correspond to the solutions of $\mathcal{ILP}(\mathcal{D}, \mathcal{AC}, D) \cup \{\lambda - \sum_{i \in \mathcal{I}}\}$, which in turn encode the *card*-minimal repairs for D w.r.t. $\mathcal{AC}$, maximizing (resp., minimizing) $\sum_{i \in \mathcal{I}(q)} (z_i - x_i)$ means evaluating the maximum (resp., minimum) value of the MAX-query q among all the "minimally"-repaired databases. As a matter of fact, the following theorem states that the boundaries of the range-CQA of a MAX-query q are the optimal values returned by the above-introduced optimization problems.

Theorem 4. *For a* MAX*-query q, either $CQA^q_{\mathcal{D},\mathcal{AC}}(D) = \emptyset$, or $CQA^q_{\mathcal{D},\mathcal{AC}}(D) = [\ell, u]$, where ℓ is the value returned by $\mathcal{OPT}^{MAX}_{glb}(\mathcal{D}, \mathcal{AC}, q, D)$ and u returned by $\mathcal{OPT}^{MAX}_{lub}(\mathcal{D}, \mathcal{AC}, q, D)$.*

4 Conclusions and Future Work

We have introduced a framework for computing range-consistent answers of MAX, MIN, and SUM queries in numerical databases violating a given set of aggregate constraints, which exploits a transformation into integer linear programming (ILP), thus allowing us to exploit well-known techniques for solving ILP problems.

Several extensions of our framework are worth investigating. Some of them are pretty straightforward. For instance, allowing multiple relations to be specified in the FROM clause in both steady aggregate constraints and aggregate queries does not affect the results stated in this paper, and in particular the correctness of our strategy for computing the range-consistent query answer. On the other hand, other extensions deserve deeper investigation. In particular, from a theoretical standpoint, it will be interesting to remove the assumption that measure attributes are bounded in value. In fact, this removal implies that the boundaries of the range-consistent answers can be $\pm\infty$: this makes it necessary to revise the strategy for computing consistent answers and make it able to detect this case.

References

1. Arenas, M., Bertossi, L.E., Chomicki, J.: Consistent query answers in inconsistent databases. In: Proc. 18th ACM Symp. on Principles of Database Systems (PODS), pp. 68–79 (1999)
2. Arenas, M., Bertossi, L.E., Chomicki, J.: Answer sets for consistent query answering in inconsistent databases. Theory and pract. of logic program (TPLP) 3(4-5), 393–424 (2003)
3. Arenas, M., Bertossi, L.E., Chomicki, J., He, X., Raghavan, V., Spinrad, J.: Scalar aggregation in inconsistent databases. Theor. Comput. Sci. (TCS) 3(296), 405–434 (2003)
4. Bertossi, L.E., Bravo, L., Franconi, E., Lopatenko, A.: The complexity and approximation of fixing numerical attributes in databases under integrity constraints. Inf. Systems 33(4-5), 407–434 (2008)
5. Bohannon, P., Flaster, M., Fan, W., Rastogi, R.: A cost-based model and effective heuristic for repairing constraints by value modification. In: Proc. Int. Conf. on Management of Data (SIGMOD), pp. 143–154 (2005)
6. Calì, A., Lembo, D., Rosati, R.: On the decidability and complexity of query answering over inconsistent and incomplete databases. In: Proc. 22nd ACM Symp. on Principles of Database Systems (PODS), pp. 260–271 (2003)
7. Chomicki, J., Marcinkowski, J., Staworko, S.: Computing consistent query answers using conflict hypergraphs. In: Proc. 13th Conf. on Information and Knowledge Management (CIKM), pp. 417–426 (2004)
8. Chomicki, J., Marcinkowski, J.: Minimal-change integrity maintenance using tuple deletions. Information and Computation (IC) 197(1-2), 90–121 (2005)
9. Fazzinga, B., Flesca, S., Furfaro, F., Parisi, F.: Dart: A data acquisition and repairing tool. In: Proc. Int. Workshop on Incons. and Incompl. in Databases (IIDB), pp. 297–317 (2006)
10. Flesca, S., Furfaro, F., Parisi, F.: Preferred database repairs under aggregate constraints. In: Prade, H., Subrahmanian, V.S. (eds.) SUM 2007. LNCS (LNAI), vol. 4772, pp. 215–229. Springer, Heidelberg (2007)
11. Flesca, S., Furfaro, F., Parisi, F.: Querying and Repairing Inconsistent Numerical Databases. ACM Transactions on Database Systems (TODS) 35(2) (2010)
12. Fuxman, A., Fazli, E., Miller, R.J.: Conquer: Efficient management of inconsistent databases. In: Proc. ACM SIGMOD Int. Conf. on Management of Data (SIGMOD), pp. 155–166 (2005)
13. Fuxman, A., Miller, R.J.: First-order query rewriting for inconsistent databases. J. Comput. Syst. Sci. 73(4), 610–635 (2007)
14. Greco, G., Greco, S., Zumpano, E.: A logical framework for querying and repairing inconsistent databases. IEEE Trans. on Knowledge and Data Engineering (TKDE) 15(6), 1389–1408 (2003)
15. Papadimitriou, C.H.: On the complexity of integer programming. Journal of the Association for Computing Machinery (JACM) 28(4), 765–768 (1981)
16. Wijsen, J.: Database repairing using updates. ACM Transactions on Database Systems (TODS) 30(3), 722–768 (2005)
17. Wijsen, J.: Consistent query answering under primary keys: a characterization of tractable queries. In: Proc. 12th Int. Conf. on Database Theory (ICDT), pp. 42–52 (2009)

Characterization, Propagation and Analysis of Aleatory and Epistemic Uncertainty in the 2008 Performance Assessment for the Proposed Repository for High-Level Radioactive Waste at Yucca Mountain, Nevada

Clifford W. Hansen, Jon C. Helton, and Cédric J. Sallaberry

Sandia National Laboratories, Albuquerque, NM 87185-1399 USA
cwhanse@sandia.gov

Abstract. The 2008 performance assessment (PA) for the proposed repository for high-level radioactive waste at Yucca Mountain (YM), Nevada, illustrates the conceptual structure of risk assessments for complex systems. The 2008 YM PA is based on the following three conceptual entities:a probability space that characterizes aleatory uncertainty; a function that predicts consequences for individual elements of the sample space for aleatory uncertainty; and a probability space that characterizes epistemic uncertainty. These entities and their use in the characterization, propagation and analysis of aleatory and epistemic uncertainty are described and illustrated with results from the 2008 YM PA.

1 Introduction

In 2008 the U.S. Department of Energy (DOE) filed an application with the U.S. Nuclear Regulatory Commission (NRC) seeking a license to construct a geologic repository for radioactive waste at Yucca Mountain (YM), Nevada, USA, which is located within the northern Mojave Desert [1]. The proposed repository comprises mined tunnels approximately 300 m underground in unsaturated volcanic tuff and 300 m above the water table. Spent nuclear fuel and high-level radioactive wastes encased in large cylindrical waste packages (WPs) would be emplaced horizontally within the tunnels underneath engineered drip shields (DSs). The regulations governing issuance of a license [2] required DOE to conduct a performance assessment (PA) to estimate, among other quantities, the mean dose to a reasonably maximally exposed individual (RMEI) (as specified in the regulations) for the time period [0, 1,000,000 yr] after repository closure.

A PA for a geologic repository for radioactive waste is a very involved and detailed analysis. However, the conceptual structure is relatively simple, comprising three basic entities: a probability space that characterizes aleatory uncertainty; a function that estimates consequences for individual elements of the sample space for aleatory uncertainty; and a probability space that characterizes epistemic uncertainty [3,4,5,6]. With this structure the conceptual and computational structure of a large PA can be described without having basic concepts

A. Deshpande and A. Hunter (Eds.): SUM 2010, LNAI 6379, pp. 177–190, 2010.

obscured by details of the analysis. In the following, these three basic entities are described and illustrated in the context of the 2008 YM PA. Specifically, the following topics are addressed: aleatory and epistemic uncertainty (Sect. 2), characterization of aleatory and epistemic uncertainty (Sect. 3), propagation of aleatory and epistemic uncertainty (Sect. 4), and analysis of aleatory and epistemic uncertainty (Sect. 5). A summary (Sect. 6) concludes the presentation.

2 Aleatory and Epistemic Uncertainty

In analyses of complex natural and engineered systems, two broad types of uncertainty are present: aleatory uncertainty, arising from an inherent variability in the behavior or properties of the system under study; and epistemic uncertainty, arising from lack of knowledge with respect to the appropriate values to use for quantities that have fixed but inexactly known values in the context of a particular analysis [7,8,9,10,11,12,13]. It is important to distinguish between these two types of uncertainty in analyses of complex systems. Classic examples of aleatory and epistemic uncertainty are the occurrence time and magnitude of future seismic events and the appropriate value to use for a spatially-averaged permeability in a ground water flow model, respectively. Alternative designations for aleatory uncertainty include stochastic, variability, irreducible, and type A; alternatives to the designation epistemic include subjective, state of knowledge, reducible and type B.

Intuitively, the analysis of a complex natural or engineered system can be viewed as an attempt to answer the following three questions about the system (i.e., Q1, Q2, Q3) and one additional question about the analysis itself (i.e., Q4): Q1, "What could happen?"; Q2, "How likely is it to happen?"; Q3, "What are the consequences if it does happen?"; and Q4, "What is the uncertainty in the answers to the first three questions?" or, equivalently, "What is the level of confidence in the answers to the first three questions?".

Formally, the answers to the first three questions can be represented by a set of ordered triples of the form

$$(\mathcal{S}_i, pS_i, \mathbf{cS}_i), i = 1, 2, \ldots, nS, \tag{1}$$

where (i) $\mathcal{S}_i$ is a set of similar occurrences, (ii) the sets $\mathcal{S}_i$ are disjoint (i.e., $\mathcal{S}_i \bigcap \mathcal{S}_j = \emptyset$ for $i \neq j$) and $\bigcup_i \mathcal{S}_i$ contains everything that could potentially occur at the particular facility under consideration, (iii) pS_i is the probability for $\mathcal{S}_i$, and (iv) $\mathbf{cS}_i$ is a vector of consequences associated with $\mathcal{S}_i$ [14]. The preceding set of ordered triples is an intuitive representation for a function f (i.e., a random variable) defined in association with a probability space $(\mathcal{A}, \mathbf{A}, p_A)$, where $\mathcal{A}$ is the set of everything that could occur in the particular universe under consideration, $\mathbf{A}$ is the collection of subsets of $\mathcal{A}$ for which probability is defined, and p_A is the function that defines probability for the elements of $\mathbf{A}$. Specifically, the sets $\mathcal{S}_i$ are elements of $\mathbf{A}$ with $\bigcup_i \mathcal{S}_i = \mathcal{A}$; $p_A(\mathcal{S}_i)$ is the probability pS_i of $\mathcal{S}_i$; and $f(\mathbf{a}_i)$ for a representative element $\mathbf{a}_i$ of $\mathcal{S}_i$ defines $\mathbf{cS}_i$. As suggested by the notational use of the letter "$\mathcal{A}$", the probability space $(\mathcal{A}, \mathbf{A}, p_A)$ characterizes aleatory uncertainty.

Question Q4 relates to uncertainty that results from a lack of knowledge with respect to the appropriateness of assumptions and/or parameter values used in an analysis. The basic idea is that an analysis has been developed to the point that it has a well-defined overall structure with identified models and parameters but uncertainty remains with respect to appropriate parameter values and possibly models selected for use in this overall structure. Many analyses use probability to characterize such uncertainty, which in turn means that there must be a corresponding probability space $(\mathcal{E}, \mathbf{E}, p_E)$. As suggested by the notational use of the letter "$\mathcal{E}$", the probability space $(\mathcal{E}, \mathbf{E}, p_E)$ is providing a characterization of epistemic uncertainty [12,15].

3 Characterization of Aleatory and Epistemic Uncertainty

Most large analyses begin with an effort to define the analysis structure and select the processes to be modeled. In PAs for radioactive waste disposal, this process is referred to as the screening of features, events and processes (FEPs). This screening process identifies what is to be included in the analysis and provides a documented justification for what is not included in the analysis. A very detailed screening of FEPs was carried out for the 2008 YM PA [16,17].

The overall structure and model components of the 2008 YM PA emerged from the indicated FEPs screening process. In particular, the FEPs screening process identified (i) aleatory uncertainties related to future occurrences, (ii) a large suite of physical processes to be represented by models in the analysis of undisturbed and/or disturbed conditions at the YM repository, and (iii) a large number of epistemically uncertain quantities that would be present in the analysis. As described below, the results of the FEPs screening process and additional supporting analyses [18] can be formally summarized in terms of the three basic analysis entities indicated in Sect. 2.

As indicated in Sect. 2, aleatory uncertainty can be formally characterized by a probability space $(\mathcal{A}, \mathbf{A}, p_A)$. The elements $\mathbf{a}$ of $\mathcal{A}$ are vectors

$$\mathbf{a} = [a_1, a_2, \ldots, a_{nA}] \tag{2}$$

characterizing individual futures that could occur at the facility under consideration. The set $\mathcal{A}$ and the individual futures $\mathbf{a}$ contained in $\mathcal{A}$ are typically defined for some specified time interval; for the proposed YM repository, time intervals of $[0, 10^4\,\text{yr}]$ and $[0, 10^6\,\text{yr}]$ are specified in different parts of the regulations ([18], App. J). In practice, the probability space $(\mathcal{A}, \mathbf{A}, p_A)$ is usually defined by specifying distributions for the individual elements of $\mathbf{a}$. For notational purposes, it is convenient to represent the distribution associated with the elements $\mathbf{a}$ of $\mathcal{A}$ with a density function $d_A(\mathbf{a})$.

For the 2008 YM PA the screening process identified three types of events that warranted inclusion in the PA structure: (i) early failure of engineered components (WPs and DSs) due to undetected manufacturing defects or errors in emplacement; (ii) igneous events (magma intrusions and eruptive conduits); and (iii) seismic events (vibratory ground motion and fault displacement). Other events were considered but were not included due to low probability of occurrence (e.g., nuclear criticality) or insignificant effect on the disposal system (e.g., glaciation).

A future involving these three types of events can be represented mathematically as

$$\mathbf{a} = [nEW, nED, nII, nIE, nSG, nSF, \mathbf{a}_{EW}, \mathbf{a}_{ED}, \mathbf{a}_{II}, \mathbf{a}_{IE}, \mathbf{a}_{SG}, \mathbf{a}_{SF}], \quad (3)$$

where nEW = number of early WP failures, nED = number of early DS failures, nII = number of igneous intrusive events, nIE = number of igneous eruptive events, nSG = number of seismic ground motion events, nSF = number of seismic fault displacement events, and $\mathbf{a}_{EW}, \mathbf{a}_{ED}, \mathbf{a}_{II}, \mathbf{a}_{IE}, \mathbf{a}_{SG}$ and $\mathbf{a}_{SF}$ are vectors defining the nEW early WP failures, nED early DS failures, nII igneous intrusive events, nIE igneous eruptive events, nSG seismic ground motion events, and nSF fault displacement events, respectively. As an example, $\mathbf{a}_{EW} = [\mathbf{a}_{EW,1}, \mathbf{a}_{EW,2}, \ldots, \mathbf{a}_{EW,nEW}]$, where $\mathbf{a}_{EW,j}$ is a vector defining early WP failure j for $j = 1, 2, \ldots, nEW$. The vectors $\mathbf{a}_{EW,j}, j = 1, 2, \ldots, nEW$, appearing in the definition of $\mathbf{a}_{EW}$ are in turn defined by $\mathbf{a}_{EW,j} = [t_j, b_j, d_j]$ and characterize the properties of failed WP j, where t_j designates WP type (i.e., $t_j = 1$ indicates a commercial spent nuclear fuel WP, $t_j = 2$ indicates a codisposed WP), b_j designates the percolation bin in which the failed WP is located (i.e., $b_j = k$ indicates that the failed WP is in percolation bin k for $k \in \{1, 2, 3, 4, 5\}$; see Fig. 6.1.4-2, Ref. [18]), and d_j designates whether the failed WP experiences nondripping or dripping conditions (i.e., $d_j = 0$ indicates nondripping conditions and $d_j = 1$ indicates dripping conditions). Distributions for each element of $\mathbf{a}_{EW,j}$ are based on assuming that the number of early failed WPs follows a binomial distribution with the failed WPs distributed randomly over WP types, percolation bins, and nondripping/dripping conditions. Definitions of the vectors $\mathbf{a}_{ED}, \mathbf{a}_{II}, \mathbf{a}_{IE}, \mathbf{a}_{SG}$ and $\mathbf{a}_{SF}$ and their associated probabilistic characterizations are given in App. J of Ref. [18].

As indicated in Sect. 2, the second entity that underlines a PA for a complex system is a function f that estimates a vector $f(\mathbf{a})$ of consequences for individual elements $\mathbf{a}$ of the sample space $\mathcal{A}$ for aleatory uncertainty. In most analyses, the function f corresponds to a sequence of models for multiple physical processes that must be implemented and numerically evaluated with one or more computer programs. However, for notational and conceptual purposes, it is useful to represent these models as a single function. Many analysis results of interest are functions of time and thus the model used to estimate system behavior can be represented by $f(\tau|\mathbf{a})$, where (i) τ corresponds to time and the indication of conditionality (i.e., $|\mathbf{a}$) emphasizes that the results at time τ depend on the particular element $\mathbf{a}$ of $\mathcal{A}$ under consideration, and (ii) $f(\tau|\mathbf{a})$ corresponds to a vector containing a large number of results. For the 2008 YM PA, the function f is an assemblage of interacting models that describe water flow, heat transfer, water chemistry, mechanical and chemical degradation of rock and metals, and contaminant transport. Model configurations used in the 2008 YM PA are illustrated in Figs. 6.1.4-1 to 6.1.4-6 and G-1 to G-6 of Ref. [18].

The third entity that underlies a PA for a complex system is a probability space $(\mathcal{E}, \mathbf{E}, p_E)$ that characterizes epistemic uncertainty. The elements $\mathbf{e}$ of $\mathcal{E}$ are vectors of the form

$$\mathbf{e} = [\mathbf{e}_A, \mathbf{e}_M] = [e_1, e_2, \ldots, e_{nE}], \tag{4}$$

where $\mathbf{e}_A$ is a vector of epistemically uncertain quantities involved in the definition of the probability space $(\mathcal{A}, \mathbf{A}, p_A)$ for aleatory uncertainty and $\mathbf{e}_M$ is a vector of epistemically uncertain quantities involved in the evaluation of the function f. The probability space $(\mathcal{E}, \mathbf{E}, p_E)$ for epistemic uncertainty and its associated density function $d_E(\mathbf{e})$ are usually developed through an expert review process that involves assigning a distribution to each element e_j of $\mathbf{e}$ [19,20,21,22].

With the introduction of the epistemically uncertain quantities that constitute the elements of $\mathbf{e} = [\mathbf{e}_A, \mathbf{e}_M]$, the notation for the density function $d_A(\mathbf{a})$ associated with the probability space $(\mathcal{A}, \mathbf{A}, p_A)$ for aleatory uncertainty becomes $d_A(\mathbf{a}|\mathbf{e}_A)$ to indicate the dependence on $\mathbf{e}_A$; similarly, the notation for the time-dependent values for the function f becomes $f(\tau|\mathbf{a}, \mathbf{e}_M)$. The model assemblage used for the 2008 YM PA included 392 epistemically uncertain quantities (i.e., $nE = 392$ in (4)). Example elements of the vector $\mathbf{e}$ are shown in Table 1, where *IGRATE* is an element of $\mathbf{e}_A$ and the remaining variables are elements of $\mathbf{e}_M$.

Table 1. Examples of the $nE = 392$ elements of the vector $\mathbf{e}$ of epistemically uncertain quantities in the 2008 YM PA ([18], Tables K3-1, K3-2, K3-3)

SCCTHRP: Residual stress threshold for stress-corrosion crack nucleation of Alloy 22 (as a percentage of yield strength in MPa) (dimensionless).
EP1LOWPU: Logarithm of the scale factor used to characterize uncertainty in plutonium solubility at an ionic strength below 1 molal (dimensionless).
IGRATE: Frequency of intersection of the repository footprint by a volcanic event (yr^{-1}).
INFIL: Pointer variable for determining infiltration conditions: 10th, 30th, 50th or 90th percentile infiltration scenario (dimensionless).
MICC14: Groundwater Biosphere Dose Conversion Factor (BDCF) for carbon-14 in modern interglacial climate ($(\mathrm{Sv/year})/(\mathrm{Bq/m^3})$).
SZFIPOVO: Logarithm of flowing interval porosity in volcanic units (dimensionless).
SZGWSPDM: Logarithm of the scale factor used to characterize uncertainty in groundwater specific discharge (dimensionless).
WDGCA22: Slope term for temperature dependence of Alloy 22 general corrosion rate (K).

The models incorporated into the function f can also be regarded as epistemically uncertain. Such uncertainty reflects questions about the degree to which the models represent the system under consideration. Extensive literature is available regarding uncertainty in models and methods for characterizing this type of uncertainty [23,24,25,26,27]. In practice, qualitative rather than quantitative methods for addressing this type of uncertainty are common [26]. In the 2008 YM PA, generally one of several candidate models is selected and the choice justified by comparisons between alternative models [18]. In a few cases, however, the 2008 YM PA employed epistemically uncertain pointer variables to select among several alternative models with the result that several models are represented in the analysis outcomes [27].

4 Propagation of Aleatory and Epistemic Uncertainty

Aleatory uncertainty in an outcome of a PA is usually summarized with a cumulative distribution function (CDF) or a complementary cumulative distribution function (CCDF). For a real-valued analysis outcome $y = f(\tau|\mathbf{a}, \mathbf{e}_M)$, the CDF and CCDF for y resulting from aleatory uncertainty in $\mathbf{a}$ are defined by

$$p_A(\tilde{y} \leq y|\mathbf{e}) = \int_{\mathcal{A}} \underline{\delta}_y[f(\tau|\mathbf{a}, \mathbf{e}_M)] d_A(\mathbf{a}|\mathbf{e}_A) \mathrm{d}\mathcal{A} \tag{5}$$

and

$$p_A(y < \tilde{y}|\mathbf{e}) = 1 - p_A(\tilde{y} \leq y|\mathbf{e}) = \int_{\mathcal{A}} \bar{\delta}_y[f(\tau|\mathbf{a}, \mathbf{e}_M)] d_A(\mathbf{a}|\mathbf{e}_A) \mathrm{d}\mathcal{A}, \tag{6}$$

respectively, where

$$\underline{\delta}_y(\tilde{y}) = \begin{cases} 1 & \text{for } \tilde{y} \leq y \\ 0 & \text{otherwise} \end{cases} \text{ and } \bar{\delta}(\tilde{y}) = 1 - \underline{\delta}_y(\tilde{y}) = \begin{cases} 1 & \text{for } y \leq \tilde{y} \\ 0 & \text{otherwise.} \end{cases} \tag{7}$$

Specifically, plots of the points $[y, p_A(\tilde{y} \leq y|\mathbf{e})]$ and $[y, p_A(y \leq \tilde{y}|\mathbf{e})]$ define the CDF and CCDF for $y = f(\tau|\mathbf{a}, \mathbf{e}_M)$, with these plots being conditional on the vector $\mathbf{e} = [\mathbf{e}_A, \mathbf{e}_M]$ of epistemically uncertain analysis inputs. In addition,

$$E_A(y|\mathbf{e}) = \int_{\mathcal{A}} f(\tau|\mathbf{a}, \mathbf{e}_M) d_A(\mathbf{a}|\mathbf{e}_A) \mathrm{d}\mathcal{A} \tag{8}$$

defines the expected value of $y = f(\tau|\mathbf{a}, \mathbf{e}_M)$ over aleatory uncertainty.

In most PAs, the integrals in Eqs. (6), (7) and (8) are too complex to estimate with formal quadrature procedures, with the result that these integrals are typically estimated with procedures based on simple random sampling or stratified sampling. The sampling procedures result in estimates of the form

$$p_A(\tilde{y} \leq y|\mathbf{e}) \cong \begin{cases} \sum_{i=1}^{m} \underline{\delta}_y[f(\tau|\mathbf{a}_i, \mathbf{e}_M)]/m \\ \sum_{i=1}^{k} \underline{\delta}_y[f(\tau|\tilde{\mathbf{a}}_i, \mathbf{e}_M)] p_A(\mathcal{A}_i|\mathbf{e}_A) \end{cases} \tag{9}$$

$$p_A(y < \tilde{y}|\mathbf{e}) \cong \begin{cases} \sum_{i=1}^{m} \bar{\delta}_y[f(\tau|\mathbf{a}_i, \mathbf{e}_M)]/m \\ \sum_{i=1}^{k} \bar{\delta}_y[f(\tau|\tilde{\mathbf{a}}_i, \mathbf{e}_M)] p_A(\mathcal{A}_i|\mathbf{e}_A) \end{cases} \tag{10}$$

and

$$E_A(y|\mathbf{e}) \cong \begin{cases} \sum_{i=1}^{m} \bar{f}(\tau|\mathbf{a}_i, \mathbf{e}_M)/m \\ \sum_{i=1}^{k} \bar{f}(\tau|\tilde{\mathbf{a}}_i, \mathbf{e}_M) p_A(\mathcal{A}_i|\mathbf{e}_A), \end{cases} \tag{11}$$

where (i) $\mathbf{a}_i, i = 1, 2, \ldots, m$ is a random sample from $\mathcal{A}$ obtained in consistency with the density function $d_A(\mathbf{a}|\mathbf{e}_A)$ for aleatory uncertainty and (ii) the sets $\mathcal{A}_i, i = 1, 2, \ldots, k$, satisfy $\bigcup_i \mathcal{A}_i = \mathcal{A}$ with $\mathcal{A}_i \bigcap \mathcal{A}_j = \emptyset$ for $i \neq j$, $\tilde{\mathbf{a}}_i$ is a representative element of $\mathcal{A}_i$, and $p_A(\mathcal{A}_i)$ is the probability of $\mathcal{A}_i$. The summations with $\mathbf{a}_i$ and $\tilde{\mathbf{a}}_i$ correspond to approximations with simple random sampling and stratified sampling, respectively. Further, use of the approximations with $\tilde{\mathbf{a}}_i$ correspond to use of the ordered triple representation for risk in (1) with $\mathcal{S}_i = \mathcal{A}_i$, $pS_i = p_A(\mathcal{A}_i)$, and $cS_i = f(\tau|\tilde{\mathbf{a}}_i, \mathbf{e}_M)$.

As an example, Fig. 1a displays CCDFs for the dose (mrem/yr) resulting from seismic ground motion events. Each CCDF is conditional on a specific realization $\mathbf{e}$ of epistemic uncertainty and was generated in the manner indicated in (10) with a random sample of size $m = 20,000$ from elements of $\mathcal{A}$ that involved seismic ground motion events prior to 10,000 yr after repository closure. In these results only the effects of the seismic ground motion events are considered; the effects of other types of events (i.e., early failures) are considered separately. Each CCDF was efficiently generated by: (i) performing detailed calculations for a small but representative set of seismic ground motion events and using appropriate interpolation and additive procedures to obtain dose to the RMEI from seismic ground motion events described by the 20,000 randomly sampled elements of $\tilde{\mathcal{A}}$; and (ii) only sampling elements of $\mathcal{A}$ that involved ground motion events and then employing an appropriate probabilistic correction to account for elements of $\mathcal{A}$ that did not involve seismic ground motion events ([18], Sect. J8.3)

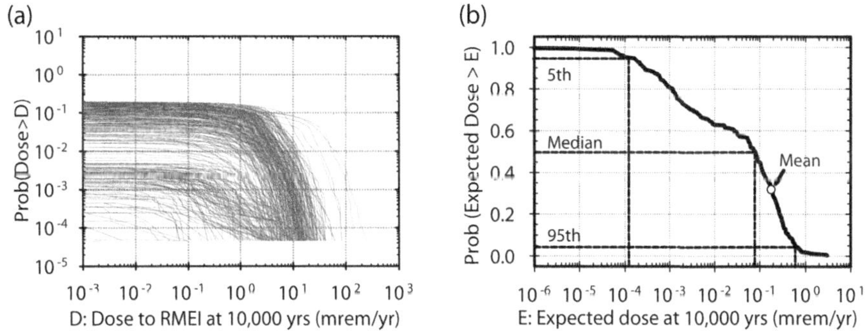

Fig. 1. Dose to the RMEI at 10,000 yr resulting from seismic ground motion events: (a) CCDFs summarizing aleatory uncertainty in dose ([18], Fig. J8.3-10), and (b) CCDF summarizing epistemic uncertainty in expected dose ([18], Fig. J8.3-5)

The individual CCDFs in Fig. 1a were generated for the elements of a Latin hypercube sample (LHS)

$$\mathbf{e}_i = [\mathbf{e}_{Ai}, \mathbf{e}_{Mi}], i = 1, 2, \ldots, n = 300 \tag{12}$$

of size 300 from the sample space $\mathcal{E}$ for epistemic uncertainty. Latin hypercube sampling is a probabilistically-based sampling procedure that is often used for the propagation of epistemic uncertainty because of its dense stratification over the range of each epistemically uncertain analysis input [15,28].

The effects of epistemic uncertainty are also usually represented with CDFs and CCDFs whose determination formally involves the evaluation of integrals over the sample space $\mathcal{E}$ for epistemic uncertainty. For example, the CCDF summarizing the epistemic uncertainty in the expected value $E_A(y|\mathbf{e})$ (expectation over aleatory uncertainty) for $y = f(\tau|\mathbf{a}, \mathbf{e}_M)$ is given by

$$\begin{aligned} p_E[\bar{y} \leq E_A(y|\mathbf{e})] &= p_E\left[\bar{y} \leq \int_{\mathcal{A}} f(\tau|\mathbf{a}, \mathbf{e}_M) d_A(\mathbf{a}|\mathbf{e}_A) \mathrm{d}\mathcal{A}\right] \\ &= \int_{\mathcal{E}} \bar{\delta}_{\bar{y}} \left[\int_{\mathcal{A}} f(\tau|\mathbf{a}, \mathbf{e}_M) d_A(\mathbf{a}|\mathbf{e}_A) \mathrm{d}\mathcal{A}\right] d_E(\mathbf{e}) \mathrm{d}\mathcal{E} \\ &\cong \begin{cases} \sum_{i=1}^{n} \bar{\delta}_{\bar{y}}[\sum_{j=1}^{m} f(\tau|\mathbf{a}_j, \mathbf{e}_{Mi})/m]/n \\ \sum_{i=1}^{n} \bar{\delta}_{\bar{y}}[\sum_{j=1}^{k} f(\tau|\tilde{\mathbf{a}}_j, \mathbf{e}_{Mi}) p_A(\mathcal{A}_j|\mathbf{e}_{A,i})]/n, \end{cases} \end{aligned} \tag{13}$$

where the samples $\mathbf{e}_i, i = 1, 2, \ldots, n$, $\mathbf{a}_j, j = 1, 2, \ldots, m$, and $\tilde{\mathbf{a}}_j, j = 1, 2, \ldots, k$, are defined the same as before. A CCDF of the form defined in (13) is illustrated in Fig. 1b for the expected dose at $\tau = 10,000$ yr resulting from seismic ground motion events.

In a similar manner, the CDF summarizing the epistemic uncertainty in $y = f(\tau|\mathbf{a}, \mathbf{e}_M)$ at time τ conditional on a specific element $\mathbf{a}$ of the sample space $\mathcal{A}$ for aleatory uncertainty is defined by the probabilities

$$p_E(\tilde{y} \leq y|\mathbf{a}) = \int_{\mathcal{E}} \underline{\delta}_y \left[f(\tau|\mathbf{a}, \mathbf{e}_M)\right] d_E(\mathbf{e}) \mathrm{d}\mathcal{E} \cong \sum_{i=1}^{n} \underline{\delta}_y \left[f(\tau|\mathbf{a}, \mathbf{e}_{Mi})\right]/n, \tag{14}$$

where $\underline{\delta}_y(\cdot)$ is defined the same as in (7) and $\mathbf{e}_i, i = 1, 2, \ldots, n$, is a simple random sample or LHS from $\mathcal{E}$ generated in consistency with the density function $d_E(\mathbf{e})$ for epistemic uncertainty. Time-dependent values for $y = f(\tau|\mathbf{a}, \mathbf{e}_M)$ are illustrated in Fig. 2a. In contrast, Fig. 2b compares results at a fixed time (i.e., 10^4 yr) for several representative elements $\mathbf{a}$ of the sample space $\mathcal{A}$. When a number of results of this form are to be presented and compared, box plots provide convenient and easily compared summaries of a number of CDFs in a compact format as illustrated in Fig. 2b.

The regulations for the YM repository [2,6] specify limits of 15 mrem/yr and 100 mrem/yr on the expected value of dose to the RMEI for the time intervals of $[0, 10^4\,\text{yr}]$ and $[10^4, 10^6\,\text{yr}]$, respectively, following repository closure. The regulations indicate that the preceding bounds apply to expected values over both aleatory and epistemic uncertainty. However, the regulations also specifically require the uncertainty in expected dose to be presented. Thus the regulations require an analysis that determines both (i) the epistemic uncertainty in estimates of the expected dose to the RMEI over aleatory uncertainty and (ii) the expectation of dose to the RMEI over both aleatory and epistemic uncertainty. For descriptive convenience, expected dose from aleatory uncertainty conditional on a specific realization of epistemic uncertainty is referred to simply as expected

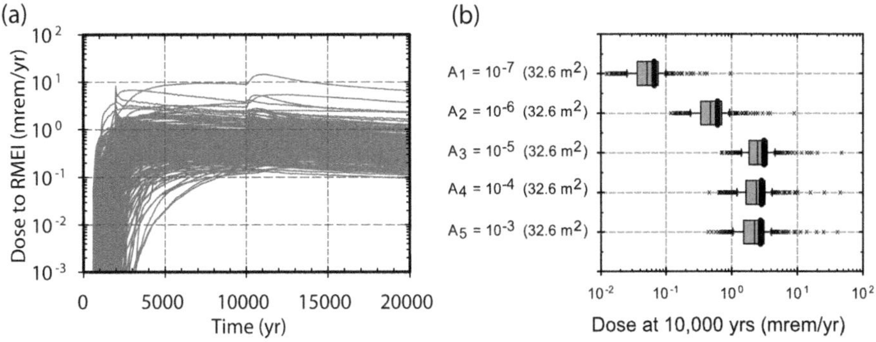

Fig. 2. Dose to the RMEI resulting from damage to CDSP WPs caused by a seismic ground motion event at 200 yr obtained with an LHS of size $nLHS = 300$: (a) time-dependent dose resulting from damaged area of $A_s = (10^{-6})(32.6\text{m}^2)$ ([18], Fig. J8.3-1) and (b) box plots summarizing epistemic uncertainty in dose to the RMEI resulting from damaged areas of $A_s = (10^{-6+s}) \times (32.6\text{m}^2)$ with $s = 1, 2, 3, 4, 5$ ([18], Fig. J8.3-2) (box extends from 0.25 to 0.75 quantile; left and right bar and whisker extend to 0.1 and 0.9 quantile, respectively; ×'s represent values outside the 0.1 to 0.9 quantile range; median and mean are represented by light and dark vertical lines, respectively)

dose, and expected dose over both aleatory and epistemic uncertainty is referred to as expected (mean) dose.

The determination of time-dependent values for both expected dose and expected (mean) dose to the RMEI required careful planning due to the complexity of (i) the probability space characterizing aleatory uncertainty, (ii) the probability space characterizing epistemic uncertainty, and (iii) the models representing physical processes associated with nominal and/or disturbed conditions at the repository and its vicinity. To deal with this complexity, the computational implementation of the 2008 YM PA was decomposed into the consideration of seven scenario classes (i.e., subsets of the sample space $\mathcal{A}$ for aleatory uncertainty) corresponding to events with differing effects: nominal (i.e., undisturbed) conditions, early WP failure, early DS failure, igneous intrusive events, igneous eruptive events, seismic ground motion events, and seismic fault displacement events ([18], App. J). For example, the seismic ground motion scenario class is defined by $\mathcal{S}_{SG} = \{\mathbf{a} \in \mathcal{A} | nSG > 0\}$ and represents futures involving one or more seismic ground motion events.

The computational implementation of the 2008 YM PA was based on the concept that (i) analyses for expected dose to the RMEI could be performed separately for the events described in each scenario class and (ii) the expected dose results for each scenario class could be summed to obtain expected and expected (mean) dose for all scenario classes. This computational strategy assumes that no synergisms are present between the effects of different types of events. The underlying justification for this decomposition and subsequent recomposition of dose results is that the more likely occurrences (e.g., early failures and seismic events) have relatively small effects and the less likely occurrences (e.g., igneous events) have large effects that dominate the effects of more likely occurrences. Thus, the error in the estimates of dose resulting from the decomposition and recomposition is small.

For each scenario class the time-dependent expected dose attributable to the events described by the scenario class can be computed as indicated in (11) and expected (mean) dose obtained. As an example, the expected and expected (mean) dose to the RMEI from seismic ground motion events and seismic fault displacement events are shown in Fig. 3a and Fig. 3b, respectively. The two indicated figures each contain 300 curves (i.e., one curve for each LHS element in (12)). The spread of these curves provides a representation of the epistemic uncertainty present in the estimation of these quantities. The individual curves could, in concept, be calculated as indicated in (10) and illustrated in Fig. 1b. However, similar to the calculation of the results in Fig. 1, it was more efficient to use a quadrature procedure with appropriate interpolations to obtain the results in Fig. 3a and Fig. 3b ([18], Sect. J8). The indicated mean and quantiles (i.e., percentiles) were obtained in the manner indicated in conjunction with (13) and provide a quantitative summary of the epistemic uncertainty present in the estimation of the expected dose for the seismic ground motion and seismic fault displacement scenario classes.

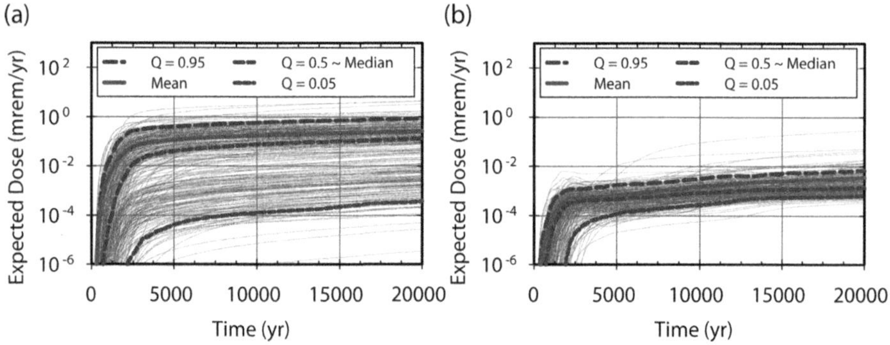

Fig. 3. Expected dose to the RMEI (*EXPDOSE*, mrem/yr) over $[0, 2\times 10^4 \text{ yr}]$ resulting from seismic events: (a) *EXPDOSE* for seismic ground motion events ([18], Fig. K7.7.1-1[a]), and (b) *EXPDOSE* for seismic fault displacement events ([18], Fig. K7.8.1-1[a])

As previously indicated, the expected dose for the seven scenario classes are summed to obtain expected and expected (mean) dose for all scenario classes. The same LHS was used in all propagations of epistemic uncertainty in the 2008 YM PA, which permitted analysis results to be combined across different parts of the analysis. The results of this calculation are shown in Fig. 4a and Fig. 4c for the time intervals $[0, 2 \times 10^4 \text{ yr}]$ and $[0, 10^6 \text{ yr}]$, respectively.

The results in Fig. 4a are smoother than the results in Fig. 4c because quadrature procedures with appropriate interpolations from precalculated results were

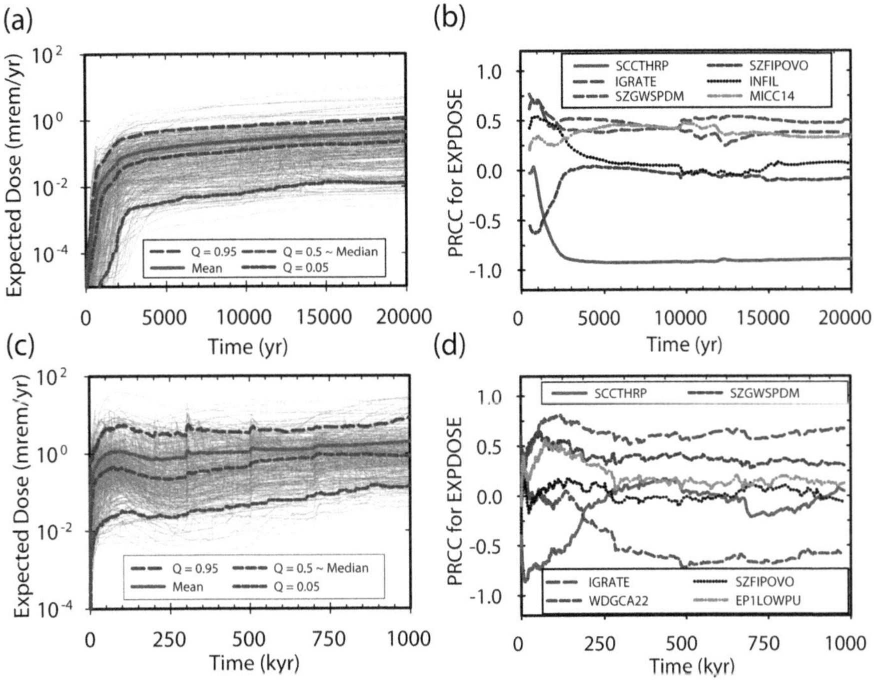

Fig. 4. Expected dose to RMEI (*EXPDOSE*, mrem/yr) for all scenario classes: (a, b) *EXPDOSE* and associated partial rank correlation coefficients (PRCCs) for $[0, 2 \times 10^4\,\text{yr}]$ ([18], Fig. K8.1-1[a]), and (c, d) *EXPDOSE* and associated PRCCs for $[0, 10^6\,\text{yr}]$ ([18], Fig. K8.2-1[a])

used to obtain the expected dose results in Fig. 4a. In contrast, the complexity of evolving repository conditions and the potential for multiple seismic ground motion events required the use of a Monte Carlo procedure to estimate the combined effects of evolving repository conditions and seismic ground motion events for the $[0, 10^6\,\text{yr}]$ time interval. Specifically, 30 random elements of the sample space $\mathcal{A}$ for aleatory uncertainty were used in the generation of the seismic ground motion component of the expected dose curves in Fig. 4c. The use of a larger sample would have reduced the choppiness in the expected dose curves for the $[0, 10^6\,\text{yr}]$ time interval but would not have changed the basic nature of the results. For both time periods, the expected (mean) dose results are substantially below the regulatory requirements of 15 and 100 mrem/yr for time intervals of $[0, 10^4\,\text{yr}]$ and $[10^4, 10^6\,\text{yr}]$, respectively.

5 Analysis of Aleatory and Epistemic Uncertainty

The analysis of aleatory and epistemic uncertainty divides into two areas. The first area involves the display and examination of the aleatory and epistemic uncertainty that is present in analysis results of interest. This is essentially the outcome of the uncertainty propagation procedures discussed in Sect. 4 and illustrated in Figs. 1 - 3, 4a and 4c. The second area is formal sensitivity analysis to determine the effects of individual epistemically uncertain analysis inputs (i.e., elements of vector $\mathbf{e}$ in (4)) on analysis results of interest.

In the 2008 YM PA, the mapping between epistemically uncertain analysis inputs and analysis results was explored for results generated with the LHS in (12) using a variety of relatively simple sensitivity analysis procedures, including examination of scatterplots, correlation analysis, regression analysis, partial correlation analysis, and rank transformations ([18], App. K). More sophisticated procedures (i.e.,statistical tests for patterns based on gridding, entropy tests for patterns based on gridding, nonparametric regression analysis, squared rank differences/rank correlation test, two dimensional Kolmogorov-Smirnov test, tests for patterns based on distance measures, top down coefficient of concordance, and variance decomposition) are available and have been used in other analyses [29,30].

Example sensitivity analyses based on partial rank correlation coefficients (PRCCs) are presented in Fig. 4b and Fig. 4d. In these examples, the PRCCs are determined by analyzing the uncertainty associated with the analysis results above individual values on the abscissas of the indicated figures, and then connecting these results to form curves of sensitivity analysis results for individual analysis inputs. As a reminder, a PRCC provides a measure of the monotonic effect of an uncertain analysis input on an analysis result after the removal of the monotonic effects of all other uncertain analysis inputs. Additional examples of sensitivity analysis results are available in App. K of Ref. [18] for the 2008 YM PA and in Ref. [31] for the 1996 Waste Isolation Pilot Plant PA.

6 Summary

The conceptual structure and computational organization of the 2008 YM PA is based on the following three basic entities: a probability space $(\mathcal{A}, \mathbf{A}, p_A)$ that characterizes aleatory uncertainty; a function f that estimates consequences for individual elements $\mathbf{a}$ of the sample space $\mathcal{A}$ for aleatory uncertainty; and a probability space $(\mathcal{E}, \mathbf{E}, p_E)$ that characterizes epistemic uncertainty. Recognition of these three basic entities makes it possible to understand the conceptual and computational basis of the 2008 YM PA without having basic concepts obscured by fine details of the analysis and leads to an analysis that produces insightful uncertainty and sensitivity results. A fuller description of the ideas and results discussed in this presentation is available in Apps. J and K of Ref. [18].

Acknowledgements. Work performed at Sandia National Laboratories (SNL), which is a multiprogram laboratory operated by Sandia Corporation, a Lockheed Martin Company, for the U.S. Department of Energy's (DOE's) National Nuclear Security Administration under Contract No. DE-AC04-94AL85000. The views expressed in this article are those of the authors and do not necessarily reflect the views or policies of the DOE or SNL.

References

1. US DOE (United States Department of Energy): Yucca Mountain Repository License Application, DOE/RW-0573, Rev 0. (2008)
2. US NRC (United States Nuclear Regulatory Commission): 10 Code of Federal Regulations Part 63: Disposal of High-Level Radioactive Wastes in a Geologic Repository at Yucca Mountain, Nevada (2009)
3. Helton, J.C., et al.: Conceptual Structure of the 1996 Performance Assessment for the Waste Isolation Pilot Plant. Reliability Engineering and System Safety 69, 151–165 (2000)
4. Helton, J.C.: Mathematical and Numerical Approaches in Performance Assessment for Radioactive Waste Disposal: Dealing with Uncertainty. In: Scott, E.M. (ed.) Modelling Radioactivity in the Environment, pp. 353–390. Elsevier Science, New York (2003)
5. Helton, J.C., et al.: Yucca Mountain 2008 Performance Assessment: Conceptual Structure and Computational Implementation. In: Proceedings of the 12th International High-Level Radioactive Waste Management Conference, pp. 524–532. American Nuclear Society, La Grange Park (2008)
6. Helton, J.C., Sallaberry, C.J.: Conceptual Basis for the Definition and Calculation of Expected Dose in Performance Assessments for the Proposed High-Level Radioactive Waste Repository at Yucca Mountain, Nevada. Reliability Engineering and System Safety 94, 677–698 (2009)
7. Parry, G.W., Winter, P.W.: Characterization and Evaluation of Uncertainty in Probabilistic Risk Analysis. Nuclear Safety 22, 28–42 (1981)
8. Parry, G.W.: The Characterization of Uncertainty in Probabilistic Risk Assessments of Complex Systems. Reliability Engineering and System Safety 54, 119–126 (1996)
9. Apostolakis, G.: The Concept of Probability in Safety Assessments of Technological Systems. Science 250, 1359–1364 (1990)
10. Hoffman, F.O., Hammonds, J.S.: Propagation of Uncertainty in Risk Assessments: The Need to Distinguish Between Uncertainty Due to Lack of Knowledge and Uncertainty Due to Variability. Risk Analysis 14, 707–712 (1994)
11. Helton, J.C.: Treatment of Uncertainty in Performance Assessments for Complex Systems. Risk Analysis 14, 483–511 (1994)
12. Helton, J.C.: Uncertainty and Sensitivity Analysis in the Presence of Stochastic and Subjective Uncertainty. Journal of Statistical Computation and Simulation 57, 3–76 (1997)
13. Paté-Cornell, M.E.: Uncertainties in Risk Analysis: Six Levels of Treatment. Reliability Engineering and System Safety 54, 95–111 (1996)
14. Kaplan, S., Garrick, B.J.: On the Quantitative Definition of Risk. Risk Analysis 1, 11–27 (1981)
15. Helton, J.C., Davis, F.J.: Latin Hypercube Sampling and the Propagation of Uncertainty in Analyses of Complex Systems. Reliability Engineering and System Safety 81, 23–69 (2003)
16. SNL (Sandia National Laboratories): Features, Events, and Processes for the Total System Performance Assessment: Methods. ANL-WIS-MD-000026 REV 00. U.S. Department of Energy Office of Civilian Radioactive Waste Management, Las Vegas (2008)

17. SNL (Sandia National Laboratories): Features, Events, and Processes for the Total System Performance Assessment: Analyses. ANL-WIS-MD-000027 REV 00. U.S. Department of Energy Office of Civilian Radioactive Waste Management, Las Vegas (2008)
18. SNL (Sandia National Laboratories): Total System Performance Assessment Model/Analysis for the License Application.MDL-WIS-PA-000005 Rev 00 AD 01. U.S. Department of Energy Office of Civilian Radioactive Waste Management, Las Vegas (2008)
19. Hora, S.C., Iman, R.L.: Expert Opinion in Risk Analysis: The NUREG-1150 Methodology. Nuclear Science and Engineering 102, 323–331 (1989)
20. Bonano, E.J., et al.: Elicitation and Use of Expert Judgment in Performance Assessment for High-Level Radioactive Waste Repositories. Sandia National Laboratories, Albuquerque (1990)
21. Cooke, R.M.: Experts in Uncertainty: Opinion and Subjective Probability in Science. Oxford University Press, New York (1991)
22. Meyer, M.A., Booker, J.M.: Eliciting and Analyzing Expert Judgment: A Practical Guide. SIAM, Philadelphia (2001)
23. Mosleh, A., Siu, N., Smidts, C., Liu, C.: Proceedings of Workshop I in Advanced Topics in Risk and Reliability Analysis, Model Uncertainty: Its Characterization and Quantification. NUREG/CP-0138. U.S. Nuclear Regulatory Commission, Washington, D.C. (1994)
24. Cullen, A., Frey, H.C.: Probabilistic Techniques in Exposure Assessment. Plenum Press, New York (1999)
25. Droguett, E.L., Mosleh, A.: Bayesian Methodology for Model Uncertainty Using Model Performance Data. Risk Analysis 28, 1457–1476 (2008)
26. Galson, D.A., Khursheed, A.: The Treatment of Uncertainty in Performance Assessment and Safety Case Development: State-of-the Art Overview. GSL/0546-WP1.2. Galson Sciences Ltd., Oakham (2007)
27. Helton, J.C., et al.: Model Uncertainty: Conceptual and Practical Issues in Risk-Informed Decision Making Context. In: Mosleh, A., et al. (eds.) Model Uncertainty, Center for Risk and Reliability, University of Maryland (2010)
28. McKay, M.D., et al.: A Comparison of Three Methods for Selecting Values of Input Variables in the Analysis of Output from a Computer Code. Technometrics 21, 239–245 (1979)
29. Helton, J.C., et al.: Survey of Sampling-Based Methods for Uncertainty and Sensitivity Analysis. Reliability Engineering and System Safety 91, 1175–1209 (2006)
30. Helton, J.C.: Uncertainty and Sensitivity Analysis Techniques for Use in Performance Assessment for Radioactive Waste Disposal. Reliability Engineering and System Safety 42, 327–367 (1993)
31. Helton, J.C., Marietta, M.G.(eds.): Special Issue: The 1996 Performance Assessment for the Waste Isolation Pilot Plant. Reliability Engineering and System Safety 69, 1–451 (2000)

Comparing Evidential Graphical Models for Imprecise Reliability

Wafa Laâmari[1], Boutheina Ben Yaghlane[1], and Christophe Simon[2]

[1] LARODEC Laboratory - Institut Supérieur de Gestion de Tunis, Tunisia
[2] CRAN Laboratory - Nancy Université - CNRS, UMR 7039, ESSTIN, France

Abstract. This paper presents a comparison of two evidential networks applied to the reliability study of complex systems with uncertain knowledge. This comparison is based on different aspects. In particular, the original structure, the graphical structures for the inference, the message-passing schemes, the storage efficiencies, the computational efficiencies and the exactness of the results are studied.

1 Introduction

The reliability analysis of systems interests today many fields since it is an important element of process control in companies. In studies of system reliability, the information on the behavior of the system components' reliability usually incorporates uncertainties [4].

The ignorance, the inconsistency and the incompleteness of available data are sources of uncertainty incorporated in systems [11]. The uncertainty presents various forms. It may be either stochastic or epistemic. The first form of uncertainty occurs when the values of available data are uncertain. Whereas the second type of uncertainty occurs in the presence of incomplete or inconsistent information [12]. The problem which arises from the uncertainty has been treated by several theories in the literature such as the probability theory, the possibility theory and the evidence theory [14, 8].

The probabilistic formalism provides different tools to handle efficiently the stochastic uncertainty. For the epistemic uncertainty, the probability theory proposes a solution which is not always efficient, it consists in using the uniform distribution to represent the total ignorance. In contrary, this type of uncertainty is well treated by the evidence theory which offers flexible and powerful tools to model it. The theory of evidence, also called Dempster-Shafer (DS)'s belief functions theory, is of a major interest for the reliability studies of complex systems which incorporate uncertain knowledge in the epistemic meaning.

Complex systems are too large to easily compute their reliability. Thus, modeling complex systems for the analysis of their reliability under uncertainty by the means of evidential network-based approaches is of a great interest.

Evidential networks are very powerful knowledge representations which allow at once graphical modeling and reasoning with the uncertainty in its random and epistemic meanings thanks to the theory of belief functions.

A. Deshpande and A. Hunter (Eds.): SUM 2010, LNAI 6379, pp. 191–204, 2010.

The aim of this paper is to present a comparison of two evidential models applied to the study of systems reliability under uncertainty. The first one which is proposed by Simon and Weber [11] is based on the junction tree (JT) inference algorithm, using an extension of the Bayes' theorem to the representation of DS's belief functions theory. The second one which is proposed by Ben Yaghlane [1] is based on the modified binary join tree (MBJT) algorithm, using the disjunctive rule of combination (DRC) and the generalized Bayesian theorem (GBT), both proposed by Smets [13]. In order to distinguish the two kinds of network, we call the first one the evidential network (EN) and we refer to the second one as the directed evidential network with conditional belief functions (DEVN).

The paper is structured as follows. In Sect.2, we first describe a system chosen for modeling its reliability based on the two models. Then, we present in Sect.3 the evidential network proposed by Simon and Weber. In Sect.4, we sketch the DEVN introduced by Ben Yaghlane. Section 5 is devoted to a comparison between these two models on the basis of criterion specified progressively.

2 The Oil Pipeline System: A Linear Consecutive-k-Out-of-n:G System

In this section, we describe a linear consecutive-k-out-of-n:G system: the oil pipeline system that we use to illustrate the two evidential networks.

The linear consecutive-k-out-of-n configuration was chosen because it is a well known configuration that has been proposed for the design and the reliability analysis of many real-world systems such as the microwave rely stations, the oil pipeline systems, the vacuum systems, and the computer ring networks [5].

A linear consecutive-k-out-of-n:G system, denoted by Lin/Con/$k$oo$n$:G system, is a sequence of n linearly connected components such that the system works if and only if at least k consecutive components out of each n consecutive components work adequately [5].

2.1 Description of the Oil Pipeline System

The oil pipeline system, designed for transporting oil from one point to another point, has been deeply studied in the literature by several authors. It consists of m pump stations (components) which are equally spaced between the two points. Each pump station Ci is able to transport oil to a distance including n pump stations and has two disjoint states ($\{Up\}$,$\{Down\}$): Up is the state of a pump station when it is working and $Down$ is its state when it has failed. Thus, the frame of discernment of each pump station is $\Omega = \{Up, Down\}$ and its corresponding power set is $2^{\Omega} = \{\emptyset, \{Up\}, \{Down\}, \{Up, Down\}\}$.

This power set 2^{Ω} offers the possibility to assign a quantity to the modality $\{Up,Down\}$ whose role is to characterize our ignorance on the real state of the pump station. It means that the pump station can be exclusively in one of the two states $\{Up\}$ or $\{Down\}$ without knowing exactly which. This ignorance characterizes the epistemic uncertainty and $\{Up,Down\}$ is the epistemic state. The elementary events on the oil pipeline system components are supposed to be independent. The system is considered homogeneous and no repair is considered.

The oil pipeline system reliability R_S which is the probability of this system to be in the state $\{Up\}$ could be evaluated applying the following formula:

$$R_S = \sum_{i=0}^{m} \begin{bmatrix} m - ik \\ i \end{bmatrix} (-1)^i (pq^k)^i - (q)^k \sum_{i=0}^{m} \begin{bmatrix} m - ik - k \\ i \end{bmatrix} (-1)^i (pq^k)^i \ . \quad (1)$$

where m is the number of pump stations in the system, k is the minimum number of consecutive pump stations whose failure causes the system failure, p is the equal probability for all the pump stations Ci to be in the state $\{Up\}$, and $q = 1 - p$ is the complementary failure probability.

2.2 DAG of the Oil Pipeline System

The oil pipeline system, modeled by the directed acyclic graph (DAG) shown in Fig.1, is a case of the Lin/Con/$koon$:G configuration with $k = 2$ and $n = 3$. The system consists of 7 pump stations. Each node Ci in the DAG represents the i-th pump station, each node Oj represents the j-th consecutive '2oo3' gate, and each node Az represents the z-th 'AND' gate. The system states are assigned to the added node R. The conditional belief mass distributions of the nodes Oj and Az are defined equivalent to the consecutive '2oo3' and 'AND' gates.

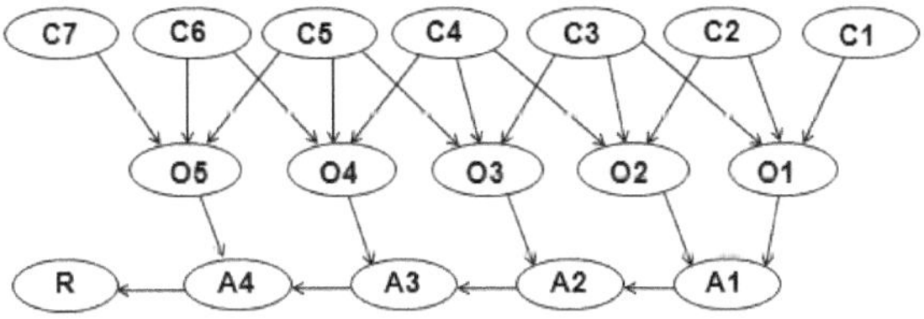

Fig. 1. Directed Acyclic Graph for the Oil Pipeline System

3 The Evidential Network (EN)

The evidential network, proposed by Simon and Weber [11], combines the DS's theory with Bayesian network (BN) to take random and epistemic uncertainties into account when modeling and reasoning. Thus, to represent the conditional dependencies between the variables in a description space integrating uncertainty, the evidential network uses conditional belief masses instead of using conditional probability functions as in BN. In the following, we briefly present the evidential network model and its corresponding computation mechanism structure.

3.1 The Evidential Network Model

The qualitative structure of an evidential network is represented by a DAG defined as a couple $G = (N, E)$, where N represents the set of nodes, and E represents the set of edges. The quantitative level of an evidential network is represented by the set of belief masses distributions M associated to each node in the graph. For each root node X (i.e. node without parent nodes), having a frame of discernment Ω constituted by q mutually exhaustive and exclusive hypotheses, an a priori belief mass distribution $M(X)$ has to be defined over the 2^q focal sets A_i^X by the following equation:

$$M(X) = [m(X \subseteq \emptyset)\ m(X \subseteq A_1^X)...m(X \subseteq A_i^X)...m(X \subseteq A_{2q-1}^X)] \ . \quad (2)$$

with

$$m(X \subseteq A_i^X) \geq 0 \quad and \quad \sum_{A_i^X, A_i^X \in 2^\Omega} m(X \subseteq A_i^X) = 1 \ . \quad (3)$$

where $m(X \subseteq A_i^X)$ is the belief that variable X verifies the hypotheses of the focal element A_i^X.

For other nodes (i.e. nodes which have got parent nodes), a conditional belief mass distribution $M[Pa(X)](X)$ is specified for each possible hypothesis A_i^X knowing the focal sets of the parents of X defined by $Pa(X)$.

The exact inference algorithm proposed by Jensen [3] is performed to compute the marginal belief mass distributions of the variables in the network. The computation mechanism is based on an extension of the Bayes' theorem to the representation of uncertain information according to the framework of DS's theory [12]. The algorithm used for inferring the beliefs is based on the construction of a secondary structure which is the junction tree.

3.2 The Junction Tree (JT)

The junction tree (JT) is a data structure which enables local computations with potentials on small domains avoiding explicitly the computation of the marginal of the joint distribution for the variables of interest [3].

The JT is a singly-connected and undirected graph whose nodes are subsets of the variables in the original network such that if a variable is in two distinct

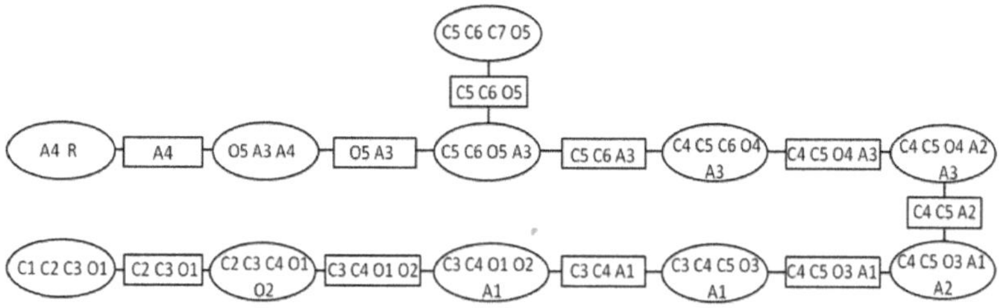

Fig. 2. The Junction Tree for the Oil Pipeline System

nodes, then the variable belongs to each node on the path between the two nodes. The nodes are called the cliques of the JT. Each separator between two adjacent nodes is labeled by the intersection of the two nodes.

To construct a JT from a directed acyclic graph G, we first create a moral graph UG, then we triangulate UG. Next we identify the cliques of the triangulated graph, and finally we organize them into a junction tree. The junction tree for the oil pipeline system is shown in Fig.2.

4 The Directed Evidential Network with Conditional Belief Functions (DEVN)

Evidential networks with conditional belief functions, called ENCs, were initially proposed by Smets for the propagation of beliefs [13]. Then these networks have been studied by Xu and Smets [15]. ENC has the same structure as BN, since it is a DAG. Nevertheless, the manner in which the conditional beliefs are defined is different from that one in which the conditional probabilities are defined in BN: each edge in the graph represents a conditional relation between the two nodes it connects. The graphical representation and the propagation algorithm proposed by Xu are related to the evidential networks which only have binary relations between the nodes [15]. Thus, in order to generalize ENC to the case where relations are given for any number of nodes, Ben Yaghlane proposed a new representation [1], called directed evidential network with conditional belief functions (DEVN).

4.1 Directed Evidential Network Model

The directed evidential network model combines the DAG and the evidence theory. Like the BN, the DEVN is represented by qualitative and quantitative levels. The qualitative level is defined by the DAG in which nodes represent random variables and directed edges describe the conditional dependencies in the model. The quantitative level is defined by a set of parameters for the DAG represented by conditional belief functions for each variable given its parents, expressing the dependence relations. Conditional belief functions are defined in DEVN like in ENC, that means in a different manner from conditional probabilities in the BN. However, as in the BN, the main objective of the reasoning process in DEVN is to compute the marginals of the global belief function for all or some variables. Thus, Ben Yaghlane proposed a propagation algorithm for the DEVN based on the use of a secondary computational structure, which permits to maintain the (in)dependence relations of the original DEVN and to perform the propagation directly with the conditional belief functions.

4.2 Modified Binary Join Tree (MBJT)

The computational data structure, proposed by Ben Yaghlane to perform the propagation process [1], is an adaptation of the binary join tree [9], called the modified binary join tree (MBJT). The concept of binary join trees was first introduced by Shenoy [9] as a significant improvement of the Shenoy-Shafer (SS) architecture described in [10]. Later, motivated by the work of Shenoy, the MBJT was proposed as a refinement of the binary join tree. It was designed to avoid the major drawback of binary join tree which is the lost of useful information about the relationship among the variables arising when transforming the original directed evidential network into a binary join tree. As for the binary join tree construction [9], the MBJT construction process is based on the fusion algorithm with some modifications used for emphasizing explicitly the conditional relations in the initial DEVN. The MBJT integrates rectangles containing these conditional relations between the variables instead of circles containing just the set of these variables [1]. Thus, by profiting efficiently of the original network's structure, the MBJT maintains the available independence relations between the variables in order to make them useful when inferring beliefs. The MBJT for the oil pipeline system is shown in Fig.3.

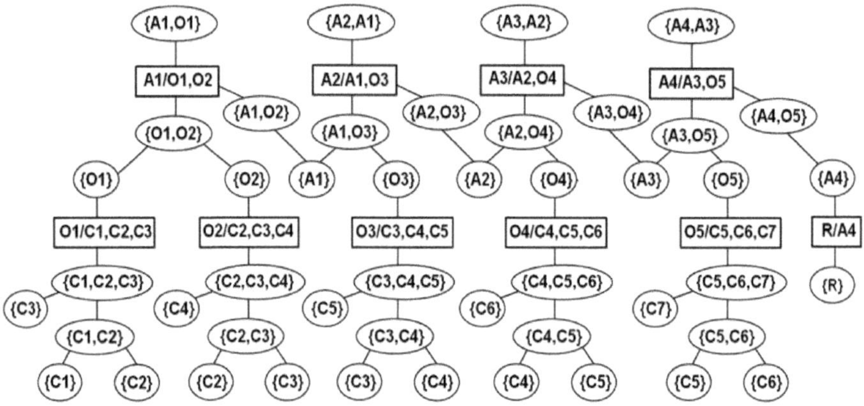

Fig. 3. The Modified Binary Join Tree for the Oil Pipeline System

5 Comparison

In this section, we compare the evidential network and the directed evidential network applied to the study of the oil pipeline system reliability. The comparison is based on different criterion specified progressively.

5.1 Original Structure

EN and DEVN are graphically two directed acyclic graphs representing the uncertainty in the knowledge using the framework of the belief functions theory.

In the two networks modeling the oil pipeline system, root nodes represent the system components and child nodes represent the logical 'AND' and consecutive

'2oo3' gates. The edges describe the dependency relations between the logical gates and their inputs. Thus, for a system reliability modeling by the means of an EN or a DEVN, the graphical representation (i.e. the DAG) is the same.

In the first network, each child node X represents a conditional relation between the variable X and its parents $Pa(X)$. This conditional relation is defined by its own conditional belief mass table $M[Pa(X)](X)$. It means that a conditional belief mass table defines the relation between the belief masses on the frame of discernment of the child node and the belief masses on the frame of discernment of the variables in the parent nodes. In the second network, each edge represents a conditional relation between the two nodes it connects. Thus, conditional beliefs are defined in the two networks in different manners.

The number of conditional relations in the EN depends on the number of child nodes. But, it depends on the number of edges in the DEVN. Nevertheless, the conditional belief function assigned to a child node given all its parents in the DEVN, can be obtained by the means of (4) proposed by Xu and Smets [15], using the initial conditional belief functions (i.e. conditional belief functions between only two nodes or variables). For example, let us consider the two spaces Ω_{O1} and Ω_{O2} associated to the two nodes O1 and O2 in the DAG of the oil pipeline system shown in Fig.1. We use $m_{And1}[o1](and1)$ to represent the conditional mass function induced on the space Ω_{And1} given $o1 \subseteq \Omega_{O1}$, and $m_{And1}[o2](and1)$ to represent the conditional mass function induced on the space Ω_{And1} given $o2 \subseteq \Omega_{O2}$.

The following formula of Xu and Smets allows to obtain $m[O1, O2](And1)$:

$$\forall o_{12} \subseteq \Omega_{O_{12}}, m[o_{12}](a) = \sum_{a1 \cap a2 = a} m_{And1}[O1](a1) m_{And1}[O2](a2) \ . \qquad (4)$$

where $\Omega_{O_{12}} = \Omega_{O1 \cup O2}$

Thus, unlike the EN in which a conditional belief mass table should be specified for each child node given all its parents, in the DEVN, we can either define this table for each child node given all its parents or simply specify one table for each child node given separately each one of its parents, and then obtain via (4) the conditional belief function of the node given all the parents.

Thus, the DEVN offers more flexibility than the EN when defining the parameters of the DAG.

5.2 Graphical Structures for Message Propagation

For the EN, propagation of beliefs is done in a JT. Whereas in the DEVN, propagation of beliefs is done in a MBJT. The nodes of the JT are the cliques identified from a triangulated moral graph of the original EN. The corresponding MBJT includes the subsets forming the hypergraph as well as singleton variables which are not already included in the hypergraph.

Let us suppose that we have a DAG with n variables. To construct a JT for this DAG, each singleton variable defines a clique in the worst case. The number of cliques in the resulting JT is n. To obtain the corresponding MBJT, we start with an hypergraph containing n subsets to which we add p singleton variables which do not occur in the hypergraph. Since the MBJT construction process is motivated by the idea of the fusion algorithm [9], k new subsets resulting from the combinations done during the construction process will be included in the MBJT. The number of nodes in the resulting MBJT is $n + p + k$.

Therefore, a MBJT modeling a system reliability has more nodes than a corresponding JT. For instance, the JT for the oil pipeline system shown in Fig.2 has 11 nodes whereas the corresponding MBJT shown in Fig.3 has 57 nodes.

The two obtained data structures used for inferring the beliefs satisfy the running intersection property [6]. It means that if a variable appears in two different nodes, it also appears on all nodes on the path linking them.

The link between two adjacent nodes in the JT is materialized by a separator which is labeled by the intersection of the two nodes, while in the MBJT, the link between two adjacent nodes is simply represented by the edge.

The construction of the first structure takes as parameters the cliques of the triangulated moral graph. The obtained model is an undirected graph showing only undirected links between nodes. It does not show any (in)dependence relations between the nodes composing the JT. In contrast, the construction of the second structure needs as parameters a directed evidential network weighted by conditional belief functions added to an hypergraph. The resulting structure is an undirected graph showing at once directed and undirected links between nodes. It emphasizes explicitly the independence relations shown in the initial DAG. These (in)dependence relations between nodes in the MBJT are represented by the conditional nodes which are going to be very useful when inferring beliefs through the MBJT [1].

5.3 Message-Passing Schemes

Once the graphical structure for message propagation is constructed, it can be initialized by assigning potentials to its nodes. In the JT, the clique potential is set to the joint density of the distributions assigned to it. The potentials for the separator nodes are set to unity. Thus, each clique and each separator in the JT stores a potential. In the MBJT, the initialization is done by the means of joint belief functions for not conditional nodes (i.e. joint nodes), and conditional belief functions for conditional nodes [1].

To begin propagating messages in the JT, a clique is selected arbitrarily as the root node. Each link in the JT is used twice during the message passing, once in the inward phase and once in the outward phase.

The inward phase, is done by collecting messages from the leaves to the root node. In reverse, distribution of messages is done in the outward phase away from the root towards each leaf in the JT.

The inference in the JT involves all the created nodes. Thus, computations are done by all the cliques and all the separators. In the JT, the inward phase is similar to the outward phase.

After completing the bidirectional message-passing scheme, we can compute the marginals for desired singleton variables. The potential at each clique and at each separator corresponds to the marginal of the joint for the variables belonging to that node. To compute the marginal distribution for a variable, we need to marginalize the smallest potential containing it over the remaining variables.

As in the JT, the propagation of messages in the MBJT is done in two stages: the propagation-up and the propagation-down. The root is determined according to the propagation-up scheme where messages are passing from the leaves over the path to the center of the tree. Unlike the inference in the JT, the inference through the MBJT involves only the joint nodes which are asked to send and to receive messages. The conditional nodes are considered as 'bridges' when inferring beliefs between joint nodes [2]. They hold an important role by allowing to determine whether the message sent from one joint node to another joint node is a 'parent' message or a 'child' message.

In the MBJT, the two phases are different: if in the propagation-up the message exchanged between two joint nodes separated by a conditional node is computed by the DRC proposed by Smets [13], then the message exchanged between the two nodes in the propagation-down is computed by the GBT proposed also by Smets [13]. In contrast, if in the propagation-up the message exchanged between two joint nodes separated by a conditional node is computed by the GBT, then in the propagation-down the message exchanged between the two nodes is computed by the DRC.

After carrying out message passing in the MBJT, we can compute the marginal of a desired variable in the corresponding singleton node, by combining its own potential with the messages received in the two phases from all the inward and the outward neighbors.

5.4 Storage Efficiencies

The potential storage memory space in the two structures depends on the number of potentials, on the domain on which each potential is specified, and on the form in which this latter is stored.

In a JT, all the cliques and the separators have local joint potentials. Since at the end of the message-passing scheme the JT yields only marginals for cliques and separators and no marginals for desired singleton variables, there is further storage memory space required to store marginals for those variables [7]. Moreover, the storage requirements for storing the input potentials are added.

In a MBJT, joint potentials are stored in the joint nodes, while conditional potentials are stored in the conditional nodes. Each conditional node stores exactly one potential, but not all the joint nodes in the MBJT store potentials.

If two adjacent joint nodes in the MBJT are separated by an edge, then either one or two potentials are stored in this edge. It depends on the number of messages exchanged between the two nodes. However, if two joint nodes are separated by a conditional node denoted by C, then the edge that connects the first joint node and C and the edge that connects C and the second joint node are regarded as one edge because the conditional node C will neither send nor receive messages. In this case, two messages are at most exchanged between the two joint nodes through the two edges, and so two potentials are at most stored.

As in the JT, a storage memory space is included for storing the input potentials in the MBJT. The marginal for each of the singleton variables is computed by combining the local potential of the corresponding node in the MBJT with all the messages received by it in the two phases. Thus, as in the JT a storage memory space is required for storing output potentials.

As mentioned in [2], storing a joint potential allocates more memory space than storing a conditional potential. The JT is storing only potentials in joint form, whereas the MBJT is storing potentials in joint form and in conditional form since it consists of both conditional and joint nodes. This offers an important memory space gain as seen in the example of the oil pipeline system:

In the JT, the memory space requirement described for the oil pipeline system in units of floating point numbers (fpn) is:

Input storage (7 root nodes, 5 child nodes with 3 parent nodes, 4 child nodes with 2 parent nodes and one child node with 1 parent node): $7*4+5*256+4*64+1*16 = 1580\ fpn$
Output storage (17 singleton variables): $17*4 = 68\ fpn$
Clique storage (6 with 5 variables, 3 with 4 variables, 1 with 3 variables and 1 with 2 variables): $6*1024+3*256+1*64+1*16 = 6992\ fpn$
Separator storage (3 with 4 variables, 5 with 3 variables, 1 with 2 variables and 1 with 1 variable): $3*256+5*64+1*16+1*4 = 1108\ fpn$

In the MBJT, the memory space used for the same system in units of fpn is:

Input storage (7 joint singleton nodes with 1 joint potential, 5 conditional nodes with 4 variables, 4 conditional nodes with 3 variables and one conditional node with 2 variables): $7*4+5*32+4*16+1*8 = 260\ fpn$
Output storage (17 singleton variables): $17*4 = 68\ fpn$
Separator storage (28 separators with 2 potentials (with 1 variable), 5 separators with 2 potentials (with 2 variables), 4 separators with 2 potentials (1 with 1 variable,1 with 2 variables), 5 separators with 2 potentials (1 with 1 variable,1 with 3 variables): $(28*(2*4)+5*(2*16)+4*(4+16)+5*(4+64)) = 804\ fpn$

Thus, the total storage space needed for the oil pipeline system is of 9748 fpn in the JT compared to 1132 fpn in the corresponding MBJT. We notice that for the oil pipeline system, the JT has more storage requirements than the MBJT. The order of magnitude of the two structures is near to 1/9 for this system.

We see clearly that by using the conditional form for storing some potentials, the MBJT requires for example less memory space to store the input potentials than the JT. Although the MBJT of the oil pipeline system consists of more nodes than the corresponding JT, the domains on which are defined the potentials associated with its nodes and its separators (edges) are smaller than the domains on which are defined the potentials assigned to cliques and separators in the corresponding JT.

Indeed, in a MBJT, the domain of the potential of the conditional node containing the maximum number of variables corresponds always to the largest domain. In contrast the domains of the clusters and of the separators, corresponding to the domains of their potentials, may be very large in the JT. The largest clique size in the JT increases exponentially with the number of variables in the initial DAG. The largest clique can consist of more variables than the largest initial conditional potential. Since the joint form is used in the JT for the clique storage and the separator storage, the memory space required in the JT is as important as the number of variables in the DAG.

For the oil pipeline system, the use of the conditional form in the MBJT adds to its storage efficiencies. The generalization of this statement regarding the storage efficiencies in the two structures requires more experimentations.

5.5 Computational Efficiencies

Both of the JT and the MBJT are structures for local computation of marginals avoiding to compute explicitly the marginal of the joint for the variables. In the first structure, the complexity is exponential in the maximum clique size, while in the second structure, the complexity is exponential in the maximum node size. Thus, it is important to construct the JT that has minimal clique size and the corresponding MBJT with the minimal node size. But for a general graph, this task is computationally difficult.

We will examine the computational efficiencies by analyzing the difference in the number of operations performed when computing the oil pipeline system reliability by the means of the two structures. Let us assume that binary multiplications are done for computing in the JT (i.e. potentials are multiplied two at time). Propagation in the JT shown in Fig.2 requires 12456 additions, 17808 multiplications and 1108 divisions, compared to only 740 additions, 2132 multiplications and 0 divisions in the corresponding MBJT in Fig.3.

For the oil pipeline system, it is clear that the MBJT is computational more efficient than the JT. We notice that the MBJT does fewer additions than the JT. In the MBJT, additions are done for marginalizing messages to the domain of the receiver, whereas in the JT, additions are done for the same reason as well as for computing the marginals of singleton variables. Since computations of the marginals for those variables are done in the separators (or in the worst case in the cliques), there are further additions required to achieve these computations in the JT. The domains on which the potentials are defined in the JT are larger than the domains on which the potentials are defined in the MBJT. Moreover, all the potentials are defined in the JT with joint form, while some of them are defined in the MBJT in the conditional form. A penalty in computations arises when using the joint form which requires more operations when computing. Thus, the JT does more additions than the MBJT. We notice also that for the oil pipeline system the MBJT does fewer multiplications than the JT.

In the JT, multiplications are done for the initialization of the structure and for updating the local potentials of the cliques during the propagation of messages, while in the MBJT, multiplications are done when combining potentials in the propagation process and when computing the marginal of the singleton variables. The JT shown in Fig.2 is composed of large cliques and separators. Thus, an important number of multiplications is done to assign potentials to nodes and to update their potentials during the propagation process. In the

MBJT, the computations of the marginals for singleton variables are done in the corresponding nodes, thus the number of multiplications done for those computations is not as important as in the JT. The gain in computation in the MBJT comes from the use of the GBT and the DRC rules which increase the computational efficiencies by avoiding many multiplications required for computing the messages to be sent.

Making general statements regarding relative computational efficiencies in the JT and in the MBJT requires more experimentations.

5.6 Comparison of Models Results

EN and DEVN are two evidential network-based approaches allowing to model a complex system in order to evaluate its reliability. The evaluation of the system reliability by the means of graphical models is based on the DAG modeling it and the reliability of each component. The system reliability can be also obtained by a mathematical formulation of reliability allowing to calculate the total reliability of the system based on its structure and the reliability of each component. To show the capacity of the two evidential networks to compute the system reliability, it is very important to compare the assessment of the system reliability obtained by the evidential models-based reasoning with the one obtained by the mathematical formula. Comparisons in this section are illustrated by the oil pipeline system. Two cases are distinguished according to whether the evidences are given under the closed or the open-world assumptions.

Before starting comparison, we remind some useful concepts:

The upper and the lower bounds of a probability interval can be obtained from a belief mass distribution. This interval contains the probability of focal sets and is bounded by two non-additives measures (the belief measure (Bel) and the plausibility measure(Pl)) [8].

Bel(Ai) is the lower bound of a focal set Ai. It gives the amount of support given to Ai. Pl(Ai) is the upper bound of a focal set Ai, representing the maximum amount of support that could be given to Ai.

The bounding property is defined by the following equation:

$$Bel(Ai) \leq Pr(Ai) \leq Pl(Ai)\ . \tag{5}$$

where $Pr(Ai)$ defines the occurrence probability of Ai which is unknown.

Uncertain Knowledge: Closed-world assumption

Let us suppose that a priori belief mass distribution M_i for each system component Ci is defined as follows:

$$M_i = [m(\emptyset) = 0 \;\; m(\{Up\}) = 0.7 \;\; m(\{Down\}) = 0.15 \;\; m(\{Up, Down\}) = 0.15]$$

The uncertainty in the knowledge induces a belief mass $m(\{Up, Down\}) > 0$. It describes a non Bayesian frame in which $Bel(Ci = \{Up\}) < Pr(Ci = \{Up\}) < Pl(Ci = \{Up\})$.

The system reliability obtained by both of EN and DEVN is $R_S = 0.487163$. $Bel(S = \{Up\}) = 0.487163$ and $Pl(S = \{Up\}) = 0.818179$. The system reliability obtained by (1) is $R_S = 0.623917$. It is between [0.487163,0.818179].

This simple example shows that in both of the EN and the DEVN, the bounding property (5) is verified when coding a priori belief masses with a closed-world assumption.

Uncertain Knowledge: Open-world assumption
Let us now suppose that a priori belief mass distribution M_i is defined for each component Ci as follows:

$$M_i = [m(\emptyset) = 0.05 \;\; m(\{Up\}) = 0.7 \;\; m(\{Down\}) = 0.1 \;\; m(\{Up, Down\}) = 0.15]$$

The non-exhaustivity of all hypotheses induces a belief mass $m(\emptyset) > 0$. It translates an open-world in which $Bel(Ci = \{Up\}) < Pr(Ci = \{Up\}) < Pl(Ci = \{Up\})$. The reliability obtained by the two evidential networks is $R_S = 0.487163$. $Bel(S = \{Up\}) = 0.487163$ and $Pl(S = \{Up\}) = 0.732888$. The system reliability obtained by (1) is $R_S = 0.623917$. It is between [0.487163, 0.732888].

Based on the exactness of the results, there are no essential differences between the two evidential networks, since they provide the same reliability evaluation.

6 Conclusion

In this paper, we have compared two evidential network-based approaches applied to the system reliability, the evidential network and the directed evidential network. This comparison was based on different aspects. Comparing the two structures applied to evaluate a specific complex system, the directed evidential network is computationally more efficient and more storage efficient than the evidential network. This statement regarding relative computational efficiencies and storage efficiencies in the two evidential networks may be generalized, but this requires more experimentations by applying them to different complex systems. In future work, the comparison of the evidential network and the directed evidential network by taking into account the dynamic aspects will be of a great interest.

References

[1] Ben Yaghlane, B., Mellouli, K.: Inference in directed evidential networks based on the transferable belief model. Int. J. of App. Reasoning 48, 399–418 (2008)
[2] Ben Yaghlane, B.: A Comparison of Architectures for Exact Inference in Evidential Networks. In: 12th Int. Conf. on Inf. Processing and Management of Uncertainty in KBS (IPMU 2008), Malaga, Spain (2008)
[3] Jensen, F.V., Lauritzen, S.L., Olesen, K.G.: Bayesian Updating in Causal Probabilistic Networks by local Computation. Computational Statistics Quarterly 4, 269–282 (1990)

[4] Kozine, I., Utkin, L.: Interval valued finite markov chaines. Reliable computing 8, 97–113 (2002)
[5] Kuo, W., Zuo, M.J.: Optimal Reliability Modeling: Principles and Applications. John Wiley & Sons, Inc., Chichester (2003)
[6] Lauritzen, S.L., Spiegelhalter, D.J.: Local Computations with Probabilities on Graphical Structures and their application to Expert Systems. Journal of Royal Statistical Society, Series B 2(50), 157–224 (1988)
[7] Lepar, V., Shenoy, P.: A comparison of Lauritzen-Spiegelhalter, Hugin, and Shenoy-Shafer Architectures for Computing Marginals of Probability Distributions. In: Cooper, G., Moral, S. (eds.) UAI 1998, pp. 328–337 (1998)
[8] Shafer, G.: A Mathematical Theory of Evidence. Princeton Univ. Press, Princeton (1976)
[9] Shenoy, P.P.: Binary join trees fo computing marginals in the Shenoy-Shafer architecture. Int. J. of App. Reasoning 17, 239–263 (1997)
[10] Shenoy, P.P., Shafer, G.: Axioms for probability and belief functions propagation. Uncertainty in Artificial Intelligence 4, 159–198 (1990)
[11] Simon, C., Weber, P., Evsukoff, A.: Bayesian networks inference algorithm to implement Dempster Shafer theory in reliability analysis. Reliability Engineering and System Safety 93, 950–963 (2008)
[12] Simon, C., Weber, P.: Evidential networks for reliability analysis and performance evaluation of systems with imprecise knowledge. IEEE Transactions on Reliability 58, 69–87 (2009)
[13] Smets, P.: Belief function: the disjunctive rule of combination and the generalized Bayesian theorem. Int. J. of App. Reasoning 9, 1–35 (1993)
[14] Walley, P.: Statistical reasoning with imprecise probabilities. Chapman & Hall, Boca Raton (1991)
[15] Xu, H., Smets, P.: Evidential Reasoning with Conditional Belief Functions. In: Heckerman, D., Poole, D., Lopez De Manatars, R. (eds.) UAI 1994, pp. 598–606. Morgan Kaufman, San Mateo (1994)

Imprecise Bipolar Belief Measures Based on Partial Knowledge from Agent Dialogues

Jonathan Lawry

Department of Engineering Mathematics,
University of Bristol,
Bristol, UK
j.lawry@bris.ac.uk

Abstract. Valuation pairs [6], [5] are proposed as a possible framework for representing assertability conventions applied to sentences from a propositional logic language. This approach is then extended by introducing imprecise valuation pairs to represent partially or imprecisely defined assertability conventions. Allowing for uncertainty then naturally results in lower and upper bipolar belief measures on the sentences of the language. The potential of this extended framework is illustrated by its application to an idealised dialogue between agents, each making assertions and condemnations based on their individual valuation pairs.

1 Introduction

In a population of communicating social agents the beliefs and opinions of individual agents must be inferred from, amongst other things, their assertions and their reactions to the assertions of other agents. For example, dialogues conducted on social networking websites, newsgroups and other web-based discussion forums often comprise of a sequence of assertions together with statements of agreement with and condemnation of, the assertions of others. Focusing in this way on assertability rather than truth follows Kyburg [3] and also Parikh [7] who states:

> Certain sentences are assertible in the sense that we might ourselves assert them and other cases of sentences which are non-assertible in the sense that we ourselves (and many others) would reproach someone who used them. But there will also be the intermediate kind of sentences, where we might allow their use.

In other words, the inherent vagueness of natural language propositions manifests itself in the bipolar nature of assertability conventions. That is to say, in a certain social context, some statements are *definitely assertable* in the sense that no one would disagree with them, while other statements are only *acceptable to assert* in the sense that, while opinions may differ as to the truth of these statements, no one would condemn their assertion. The bipolarity of assertability would seem to be a special case of what Dubois and Prade [1] refer to as *symmetric bivariate*

A. Deshpande and A. Hunter (Eds.): SUM 2010, LNAI 6379, pp. 205–218, 2010.

unipolarity, whereby judgments are made according to two distinct evaluations on unipolar scales. In the current context, we have a strong and a weak evaluation criterion where the former corresponds to definite assertability and the latter to acceptable assertability. As with many examples of this type of bipolarity there is a natural duality between the two evaluation criterion in that a proposition is definitely assertable if and only if it is not acceptable to assert its negation.

Lawry and Gonzalez-Rodrigues [5], [6] proposed a framework for quantifying beliefs about the assertability of propositions. In this paper we extend this framework to allow for only partial knowledge of the assertability conventions of the population as may be inferred from a dialogue between agents. An outline of the paper is as follows: Section two gives an introduction to *valuation pairs* as a formal representation of assertability conventions in a simple propositional logic language. In this section we also describe the bipolar belief measures obtained when there is uncertainty about the actual assertability convention being employed by the population. Section three introduces *imprecise valuation pairs* to represent assertability conventions which are only partially defined. In the presence of uncertainty this approach naturally results in lower and upper bipolar belief measures. Section four describes how imprecise valuation pairs together with lower and upper bipolar belief measures can be used to represent the kind of knowledge about underlying assertability conventions which can be inferred from a simple idealised dialogue between communicating agents. Finally section five gives some conclusions and outlines future work.

2 Valuation Pairs and Bipolar Belief

Let L be a language of the propositional logic with connectives $\wedge$, $\vee$ and $\neg$ and a finite set of propositional variables $P = \{p_1, \ldots, p_n\}$. Let SL denote the sentences of L. To model the bipolarity of assertability we introduce the notion of a valuation pair defined on SL. This consists of two binary functions v^+ and v^- linked, through negation, by a duality relationship. The underlying idea is that v^+ represents the strong criteria of definite assertability while v^- represents the weaker criteria of acceptable assertability.

Definition 1. *Coherent Valuations Pairs*
A coherent valuation pair for L is a pair of functions $\boldsymbol{v} = (v^+, v^-)$ *such that* $v^+ : SL \rightarrow \{0,1\}$, $v^- : SL \rightarrow \{0,1\}$ *and where* $\forall \theta \in SL$, $v^+(\theta) = 1$ *iff it is* definitely correct *to assert* θ *and* $v^-(\theta) = 1$ *iff it is* acceptable *to assert* θ. *Furthermore,* v^+ *and* v^- *satisfy the following properties:* $\forall \theta, \varphi \in SL$ $v^+(\theta) \leq v^-(\theta)$[1] *and*

- $v^+(\theta \wedge \varphi) = \min(v^+(\theta), v^+(\varphi))$, $v^-(\theta \wedge \varphi) = \min(v^-(\theta), v^-(\varphi))$
- $v^+(\theta \vee \varphi) = \max(v^+(\theta), v^+(\varphi))$, $v^-(\theta \vee \varphi) = \max(v^-(\theta), v^-(\varphi))$

[1] The condition that $v^+ \leq v^-$ is referred to as coherence. In [6] this is presented as an additional constraint on valuation pairs. However, in this paper we shall only consider coherent valuation pairs.

– $v^+(\neg\theta) = 1 - v^-(\theta)$, $v^-(\neg\theta) = 1 - v^+(\theta)$

The last rule is motivated by the assumption that it is definitely correct *to assert* $\neg\theta$ *if and only if it is* not acceptable *to assert* θ. *Let* CV *denote the set of coherent valuation pairs defined on* L.

Viewed as function from SL into $\{0,1\}^2$ a valuation pair represents a complete description of an assertability convention for the language L. The difference between v^- and v^+ is due to inherent vagueness in the sentences of L. So that for $\theta \in SL$, $v^+(\theta) = 0$ and $v^-(\theta) = 1$ means that θ expresses a borderline case, which while not definitely assertable, is nonetheless acceptable to assert. Lawry and Gonzalez-Rodrigues [6] also point out a clear relationship between valuation pairs and three-valued logic. To see this consider, for $\theta \in SL$, the three possible values of $\boldsymbol{v}(\theta)$, $(0,0)$, $(0,1)$ and $(0,1)$, as corresponding to the truth values *true*, *borderline* and *false* respectively. Translating the combination rules from definition 1 so that they relate to these truth-values, then results in Kleene's strong three valued logic [2].

Two sentences from SL are deemed to be assertably equivalent if both their strong and weak assertability valuations agree for all valuation pairs.

Definition 2. *Equivalence*
For $\theta, \varphi \in SL$, θ *and* φ *are equivalent, denoted* $\theta \equiv \varphi$, *iff* $\forall \boldsymbol{v} \in CV$ $v^+(\theta) = v^+(\varphi)$ *and* $v^-(\theta) = v^-(\varphi)$

Important equivalences include; $\forall \theta, \varphi, \psi \in SL$ (Double Negation) $\neg\neg\theta \equiv \theta$, (Idempotence) $\theta \wedge \theta \equiv \theta$ and $\theta \vee \theta \equiv \theta$, (Associativity) $\theta \vee (\varphi \vee \psi) \equiv (\theta \vee \varphi) \vee \psi$ and $\theta \wedge (\varphi \wedge \psi) \equiv (\theta \wedge \varphi) \wedge \psi$, (Commutativity) $\theta \vee \varphi \equiv \varphi \vee \theta$ and $\theta \wedge \varphi \equiv \varphi \wedge \theta$, (Distributivity) $\theta \vee (\varphi \wedge \psi) \equiv (\theta \vee \varphi) \wedge (\theta \vee \psi)$ and $\theta \wedge (\varphi \vee \psi) \equiv (\theta \wedge \varphi) \vee (\theta \wedge \psi)$, (De Morgan) $\neg(\theta \wedge \varphi) \equiv \neg\theta \vee \neg\varphi$ and $\neg(\theta \vee \varphi) \equiv \neg\theta \wedge \neg\varphi$.

Within the proposed bipolar framework, uncertainty concerning the sentences in L effectively corresponds to uncertainty as to which is the true valuation pair (assertability convention) for L. In general we view this uncertainty as being epistemic in nature, resulting from a lack of knowledge concerning either, the domain of discourse to which propositions refer (e.g. Bill's height in the proposition 'Bill is tall'), or the linguistic conventions governing the assertability of propositions as part of communications (e.g. whether the proposition 'Bill is tall' is assertable when Bill's height is known to be 1.8 metres). The latter requires an underlying assumption on the part of each agent that there exists a coherent set of rules governing assertability to which they should adhere if they wish to be understood by other agents (i.e. that there is a *true* valuation pair). In previous work we have referred to this assumption as the *epistemic stance* [4]. In the following definition we assume that this uncertainty about valuation pairs is quantified by a probability measure w, although clearly other epistemic uncertainty measures (e.g. possibility measures) could also be considered. This assumption naturally results in two bipolar measures of belief in the assertability of sentences in SL.

Definition 3. *Bipolar Belief Measures*

- *Let w be a probability distribution defined on CV so that $w(v^+, v^-)$ is the agent's subjective belief that (v^+, v^-) is the true valuation pair for L.*
- *Let $\mu^+ : SL \to [0,1]$ such that $\forall \theta \in SL\ \mu^+(\theta) = w(\{\boldsymbol{v} \in V : v^+(\theta) = 1\})$*
- *Let $\mu^- : SL \to [0,1]$ such that $\forall \theta \in SL\ \mu^-(\theta) = w(\{\boldsymbol{v} \in V : v^-(\theta) = 1\})$*

Intuitively speaking, $\mu^+(\theta)$ is the level of belief that θ is definitely assertable, while $\mu^-(\theta)$ is the level of belief that θ is acceptable to assert.

The following theorem gives a number of general properties satisfied by μ^+ and μ^-.

Theorem 1. *[6]*
- $\mu^+ \leq \mu^-$
- $\forall \theta \in SL\ \mu^+(\neg\theta) = 1 - \mu^-(\theta)$ *and* $\mu^-(\neg\theta) = 1 - \mu^+(\theta)$
- $\forall \theta, \varphi \in SL$, *if* $\theta \equiv \varphi$ *then* $\mu^+(\theta) = \mu^+(\varphi)$ *and* $\mu^-(\theta) = \mu^-(\varphi)$

We now introduce a characterisation of μ^+ and μ^- in terms of a mass function $m : 2^P \times 2^P \to [0,1]$, which gives mass values to pairs of sets of propositional variables.

Definition 4. *Mass Functions*
Given a valuation pair $\boldsymbol{v} \in V$
- *Let $\mathcal{D}^{\boldsymbol{v}} = \{p_i \in P : v^+(p_i) = 1\}$ and let $\mathcal{C}^{\boldsymbol{v}} = \{p_i \in P : v^+(\neg p_i) = 1\}$*
- *Let $m : 2^P \times 2^P \to [0,1]$ such that for $F, G \subseteq P$*
  $m(F,G) = w(\{\boldsymbol{v} \in V : \mathcal{D}^{\boldsymbol{v}} = F, \mathcal{C}^{\boldsymbol{v}} = G\})$

Theorem 2. *[6] For all coherent valuation pairs $\boldsymbol{v} \in CV$ $\mathcal{C}^{\boldsymbol{v}} \subseteq (\mathcal{D}^{\boldsymbol{v}})^c$*

$m(F,G)$ is the belief that the set of definitely assertable propositions is F and that the set of definitely assertable negated propositions is G. Notice, that given the duality between v^+ and v^-, $m(F,G)$ is also the belief that G^c is the set of acceptable propositions and that F^c is the set of acceptable negated propositions. In fact, any valuation pair $\boldsymbol{v}$ is completely characterised by the pair of sets $(\mathcal{D}^{\boldsymbol{v}}, \mathcal{C}^{\boldsymbol{v}})$. For example, if $P = \{p_1, p_2, p_3, p_4\}$ and $(\mathcal{D}^{\boldsymbol{v}}, \mathcal{C}^{\boldsymbol{v}}) = (\{p_1, p_2\}, \{p_4\})$ then $\boldsymbol{v}$ is such that $\boldsymbol{v}(p_1) = \boldsymbol{v}(p_2) = (1,1)$, $\boldsymbol{v}(p_3) = (0,1)$ and $\boldsymbol{v}(p_4) = (0,0)$. Hence, it holds that for any pair of sets $(F,G) \in 2^P \times 2^P$ there is a unique valuation pair $\boldsymbol{v} \in V$, for which $\mathcal{D}^{\boldsymbol{v}} = F$ and $\mathcal{C}^{\boldsymbol{v}} = G$. Let this valuation pair be denoted $\boldsymbol{v}_{(F,G)} = (v^+_{(F,G)}, v^-_{(F,G)})$ and which is then defined as follows:

Definition 5

$v^+_{(F,G)}(p_i) = 1$ *iff* $p_i \in F$, $v^+_{(F,G)}(\neg p_i) = 1$ *iff* $p_i \in G$

$v^-_{(F,G)}(p_i) = 1$ *iff* $p_i \notin G$, $v^-_{(F,G)}(\neg p_i) = 1$ *iff* $p_i \notin F$

From this we can see that definition 4 can be simplified so that $m(F,G) = w(\boldsymbol{v}_{(F,G)})$.

Definitions 4 and 5 allow us to represent any coherent valuation pair by a tuple (F,G) where $F, G \subseteq P$ and $G \subseteq F^c$ such that $\mathcal{D}^{\boldsymbol{v}} = F$ and $\mathcal{C}^{\boldsymbol{v}} = G$. In the following section we shall employ this representation in order to generalise from valuation pairs to imprecise valuation pairs.

The following results show that μ^+ and μ^- can be represented as the sums of certain mass values. In particular, for $\theta \in SL$, we define $\lambda(\theta)$ and $\kappa(\theta)$, such that $\mu^+(\theta)$ is the sum of mass values across $\lambda(\theta)$ and $\mu^-(\theta)$ is the sum of mass values across $\kappa(\theta)$.

Definition 6. *λ-mapping [6]*
$\lambda : SL \to 2^{2^P \times 2^P}$ *be defined recursively as follows:* $\forall p_i \in P$, $\forall \theta, \varphi \in SL$
- $\lambda(p_i) = \{(F, G) : p_i \in F\}$
- $\lambda(\theta \wedge \varphi) = \lambda(\theta) \cap \lambda(\varphi)$
- $\lambda(\theta \vee \varphi) = \lambda(\theta) \cup \lambda(\varphi)$
- $\lambda(\neg\theta) = \{(G^c, F^c) : (F, G) \in \lambda(\theta)\}^c$

Definition 7. *κ-mapping [6]*
$\kappa : SL \to 2^{2^P \times 2^P}$ *such that* $\forall \theta \in SL$ $\kappa(\theta) = \lambda(\neg\theta)^c$

Theorem 3. *Characterisation Theroem [6]*
$\forall \theta \in SL$ *and* $\forall \boldsymbol{v} \in CV$ $(\mathcal{D}^{\boldsymbol{v}}, \mathcal{C}^{\boldsymbol{v}}) \in \lambda(\theta)$ *iff* $v^+(\Psi) = 1$ *and* $(\mathcal{D}^{\boldsymbol{v}}, \mathcal{C}^{\boldsymbol{v}}) \in \kappa(\theta)$ *iff* $v^-(\theta) = 1$

Corollary 1. *[6]* $\forall \theta \in SL$ $\mu^+(\theta) = \sum_{(F,G) \in \lambda(\theta)} m(F, G)$ *and* $\mu^-(\theta) = \sum_{(F,G) \in \kappa(\theta)} m(F, G)$

Theorem 4. *[6]* $\forall \theta \in SL$ *if* $(F, G) \in \lambda(\theta)$ *then* $(F', G') \in \lambda(\theta)$ *where* $F' \supseteq F$ *and* $G' \supseteq G$.

Corollary 2. $\forall \theta \in SL$ *if* $(F, G) \in \kappa(\theta)$ *then* $(F', G') \in \kappa(\theta)$ *where* $F' \subseteq F$ *and* $G' \subseteq G$.

Proof. By definition $\kappa(\theta) = \lambda(\neg\theta)^c$. Therefore $(F, G) \in \kappa(\theta)$ iff $(F, G) \notin \lambda(\neg\theta)$. Now suppose for $F' \subseteq F$ and $G' \subseteq G$ $(F', G') \in \lambda(\neg\theta)$ then by theorem 4 this implies that $(F, G) \in \lambda(\neg\theta)$ which is a contradiction. Therefore, $(F', G') \notin \lambda(\neg\theta)$ which implies $(F', G') \in \kappa(\theta)$ as required.

In [6] it is shown that, given additional assumptions, μ^+ and μ^- can satisfy the combination rules for interval fuzzy logic (and consequently the isomorphic rules of intuitionistic fuzzy logic). Furthermore, in [6] we also discuss the relationship between bipolar belief measures and interval-set methods [8].

3 Belief Measures from Imprecise Valuation Pairs

In many scenarios knowledge regarding the assertability conventions of a population will be partial or imprecise. For example, monitoring an individual agent involved in a dialogue is only likely to provide partial information about his/her use of assertability conventions. From the perspective of valuation pairs this can mean that for certain propositional variables p_j, the values of $v^+(p_j)$ and $v^-(\neg p_j)$ are unknown or the values of $v^+(\neg p_j)$ and $v^-(p_j)$ are unknown. Consequently only lower and upper bounds (relative to $\subseteq$) can be determined for $\mathcal{D}^{\boldsymbol{v}}$ and $\mathcal{C}^{\boldsymbol{v}}$ which motivates the following definition of imprecise valuation pairs.

Definition 8. *Imprecise Valuation Pairs*
An imprecise valuation is a pair of lower and upper sets $\Theta = (\langle \underline{F}, \overline{F}\rangle, \langle \underline{G}, \overline{G}\rangle)$ *where* $\underline{F} \subseteq \overline{F} \subseteq P$ *and* $\underline{G} \subseteq \overline{G} \subseteq P$*, representing the following constraints on a coherent valuation pair* $\boldsymbol{v}$*:*

$\underline{F} \subseteq \mathcal{D}^{\boldsymbol{v}} \subseteq \overline{F}$ *and* $\underline{G} \subseteq \mathcal{C}^{\boldsymbol{v}} \subseteq \overline{G}$

Example 1. Let $P = \{p_1, p_2, p_3\}$ and suppose we have the following partial knowledge concerning valuation pair $\boldsymbol{v}$; $v^+(p_1) = 1$ and $v^-(p_2) = 0$. From this we can infer the following:

$$v^+(p_1) = 1 \Rightarrow v^-(p_1) = 1 \Rightarrow v^+(\neg p_1) = 0$$
$$v^-(p_2) = 0 \Rightarrow v^+(\neg p_2) = 1$$
$$v^-(p_2) = 0 \Rightarrow v^+(p_2) = 0$$

Consequently we have the constraints $\{p_1\} \subseteq \mathcal{D}^{\boldsymbol{v}} \subseteq \{p_1, p_3\}$ and $\{p_2\} \subseteq \mathcal{C}^{\boldsymbol{v}} \subseteq \{p_2, p_3\}$ as represented by the imprecise valuation pair
$\Theta = (\langle \{p_1\}, \{p_1, p_2\}\rangle, \langle \{p_2\}, \{p_2, p_3\}\rangle)$.

An imprecise valuation pair Θ then naturally defines a set of valuation pairs corresponding to those $\boldsymbol{v}$ consistent with the constraints $\underline{F} \subseteq \mathcal{D}^{\boldsymbol{v}} \subseteq \overline{F}$ and $\underline{G} \subseteq \mathcal{C}^{\boldsymbol{v}} \subseteq \overline{G}$. This set is characterised by the following set, K^Θ, of tuples (F, G).

Definition 9
$K^\Theta = \{(F, G) : \underline{F} \subseteq F \subseteq \overline{F},\ \underline{G} \subseteq G \subseteq \overline{G},\ \textit{and}\ G \subseteq F^c\}$

K^Θ is the set of possible values for $(\mathcal{D}^{\boldsymbol{v}}, \mathcal{C}^{\boldsymbol{v}})$, and consequently the set of possible valuation pairs $\boldsymbol{v}$, consistent with imprecise valuation pair $\Theta = (\langle \underline{F}, \overline{F}\rangle, \langle \underline{G}, \overline{G}\rangle)$.

Theorem 5. *$K^\Theta \neq \emptyset$ if and only if $\underline{G} \subseteq \underline{F}^c$*

Proof. ($\Leftarrow$) Trivial since $(\underline{F}, \underline{G}) \in K^\Theta$
($\Rightarrow$) Suppose $\underline{G} \not\subseteq \underline{F}^c$ and $K^\Theta \neq \emptyset$. In this case, $\exists p_j \in \underline{G}$ and $p_j \notin \underline{F}^c$. Let $(F, G) \in K^\Theta$ then $G \supseteq \underline{G} \Rightarrow p_j \in G$. Also, $G \subseteq F^c$ and therefore $p_j \in F^c$. But $F^c \subseteq \underline{F}^c \Rightarrow p_j \in \underline{F}^c$ which is a contradiction as required.

Theorem 6. *If $(F, G) \in K^\Theta$ then $F \subseteq \overline{F} \cap \underline{G}^c$ and $G \subseteq \overline{G} \cap \underline{F}^c$*

Proof. $G \subseteq F^c \subseteq \underline{F}^c \Rightarrow G \subseteq \overline{G} \cap \underline{F}^c$
$F \subseteq G^c \subseteq \underline{G}^c \Rightarrow F \subseteq \overline{F} \cap \underline{G}^c$

Definition 10. *Consistency and Tightness*
- *An imprecise valuation is* consistent *if* $\underline{G} \subseteq \underline{F}^c$
- *An imprecise valuation is* tight *if* $\overline{G} \subseteq \underline{F}^c$ *and* $\overline{F} \subseteq \underline{G}^c$

The imprecise valuation pair given in example 1 is both tight and consistent, and $K^\Theta = \{(\{p_1\}, \{p_2\}), (\{p_1, p_3\}, \{p_2\}), (\{p_1\}, \{p_2, p_3\})\}$. Notice that for any consistent imprecise valuation Θ there is a tight consistent valuation Θ' for which $K^{\Theta} = K^{\Theta'}$, where $\Theta' = (\langle \underline{F}, \overline{F} \cap \underline{G}^c \rangle, \langle \underline{G}, \overline{G} \cap \underline{F}^c \rangle)$. In the sequel we shall only consider tight consistent imprecise valuation pairs.

The following theorems provide some insight into the extent to which the bounding sets $\underline{F}$, $\overline{F}$ and $\underline{G}$, $\overline{G}$ constrain the values of v^+ and v^- for sentences in SL.

Theorem 7. *$\forall \theta \in SL$, and for any consistent imprecise valuation pair Θ, $v^+_{(\underline{F},\underline{G})}(\theta) = 1$ if and only if $\forall (F,G) \in K^{\Theta}$, $v^+_{(F,G)}(\theta) = 1$*

Proof. ($\Rightarrow$) Suppose $(\underline{F},\underline{G}) \in \lambda(\theta)$ then $\forall (F,G) \in K^{\Theta}$, $F \supseteq \underline{F}$ and $G \supseteq \underline{G}$. By theorem 4 this implies that $(F,G) \in \lambda(\theta)$ as required.
($\Leftarrow$) Suppose $\forall (F,G) \in K^{\Theta}$, $(F,G) \in \lambda(\theta)$, then this implies $(\underline{F},\underline{G})$ since $(\underline{F},\underline{G}) \in K^{\Theta}$

Theorem 8. *$\forall \theta \in SL$ and for any consistent imprecise valuation pair Θ, $v^-_{(\underline{F},\underline{G})}(\theta) = 1$ if and only if $\exists (F,G) \in K^{\Theta}$ such that $v^-_{(F,G)}(\theta) = 1$.*

Proof. ($\Rightarrow$) Trivial since $(\underline{F},\underline{G}) \in \kappa(\theta)$.
($\Leftarrow$) Suppose $\exists (F,G) \in K^{\Theta}$ such that $(F,G) \in \kappa(\theta)$. Now since $\underline{F} \subseteq F$ and $\underline{G} \subseteq G$ then by corollary 2 it follows that $(\underline{F},\underline{G}) \in \kappa(\theta)$.

Theorem 9. *$\forall \theta \in SL$ and for any consistent imprecise valuation pair Θ, if $v^-_{(\overline{F},\overline{G})}(\theta) = 1$ then $\forall (F,G) \in K^{\Theta}$, $v^-_{(F,G)}(\theta) = 1$.*

Proof. $\forall (F,G) \in K^{\Theta}$ $F \subseteq \overline{F}$ and $G \subseteq \overline{G}$. Therefore, since since $(\overline{F},\overline{G}) \in \kappa(\theta)$ it follows by corollary 2 that $(F,G) \in \kappa(\theta)$ as required.

Theorem 10. *$\forall \theta \in SL$, and for any consistent imprecise valuation pair Θ, if $\exists (F,G) \in K^{\Theta}$ such that $v^+_{(F,G)}(\theta) = 1$ then $v^+_{(\overline{F},\overline{G})}(\theta) = 1$*

Proof. Let $(F,G) \in K^{\Theta}$ and $(F,G) \in \lambda(\theta)$. Then $\overline{F} \supseteq F$ and $\overline{G} \supseteq G$ and by theorem 4 it follows that $(\overline{F},\overline{G}) \in \lambda(\theta)$.

The following example shows that the converses of theorems 9 and 10 do not hold in general.

Example 2. Let $\Theta = \langle (\emptyset, \{p_1\}), (\emptyset, \{p_1\}) \rangle$ then

$$K^{\Theta} = \{(\emptyset,\emptyset), (\emptyset,\{p_1\}), (\{p_1\},\emptyset)\} \text{ and } (\overline{F},\overline{G}) = (\{p_1\},\{p_1\})$$

Let $\theta = p_1 \vee \neg p_1$ then $\kappa(\theta) = \{(F,G) : p_1 \notin F \text{ or } p_1 \notin G\}$. Hence $\forall (F,G) \in K^{\Theta}$ $(F,G) \in \kappa(\theta)$ while $(\{p_1\},\{p_1\}) \notin \kappa(\theta)$. Alternatively, let $\theta = p_1 \wedge \neg p_1$ then $\lambda(\theta) = \{(F,G) : p_1 \in F, p_1 \in G\}$. Clearly in this case $(\{p_1\},\{p_1\}) \in \lambda(\theta)$ while $\forall (F,G) \in K^{\Theta}$ $(F,G) \notin \lambda(\theta)$.

The following theorems show that if we restrict ourselves to literals then the converses of theorems 9 and 10 do in fact hold.

Theorem 11. *Let $l = p_i$ or $l = \neg p_i$ where $p_i \in P$ then for any consistent tight imprecise valuation pair Θ, $v^-_{(\overline{F},\overline{G})}(l) = 1$ if and only if $\forall (F,G) \in K^\Theta$, $v^-_{(F,G)}(l) = 1$*

Proof. For $l = p_i$ we need only show that, if $\forall (F,G) \in K^\Theta$ $(F,G) \in \kappa(p_i)$ then $(\overline{F},\overline{G}) \in \kappa(p_i)$. Suppose that $\forall (F,G) \in K^\Theta$ $(F,G) \in \kappa(p_i)$ then $\forall (F,G) \in K^\Theta$ $p_i \notin G$. Now since Θ is tight $(\underline{F},\overline{G}) \in K^\Theta$ and hence $p_i \notin \overline{G}$. Therefore $(\overline{F},\overline{G}) \in \kappa(p_i)$ as required. The result for $l = \neg p_i$ follows similarly.

Theorem 12. *Let $l = p_i$ or $l = \neg p_i$ where $p_i \in P$ then for any consistent tight imprecise valuation pair Θ, $\exists (F,G) \in K^\Theta$ such that $v^+_{(F,G)}(l) = 1$ if and only if $v^+_{(\overline{F},\overline{G})}(l) = 1$*

Proof. For $l = p_i$ we need only show that, if $(\overline{F},\overline{G}) \in \lambda(p_i)$ then $\exists (F,G) \in K^\Theta$ such that $(F,G) \in \lambda(p_i)$. Suppose $(\overline{F},\overline{G}) \in \lambda(p_i)$ then $p_i \in \overline{F}$. Then since Θ is tight $(\overline{F},\underline{G}) \in K^\Theta$ and $(\overline{F},\underline{G}) \in \lambda(p_i)$. Hence, $\exists (F,G) \in K^\Theta$ such that $(F,G) \in \lambda(p_i)$ as required. The result for $l = \neg p_i$ follows similarly.

We now consider a situation where we have a number of possible partial descriptions of assertability conventions as represented by a set $\mathcal{IV} = \{\Theta_1, \ldots, \Theta_k\}$ of imprecise valuation pairs. For example, each member of $\mathcal{IV}$ might represent the knowledge obtained by monitoring a certain agent involved in a dialogue. If we then assume a weighting on the members of $\mathcal{IV}$ as represented by a probability distribution, then this naturally generates lower and upper bipolar belief measures as follows:

Definition 11. *Lower and Upper Bipolar Belief Measures*
Let $\mathcal{IV} = \{\Theta_1, \ldots \Theta_k\}$ where $\Theta_i = \langle (\underline{F}_i, \overline{F}_i), (\underline{G}_i, \overline{G}_i) \rangle$, be the set of possible consistent tight imprecise valuations pairs and let w be a probability distribution on $\mathcal{IV}$. Then we define $\forall \theta \in SL$:

- $\underline{\mu^+}(\theta) = w(\{\Theta_i : \forall (F,G) \in K^{\Theta_i}\ v^+_{(F,G)}(\theta) = 1\})$
- $\overline{\mu^+}(\theta) = w(\{\Theta_i : \exists (F,G) \in K^{\Theta_i}\ v^+_{(F,G)}(\theta) = 1\})$
- $\underline{\mu^-}(\theta) = w(\{\Theta_i : \forall (F,G) \in K^{\Theta_i}\ v^-_{(F,G)}(\theta) = 1\})$
- $\overline{\mu^-}(\theta) = w(\{\Theta_i : \exists (F,G) \in K^{\Theta_i}\ v^-_{(F,G)}(\theta) = 1\})$

Trivially, if $\theta, \varphi \in SL$ are equivalent as given in definition 2 then it holds that $\underline{\mu^+}(\theta) = \underline{\mu^+}(\varphi)$, $\overline{\mu^+}(\theta) = \overline{\mu^+}(\varphi)$, $\underline{\mu^-}(\theta) = \underline{\mu^-}(\varphi)$ and $\overline{\mu^-}(\theta) = \overline{\mu^-}(\varphi)$. Also, the next two corollaries follow from theorems 7, 8, 9 and 10, and theorems 11 and 12 respectively.

Corollary 3.
- $\underline{\mu^+}(\theta) = w(\{\Theta_i : v^+_{(\underline{F}_i,\underline{G}_i)}(\theta) = 1\})$
- $\underline{\mu^-}(\theta) \geq w(\{\Theta_i : v^-_{(\overline{F}_i,\overline{G}_i)}(\theta) = 1\})$
- $\overline{\mu^+}(\theta) \leq w(\{\Theta_i : v^+_{(\overline{F}_i,\overline{G}_i)}(\theta) = 1\})$
- $\overline{\mu^-}(\theta) = w(\{\Theta_i : v^-_{(\underline{F}_i,\underline{G}_i)}(\theta) = 1\})$

Corollary 3 shows that to determine $\underline{\mu^+}$ and $\overline{\mu^-}$ we need only consider the bounding set of each imprecise valuation pair $\mathcal{IV}$. This significantly improves the tractability of such calculations in the case that the number of propositional variables in L is large. Unfortunately, in general, for $\underline{\mu^-}$ and $\overline{\mu^+}$ the bounding sets only provide sufficient information to determine lower and upper bounds on these two measures respectively. However, the following corollary shows that if we restrict ourselves to literals then $\underline{\mu^-}$ and $\overline{\mu^+}$ can in fact be determined from only the bounding sets.

Corollary 4. *If l is a literal corresponding to p_i or $\neg p_i$ for some $p_i \in P$, then*

- $\underline{\mu^-}(l) = w(\{\Theta_i : v^-_{(\overline{F_i},\overline{G_i})}(l) = 1\})$
- $\overline{\mu^+}(l) = w(\{\Theta_i : v^+_{(\overline{F_i},\overline{G_i})}(l) = 1\})$

In some circumstances it may not be sufficient to determine lower and upper measures from $\mathcal{IV}$ and instead precise values of μ^+ and μ^- may be needed. This requires us to make additional assumptions above and beyond the information provided by each imprecise valuation pair. In the following we propose three such additional assumptions in order to generate, maximally conservative, maximally positive and maximally negative measures.

The underlying assumption behind maximally conservative measures is that given no additional information than that provided by an imprecise valuation pair Θ we should adopt the *least decisive* valuation pair in K^Θ. In other words, for any literal l we should assume, where possible, that l is not definitely assertable (i.e. $v^+(l) = 0$) but that it is acceptable to assert l (i.e. $v^-(l) = 1$).

Definition 12. *Maximally Conservative Measures*
The maximally conservative measures generated by $\mathcal{IV}$ and w are defined by:

- $\mu^+_{mc}(\theta) = w(\{\Theta_i : v^+_{(\underline{F}_i,\underline{G}_i)}(\theta) = 1\})$
- $\mu^-_{mc}(\theta) = w(\{\Theta_i : v^-_{(\underline{F}_i,\underline{G}_i)}(\theta) = 1\})$

Notice that $\boldsymbol{v}_{(\underline{F_i},\underline{G_i})}$ is the valuation pair consistent with Θ_i which minimizes *both the number of propositional variables and negated propositional variables deemed to be definitely assertable. Furthermore, given the duality between v^+ and v^-, this same valuation pair also* maximizes *the number of propositional and negated propositional variables which are deemed acceptable to assert. μ^+_{mc} and μ^-_{mc} can be determined from the mass function m_{mc} defined such that:*

$$\forall F, G \subseteq P \; m_{mc}(F,G) = w(\{\Theta_i : \underline{F}_i = F, \underline{G}_i = G\})$$

Theorem 13. $\forall \theta \in SL$, $\mu^+_{mc}(\theta) = \underline{\mu^+}(\theta)$ *and* $\mu^-_{mc}(\theta) = \overline{\mu^-}(\theta)$

Corollary 5. $\forall \theta \in SL$, $\underline{\mu^-}(\neg\theta) = 1 - \overline{\mu^+}(\theta)$ *and* $\overline{\mu^-}(\neg\theta) = 1 - \underline{\mu^+}(\theta)$

For maximally positive measures it is assumed that, amongst literals, propositional variables are inherently more assertable than negated propositional variables. Perhaps this might follow from the status of propositional variables as primitives of the language L, so that typically only assertable statements would be identified as propositions, with negation then being effectively used to express non-assertability. This assumption requires us to select the valuation pair in K^{Θ} for which $v^{+}(p_j) = v^{-}(p_j) = 1$ for the maximal number of propositional variables p_j.

Definition 13. *Maximally Positive Measures*

- $\mu^{+}_{mp}(\theta) = w(\{\Theta_i : v^{+}_{(\overline{F}_i,\underline{G}_i)}(\theta) = 1\})$
- $\mu^{-}_{mp}(\theta) = w(\{\Theta_i : v^{-}_{(\overline{F}_i,\underline{G}_i)}(\theta) = 1\})$

Notice that $\boldsymbol{v}_{(\overline{F_i},\underline{G_i})}$ *is the valuation pair consistent with* Θ_i *which simultaneously* maximizes *both the number of propositional variables deemed to be definitely assertable and the number of propositional variables deemed to be acceptable to assert. Due to the duality between* v^{+} *and* v^{-} *this valuation pair also* minimizes *the number of negated propositional variables deemed to be definitely assertable.* μ^{+}_{mp} *and* μ^{-}_{mp} *can be determined from the mass function* m_{mp} *defined such that:*

$$\forall F,G \subseteq P \ m_{mp}(F,G) = w(\{\Theta_i : \overline{F}_i = F, \underline{G}_i = G\})$$

Maximally negative measures are based on the assumption that assertability conventions favour negative statements. In effect this is the dual of maximal positivity in that, amongst literals, it is assumed that negated propositional variables are more assertable than propositional variables. Making this assumption corresponds to identifying the valuation pair in K^{Θ} for which $v^{+}(\neg p_j) = v^{-}(\neg p_j) = 1$ for the maximal number of propositional variables p_j.

Definition 14. *Maximally Negative Measures*

The maximally negative measures generated by $\mathcal{IV}$ *and* w *are defined by:*

- $\mu^{+}_{mn}(\theta) = w(\{\Theta_i : v^{+}_{(\underline{F}_i,\overline{G}_i)}(\theta) = 1\})$
- $\mu^{-}_{mn}(\theta) = w(\{\Theta_i : v^{-}_{(\underline{F}_i,\overline{G}_i)}(\theta) = 1\})$

Notice that $\boldsymbol{v}_{(\underline{F}_i,\overline{G}_i)}$ *is the valuation pair consistent with* Θ_i *which simultaneously* maximizes *both the number of negated propositional variables deemed to be definitely assertable and the number of negated propositional variables deemed to be acceptable to assert. Due to the duality between* v^{+} *and* v^{-} *this valuation pair also* minimizes *the number of propositional variables deemed to be definitely assertable.* μ^{+}_{mn} *and* μ^{-}_{mn} *can be determined from the mass function* m_{mn} *defined such that:*

$$\forall F,G \subseteq P \ m_{mn}(F,G) = w(\{\Theta_i : \underline{F}_i = F, \overline{G}_i = G\})$$

From corollaries 3 and 4 and definitions 12, 13 and 14 it follows immediately that: $\forall p_j \in P$

$$\underline{\mu^{+}}(p_j) = \mu^{+}_{mc}(p_j) = \mu^{+}_{mn}(p_j) = w(\{\Theta_i : p_j \in \underline{F}_i\})$$

$$\overline{\mu^{+}}(p_j) = \mu^{+}_{mp}(p_j) = w(\{\Theta_i : p_j \in \overline{F}_i\})$$

$$\underline{\mu^{-}}(p_j) = \mu^{-}_{mn}(p_j) = w(\{\Theta_i : p_j \notin \overline{G}_i\})$$

$$\overline{\mu^{-}}(p_j) = \mu^{-}_{mc}(p_j) = \mu^{-}_{mp}(p_j) = w(\{\Theta_i : p_j \notin \underline{G}_i\})$$

Similarly, it also holds that: $\underline{\mu^+}(\neg p_j) = \mu^+_{mc}(\neg p_j) = \mu^+_{mp}(\neg p_j)$, $\overline{\mu^+}(\neg p_j) = \mu^+_{mn}(\neg p_j)$, $\underline{\mu^-}(\neg p_j) = \mu^-_{mp}(\neg p_j)$ and $\overline{\mu^-}(\neg p_j) = \mu^-_{mc}(\neg p_j) = \mu^-_{mn}(\neg p_j)$.

4 Learning Bipolar Belief Measures from Agent Assertions

In this section we consider a simplified dialogue between a population of agents in order to illustrate the potential use of imprecise valuation pairs as a representation framework to capture partial knowledge about assertability conventions. Let $\mathcal{A} = \{a_1, \ldots, a_m\}$ be a population of communicating agents. In this simplified model each agent in $\mathcal{A}$ is able to *assert* propositions or negated propositions from P and to *condemn* or to *agree with* the assertions of others. Communications between these agents are recorded for a fixed time period T. For sentence $\theta \in SL$ and agents $a_i, a_j \in \mathcal{A}$ let $AS(a_i, \theta)$, $CD(a_i, a_j, \theta)$ and $AG(a_i, a_j, \theta)$ denote the actions 'During T a_i asserts θ' 'During T a_i condemns a_j for asserting θ' and 'During T a_i agrees with a_j for asserting θ' respectively. Furthermore we define the following subsets of P:

- $PAS_i = \{p_j \in P : AS(a_i, p_j)\}$ denoting the positive assertions made by a_i.
- $NAS_i = \{p_j \in P : AS(a_i, \neg p_j)\}$ denoting the negative assertions made by a_i.
- $PCD_i = \{p_j \in P : \exists a_k \in \mathcal{A},\ CD(a_i, a_k, p_j)\}$ denoting the propositions condemned by a_i.
- $NCD_i = \{p_j \in P : \exists a_k \in \mathcal{A},\ CD(a_i, a_k, \neg p_j)\}$ denoting the negated propositions condemned by a_i.
- $PAG_i = \{p_j \in P : \exists a_k \in \mathcal{A},\ AG(a_i, a_k, p_j)\}$ denoting the propositions agreed with by a_i.
- $NAG_i = \{p_j \in P : \exists a_k \in \mathcal{A},\ AG(a_i, a_k, \neg p_j)\}$ denoting the negated propositions agreed with by a_i.

It is assumed that the assertions, agreements and condemnations made by each agent a_i is consistent with a consistent valuation pair $\boldsymbol{v}_i$ in the following sense: For literal $l = \pm p_j$

- If $AS(a_i, l)$ then $v_i^-(l) = 1$
- If $CD(a_i, a_k, l)$ then $AS(a_k, l)$ and $v_i^+(\neg l) = 1$
- If $AG(a_i, a_k, l)$ then $AS(a_k, l)$ and $v_i^+(l) = 1$

The intuition here is that the acts of condemnation and agreement indicate stronger beliefs about literals than the act of assertion. Indeed, from the assertion of a statement it is assumed that we can only infer that the agent in question believes it to be at least acceptable to assert that statement.

Theorem 14. $\forall a_i \in \mathcal{A}$

$$NCD_i \cup PAG_i \subseteq \mathcal{D}^{\boldsymbol{v}_i} \subseteq (NAS_i)^c \cap (PCD_i)^c \cap (NAG_i)^c$$
$$PCD_i \cup NAG_i \subseteq \mathcal{C}^{\boldsymbol{v}_i} \subseteq (PAS_i)^c \cap (NCD_i)^c \cap (PAG_i)^c$$

Proof.

$NAS_i \subseteq \{p_j : v_i^-(\neg p_j) = 1\} = \{p_j : v_i^+(p_j) = 0\} = (\mathcal{D}^{\boldsymbol{v}_i})^c \Rightarrow \mathcal{D}^{\boldsymbol{v}_i} \subseteq (NAS_i)^c$
$PAS_i \subseteq \{p_j : v_i^-(p_j) = 1\} = \{p_j : v_i^+(\neg p_j) = 0\} = (\mathcal{C}^{\boldsymbol{v}_i})^c \Rightarrow \mathcal{C}^{\boldsymbol{v}_i} \subseteq (PAS_i)^c$
Also $p_j \in PCD_i \Rightarrow v_i^+(\neg p_j) = 1 \Rightarrow p_j \in \mathcal{C}^{\boldsymbol{v}_i} \Rightarrow PCD_i \subseteq \mathcal{C}^{\boldsymbol{v}_i}$
and $p_j \in NCD_i \Rightarrow v_i^+(p_j) = 1 \Rightarrow p_j \in \mathcal{D}^{\boldsymbol{v}_i} \Rightarrow NCD_i \subseteq \mathcal{D}^{\boldsymbol{v}_i}$
and $p_j \in PAG_i \Rightarrow v_i^+(p_j) = 1 \Rightarrow p_j \in \mathcal{D}^{\boldsymbol{v}_i} \Rightarrow PAG_i \subseteq \mathcal{D}^{\boldsymbol{v}_i}$
and $p_j \in NAG_i \Rightarrow v_i^+(\neg p_j) = 1 \Rightarrow p_j \in \mathcal{C}^{\boldsymbol{v}_i} \Rightarrow NAG_i \subseteq \mathcal{C}^{\boldsymbol{v}_i}$

Therefore we have shown that

$$NCD_i \cup PAG_i \subseteq \mathcal{D}^{\boldsymbol{v}_i} \subseteq (NAS_i)^c \text{ and } PCD_i \cup NAG_i \subseteq \mathcal{C}^{\boldsymbol{v}_i} \subseteq (PAS_i)^c$$

However, by definition of a coherent valuation pair $\boldsymbol{v}_i$ it also follows from the above that:

$$\mathcal{C}^{\boldsymbol{v}_i} \subseteq (\mathcal{D}^{\boldsymbol{v}_i})^c \subseteq (NCD_i \cup PAG_i)^c \text{ and } \mathcal{D}^{\boldsymbol{v}_i} \subseteq (\mathcal{C}^{\boldsymbol{v}_i})^c \subseteq (PCD_i \cup NAG_i)^c$$

Combining these constraints then gives:

$$NCD_i \cup PAG_i \subseteq \mathcal{D}^{\boldsymbol{v}_i} \subseteq (NAS_i)^c \cap (PCD_i)^c \cap (NAG_i)^c$$
$$PCD_i \cup NAG_i \subseteq \mathcal{C}^{\boldsymbol{v}_i} \subseteq (PAS_i)^c \cap (NCD_i)^c \cap (PAG_i)^c$$

as required.

Hence, we see that the assertions and condemnations of agent a_i determines a consistent tight imprecise valuation $\Theta_i = (\langle \underline{F}_i, \overline{F}_i \rangle, \langle \underline{G}_i, \overline{G}_i \rangle)$ where

$$\underline{F}_i = NCD_i \cup PAG_i,\ \overline{F}_i = (NAS_i)^c \cap (PCD_i)^c \cap (NAG_i)^c$$
$$\underline{G}_i = PCD_i \cup NAG_i,\ \overline{G}_i = (PAS_i)^c \cap (NCD_i)^c \cap (PAG_i)^c$$

Example 3. Consider a population of four agents $\mathcal{A} = \{a_1, a_2, a_3, a_4\}$ and let L be a language with four propositional variables $P = \{p_1, p_2, p_3, p_4\}$. The agents' communications over a fixed time period are summarized in table 1. From this data we generate the following imprecise valuation pairs according to theorem 14:

Table 1. Table summarizing the communications between agents

Agent	PAS_i	NAS_i	PCD_i	NCD_i	PAG_i	NAG_i
a_1	$\{p_1, p_3\}$	$\emptyset$	$\{p_4\}$	$\emptyset$	$\{p_3\}$	$\emptyset$
a_2	$\{p_2\}$	$\{p_4\}$	$\{p_4\}$	$\emptyset$	$\emptyset$	$\{p_4\}$
a_3	$\{p_1\}$	$\{p_4\}$	$\{p_3\}$	$\emptyset$	$\emptyset$	$\{p_2\}$
a_4	$\{p_2, p_4\}$	$\emptyset$	$\emptyset$	$\{p_4\}$	$\{p_4\}$	$\emptyset$

$$\Theta_1 = (\langle\{p_3\}, \{p_1, p_2, p_3\}\rangle, \langle\{p_4\}, \{p_2, p_4\}\rangle)$$
$$\Theta_2 = (\langle\emptyset, \{p_1, p_2, p_3\}\rangle, \langle\{p_4\}, \{p_1, p_3, p_4\}\rangle)$$
$$\Theta_3 = (\langle\emptyset, \{p_1\}\rangle, \langle\{p_2, p_3\}, \{p_2, p_3, p_4\}\rangle)$$
$$\Theta_4 = (\langle\{p_4\}, \{p_1, p_2, p_3, p_4\}\rangle, \langle\emptyset, \{p_1, p_3\}\rangle)$$

Now assuming equal weighting to agents results in a uniform distribution w on $\{\Theta_1, \Theta_2, \Theta_3, \Theta_4\}$. From this we obtain the following lower and upper bipolar belief measures as shown for the propositional variables in table 2.

Table 2. Table summarizing $\underline{\mu}^+$, $\overline{\mu}^+$, $\underline{\mu}^-$, $\overline{\mu}^-$

Proposition	$\underline{\mu}^+$	$\overline{\mu}^+$	$\underline{\mu}^-$	$\overline{\mu}^-$
p_1	0	1	$\frac{1}{2}$	1
p_2	0	$\frac{3}{4}$	$\frac{1}{2}$	$\frac{3}{4}$
p_3	$\frac{1}{4}$	$\frac{3}{4}$	$\frac{1}{4}$	$\frac{3}{4}$
p_4	$\frac{1}{4}$	$\frac{1}{4}$	$\frac{1}{4}$	$\frac{1}{2}$

From definitions 12, 13 and 14 we have that mass functions m_{mc}, m_{mp} and m_{mn} are given by:

$$m_{mc} := (\{p_3\}, \{p_4\}) : \frac{1}{4},\ (\emptyset, \{p_4\}) : \frac{1}{4},\ (\emptyset, \{p_2, p_3\}) : \frac{1}{4},\ (\{p_4\}, \emptyset) : \frac{1}{4}$$
$$m_{mp} := (\{p_1, p_2, p_3\}, \{p_4\}) : \frac{1}{2},\ (\{p_1\}, \{p_2, p_3\}) : \frac{1}{4},\ (\{p_1, p_2, p_3, p_4\}, \emptyset\}) : \frac{1}{4}$$
$$m_{mn} := (\{p_3\}, \{p_2, p_4\}) : \frac{1}{4},\ (\emptyset, \{p_1, p_3, p_4\}) : \frac{1}{4},\ (\emptyset, \{p_2, p_3, p_4\}) : \frac{1}{4},\ (\{p_4\}, \{p_1, p_3\}) : \frac{1}{4}$$

The resulting bipolar belief measures for these mass functions are then summarised in table 3 for the propositional variables.

Table 3. Table summarizing μ^+_{mc},μ^-_{mc},μ^+_{mp},μ^-_{mp},μ^+_{mn},μ^-_{mn}

Proposition	μ^+_{mc}	μ^-_{mc}	μ^+_{mp}	μ^-_{mp}	μ^+_{mn}	μ^-_{mn}
p_1	0	1	1	1	0	$\frac{1}{2}$
p_2	0	$\frac{3}{4}$	$\frac{3}{4}$	$\frac{3}{4}$	0	$\frac{1}{2}$
p_3	$\frac{1}{4}$	$\frac{3}{4}$	$\frac{3}{4}$	$\frac{3}{4}$	$\frac{1}{4}$	$\frac{1}{4}$
p_4	$\frac{1}{4}$	$\frac{1}{2}$	$\frac{1}{4}$	$\frac{1}{2}$	$\frac{1}{4}$	$\frac{1}{4}$

5 Conclusions and Future Work

In this paper we have proposed valuation pairs as a possible framework for representing assertability conventions applied to sentences from a propositional logic language. This approach has then been extended by introducing imprecise valuation pairs to represent partially or imprecisely defined assertability conventions. Allowing for uncertainty then naturally results in lower and upper bipolar belief measures on the sentences of the language. The potential of this extended framework has then been illustrated by its application to an idealised dialogue between agents, each making assertions, agreements and condemnations based on their individual valuation pairs. Information inferred from this dialogue can then be represented by imprecise valuation pairs together with lower and upper bipolar belief measures.

Future work will investigate extending the form of dialogue described in section 4 to allow more complex communications between agents. Another focus of research may be to consider methods for combining (fusing) imprecise valuation pairs, and the related issue of consistency. For instance, one might define a conjunction of two imprecise valuation pairs Θ_i and Θ_j by;

$$\Theta_i \sqcap \Theta_j = \left(\left\langle \underline{F}_i \cup \underline{F}_j, \overline{F}_i \cap \overline{F}_j \right\rangle, \left\langle \underline{G}_i \cup \underline{G}_j, \overline{G}_i \cap \overline{G}_j \right\rangle\right)$$

provided $\underline{F}_i \cup \underline{F}_j \subseteq \overline{F}_i \cap \overline{F}_j$, $\underline{G}_i \cup \underline{G}_j \subseteq \overline{G}_i \cap \overline{G}_j$, $\underline{G}_i \cup \underline{G}_j \subseteq (\underline{F}_i \cup \underline{F}_j)^c$

and is undefined otherwise

Θ_i and Θ_j would then be deemed consistent provided that $\Theta_i \sqcap \Theta_j$ was defined.

Currently, individual agents are viewed as providing alternative potential assertability conventions and a probability distribution is then employed to provide a weighting to each of these possibilities. However, this somewhat ignores the fact that each agent's valuation pair has been inferred from its interactions and dialogue with other agents in the population. Consequently, one might expect to gain insight into the shared assertability conventions of the whole population by comparing the imprecise valuation pairs inferred from the different agents. For example, one approach might be to investigate maximally consistent sets of imprecise valuation pairs where consistency is defined as outlined above.

References

1. Dubois, D., Prade, H.: An Introduction to Bipolar Representations of Information and Preference. International Journal of Intelligent Systems 23, 866–877 (2008)
2. Kleene, S.C.: Introduction to Metamathematics. D. Van Nostrand Company Inc., Princeton (1952)
3. Kyburg, A.: When Vague Sentences Inform: A Model of Assertability. Synthese 124, 175–192 (2000)
4. Lawry, J.: Appropriateness Measures: An uncertainty model for vague concepts. Synthese 161(2), 255–269 (2008)
5. Lawry, J., Gonzalez Rodriguez, I.: Generalised Label Semantics as a Model of Epistemic Vagueness. In: Sossai, C., Chemello, G. (eds.) ECSQARU 2009. LNCS, vol. 5590, pp. 626–637. Springer, Heidelberg (2009)
6. Lawry, J., Gonzalez Rodrigues, I.: A Bipolar Model of Assertability and Belief. International Journal of Approximate Reasoning (in Press, 2010)
7. Parikh, R.: Vague Predicates and Language Games. Theoria (Spain) XI(27), 97–107 (1996)
8. Yao, Y.Y., Li, X.: Uncertain Reasoning with Interval-set Algebra. In: Ziarko, W.P. (ed.) Rough Sets, Fuzzy Sets and Knowledge Discovery, pp. 178–185. Springer, Heidelberg (1994)

Kriging with Ill-Known Variogram and Data

Kevin Loquin and Didier Dubois

Institut TELECOM ParisTech,
TSI department, 46, rue Barrault
75013 Paris, France
kevin.loquin@gmail.com
Institut de Recherche en Informatique de Toulouse,
RPDMP department, 118 Route de Narbonne
31062 Toulouse Cedex 9, France
dubois@irit.fr

Abstract. Kriging consists in estimating or predicting the spatial phenomenon at non sampled locations from an estimated random function. Although information necessary to properly select a unique random function model seems to be partially lacking, geostatistics in general, and the kriging methodology in particular, does not account for the incompleteness of the information that seems to pervade the procedure. On the one hand, the collected data may be tainted with errors that can be modelled by intervals or fuzzy intervals. On the other hand, the choice of parameter values for the theoretical variogram, an essential step, contains some degrees of freedom that is seldom acknowledged. In this paper we propose to account for epistemic uncertainty pervading the variogram parameters, and possibly the data set, by means of fuzzy interval uncertainty. We lay bare its impact on the kriging results, improving on previous attempts by Bardossy and colleagues in the late 1980's.

Keywords: geostatistics, kriging, variogram, random function, epistemic uncertainty, fuzzy subset, possibility theory, optimisation.

1 Introduction

Epistemic uncertainty is uncertainty that stems from a lack of knowledge, from insufficient available information, about a phenomenon. It is different from uncertainty due to the variability of the phenomenon. In risk analysis, there is a growing consensus that epistemic uncertainty and variability should not be modelled in the same way [5,21,22,27] because they do not lead to the same kind of decision. When epistemic uncertainty is prevailing, the natural approach is to try to collect more information, so as to reduce it. On the contrary, in the face of potentially dangerous variability, it is important to confine it by a suitable course of action. These kinds of actions are called epistemic and ontic actions, respectively, in artificial intelligence.

In geostatistics and related areas of applications (such as underground CO2 storage), one source of information consists of maps featuring the variation of

A. Deshpande and A. Hunter (Eds.): SUM 2010, LNAI 6379, pp. 219–235, 2010.

quantities of interest measuring properties across a certain geographical area, on the basis of some local measurements. Constructing such maps requires interpolation techniques that may account for observed spatial variability. The most well-known of them is called kriging.

Kriging methods have been studied and applied extensively since 1970 and later on adapted, extended, and generalized [8,28]. The approach is based on the use a random field, and a number of assumptions such as stationarity and spatial ergodicity, so as the reduce the needed information to a so-called variogram that can be estimated from the available measurements. This variogram enables least-square like equations to be derived, whose solutions yield the interpolation coefficients.

Uncertainty present in the process may cast some doubts on the validity of the resulting interpolation: physical measurements can be inaccurate, and the choice of a theoretical variogram is partially arbitrary, as it relies on the skills of the geostatistician. Moreover, the statistical validity of the variogram conflicts with the local relevance of the obtained estimates: the larger the considered area, the more data, the more valid is the variogram, and the less locally relevant is the interpolated estimate at a given location. Conversely, a local analysis will have poor statistical validity as available data will be very scarce. However, very few scholars discussed the nature of the uncertainty that underlies the standard Matherionan geostatistics except G. Matheron himself [36] and even fewer considered theories alternative to probability theory that could more reliably handle epistemic uncertainty in geostatistics (See Loquin and Dubois [32] for a survey).

Typically, possibility distributions [18] in the form of intervals or fuzzy intervals are supposed to handle epistemic uncertainty, while probability distributions are supposed to properly quantify variability. This paper proposes an approach to evaluate the impact, on kriging results, of epistemic uncertainty pervading the data and the theoretical variogram represented by fuzzy intervals.

2 (Ordinary) Kriging

Geostatistics, is commonly viewed as the application of statistics to the study of spatially distributed data. This theory is not new and borrows most of its models and tools from the concept of random function. Consider a spatial phenomenon, such as a soil pollutant or permeability, which can be viewed as a deterministic spatial function $z : \mathcal{D} \longrightarrow \Gamma \subseteq \mathbb{R}$, where $\mathcal{D}$ is a compact subset of $\mathbb{R}^\ell$ (generally of $\mathbb{R}^2$). Available information about this spatial entity are n observations $\mathbf{z} = \{z(x_i), i = 1, \ldots, n\}$, located at n known distinct sampling positions $\{x_1, \ldots, x_n\}$ in $\mathcal{D}$. Geostatistics aim at extracting information about z from the unique spatially distributed statistic $\mathbf{z}$. To do so, it makes an extensive use of a real valued random function $Z = \{Z(x), x \in \mathcal{D}\}$, which allegedly captures uncertainty about the spatial phenomenon under study.

2.1 From Random Functions to Variograms

A particular case of random function commonly used in geostatistics is the *intrinsic* random function [8], which is characterized by a mean value and a variance of increments of $Z(x)$:

$$\begin{cases} \mathbb{E}[Z(x)] = m, \\ \mathbb{V}[Z(x+h) - Z(x)] = 2\gamma(h). \end{cases} \tag{1}$$

$\forall x, h \in \mathcal{D}$. The function $\gamma(h)$ is called a *(semi-)variogram.* The variogram is a key concept in geostatistics. It quantifies the dependence or interaction between $Z(x)$ and $Z(x')$ at any two locations x and x' of $\mathcal{D}$, as a function of their separation $h = x - x'$.

Mathematically, variogram and variance are strongly linked and some properties of the variance are propagated to the variogram. Namely, the so-called conditional negative definiteness property of the variogram is inherited from the positivity of the variance.

Definition 1 (Conditionally negative definite function). *A function $\gamma(h)$, defined for any $h \in \mathbb{R}^\ell$, is conditionally negative definite if, for any choice of p, $\{x_i, i = 1, \dots, p\}$ and $\{\mu_i, i = 1, \dots, p\}$, conditionally to the fact that $\sum_{i=1}^{p} \mu_i = 0$,*

$$\sum_{i=1}^{p} \sum_{j=1}^{p} \mu_i \mu_j \gamma(x_j - x_i) \leq 0.$$

This conditionally negative definiteness requirement is necessary for any function to qualify as the variogram of some random function. However, it forbids any empirical derivation of the variogram and reduces the set of possible variogram models to some specific functions (e.g. spherical or Gaussian models). Explicit formulations of many popular variogram models are surveyed in [8]. Generally, a linear combination, which preserves definiteness of variograms is used. Hence in its most general formulation, the theoretical variogram is a function with p parameters, denoted by $\mathbf{a} = \{a_1, \dots, a_p\}$.

In most cases only three variogram parameters are specified. The **sill** is the asymptotic value of the variogram when h increases. It means that there is a distance, called the **range**, beyond which $Z(x)$ and $Z(x+h)$ are uncorrelated. In some sense, the range gives some meaning to the concept of area of influence. Another parameter of a variogram that can be physically interpreted is the **nugget effect**: it is the value taken by the variogram when h tends to 0. This discontinuity at the origin is generally due to geological discontinuity, measurement noise or positioning errors.

2.2 Ordinary Kriging

In the kriging framework, it is assumed that the estimate $z^*(x_0)$ of the deterministic function z at any location x_0 of $\mathcal{D}$, is formed by a linear combination of the n collected data $\mathbf{z} = \{z(x_i), i = 1, \dots, n\}$. The *kriging estimate* is given by:

$$z^*(x_0) = \sum_{i=1}^{n} \lambda_i(x_0) z(x_i). \quad (2)$$

The computation of $z^*(x_0)$ depends on the estimation of the kriging weights $\Lambda_0 = \{\lambda_i(x_0), i = 1, \ldots, n\}$ at location x_0. For short, we shall denote the kriging weights by λ_i instead of $\lambda_i(x_0)$ without any ambiguity for the rest of the paper. Each weight λ_i corresponds to the *influence* of the value $z(x_i)$ in the computation of $z^*(x_0)$.

Kriging weights can be obtained by minimizing the variance of the estimation error $\mathbb{V}(Z(x_0) - Z^*(x_0))$ under the unbiasedness condition $\mathbb{E}(Z^*(x_0)) = \mathbb{E}(Z(x_0))$. In this framework, $Z(x_0)$ is the random variable at location x_0 and $Z^*(x_0) = \sum_{i=1}^{n} \lambda_i Z(x_i)$ is the random counterpart to the kriging estimate (2).

For any intrinsic random function Z with a known variogram function and an unknown constant mean value m, ordinary kriging can be viewed as a least squares method. The kriging weights Λ_0 minimizing the estimation variance or equivalently the mean squared error $\mathbb{E}\big(Z(x_0) - Z^*(x_0)\big)^2$ are obtained by solving the linear system [8,24]:

$$\begin{cases} \sum_{j=1}^{n} \lambda_j \gamma(x_i - x_j) + \mu = \gamma(x_0 - x_i), \forall i = 1, \ldots, n, \\ \sum_{i=1}^{n} \lambda_i = 1, \end{cases} \quad (3)$$

where μ is a Lagrange multiplier that makes the ordinary kriging system solvable. The $n+1$-th equation $\sum_{i=1}^{n} \lambda_i = 1$ is inherited from the unbiasedness condition.

The ordinary kriging variance is then given by:

$$\sigma_O^2(x_0) = \sum_{i=1}^{n} \lambda_i \gamma(x_0 - x_i) + \mu. \quad (4)$$

See the monograph by Chilès and Delfiner [8] and the one by Goovaerts [24] for deeper presentations of many kriging methods.

3 Epistemic Uncertainty in Kriging

The traditional kriging method delivers precise interpolation results based on a probabilistic analysis of the problem. It may look paradoxical since, on the one hand, the only form of variability that makes sense in the problem is spatial and reflected in the data. The random function setting appears as a mathematical artefact. On the other hand, when data is scarce, information about the variogram is incomplete, and besides the data itself may be inaccurate to some extent. In this situation, one should expect the interpolation results to be imprecise in areas where few data have been collected.

3.1 Variability in Kriging: A Discussion

Probabilistic models are natural representations of phenomena displaying some form of variability. Repeatability is the central feature of the idea of probability as pointed out by Shafer and Vovk [39]. A random variable $V(\omega)$ is a mapping from a sample space Ω to the real line, and variability is captured by binding the value of V to the repeated choices of $\omega \in \Omega$, and the probability measure that equips Ω summarizes the repeatability pattern.

In the case of the random function approach to geostatistics, the role of this scenario is not quite clear. Geostatistics is supposed to handle spatial variability of a numerical quantity $z(x)$ over some geographical area $\mathcal{D}$. Taken at face value, spatial variability means that when the location $x \in \mathcal{D}$ changes, so does $z(x)$. However, when x is fixed $z(x)$ is a precise deterministic value. Strictly speaking, these considerations would lead us to identify the sample space with $\mathcal{D}$, equipped with the Lebesgue measure. In that case, one way of interpreting such a random function $Z(x)$ is to consider that it accounts for the spatial variability of $z(x)$ in a neighborhood of x (this is the so-called transitive model of Matheron [8,35] or the deterministic view of kriging described by Journel [29]).

In order to justify the choice of a random function, whereby spatial variability is replaced by a thought repeatable experiment at each location, a spatial ergodicity assumption is advocated. It claims that the spatial moments of the deterministic spatial entity z are equal to the statistical moments of any random variable $Z(x)$ pertaining to the random function Z. This is clearly an adaptation, to spatial phenomena, of an assumption made on random functions for temporal phenomena. Thus the use of a random field as a mathematical model for the problem looks like a clever trick to formally derive the least-square like equations whose solutions yield the interpolation coefficients.

3.2 Epistemic Uncertainty in the Variogram and the Data

Another source of uncertainty possibly pervading any information processing method is epistemic uncertainty, which stems from a lack of knowledge, from insufficient available information, about a phenomenon. In the kriging estimation procedure, epistemic uncertainty clearly lies in two places of the process: the knowledge of data points and the choice of the theoretical variogram.

Empirical tools like variogram clouds or sample variograms [32] are considered by geostatisticians as visualization or preliminary guiding tools. They cannot be used in the kriging computation because they generally do not fulfil the conditional negative definite requirement. In order to overcome this difficulty, two methods are generally considered : either an automated fitting (by means of a regression analysis) of a chosen theoretical variogram model to the empirical variogram is performed, or even a manual fitting is made at a glance by the experts.

Whatever the chosen method, the fitting of a variogram involves an important epistemic transfer. Indeed, the practitioner tries to extrapolate from some lacunary objective information (the sample variogram built from the data), by means of a unique subjectively chosen dependence model, the theoretical variogram. As pointed out by A. G. Journel [30]:

> *Any serious practitioner of geostatistics would expect to spend a good half of his or her time looking at all faces of a data set, relating them to various geological interpretations, prior to any kriging.*

Except in the papers by Bardossy *et al.* [2,3,4] discussed later on, this fundamental step of the kriging method is never considered in terms of the epistemic uncertainty it creates. Intuitively, however, there is a certain degree of freedom in the choice of a single variogram, expressing a lack of information. This lack of information is one source of epistemic uncertainty, by definition [27].

As the variogram model plays a critical role in the calculation of the reliability of a kriging estimate, the epistemic uncertainty of its fit should not be neglected. Not considering the epistemic uncertainty pervading the variogram parameters, as propagated to the kriging estimate, may result in underestimated risks and a false confidence in the results. Even in the presence of a precise dataset, one may argue that the chosen variogram is tainted with epistemic uncertainty that only the expert who selects it could estimate.

One may also challenge the precision or accuracy of the measurements. Especially, geological measurements are often highly imprecise. Let us take a simple example: the measurement of permeability in an aquifer. It results from the interpretation of a pumping test: if we pump water in a well, the water level will decrease in that well and also in neighboring wells. The local permeability is obtained by fitting theoretical draw-down curves to experimental ones. There is obviously some imprecision in that fitting that is based on assumptions on the environment (e.g., homogeneous substrate). Epistemic uncertainty due to measurement imperfections should pervade the measured permeability data. For the inexact (imprecise) information resulting from unique assessments of deterministic values, a nonfrequentist or subjective approach reflecting imprecision could be used.

3.3 The Limited Usefulness of the Kriging Variance

Note that epistemic uncertainty has little to do with the kriging variance. Indeed, this variance remains a theoretical artifact that only depends on the sampling positions $\{x_1, \ldots, x_n\}$ and not on the values $z(x_i)$. The only link between estimation variance and data values is through the variogram, which is global rather than local in its definition. A variogram is not supposed to handle variability in the measurements, only mutual influence between measurements. Even the nugget effect, which is supposed to capture, to some extent, measurement errors, contributes to the kriging variance (4) only for measurements inside a very small neighborhood of x_0. Once parameters of the theoretical variogram are decided, the kriging variance no longer depends on the measured values, but merely on the geometry of the measurement location sample. Thus, it is clear that the usual kriging variance does not produce an estimation of the kriging error or imprecision due to incomplete information. Other techniques are required.

4 Modelling and Propagating Epistemic Uncertainty

Uncertainty theories such as possibility theory [18], belief functions [38] or imprecise previsions [42] are supposed to jointly handle variability and epistemic uncertainty. These approaches share the same basic principle: when facing a lack of knowledge or insufficient available information, it is safer to work with a family of probability measures. Such models are generically called imprecise probability models. The simplest of these approaches consists in using intervals or fuzzy intervals.

4.1 Fuzzy Intervals

A fuzzy subset F of the real line [18,45] can model the available knowledge (an epistemic state) about a deterministic value, for instance the value $z(x)$ measured at location x, in the sense that the membership degree $F(r)$ is a gradual estimation of the adequation of r to the value $z(x)$ according to the expert knowledge. The membership grade $F(r)$ is then interpreted as a degree of possibility of $z(x) = r$ according to the expert [46]. In our problem, fuzzy sets are representations of knowledge about underlying precise data and variogram parameters.

Possibility distributions can often be viewed as nested sets of intervals [14]. Let $F_\alpha = \{r \in \mathbb{R} : F(r) > \alpha\}$ be called an α-cut. F is called a fuzzy interval if and only if $\forall 0 < \alpha \leq 1, F_\alpha$ is an interval. If the membership function is continuous, the degree of certainty of $z(x) \in F_\alpha$ is equal to $N(F_\alpha) = 1 - \alpha$, in the sense that any value outside F_α has possibility degree at most α. So it is sure that $z(x) \in S(F) = lim_{\alpha \to 0} F_\alpha$ (the support of F), while there is no certainty that the most plausible values in F_1 contain the actual value. Note that the membership function can be retrieved from its α-cuts, by means of the relation:

$$F(r) = \sup_{r \in F_\alpha} \alpha.$$

Therefore, suppose that the available knowledge supplied by an expert comes in the form of nested intervals $\mathcal{F} = \{I_k, k = 1, \ldots, K\}$ such that $I_1 \subset I_2 \subset \cdots \subset I_K$ with increasing confidence levels $c_k > c_{k'}$ if $k > k'$, the possibility distribution defined by

$$F(r) = \min_{k=1,\ldots,K} \max(1 - c_k, I_k(r)),$$

where function $I_k(\cdot)$ is the characteristic function of set I_k, is a faithful representation of the supplied information. The nestedness assumption is natural in the case of expert knowledge: we expect an expert to be coherent when providing information, while we cannot demand full precision. On the contrary, we cannot expect a series of imprecise sensor measurements to be coherent, due to noise, while we expect them to be as precise as possible.

Viewing a necessity degree as a lower probability bound [19], F is an encoding of the probability family $\mathcal{P}_\mathcal{F} = \{P : P(I_k) \geq c_k\}$. If $c_K = 1$ then the support of this fuzzy interval is I_K. The continuous fuzzy interval is obtained in the limit using an infinite number of confidence intervals. If an expert only provides a modal value c and a support $[a, b]$, it makes sense to represent this information as the triangular fuzzy interval with mode c and support $[a, b]$ [15]. Indeed F then encodes a family of (subjective) probability distributions containing all the unimodal ones with the same mode and support included in $[a, b]$.

Note that representing data or model parameters by intervals or fuzzy intervals does not mean a change in the nature of the data or the parameters. The best model and the actual measured value are still precise but they are ill-known. Intervals and fuzzy sets cannot be handled as existing entities, they are epistemic constructs: they are models of our knowledge of reality, not models of reality per se.

4.2 The Extension Principle

The extension principle introduced by Lotfi Zadeh [18,46] provides a general method for extending non fuzzy models or functions in order to deal with fuzzy parameters. For instance, fuzzy interval arithmetic [16] that generalizes interval arithmetic has been developed by applying the extension principle to standard arithmetic operations like addition, subtraction, etc. The extension principle applied to any function $f : X \to Y$, of m precise parameters $\mathbf{x} = \{x_k, k = 1, \ldots, m\}$, to a set of m fuzzy parameters $\hat{\mathbf{x}} = \{\hat{x}_k, k = 1, \ldots, m\}$ modelling epistemic uncertainty about $\mathbf{x}$ leads to a fuzzy set-valued result $f(\hat{\mathbf{x}})$, whose membership function is given, for any $y \in Y$, by:

$$\mu_{f(\hat{\mathbf{x}})}(y) = \sup_{\mathbf{x}:y=f(\mathbf{x})} \min_{k=1,\ldots,m} \mu_{\hat{x}_k}(x_k). \tag{5}$$

This fuzzy result can be expressed as the possibility measure of the set $f^{-1}(y)$ based on the possibility distribution $\pi(\mathbf{x}) = \min_{k=1,\ldots,m} \mu_{\hat{x}_k}(x_k)$, i.e. $\mu_{f(\hat{\mathbf{x}})}(y) = \Pi(f^{-1}(y))$.

5 A New Kriging Method Integrating Epistemic Uncertainty

5.1 Principle

The kriging estimate can be expressed as a function f_0 having $n + p$ variables, namely, the observations $\mathbf{z}$ and the parameters of the variogram model $\mathbf{a}$:

$$z^*(x_0) = \sum_{i=1}^{n} \lambda_i z(x_i) = f_0(\mathbf{a}, \mathbf{z}), \tag{6}$$

The epistemic uncertainty that the geostatistician expresses on the theoretical variogram parameters and the epistemic uncertainty pervading the data can be modeled by sets of fuzzy intervals denoted by $\hat{\mathbf{a}} = \{\hat{a}_j, j = 1, \ldots, p\}$ and $\hat{\mathbf{z}} = \{\hat{z}(x_i), i = 1, \ldots, n\}$, respectively. They can be propagated to the kriging estimate f_0, by means of the extension principle, which leads to a fuzzy set-valued result, that will be denoted by $\hat{z}^0$ in the rest of the paper. The idea of applying the extension principle to the global kriging resolution dates back from the work of Bardossy *et al.*. The paper [2] considers fuzzy quantities to model imprecision of hard measurements. An empirical fuzzy variogram is obtained by using the usual fuzzy extension of interval arithmetic. However, the kriging computation is performed thanks to a precise theoretical variogram evaluated at a glance by the experts. At the final stage, the fuzzy nature of their kriging result is inherited from an application of the fuzzy arithmetic to the kriging equation 2 for precise kriging weights and fuzzy data. The next couple of papers [3,4] goes by proposing a brute force application of the extension principle to the propagation of the imprecision about the variogram parameters and the data to the kriging result. The kriging resolution, without epistemic uncertainty handling, is computationally expensive. Incorporating epistemic uncertainty in the system makes it even more complex, computationally, but also conceptually. These papers do not propose any algorithm or computational trick, which is what we propose in this section.

5.2 Detailed Algorithm

From fuzzy to interval analysis. A first simplication is to assume that the involved fuzzy parameters are *epistemically* bound, i.e. there is a unique source of information setting a uniform confidence level. Under this assumption, fuzzy interval analysis or propagation reduces to performing standard interval analysis on the set of α-cuts of the involved fuzzy intervals, as already pointed out by Bardossy *et al.* [3]. This modeling choice is obviously debatable since it does not seem natural to consider that the epistemic uncertainty pervading the variogram parameters has the same origin as the epistemic uncertainty pervading the data. What the use of a single α threshold suggests is that if data is considered precise to some extent, variogram parameters should be considered precise to the same extent.

In that case, the output fuzzy interval is just a superposition of the nested crisp interval results. The involved algorithms can be optimised by performing crisp interval propagation starting with small intervals, i.e., high membership grades. The nested nature of fuzzy intervals enables the exploration of the solution domain at level α, to be used when exploring larger domains at levels $\alpha' < \alpha$. Such an idea is at work for instance in the transformation method of Michael Hanss [26] for mechanical engineering computations under uncertainty modelled by fuzzy sets.

On this basis let us focus our attention on the case where uncertainty pervading the variogram parameters $\mathbf{a}$ and the data $\mathbf{z}$ take the form of crisp intervals $\overline{\underline{\mathbf{a}}} = \{[\underline{a}_j, \overline{a}_j], j = 1, \ldots, p\}$ and $\overline{\underline{\mathbf{z}}} = \{[\underline{z}(x_i), \overline{z}(x_i)], i = 1, \ldots, n\}$, fixing the membership level. The extension principle problem comes down to a sequence of two-fold (minimisation and maximisation) global optimisation problems:

$$\underline{z}^0 = \min_{(\mathbf{a},\mathbf{z}) \in \overline{\underline{\mathbf{a}}} \times \overline{\underline{\mathbf{z}}}} f_0(\mathbf{a}, \mathbf{z}); \quad \overline{z}^0 = \max_{(\mathbf{a},\mathbf{z}) \in \overline{\underline{\mathbf{a}}} \times \overline{\underline{\mathbf{z}}}} f_0(\mathbf{a}, \mathbf{z}), \tag{7}$$

where $\overline{\underline{\mathbf{a}}} \times \overline{\underline{\mathbf{z}}}$ is the Cartesian product of the domains $\overline{\underline{\mathbf{a}}}$ and $\overline{\underline{\mathbf{z}}}$.

Separation of the constraints. Now instead of handling the set of $n+p$ constraints as a whole, a separation between variogram parameter constraints $\overline{\underline{\mathbf{a}}}$ and data constraints $\overline{\underline{\mathbf{z}}}$ leads to an important computational gain. In our problem, it is clear that the constraints, represented by domains to be explored, $\overline{\underline{\mathbf{a}}}$ and $\overline{\underline{\mathbf{z}}}$ are non interactive. It means [17] that the underlying variables ($\mathbf{a}$ and $\mathbf{z}$) are not linked by any mathematical relation. It reflects the jump from empirical to theoretical variogram. In that case, it can easily be shown that optimising a function (here f_0) over a Cartesian product domain (here $\overline{\underline{\mathbf{a}}} \times \overline{\underline{\mathbf{z}}}$) is equivalent to separately and sequentially optimising this function over the domains forming the Cartesian product (here $\overline{\underline{\mathbf{a}}}$ and $\overline{\underline{\mathbf{z}}}$). This remark is useful when an explicit analytical expression of the optimal range of the function to optimise is available for one constraint domain when the other is fixed. In our case, we can easily show that for a fixed set of precise variogram parameters $\mathbf{a}$, an analytical formulation of the kriging estimate bounds can be expressed.

First, from now on, the kriging weights will be denoted by $\lambda_i(\mathbf{a})$ for the sake of clarity in this technical presentation because they only depend on the variogram parameters $\mathbf{a}$. When the parameters $\mathbf{a}$ are fixed, the kriging weights $\lambda_i(\mathbf{a})$ are precise. Then, for fixed variogram parameters, when the data lie in intervals $[\underline{z}_i, \overline{z}_i]$, the function f_0, with $n+p$ variables $\mathbf{a}$ and $\mathbf{z}$, can be turned into a function with n variables $\mathbf{z}$ and p fixed parameters $\mathbf{a}$. This function is locally monotonic [23] over the domains of the n variables $\mathbf{z}$ represented by the components of the vector $\overline{\underline{\mathbf{z}}}$. In other words, f_0 is monotonic when we fix all variables $z(x_i)$ but one. But the direction of the monotonicity with respect to each $z(x_i)$ depends on the sign of the corresponding $\lambda_i(\mathbf{a})$. For such locally monotonic functions, we know that the supremum and the infimum of f_0 (with fixed $\mathbf{a}$) are reached for n-tuples $\overline{\mathbf{z}}^{\mathbf{a}}$ and $\underline{\mathbf{z}}^{\mathbf{a}}$ consisting of boundaries of the domains $[\underline{z}(x_i), \overline{z}(x_i)]$. Actually, it is trivial to see that

$$\underline{\mathbf{z}}_i^{\mathbf{a}} = \begin{cases} \underline{z}(x_i), & \text{if } \lambda_i \geq 0, \\ \overline{z}(x_i), & \text{otherwise,} \end{cases} \qquad \overline{\mathbf{z}}_i^{\mathbf{a}} = \begin{cases} \overline{z}(x_i), & \text{if } \lambda_i \geq 0, \\ \underline{z}(x_i), & \text{otherwise.} \end{cases} \tag{8}$$

Formally, it leads to the kriging estimate bounds:

$$\underline{z}_{\mathbf{a}}^0 = \min_{\mathbf{z} \in \overline{\underline{\mathbf{z}}}} f_0(\mathbf{a}, \mathbf{z}) = \sum_{i=1}^{n} \lambda_i(\mathbf{a}) \underline{\mathbf{z}}_i^{\mathbf{a}}; \ \overline{z}_{\mathbf{a}}^0 = \max_{\mathbf{z} \in \overline{\underline{\mathbf{z}}}} f_0(\mathbf{a}, \mathbf{z}) = \sum_{i=1}^{n} \lambda_i(\mathbf{a}) \overline{\mathbf{z}}_i^{\mathbf{a}}.$$

Finally, the global optimisation problem is reduced to:

$$\underline{z}^0 = \min_{\mathbf{a} \in \overline{\underline{\mathbf{a}}}} \sum_{i=1}^{n} \lambda_i(\mathbf{a}) \underline{\mathbf{z}}_i^{\mathbf{a}}; \ \overline{z}^0 = \max_{\mathbf{a} \in \overline{\underline{\mathbf{a}}}} \sum_{i=1}^{n} \lambda_i(\mathbf{a}) \overline{\mathbf{z}}_i^{\mathbf{a}}. \tag{9}$$

It consists of optimising the modified kriging estimates taken for the n-tuples $\underline{\mathbf{z}}^{\mathbf{a}}$ and $\overline{\mathbf{z}}^{\mathbf{a}}$ when the variogram parameters span the domains of $\overline{\underline{\mathbf{a}}}$.

This separation of constraints leads to an optimisation problem with p constraints instead of $n+p$. This problem size reduction is very significant in applications of kriging with large datasets, i.e. when n is high. Besides, in most kriging problems, the theoretical variogram parameters are reduced to three, i.e., the nugget effect, the sill and the range. Since in any kriging application we have that $n >> 3$ (and generally $n >> p$), the computational cost of our approach is greatly reduced compared to any other global brute force approach.

Optimisation scheme. Now the actual efficiency of our approach relies on the choice of an algorithm for solving the global optimisation problem on the variogram parameters constraints (9). In opposition to the optimisation of the kriged estimates when $\mathbf{a}$ is fixed and the only constraints are due to the data $\overline{\underline{\mathbf{z}}}$, it does not seem possible to find a simple procedure for directly solving the problem when the variogram parameters are imprecise, i.e. are constrained. Indeed, the kriging weights $\lambda_i(\mathbf{a})$'s are obtained by solving a linear system of equations (3). Therefore, the $\lambda_i(\mathbf{a})$'s non-monotonically depend on the variogram parameters.

The main task of any global function optimisation technique is to compute the value of the function many times, and to compare the obtained results. What is at stake is to obtain the best solution with a minimal number of computations of the function f_0. To this aim, we try to decrease the number of computations by the following procedure: instead of performing the minimization and the maximization separately, they are done at the same time. Indeed, the computations of $\underline{z}^0_{\mathbf{a}}$ and $\overline{z}^0_{\mathbf{a}}$, for the same variogram parameters $\mathbf{a}$ share the same kriging weights $\lambda_i(\mathbf{a})$'s. Therefore, the kriging resolution made for a minimization can be tested for the maximization without any additional kriging system resolution and conversely. We thus increase the number of results without increasing the number of calls to the kriging routine.

Now in order to increase the chance of reaching the exact bounds with a small computational cost increase, we propose the following procedure. Empirically we observed that, very often, the bounds $\underline{z}^0$ and $\overline{z}^0$ are reached for p-tuples formed with the bounds of $\overline{\underline{\mathbf{a}}}$. Therefore, we propose a preliminary combinatorial exploration which consists of testing the 2^p possible p-tuples $\mathbf{a}$ of variogram parameters such that $a_j \in \{\underline{a}_j, \overline{a}_j\}$, $\forall j \in \{1, \ldots, p\}$ and storing the least value of $\underline{z}^0_{\mathbf{a}}$ and the greatest value of $\overline{z}^0_{\mathbf{a}}$, computed with those 2^p possible p-tuples. The parallel computation presented above can obviously be used with this procedure.

This preliminary step is useful since, in many cases, it will provide the exact bounds $\underline{z}^0$ and $\overline{z}^0$ of the kriging estimate. However, this step is not sufficient at locations x_0 where the exact bounds $\underline{z}^0$ and $\overline{z}^0$ are reached for p-tuples $\mathbf{a}$ inside the variogram parameter domain, i.e., such that $a_j \in]\underline{a}_j, \overline{a}_j[$, for some $j \in \{1, \ldots, p\}$.

Since we cannot predict in advance if the bounds of the kriged estimates will be reached for extreme variogram parameter values, we need to combine this preliminary combinatorial step with a probabilistic metaheuristic method: simulated annealing. The idea of this random search method is to "cleverly" explore the parameters domains $\overline{\underline{\mathbf{a}}}$ in order to be more effective than with an exhaustive combinatorial method and to avoid being stuck in local minima or maxima of the function to optimise. We still perform a joint computation of the minimum and maximum.

6 Experiment

The long term storage of CO_2 into geological underground formations is a relatively new concept expected to play an important role in longterm management of CO_2 produced by human activity. Worldwide sites and their associated risks[1] have to be studied as candidates for capturing CO_2. Permeability measures the capability of a material (in our case a geological formation such as an aquifer) to transfer fluids inside an underground spatial domain. Describing its geographical distribution is a fundamental issue for deciding about a storing site. Highly permeable sites such as aquifers, which are underground layers of unconsolidated materials (generally sand, silt, or clay) covered by highly impermeable caprocks, that would prevent CO_2 from escaping to the surface, have to be identified. Therefore, permeability is the basic parameter for characterizing a site performance and reliability.

Epistemic uncertainty due to measurement imperfections pervades the measured permeability data. A characteristic feature of the problem of kriging permeability measurements is the general unavailability of additional expert information about the studied field so that permeability measurements do not allow any credible geostatistical inference of a precise variogram model. Therefore, imprecise variogram parameters and data can be used in such a context.

Figure 1 presents the results of our kriging approach integrating epistemic uncertainty on a 40 x 60 km domain with 41 imprecise permeability measurements. We consider here that the domain $[590, 630]$ is the X component and that the domain $[2390, 2450]$ is the Y component of the studied area. This domain is part of the Dogger of Paris, which is an aquifer located at an altitude between -1500m and -2000m containing water at a temperature between 65 degrees Celsius and 85 degrees Celsius. The available measurements all lie between 10^{-11} m^2 and 10^{-13} m^2. In the rest of this section we will work with the negative logarithm of these measurements in order to deal with quantities ranging between 11 and 13.

In the bottom left box of Figure 1 the distribution of measurement locations over this domain is shown. On the kriging map of Figure 1, the lower and upper measurement values are represented by oriented triangles. Note that the lower and upper maps do not go through all these measurements only because of the sampling of the grid. If we check the imprecise kriging value at a measurement location, we retrieve the measurement value exactly as in the usual kriging approach.

This analysis is made with imprecise prior information on nugget effect η, sill s, and range r of a theoretical spherical variogram [8]. The chosen imprecise variogram parameters are $\overline{\underline{\eta}} = [0, 0.06]$, $\overline{\underline{s}} = [0.06, 0.11]$ and $\overline{\underline{r}} = [10, 15]$.

Figure 1 is a good illustration of what imprecise kriging looks like for intervallist data and parameters. However, such a graph is difficult to interpret and analyse. Therefore we decided not to try to represent a more complex fuzzy kriged surface. But, in order to analyse more thoroughly our obtained results,

[1] Proposing routines and criteria for evaluating the risks of a candidate site is the aim of the project CRISCO2 in which the authors are involved.

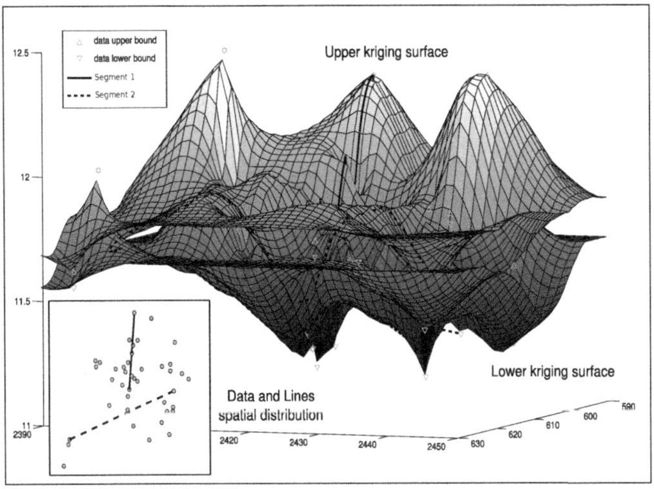

Fig. 1. Illustration of our kriging method on imprecise permeability measurements

we now focus our attention on the results obtained along two straight lines (Segment 1: the red segment and Segment 2: the blue segment on the picture), each one relating two measurement locations: Segment 1 is between (594.097, 2418.75) and (609.42, 2419.22) and Segment 2 is between (608.712, 2439.16) and (620.824, 2394.28).

Figure 2 presents the results of our kriging algorithm on Segment 1 and Segment 2 with the same parameters as for Figure 1, i.e. for the same dataset and for the same variogram parameters: $\overline{\eta} = [0, 0.06]$, $\overline{\underline{s}} = [0.06, 0.11]$ and $\overline{\underline{r}} = [10, 15]$. For both Segment 1 and Segment 2, the lower and upper kriging results with these intervallist parameters are the lines. Figure 2 also presents the results, on Segment 1 and Segment 2, of the simple generalization of kriging to imprecise measurements but with precise variogram parameters: $\eta = 0.01822$, $s = 0.0682$ and $r = 11.3$. For both Segment 1 and Segment 2, the lower and upper kriging results with such precise parameters are the dashed lines. Both graphs plot the kriging estimate as a function of the X domain.

With this experiment, we aim at separating the influence of the epistemic uncertainty pervading the data from the influence of the epistemic pervading the variogram parameters on the kriging results. To have a clear view of this experiment, reader should remember the data distribution over the domain and especially over Segment 1 and Segment 2 presented in the left bottom box of Figure 1.

Whatever the studied segment (Segment 1 or Segment 2) a common observation that can be made is that when the data surrounding the kriging locations are dense (i.e. for X between 600 and 610 for Segment 1 and for X around 614 for Segment 2 (approximately) and their extremes), kriging with precise or imprecise variogram parameters lead to similar estimations. The differences between the results of these different kriging approaches become significant for sparsely

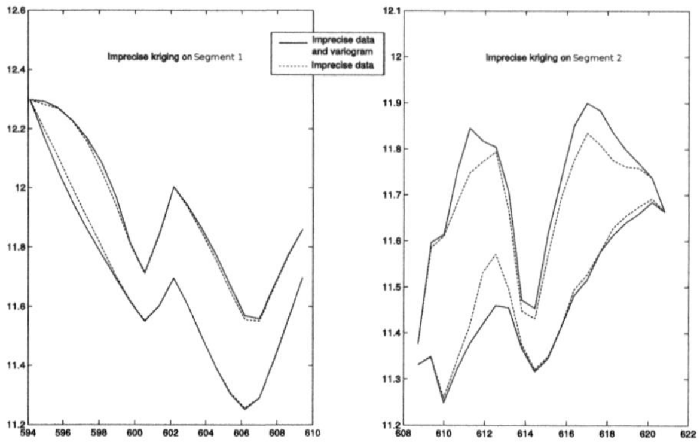

Fig. 2. Our kriging method on imprecise permeability measurements with precise vs. imprecise variogram parameters

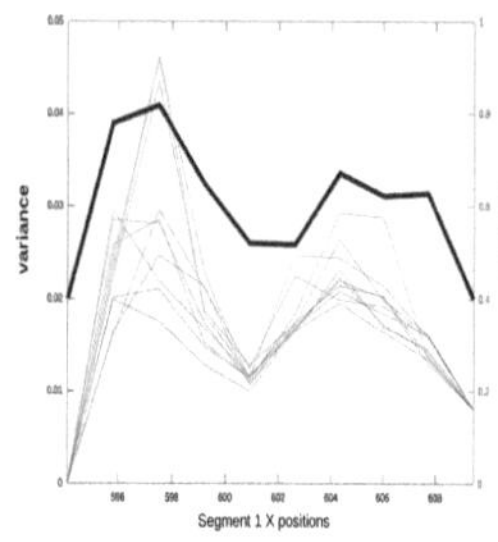

Fig. 3. Kriging variance and imprecision on Segment 1 with changing measurement values

located measurements. Therefore we can conclude that the epistemic uncertainty pervading the data have more influence than the epistemic uncertainty pervading the variogram parameters on the kriging result when data are dense around the kriging location. When the data are sparse around a kriged location, the epistemic uncertainty pervading the variogram parameters thus significantly affects the kriged result.

Now, we propose to qualitatively compare the imprecision of our kriging approach to the usual kriging variance. In that aim, we made a series of ten artificial random modifications in the imprecise measurement values (without changing the imprecision of each data). We then compared, on Segment 1, the resulting kriging imprecision of our method (the thin curves of Figure 3), for variogram parameters: $\overline{\eta} = [0, 0.04]$, $\overline{s} = [0.06, 0.1]$ and $\overline{r} = [10, 12]$, to the usual kriging variance (the thick curve of Figure 3), for variogram parameters: $\eta = 0.02$, $s = 0.08$ and $r = 11$, obtained with precise data taken as the centers of the modified imprecise measurements.

Figure 3 shows that the kriging variance (the thick line) only depends on the measurement locations and not on the measurement values, which is not a correct feature of an error index, as is supposed to be the kriging variance. Suppose that a sudden unexpected observation (an outlier) occurs, the kriging variance does not reflect this aberration. With the imprecise kriging approach, we can notice that the imprecision magnitude (the thin lines) reflect the gradients between observations in the neighborhood of the kriging location. Thus an outlier will result in an imprecision increase.

This remark can be extrapolated to the use of sensitivity analysis in any quantitative data processing method, i.e. the imprecision of a sensitivity analysis on a method is a marker of the error or the noise in the data [33]. Indeed, it is obvious to notice that, even for only two parameterizations, significant local variations in the data (generally considered as due to error or noise) will lead to more significant variations in the results than for small local variations. Hence, extrapolated to imprecise parameterizations, we can conclude that the resulting imprecision is a marker of the measurement errors (or noise).

7 Conclusion

This paper tries to revive the fuzzy kriging technique by putting it in the proper perspective, namely the one of handling epistemic uncertainty. Uncertainty pervading the results of kriging are due not only to the possible scarcity and imprecision of available measurements, but also to the difficulty to relate the empirical and the theoretical variograms. The latter is often chosen by an expert that relies at least as much on his or her experience in the area to be investigated as on the measurements themselves. So, we argue that accounting for epistemic uncertainty about measurements and variogram parameters makes the kriging procedure more cogent. A major contribution of this paper is to propose a computationally tractable method for fuzzy kriging. Unsurprisingly, the major source of uncertainty seems to be the ill-knowledge about measurements. However, our technique yields fuzzy set-valued results even if measurements are precise, due to epistemic uncertainty about the variogram parameters that exists even with precise data. We point out that this epistemic uncertainty affects the resulting kriged surface all the more as the data are sparser. In dense data zones, any interpolation is good enough. Our paper tries to address the intuitive consideration that interpolation techniques, insofar as they account for actual knowledge about an incompletely known function, should deliver imprecise results in areas located far away from the data points.

Acknowledgements

This work is supported by the French Research National Agency (ANR) through the CO_2 program (project CRISCO2 ANR-06-CO2-003). The authors wish to thank Jean-Paul Chilès and Nicolas Desassis for their comments on a first draft of this paper and their support during the project.

References

1. Aumann, R.J.: Integrals of set-valued functions. J. Math. Anal. Appl. 12, 1–12 (1965)
2. Bardossy, A., Bogardi, I., Kelly, W.E.: Imprecise (fuzzy) information in geostatistics. Math. Geol. 20, 287–311 (1988)
3. Bardossy, A., Bogardi, I., Kelly, W.E.: Kriging with imprecise (fuzzy) variograms. I: Theory. Math. Geol. 22, 63–79 (1990)
4. Bardossy, A., Bogardi, I., Kelly, W.E.: Kriging with imprecise (fuzzy) variograms. I: Application. Math. Geol. 22, 81–94 (1990)
5. Baudrit, C., Dubois, D., Couso, I.: Joint propagation of probability and possibility in risk analysis: Towards a formal framework. International Journal of Approximate Reasoning 45(1), 82–105 (2007)

6. Baudrit, C., Dubois, D., Perrot, N.: Representing parametric probabilistic models tainted with imprecision. Fuzzy Sets and Systems 159, 1913–1928 (2008)
7. Berger, J.O., de Oliveira, V., Sanso, B.: Objective Bayesian analysis of spatially correlated data. Journal of the American Statistical Association 96, 1361–1374 (2001)
8. Chilès, J.P., Delfiner, P.: Geostatistics, Modeling Spatial Uncertainty. Wiley, Chichester (1999)
9. Couso, I., Dubois, D.: On the variability of the concept of variance for fuzzy random variables. IEEE Transactions on Fuzzy Systems 17(5), 1070–1080 (2009)
10. Cressie, N.A.C.: The origins of kriging. Math. Geol. 22, 239–252 (1990)
11. Diamond, P.: Fuzzy least squares. Inf. Sci. 46, 141–157 (1988)
12. Diamond, P.: Interval-valued random functions and the kriging of intervals. Math. Geol. 20, 145–165 (1988)
13. Diamond, P.: Fuzzy kriging. Fuzzy Sets and Systems 33, 315–332 (1989)
14. Dubois, D.: Possibility theory and statistical reasoning. Computational Statistics & Data Analysis 51, 47–69 (2006)
15. Dubois, D., Foulloy, L., Mauris, G., Prade, H.: Probability-possibility transformations, triangular fuzzy sets, and probabilistic inequalities. Reliable Computing 10, 273–297 (2004)
16. Dubois, D., Kerre, E., Mesiar, R., Prade, H.: Fuzzy interval analysis. In: Dubois, D., Prade, H. (eds.) The Handbook of Fuzzy Sets. Fundamentals of Fuzzy Sets, vol. I, pp. 483–581. Kluwer Academic Publishers, Dordrecht (2000)
17. Dubois, D., Prade, H.: Additions of fuzzy interactive numbers. IEEE Transactions on Automatic Control 26, 926–936 (1981)
18. Dubois, D., Prade, H.: Possibility Theory. Plenum Press, New York (1988)
19. Dubois, D., Prade, H.: When upper probabilities are possibility measures. Fuzzy Sets and Systems 49, 65–74 (1992)
20. Dubrule, O.: Comparing splines and kriging. Comp. Geosci. 10, 327–338 (1984)
21. Ferson, S.: What Monte Carlo methods cannot do. Human and Ecological Risk Assessment: An International Journal 2(4), 990–1007 (1996)
22. Ferson, S., Troy Tucker, W.: Sensitivity analysis using probability bounding. Reliability Engineering and System Safety 91, 1435–1442 (2006)
23. Fortin, J., Dubois, D., Fargier, H.: Gradual numbers and their application to fuzzy interval analysis. IEEE Transactions on Fuzzy Systems 16, 388–402 (2008)
24. Goovaerts, P.: Geostatistics for Natural Resources Evaluation. Oxford Univ. Press, New-York (1997)
25. Handcock, M.S., Stein, M.L.: A Bayesian analysis of kriging. Technometrics 35, 403–410 (1993)
26. Hanss, M.: The transformation method for the simulation and analysis of systems with uncertain parameters. Fuzzy Sets and Systems 130, 277–289 (2002)
27. Helton, J.C., Oberkampf, W.L.: Alternative representations of epistemic uncertainty. Reliability Engineering and System Safety 85, 1–10 (2004)
28. Journel, A.G., Huijbregts, C.J.: Mining Geostatistics. Academic Press, New York (1978)
29. Journel, A.G.: The deterministic side of geostatistics. Math. Geol. 17, 1–15 (1985)
30. Journel, A.G.: Geostatistics: Models and Tools for the Earth Sciences. Math. Geol. 18, 119–140 (1986)
31. Krige, D.G.: A statistical approach to some basic mine valuation problems on the Witwatersrand. Journal of the Chemical, Metallurgical and Mining Society of South Africa 52, 119–139 (1951)

32. Loquin, K., Dubois, D.: Kriging and epistemic uncertainty: a critical discussion. In: Jeansoulin, R., Papini, O., Prade, H., Schockaert, S. (eds.) Methods for Handling Imperfect Spatial Information. Springer, Heidelberg (2009)
33. Loquin, K., Strauss, O., Crouzet, J.F.: Possibilistic signal processing: How to handle noise? International Journal of Approximate Reasoning, special issue selected papers ISIPTA 2009 (2009) (in Press)
34. Matheron, G., Blondel, F.: Traité de géostatistique appliquée. Editions Technip (1962)
35. Matheron, G.: Le krigeage Transitif. Unpublished note, Centre de Morphologie Mathmatique de Fontainebleau (1967)
36. Matheron, G.: Estimer et choisir: essai sur la pratique des probabilités. ENSMP (1978)
37. Puri, M.L., Ralescu, D.A.: Fuzzy random variables. J. Math. Anal. Appl. 114, 409–422 (1986)
38. Shafer, G.: A mathematical theory of evidence. Princeton University Press, Princeton (1976)
39. Shafer, G., Vovk, V.: Probability and Finance: It's Only a Game! Wiley, New York (2001)
40. Srivastava, R.M.: Philip and Watson–Quo vadunt? Math. Geol. 18, 141–146 (1986)
41. Taboada, J., Rivas, T., Saavedra, A., Ordóñez, C., Bastante, F., Giráldez, E.: Evaluation of the reserve of a granite deposit by fuzzy kriging. Engineering Geol. 99, 23–30 (2008)
42. Walley, P.: Statistical reasoning with imprecise probabilities. Chapman and Hall, Boca Raton (1991)
43. Watson, G.S.: Smoothing and interpolation by kriging and with splines. Math. Geol. 16, 601–615 (1984)
44. Yaglom, A.M.: An introduction to the theory of stationary random functions. Courier Dover Publications (2004)
45. Zadeh, L.A.: Fuzzy sets. Inf. Control 8, 338–353 (1965)
46. Zadeh, L.A.: Fuzzy sets as a basis for a theory of possibility. Fuzzy Sets and Systems 1, 3–28 (1978)

Event Modelling and Reasoning with Uncertain Information for Distributed Sensor Networks

Jianbing Ma, Weiru Liu, and Paul Miller

School of Electronics, Electrical Engineering and Computer Science,
Queen's University Belfast, Belfast BT7 1NN, UK
{jma03,w.liu}@qub.ac.uk, p.miller@ecit.qub.ac.uk

Abstract. CCTV and sensor based surveillance systems are part of our daily lives now in this modern society due to the advances in telecommunications technology and the demand for better security. The analysis of sensor data produces semantic rich events describing activities and behaviours of objects being monitored. Three issues usually are associated with events descriptions. First, data could be collected from multiple sources (e.g., sensors, CCTVs, speedometers, etc). Second, descriptions about these data can be poor, inaccurate or uncertain when they are gathered from unreliable sensors or generated by analysis non-perfect algorithms. Third, in such systems, there is a need to incorporate domain specific knowledge, e.g., criminal statistics about certain areas or patterns, when making inferences. However, in the literature, these three phenomena are seldom considered in CCTV-based event composition models. To overcome these weaknesses, in this paper, we propose a general event modelling and reasoning model which can represent and reason with events from multiple sources including domain knowledge, integrating the Dempster-Shafer theory for dealing with uncertainty and incompleteness. We introduce a notion called event cluster to represent uncertain and incomplete events induced from an observation. Event clusters are then used in the merging and inference process. Furthermore, we provide a method to calculate the mass values of events which use evidential mapping techniques.

Keywords: Bus Surveillance; Active System; Event Composition; Event Reasoning; Inference.

1 Introduction

CCTV-based[1] surveillance is an inseparable part of our society now – everywhere we go we see CCTV cameras (e.g. [2,11,5,13], etc). The role of such systems has shifted from purely passively recording information for forensics to proactively providing analytical information about potential threats/dangers in real-time fashion. This shift poses some dramatic challenges on how information collected in such a network shall be

[1] This paper is an extended version of [9] in which we have included a set of running examples, a method (summarized by an algorithm) to calculating mass values of events which uses evidential mapping techniques, and the newly introduced notion *rule clusters*. Furthermore, we also demonstrate in this paper how to interpret and use the background knowledge, or domain knowledge, which was only preliminarily introduced in [9].

A. Deshpande and A. Hunter (Eds.): SUM 2010, LNAI 6379, pp. 236–249, 2010.

exchanged, correlated, reasoned with and ultimately be used to provide significantly valuable predictions for threats or actions that may lead to devastating consequences.

Central to this is the ability to deal with a large collection of meaningful events derived from sensor/camera data analysis algorithms. An event can be understood as something that happened somewhere at a certain time (or time interval). Typically, a life cycle of event includes detection, storage, reasoning, mining, exploration and actions.

In this paper, we focus on a real-time event modelling and reasoning framework for supporting the instant recognition of emergent events based on uncertain or imperfect information from multiple sources. This framework has many potential uses in various applications, e.g., active databases, smart home projects, bus/airport surveillance, and stock trading systems, etc.

Various event reasoning systems have been proposed in the literature, e.g., an event language based on *event expressions* for active database systems in [6], the Semantic Web Rule Language (SWRL) for semantic web applications and the situation manager rule language [1] for general purposes, etc. These systems provide both event representation and deterministic event inference in the form of rules. However, these systems do not take into account uncertainties which are usually associated with real-world events. To remedy this weakness, in [15,16,17], an event composition model was proposed with uncertainties represented by probability measures.

However, this model cannot deal with the problem of incomplete information in event reasoning. For example, in the case of monitoring a person entering an building, the person may be classified as male with a certainty of 85% by an event detection algorithm (an event here is to identify a person's gender). However, the remainder does not imply that the person is female with a 15% certainty, rather, it is unknown. That is, we do not know how the remaining 15% shall be distributed on alternatives $\{male\}$ or $\{female\}$. Hence with probability theory, this information can only be represented as $p(male) \geq 0.85$ and $p(female) \leq 0.15$ which is difficult for subsequent reasoning (e.g., a Bayesian network).

In distributed sensor networks, events are more often gathered from multiple heterogeneous sources, e.g., the same event can be obtained from video or audio data analysis, or from speedometers. We assume that each source channels its information via event descriptions, hence a practical event model should consider combining information about the same event from multiple sources. As different sources may provide possibly conflicting descriptions on the same event, the event composition model should also be able to deal with such conflict between multiple sources. Unfortunately, to the best of our knowledge, this issue is hardly mentioned in the literature on event composition models. In [1], although events can come from multiple sources, a particular event can only be from one source, so this model cannot deal with multiple events from different sources relating to the same situation (scenario).

Furthermore, when an event reasoning system receives events descriptions from multiple sources, it also needs to consider the reliabilities of these information sources. For instance, in surveillance applications, sensors/cameras, etc, are frequently used. However, since sensors/cameras can be malfunctioning such as a camera may have been tampered with, illumination could be poor, or the battery is low, etc, they may give imprecise information which cannot be simply represented by probability measures, either.

Dempster-Shafer (DS) Theory [4,12] is a popular framework to deal with uncertain or incomplete information from multiple sources. This theory is capable of modelling incomplete information through ignorance as well as considering the reliabilities of sources by using the discounting function. In this paper, we propose an event model integrating DS theory that can represent and reason with possibly conflicting information (recorded as events) from multiple sources which may be uncertain or incomplete. We also deploy the *discounting* function [8] to resolve imprecise information due to unreliable sources.

Furthermore, it is also a key requirement for an event model to have the ability to represent and manage domain knowledge [14]. Because domain knowledge does not fit into the usual definitions of events in the literature, it is not surprising that it is generally ignored by the existing event models, e.g., [1,15,16,17], etc. In our event model, however, domain knowledge is treated as a special kind of event and is managed the same way as other types of events.

To summarize, the main contributions of our event composition model are

1. a general model for representing uncertain and incomplete information (events),
2. a combination framework for dealing with events from multiple sources,
3. utilization of domain knowledge for assisting inferences,
4. using evidential mapping technique to calculate the event mass.

The framework has been implemented and tested with a set of events acquired from the Intelligent Sensor Information System (ISIS) project which aims at developing a state-of-the-art surveillance sensor network concept demonstrator for public transport. A set of domain specific rules are constructed with the help of criminologist working on the project.

The rest of the paper is organized as follows. In Section 2, we provide the preliminaries on Dempster-Shafer theory. In Section 3, formal definitions of event model are given including the definitions of events, multi-source events combination, event flow and event inference. We then provide an algorithm for calculating mass values of events in Section 4. Finally, we discuss related work and conclude the paper in Section 5 and Section 6 respectively.

2 Dempster-Shafer Theory

For convenience, we recall some basic concepts of Dempster-Shafer's theory of evidence (DS theory). Let Ω be a finite, non-empty set called the frame of discernment, denoted as, $\Omega = \{w_1, \cdots, w_n\}$.

Definition 1. *A mass function is a mapping* $m : 2^{\Omega} \to [0,1]$ *such that* $m(\emptyset) = 0$ *and* $\sum_{A \subseteq \Omega} m(A) = 1$.

If $m(A) > 0$, then A is called a focal element of m. Let $\mathscr{F}_m$ denote the set of focal elements of m. From a mass function, m, *belief* function (Bel) and *plausibility* function (Pl) can be defined to represent the lower and upper bounds of the beliefs implied by m as follows.

$$Bel(A) = \sum_{B \subseteq A} m(B) \text{ and } Pl(A) = \sum_{C \cap A \neq \emptyset} m(C). \tag{1}$$

One advantage of DS theory is that its has the ability to accumulate and combine evidence from multiple sources by using *Dempster's rule of combination*. Let m_1 and m_2 be two mass functions from two distinct sources over Ω. Combining m_1 and m_2 gives a new mass function m as follows:

$$m(C) = (m_1 \oplus m_2)(C) = \frac{\sum_{A \cap B = C} m_1(A) m_2(B)}{1 - \sum_{A \cap B = \emptyset} m_1(A) m_2(B)} \quad (2)$$

In practice, sources may not be completely reliable, to reflect this, in [12], a *discount rate* was introduced by which the mass function may be discounted in order to reflect the reliability of a source. Let r $(0 \leq r \leq 1)$ be a discount rate, a discounted mass function using r is represented as:

$$m^r(A) - \begin{cases} (1-r)m(A) & A \subset \Omega \\ r + (1-r)m(\Omega) & A = \Omega \end{cases} \quad (3)$$

When $r = 0$ the source is absolutely reliable and when $r = 1$ the source is completely unreliable. After discounting, the source is treated as totally reliable.

In our event composition and inference model, we use a set of rules (with degrees of certainty) to describe which collection of events could imply what other events to a particular degree. A simplified form of a rule of this kind[2] is as *if E then H_1 with degree of belief f_1, ..., H_n with degree of belief f_n*. These rules are called heuristic in [7], in which a modelling and propagation approach was proposed to represent a set of heuristic rules and to propagate degrees of beliefs along these rules, through the notion *evidential mapping Γ^**.

An evidential mapping is to establish relationships between two frames of discernment Ω_E, Ω_H such that $\Gamma^* : 2^{\Omega_E} \to 2^{2^{\Omega_H} \times [0,1]}$ assigning a subset $E_i \subseteq \Omega_E$ to a set of subset-mass pairs in the following way:

$$\Gamma^*(E_i) = \big((H_{ij}, f(E_i \to H_{ij})), \ ..., \ (H_{it}, f(E_i \to H_{it}))\big) \quad (4)$$

where $H_{ij} \subseteq \Omega_H, i = 1, \ ..., \ n, j = 1, \ ..., \ t$, and $f : 2^{\Omega_E} \times 2^{2^{\Omega_H}} \to [0,1]$ satisfying[3]
(a) $H_{ij} \neq \emptyset, \ j = 1, \ ..., \ t$;
(b) $f(e_i \to H_{ij}) \geq 0, \ j = 1, \ ..., \ t$;
(c) $\sum_{j=1}^{n} f(e_i \to H_{ij}) = 1$;
(d) $\Gamma^*(\Omega_E) = ((\Omega_H, 1))$;

A piece of evidence on Ω_E can then be propagated to Ω_H through evidential mapping Γ^* as follows:

$$m_{\Omega_H}(H_j) = \sum_i m_{\Omega_E}(E_i) f(E_i \to H_{ij}). \quad (5)$$

To calculate the mass values of inferred events based on the premise events of inference rules, we integrate *evidential mapping Γ^** technique [7] into our event composition and reasoning model, which is detailed in Section 4.

---

[2] The definition of rules is given in Section 3.
[3] For the sake of clear illustration, instead of writing $f(E_i, H_{ij})$, we write $f(E_i \to H_{ij})$.

3 A General Framework for Event Modelling

3.1 Event Definition

For an event model, the first issue we should address is the definition of events. The definition of an event should be expressive enough to deliver all the information of interest for an application and also be as simple and clear as possible.

Definitions of an event from different research fields are very diverse and tend to reflect the content of the designated application. For instance, in text topic detection and track, an event is something that happened somewhere at a certain time; in pattern recognition, an event is defined as a pattern that can be matched with a certain class of pattern types, and in signal processing, an event is triggered by a status change in the signal, etc.

In this paper, to make our framework more general, we define the events as follows: an event is *an occurrence that is instantaneous (event duration is* 0*, i.e., takes place at a specific point of time)*[4] *and atomic (it happens or not)*. The atomic requirement of an event does not exclude uncertainty. For instance, when there is a person boarding a bus and this person can be a male or a female (suppose we only focus on the gender), then whether it is a male/female that boards the bus is an example of uncertainty. But *a male (resp. a female) is boarding the bus* is an atomic event which either occurs completely or does not occur at all. To represent uncertainty encountered during event detection, in the following, we distinguish an *observation* (with uncertainty) from *possible events associated with the observation* (because of the uncertainty). This can be illustrated by the above example: an observation is that *a person is boarding the bus* and the possible events are *a male is boarding the bus* and *a female is boarding the bus*. An observation says that something happened, but the *entity* being observed is not completely certain yet, so we have multiple events listing what that *entity* might be.

This definition of events is particularly suitable for surveillance problems, where the objects being monitored are not complete clear to the observer.

In the literature, there are two types of events, one type contains *external events* [1] or *explicit events* [15,16] and the other consists of *inferred events*. External events are events directly gathered from external sources (within the application) while inferred events are the results of the inference rules of an event model. In addition, to make use of domain knowledge, we introduce the third type of events, *domain events*, which are usually extracted from experts's opinions or background knowledge about this application the domain. Intuitively, domain knowledge is not from observed facts while external events are. Examples of these events can be seen in the next subsection.

3.2 Event Representation

Intuitively, a concrete event definition is determined by the application domain which contains all the information of interest for the application (including data relevant to the application and some auxiliary data). But there are some common attributes that every event shall possess, such as

[4] Domain events introduced later in this subsection may have a nonzero duration. A domain event can be seen as a series of instantaneous events.

1. $EType$: describing the type of an event, such as, *Person Boarding Vehicle* abbreviated as PBV.
2. $occT$: the point in time that an event occurred.
3. ID: the ID of a source from which an event is detected.
4. rb: the degree of reliability of a source.
5. sig: the degree of significance of an event.

Formally, we define an event e as follows.

$$e = (EType, occT, ID, rb, sig, v_1, \cdots, v_n)$$

where v_is are any additional attributes required to define event e based on the application. Attribute v_i can either have a single or a set of elements as its value, e.g., for attribute *gender*, its value can be *male*, or *female*, or {*male, female*} (however, it is not possible to tell the gender of a person when their face is obscured, so we introduce a value *obscured* as an unknown[5] value for gender). Any two events with the same event type, source ID and time of occurrence (Typically the occurrence time is like $21:05:31pm12/2/09$, for simplicity we only use the hours) are from the set of possible events related to a single observation. For example, $\mathsf{e}_1 = (\mathsf{PBV}, \mathsf{20pm}, 1, 0.8, 0.7, \mathsf{male}, \cdots)$ and $\mathsf{e}_2 = (\mathsf{PBV}, \mathsf{20pm}, 1, 0.8, 0.7, \{\mathsf{male}, \mathsf{female}\}, \cdots)$ are two events with v_1 for *gender* (we have omitted other attributes for simplicity).

Events of the same type have the same set of attributes.

Example 1. *Suppose we are monitoring passengers boarding a bus through the front door. Then we have an event type* PBV *(Person Boarding Vehicle) which may include related attributes such as source ID, occurrence time, reliability, significance, person gender, person ID, person age, front/back door, vehicle type, vehicle ID, bus route (we omit the bus position for simplicity). An instance of* PBV *is* (PBV, 21pm, 1, 0.9, 0.7, male, 3283, young, fDoor, double decker bus, Bus1248, 45).

An event is always attached with a mass value. Semantically, for a particular event type with each of its event represented as $(EType, occT, ID, rb, sig, v_1, \cdots, v_n)$, we use Dom_i to denote the domain of v_i, and $\mathcal{V} = \prod_{i=1}^{n} Dom_i$ to denote the frame of discernment (domain of tuple $(v_1, \cdots, v_n)$), and m to denote a mass function over $2^{\mathcal{V}}$.

To represent an observed fact with uncertainty, we introduce concept *event cluster*. An *event cluster* EC is a set of events which have the same event type ($EType$), occurrence time ($occT$) and source ID (ID) , but with different $v_1, \cdots, v_n$ values. Events e_1 and e_2 above form an event cluster for the observed fact *someone is boarding the bus*. Note that as the reliability is based on the source, events in a specified event cluster EC will have the same reliability.

For an event e in event cluster EC, we use $e.EType$ (resp. $e.occT$, etc) to denote the event type (resp. time of occurrence, etc) of e, $e.v$ to denote $(v_1, \cdots, v_n)$, and $e.m$ to denote the value $m(e.v)$. By abuse of notations, we also write $EC.EType$ (resp.

[5] Note that *obscured* is not the same as {male, female} since it may indicate some malicious event. In fact, here the attribute *gender* can be seen as an output from a face recognition program in which *obscured* means that information about face recognition is not available. In this sense, *gender* is an abbreviation of *gender recognition result*.

$EC.ID$, $EC.occT$, $EC.rb$) to denote the event type (resp. source ID, time of occurrence, reliability) of any event in EC since all the events in EC have the same values for these attributes.

It should be noted that within a particular application, the degree of significance of an event is self-evident (i.e., a function over $e.v$). For example, in bus surveillance, the event *a young man boards a bus around 10pm in an area with high crime statistics* is more significant than the event *a middle-aged woman boards a bus around 6pm in an area of low-crime*. However, due to space limitation, we will not discuss it further.

A mass function m over $\mathcal{V}$ for event cluster EC should satisfy the normalization condition: $\sum_{e \in EC} e.m = 1$. That is, EC does contain an event that really occurred. For example, for the two events, e_1 and e_2, introduced above, a mass function m can be defined as $m(\mathsf{male}, \cdots) = 0.85$ and $m(\{\mathsf{male}, \mathsf{female}\}, \cdots) = 0.15$.

An event cluster hence gives a full description of an observed fact with uncertainty from the perspective of one source.

Example 2. *(Example 1 continued) Consider a camera overlooking the front entrance to a bus. There are several possible events relating to a camera recording: a male boards the bus, a female boards the bus, a person that we cannot distinguish it is a man or woman boards the bus, a person boards the bus but its face is obscured. Among these the last is the most significant as someone who boards the bus with its face obscured is likely to be up to no good. Most vandals and criminals will take steps to ensure their faces are not caught by the cameras. Therefore, we have the following event cluster with a set of events as (we omitted other details for simplicity)*
$\{(\mathsf{PBV}, 21:05:31, 1, 0.9, 0.7, \mathsf{male}, 3283)$, $(\mathsf{PBV}, 21:05:31, 1, 0.9, 0.4, \mathsf{female}, 3283)$,
$(\mathsf{PBV}, 21:05:31, 1, 0.9, 0.7, \{\mathsf{male}, \mathsf{female}\}, 3283)$,
$(\mathsf{PBV}, 21:05:31, 1, 0.9, 1, \mathsf{obscured}, 3283)\}$.

A sample mass function assigning mass values to focal elements from frame $\Omega = \{male, female, obscurred\}$ *can be* $m(\mathsf{male}, 3283) = 0.4$ $m(\mathsf{female}, 3283) = 0.3$, $m(\{\mathsf{male}, \mathsf{female}\}, 3283) = 0.2$, *and* $m(\mathsf{obscured}, 3283) = 0.1$.

Observe that if some E^* in ET s.t., $E^*.m = 1$, then the event cluster ET simply reduces to a single event, i.e., E^* (other events with mass values 0 are ignored).

Domain knowledge[6] can be represented as a special event cluster in which an event (called a *domain event*) is in the same form of the external/inferred events except that the time of occurrence can be an interval.

Example 3. *(Example 2 con't) After a survey, we obtained a distribution (with reliability* 0.8*) on person boarding bus 1248 at route 45 between* $20:00$ *and* $22:00$ *as* $\mathsf{male} : \mathsf{female} : \mathsf{obscured} = 5:4:1$. *Then this piece of domain knowledge produces the following event cluster with events*
$(\mathsf{PBV}, [20:00:00, 22:00:00], 0, 0.8, 0.7, \mathsf{male}, \mathsf{double\ decker\ bus}, \mathsf{Bus\ 1248}, 45)$,
$(\mathsf{PBV}, [20:00:00, 22:00:00], 0, 0.8, 0.4, \mathsf{female}, \mathsf{double\ decker\ bus}, \mathsf{Bus\ 1248}, 45)$, *and*
$(\mathsf{PBV}, [20:00:00, 22:00:00], 0, 0.8, 1, \mathsf{obscured}, \mathsf{double\ decker\ bus}, \mathsf{Bus\ 1248}, 45)$.

The mass values given by the domain knowledge are (we omitted other details here) $m(\mathsf{male}, \cdots) = 0.5$, $m(\mathsf{female}, \cdots) = 0.4$, *and* $m(\mathsf{obscured}, \cdots) = 0.1$.

6 In our event model, we reserve source 0 for domain knowledge.

A domain event has a *series interpretation* such that it contains a series of external events, the occurrence time of each such external event is one time point in the time interval of the domain event while other attributes are unchanged. For instance, event (bus details omitted) $(\mathsf{PBV}, [20:00:00, 22:00:00], 0, 0.8, 0.7, \mathsf{male}, \cdots)$ can be seen as a series of events
$(\mathsf{PBV}, 20:00:00, 0, 0.8, 0.7, \mathsf{male}, \cdots)$, $(\mathsf{PBV}, 20:00:01, 0, 0.8, 0.7, \mathsf{male}, \cdots)$, $\cdots$,
$(\mathsf{PBV}, 22:00:00, 0, 0.8, 0.7, \mathsf{male}, \cdots)$.

With the series interpretation of domain events, intuitively we do not allow two domain events having the same event attribute values except that their time intervals are overlapped. To illustrate, if one domain event gives $e_3 = (\mathsf{PBV}, [20:00:00, 22:00:00], 0, 0.8, 0.7, \mathsf{male}, \cdots)$ with $e_3.m = 0.5$ (i.e., $m(\mathsf{male}, \cdots) = 0.5$) and another domain event provides $e_4 = (\mathsf{PBV}, [21:00:00, 23:00:00], 0, 0.8, 0.7, \mathsf{male}, \cdots)$ with $e_4.m' = 0.9$ (i.e., $m(\mathsf{male}, \cdots) = 0.9$), then they contradict each other during the time interval $[21:00:00, 22:00:00]$. If the second domain event gives the same mass value (i.e., $m'(\mathsf{male}, \cdots) = 0.5$), then in fact these two event clusters can be merged with a time interval $[20:00:00, 23:00:00]$. Similarly, two domain event clusters with their events pairwise having the same attributes values except overlapping time intervals either contradict each other or can be merged into one domain event cluster. Hence hereafter we assume that there does not exist two domain event clusters having the same event type and with overlapping time intervals.

3.3 Event Combination

When a set of event clusters have the same event type and time of occurrence but different source IDs, we call them *concurrent* event clusters[7]. This means that multi-model sensors may have been used to monitor the situation. Therefore, we need to combine these event clusters since they refer to the same observed fact from different perspectives. The combined result is a new event cluster with the same event type and time of occurrence, but the source ID of the combined event will be the union of the original sources. The combination of event clusters is realized by applying Dempster's combination rule on discounted mass functions. That is, the mass function of an event cluster is discounted with the discount rate defined as the reliability of a source.

Definition 2. *Let $EC_1, \cdots, EC_k$ be a set of concurrent event clusters, and $m_1^r, \cdots, m_k^r$ be the corresponding discounted mass functions over $2^{\mathcal{V}}$, m be the mass function obtained by combining $m_1^r, \cdots, m_k^r$ using the Dempster's combination rule, then we get the combined event cluster $EC = \oplus_{j=1}^k EC_j$ such that $\forall e \in EC$, we have $e.v \in \mathscr{F}_m$, $e.EType = EC_1.EType$, $e.occT = EC_1.occT$, $e.ID = \{EC_1.ID, \cdots, EC_k.ID\}$, $e.rb = 1$, and $e.m = m(e.v)$. Conversely, for each focal element A in $\mathscr{F}_m$, there exists a unique $e \in EC$, s.t., $e.v = A$.*

As stated earlier, $e.sig$ (event significance) is a function on $e.v$.

Example 4. *(Example 3 continued) Let EC_0 be the event cluster given in Example 2, and EC_1 be the event cluster given in Example 3, then the combined event cluster*

[7] Due to the series interpretation of domain event clusters, a domain event cluster and an external event cluster are called concurrent iff the time of occurrence of the latter is within the time interval of the former.

$EC = EC_0 \oplus EC_1$ *is* { $(\mathsf{PBV}, 21:05:31, \{0,1\}, 1, 0.7, \mathsf{male}, 3283)$, $(\mathsf{PBV}, 21:05:31, \{0,1\}, 1, 0.4, \mathsf{female}, 3283)$, $(\mathsf{PBV}, 21:05:31, \{0,1\}, 1, 0.7, \{\mathsf{male}, \mathsf{female}\}, 3283)$, $(\mathsf{PBV}, 21:05:31, \{0,1\}, 1, 1, \mathsf{obscured}, 3283)$, $(\mathsf{PBV}, 21:05:31, \{0,1\}, 1, 1, \{\mathsf{male}, \mathsf{female}, \mathsf{obscured}\}, 3283)$ }, *and the corresponding mass values are* $m(\mathsf{male}, 3283) = 0.478$, $m(\mathsf{female}, 3283) = 0.376$, $m(\{\mathsf{male}, \mathsf{female}\}, 3283) = 0.059$, $m(\mathsf{obscured}, 3283) = 0.054$, *and* $m(\{\mathsf{male}, \mathsf{female}, \mathsf{obscured}\}, 3283) = 0.033$.

3.4 Event Flow

Event models usually use the concept *Event History* (EH) to describe the set of all events whose occurrences fall between a certain period of time. However, in our framework, given a set of event clusters, we first carry out events combination, and then retain only the combined event clusters. So what we have is not a *history*, because of this, we call it an *event flow* and denote it as EF. We use $EF_{t_1}^{t_2}$ to represent a set of combined event clusters whose occurrences fall between t_1 and t_2. Since an event flow contains the combined events, to some extent, we have already considered the *opinions* (of the original events) from different sources.

Example 5. *(Example 4 continued) Let* EC_0 *and* EC_1 *be the event clusters given in Example 2 and Example 3, respectively,* EC *be the combined event cluster of* EC_0 *and* EC_1 *in Example 4. Let* EC_2 *be the event cluster for describing a person loitering in a bus (for simplicity we also omitted other details) given by source 2 as* $(\mathsf{PL}, 21:05:37, 2, 1, 1, 3283, \mathsf{DriveCabin})$, $(\mathsf{PL}, 21:05:37, 2, 1, 0.7, 3283, \mathsf{StairWay})$, *and* $(\mathsf{PL}, 21:05:37, 2, 1, 0.3, 3283, \mathsf{Seated})$ *and the mass values given by source 2 are* $m_2(3283, \mathsf{DriveCabin}) = 0.2$, $m_2(3283, \mathsf{Stairway}) = 0.1$, *and* $m_2(3283, \mathsf{Seated}) = 0.7$. *Then the event flow is* $[EC, EC_2]$.

3.5 Event Inference

Event inferences are expressed as a set of inference rules which are used to represent the relationships between events. In the literature of event models, most rules were defined in a deterministic manner without uncertainty except [15], where rules are defined in a probabilistic way. Simply speaking, rules in [15] are defined as follows: if some conditions of a rule are satisfied, then a certain event E occurs with a probability p, and does not occur with a probability $1 - p$. This type of inference rules is an uncertainty-based extension to the Event-Condition-Action (ECA) paradigm proposed in active databases. However, this approach ignores situations where a set of events can be inferred due to uncertainty or incompleteness[8].

In this paper, we define our event inference rules which can resolve uncertainty and incompleteness. An inference rule R is defined as a tuple $(LS, EType, Premise, Condition, m_{IEC})$ where:

LS, abbreviated for Life Span, is used to determine the temporal aspect of a rule R [3,1,16]. LS is an interval determined by a starting point and an end point, or an

[8] In fact, the event model in [15] could be extended so that more than one target event can be inferred, see [17]. However, rules like this still cannot infer events with incomplete information due to the different expressive power between probability theory and DS theory.

initiator and a terminator, respectively. The starting point and the end point are two points in time which can be determined by the event flow that is known at the time a rule is executed. For instance, a starting time point may refer to the occurrence time of a specific event, a prior given time, etc, and an end time point can be the occurrence time of another event, a prior given time, or a time period plus the starting point, etc.

$EType$ is the event type of the inferred event cluster. For example, SAD standing for Shout At Driver is an inferred event type.

$Premise$ is a set of $ETypes$ that a set of events of such types are used by the rule as prerequisites. For example, to induce an SAD event, we need to have the corresponding PBV, PL (Person Loiter), PS (Person Shout) events[9], hence $Premise = \{PBV, PL, PS\}$. $Premise$ is used to select the premise events for a rule.

$Condition$ is a conjunction of a set of conditions used to select appropriate events from the event flow to infer other events. The conditions in $Condition$ can be any type of assertions w.r.t the attributes of events. For example, let e_1 and e_2 both denote a male loitering event and e_3 denote a person shouting event, then
"$e_1.pID = e_2.pID \wedge e_1.gender = male \wedge e_1.location = e_2.location = DriverCabin$
$\wedge e_2.occT - e_1.occT \geq 10s \wedge e_1.occT \leq e_3.occT \leq e_2.occT \wedge e_3.volume = shouting$"
is a valid $Condition$. Note that for each inference rule, we only select events in the event flow within the lifespan LS (denoted by $LS(EF^t_{t'})$). In addition, the types of events used in the $Condition$ belong to $Premise$. Let the events used in $Condition$ be denoted as $Evn(Condition)$, then $Evn(Condition)$ is an instantiation of $Premise$.

m_{IEC} is the mass function for the inferred event cluster and it is in the form of $(< v^1_1, \cdots, v^1_n, mv_1 >, < v^2_1, \cdots, v^2_n, mv_2 >, \cdots, < v^k_1, \cdots, v^k_n, mv_k >)$ where each mv_i is a mass value and $\sum_{i=1}^k mv_i = 1$. We will explain this in detail when discussing rule semantics next.

To differentiate inferred events from other events, we use -1 to denote the source ID of an inferred event cluster and the occurrence time is set as the point in time an inference rule is executed. Moreover, the reliability is set to 1 as we assume that the inference rules are correct.

The semantics of using an inference rule R is interpreted as follows. Given an event flow $EF^t_{t'}$, if $Condition$ of any rule R is true at some time point $t^* > t'$, then an event cluster is inferred from rule R with mass function m_{IEC}. Otherwise, no events are inferred.

Formally, for any vector $< v^i_1, \cdots, v^i_n, mv_i >$, if $Condition(LS(EF^t_{t'})) = true$, we in fact generate an event E_i whose event type is $EType$, source ID is -1, occurrence time is the time of rule execution, reliability is 1, $E_i.v = (v^i_1, \cdots, v^i_n)$ (and $E.sig$ is a function over $E_i.v$), and

$$m_{IEC}(E_i.v) = mv_i, 1 \leq i \leq k$$

For any two rules having the same $Premise$, we consider them from a single *rule cluster*. Intuitively, rules in a rule cluster describe inferences based on the same observations, hence these rules have the same lifespan and the same inferred event type but with different $Condition$ and m_{IEC} values due to the different premise events induced

[9] Simply speaking, to get a SAD event, we need to check that a person X entered the bus, went to the driver's cabin and then a shout at the cabin was detected.

from the observations. In addition, in this framework, if two rules do not have the same $Premise$, then they will not infer the same type of events.

Example 6. *An inference rule* R_1 *which reports an obscured person loiters at the driver's cabin can be defined as* $(LS, EType, Premise, Condition, m_{IEC})$ *where*
$LS = [0, E.occT]$ *where* E *is an event of* PL,
$EType$ *is* PD *abbreviated for Passenger-Driver,*
$Premise$ *is* $\{PBV, PL, PL\}$,
$Condition$ *is* "$e_1.gender = obscured \wedge e_1.pID = e_2.pID = e_3.pID \wedge e_2.location = e_3.location = DriverCabin \wedge e_3.occT \geq e_2.occT + 10s$",
and m_{IEC} is $< \{Stand, Talk\}, 0.7 >$, $< Leaving, 0.2 >$, $< hasThreat, 0.1 >$.
A similar inference rule R_2 *can be used to report a male loiters at the driver's cabin with* $R_2 = (LS', EType', Premise', Condition', m'_{IEC})$ *where*
$LS' = LS$, $EType' = EType$, $Premise' = Premise$,
$Condition'$ *is* "$e_1.gender = male \wedge e_1.pID = e_2.pID = e_3.pID \wedge e_2.location = e_3.location = DriverCabin \wedge e_3.occT \geq e_2.occT + 10s$",
and m'_{IEC} is $< \{Stand, Talk\}, 0.7 >$, $< Leaving, 0.22 >$, $< hasThreat, 0.08 >$.
Hence R_1, R_2 *are in a rule cluster.*

4 Calculation of Event Mass Values

Since events in $Evn(Condition)$ are themselves uncertain, to get the mass value of an inferred event, we need to consider both m_{IEC} and the mass values of events in $Evn(Condtion)$. Here the mass value can be seen as a joint degree of certainty of all events involved (similar to the joint probability in Bayesian networks).

To proceed, first, it is necessary to ensure that the execution of the event model in an application is guaranteed to terminate. That is, in a finite time period, there would be only finite external events, finite domain events, finite inference rules to be triggered (hence finite inferred events). Second, it is also necessary that the execution of the event model is guaranteed to be deterministic. That is, with the same time period, same input events and same set of rules, the resultant event flow (after applying all rules) should be unique. These two issues are discussed in [10] where they can be solved by avoiding cycles in rule definitions and by ranking the rules, respectively.

Now assume that there are no cycles in rules and the rules are ranked. Typically, for a specific inferred event, it can be inferred from more than one rule in a rule cluster (e.g., a PD event with value $hasThreat$ in Example 6). Hence the mass value should consider all these rules in that rule cluster RC. Since each rule in RC has the same $Premise$, let $Premise = \{ET_1, \ldots, ET_t\}$ be a set of event types, and let $\mathscr{V}_i$ be the corresponding frames of discernment of ET_i. Let $\Omega_{RCE} = \prod_{i=1}^{t} \mathscr{V}_i$ be a joint frame of discernment for the premise event types and Ω_{RCH} be the frame of discernment of the inferred event type. Then we can use evidential mapping to get the mass value of an inferred event. Formally, for each rule R in RC, we set $\Gamma^*_R(e^R_1.v, \cdots, e^R_t.v) = ((v^1_1, \cdots, v^1_n), mv_1), \cdots, ((v^k_1, \cdots, v^k_n), mv_k)$ where $(e^R_1, \cdots, e^R_t) = Evn^R(Condition^R)$ and $((v^i_1, \cdots, v^i_n), mv_i) \in m^R_{IEC}$. For other $(e'_1, \cdots, e'_t)$ where $(e'_1, \cdots, e'_t)$ is an instantiation of $Premise$ but there does not exist a rule R in RC, s.t.,

$(e_1^R, \cdots, e_t^R) = (e'_1, \cdots, e'_t)$, we set $\Gamma_R^*(e'_1.v, \cdots, e'_t.v) = (\emptyset, 1)$. Therefore, the mass value of inferred event can be obtained using Equation 5.

We can also use Bel and Pl functions to get a plausible interval of inferred events.

Based on the above, here we give an algorithm to calculate event mass in **Algorithm 1**. Note that this algorithm executes when a new observation (hence a set of event clusters are gathered from multiple sources, and a combination is carried out) is obtained.

Algorithm 1. Event Mass Calculation

Input: An event flow EF^t, most recent combined event cluster $EC^{t'}$, all rule clusters RCs in a pre-specified order.

Output: Output a mass value of each inferred event when some rules are triggered.

$EF^{t'} \leftarrow EF^t \cup \{EC^{t'}\}$;
for each rule cluster RC **do**
  calculate $LS^{RC}(EF^{t'})$;
  select event clusters in $LS^{RC}(EF^{t'})$ according to $Premise^{RC}$;
  for each selected event clusters $EC_1, \cdots, EC_t$ **do**
    construct the frames of discernment Ω_{RCE} and Ω_{RCH};
    for each rule R in RC **do**
      for each list of events $e_1, \cdots, e_t$, s.t., $e_i \in EC_i$ **do**
        if $Condition^R$ is satisfied **then**
          add the contents of m_{IEC}^R to $\Gamma_R^*(e_1, \cdots, e_t)$;
        else
          set $\Gamma_R^*(e_1, \cdots, e_t)$ to $(\emptyset, 1)$;
        end if
      end for
    end for
    calculate the mass values of focal elements in Ω_{RCH} using Equation 5;
    set the mass value of a focal element to the mass value of an inferred event whose $e.v$ is the focal element;
  end for
end for
return the mass values for the inferred events.

5 Related Work

Our event definition is similar to that considered in [1,15,16] where events are considered significant (w.r.t the specified domain of the application), instantaneous and atomic. The reason why we do not require the events to be significant is that in real applications, we also need to model insignificant events (otherwise we may lose information). For instance, in surveillance applications, up to 99% of the events are just trivial events. Hence, instead of defining events as significant, we introduce a built-in significance value in the representation of events to facilitate subsequent processing.

For the inference rules, in [15], a rule is defined as $(sel^n, pattern^n, eventType, mappingExpressions, prob)$ where sel^n is used to get n events, $pattern^n$ is a conjunction of a set of conditions. However, conditions for $pattern^n$ can only be an equality form as $e.attr_i = e'.attr_j$ or temporal conditions of the forms, $a \leq e.occT \leq b$

or $e.occT < e'.occT$ or $e.occT \leq e'.occT \leq e.occT + c$. Obviously, it can not express conditions like $E.gender = obscured$, $E_1.speed < E_2.speed$, etc, while our $Condition$ can. In addition, a rule in [15] can only provide a single inferred event with a probability $prob$ whilst a rule in our model can provide a set of possibilities with mass values.

Classical deterministic rules are special cases of our rule definition with the inferred event cluster having only one event with a mass value 1 and probabilistic rules are also special cases of our rule definition.

Furthermore, in our model, the notion of event history or event flow is also different from those used in [6,1,15,16] such that our event history/flow takes embedded uncertainty (in fact it contains observations (event clusters) which consist of multiple possible events) while in those models an event history itself is considered deterministic and the uncertainty on event history is expressed as there can be multiple possible event histories. Due to this difference, the rule semantics is totally different from the conditional representation in [15].

Finally, when our event model reduces to the situation considered in [15], it is easy to find that our calculation of event mass reduces to the probability calculation by Bayesian networks in [15].

Proposition 1. *If we consider only one-source, one-inference-target probabilistic case as in [15], then the mass value of inferred event is equivalent to the joint probability obtained by the Bayesian network approach in [15].*

6 Conclusion

In this paper, we proposed an event model which can represent and reason with events from multiple sources, events from domain knowledge, and have the ability to represent and deal with uncertainty and incompleteness. For events obtained from multiple sources, we combined them using Dempster-Shafer theory. We introduced a notion called event cluster to represent events induced from an uncertain observation. In addition, in our model, inference rules can also be uncertain. Furthermore, we discussed how to calculate the mass values of a set of events. This framework has been implemented and evaluated by a bus surveillance case study.

Since in real-world applications, information is frequently gathered from multiple sources, and uncertainties can appear in any part of the applications, our event model can serve as an important foundation for these applications. Domain knowledge is very useful in many active systems, however, it is somehow ignored in the existing event reasoning systems. Our model can also represent and deal with it.

For future work, we want to extend this event model to include temporal aspects of events. First, in some active systems, there is no accurate occurrence time attached with events. Second, some behaviours associated with a time interval such as *a person is holding a knife* is hard to be represented in this event model. In fact, as the instantaneous nature of events in this event model, we can only tell at a certain time point (or a set of successive time points), the person is holding a knife.

Acknowledgement. This research work is partially sponsored by the EPSRC projects EP/D070864/1 and EP/E028640/1 (the ISIS project).

References

1. Adi, A., Etzion, O.: Amit - the situation manger. VLDB J. 13(2), 177–203 (2004)
2. Bsia. Florida school bus surveillance, `http://www.bsia.co.uk/LY8VIM18989_action;displaystudy_sectorid;LYCQYL79312_caseid;NFLEN064798`
3. Chakravarthy, S.S., Mishra, D.: Snoop: an expressive event specification language for active databases. Data and Knowledge Engineering 14(1), 1–26 (1994)
4. Dempster, A.P.: Upper and lower probabilities induced by a multivalued mapping. The Annals of Statistics 28, 325–339 (1967)
5. Abreu, B., et al.: Video-Based Multi-Agent Traffic Surveillance System. In: Proc. IEEE Intel. Vehi. Symp. LNCS, pp. 457–462. SPIE, Bellingram (2000)
6. Gehani, N.H., Jagadish, H.V., Shmueli, O.: Compostite event specification in active databases: Model & implementation. In: Proc. of VLDB, pp. 327–338 (1992)
7. Liu, W., Hughes, J.G., McTear, M.F.: Representating heuristic knowledge in d-s theory. In: Proc. of UAI, pp. 182–190 (1992)
8. Lowrance, J.D., Garvey, T.D., Strat, T.M.: A framework for evidential reasoning systems. In: Proc. of 5th AAAI, pp. 896–903 (1986)
9. Ma, J., Liu, W., Miller, P., Yan, W.: Event composition with imperfect information for bus surveillance. In: Procs. of 6th IEEE Inter. Conf. on Advanced Video and Signal Based Surveillance (AVSS 2009), pp. 382–387. IEEE Press, Los Alamitos (2009)
10. Patton, N.W.: Active Rules in Database Systems. Springer, Heidelberg (1998)
11. Gardiner Security. Glasgow transforms bus security with ip video surveillance, `http://www.ipusergroup.com/doc-upload/Gardiner-Glasgowbuses.pdf`
12. Shafer, G.: A Mathematical Theory of Evidence. Princeton University Press, Princeton (1976)
13. Shu, C.F., Hampapur, A., Lu, M., Brown, L., Connell, J., Senior, A., Tian, Y.: Ibm smart surveillance system (s3): a open and extensible framework for event based surveillance. In: Proc. of IEEE Conference on AVSS, pp. 318–323 (2005)
14. Snidaro, L., Belluz, M., Foresti, G.L.: Domain knowledge for surveillance applications. In: Proc. of 10th Intern. Conf. on Information Fusion (2007)
15. Wasserkrug, S., Gal, A., Etzion, O.: A model for reasoning with uncertain rules in event composition. In: Proc. of UAI, pp. 599–608 (2005)
16. Wasserkrug, S., Gal, A., Etzion, O.: Inference of security hazards from event composition based on incomplete or uncertain information. IEEE Transactions on Knowledge and Data Engineering 20(8), 1111–1114 (2008)
17. Wasserkrug, S., Gal, A., Etzion, O., Turchin, Y.: Complex event processing over uncertain data. In: Proc. of DEBS, pp. 253–264 (2008)

Uncertainty in Decision Tree Classifiers*

Matteo Magnani and Danilo Montesi

Dept. of Computer Science, University of Bologna
Mura A. Zamboni 7, 40100 Bologna
{matteo.magnani,montesi}@cs.unibo.it

Abstract. One of the current challenges in the field of data mining is to develop techniques to analyze uncertain data. Among these techniques, in this paper we focus on decision tree classifiers. In particular, we introduce a new data structure that can be used to represent multiple decision trees generated from uncertain datasets.

1 Introduction

Today more than ever it is fundamental to be able to extract relevant information and knowledge from very large amounts of data available in several different contexts like sensor and satellite databases, user generated repositories like Flickr, YouTube and Facebook, and company data warehouses. While data analysts are usually necessary to guide the knowledge discovery process, the size and complexity of these data require the application of automated or semi-automated data analysis (or mining) techniques.

One of the current challenges regarding the development of data mining techniques is the ability to manage uncertainty. This is particularly important when data have been obtained through information extraction tasks, like the creation of a user profile based on her on-line public identities or the collection of opinions regarding a given brand. In fact, while the persistence of user generated content on the Web enables its later retrieval and analysis making it a very valuable source of information, this is often of low quality, containing missing, wrong or inconsistent data. As an example, we may think of a dataset containing age and income of some individuals based on the information available on the Internet: different pages may provide different values, and some values may be missing and thus guessed using additional information on lifestyles or connections.

In the last decades many data analysis techniques have been developed, grouped in two main families: descriptive methods, e.g., clustering, and predictive methods like classification. *Classification* tasks, that are the general object of this paper, consist in building (aka *learning*) a model from a set of records (aka *statistical units*) annotated with a class so that previously unseen and unclassified records may be assigned to the most likely class. One well known and widely used kind of classifier is the *decision tree*, which we extend in this work to handle uncertain information.

* This work has been partly funded by Telecom Italia and PRIN project *Tecniche logiche e operazionali per interazione tra componenti.*

A. Deshpande and A. Hunter (Eds.): SUM 2010, LNAI 6379, pp. 250–263, 2010.

The choice of extending decision trees is justified by their simplicity and effectiveness: they produce understandable models (they can be translated into classification rules), do not require prior assumptions on data distributions, and are very fast to build (using heuristics) and to classify. In addition, their accuracy is comparable to those of other classifiers except for specific datasets. Obviously, this does not mean that decision trees are the best classification tools: given the nature and the complexity of real classification problems, it is fundamental to have several classification techniques. Among these, decision trees are certainly a strong option, and it is thus important to be able to apply them also to uncertain data.

This paper is organized as follows. In the next section we briefly summarize the main related works, highlighting our contributions. In Section 3 we review the main concepts of traditional decision trees. Specifically, we focus only on those features that are necessary to understand our extension. Then, in Section 4 we extend this traditional classification scenario using the semantics of possible worlds, showing that decision tree induction over uncertain data corresponds to multiple inductions on certain alternative datasets. This poses the problem of providing compact representations of multiple classifiers. A novel data structure to address this problem is introduced in Section 5, where we also prove that it can always be constructed from a set of alternative trees providing an algorithm to build uncertain trees — however, this algorithm is only used as a constructive proof, and is not intended to be used directly to build real uncertain trees. We conclude the paper with a summary of our work and a brief discussion of future research directions.

2 Related Work

Several families of data mining techniques have been extended to work on uncertain data: for a recent survey, see [1]. In the specific sub-field of decision tree classifiers probabilities have been used with several different objectives. First, there are data structures usually called *probabilistic decision trees* [2] which are however different from the classification models used in data mining and should not be confused with them: they represent decision processes, and not data. Second, even traditional decision tree classifiers often produce probabilistic results, either because training data contain records with the same values on the independent attributes but different classes or because trees are pruned generating heterogeneous leaves with different records in multiple classes. However, in both cases the input data is not uncertain. Third, probabilistic trees have been used to deal with missing values in the training or test records [3,4,5]. Also in this case, though very important, the problem of building decision trees from probabilistic data has not been considered. Among non-probabilistic approaches, fuzzy sets have been used to extend decision tree classifiers [6]. Finally, there are a few recent proposals of approaches to build decision trees from uncertain data [7,8], for which a more precise comparison with our work is necessary.

In [7] uncertainty is represented using probability distribution functions instead of certain feature attributes, while the class attribute is certain. In fact, the focus of the paper is on the development of an algorithm for building decision trees from uncertain numerical data and in particular to efficiently identify good split intervals — a topic not treated in our work, where we allow uncertain categorical and class attributes but with discrete probabilities. In [8] the authors consider uncertainty on both numerical and categorical data (also in this case the class attribute is certain) and introduce the concept of probabilistic entropy to be used in an extended algorithm for tree induction. The result is a traditional decision tree whose structure and classification outcomes depend on the uncertainty in the input data source.

The main contribution of our work complementing these approaches consists in applying the possible worlds semantics to decision trees for uncertain data. To the best of our knowledge, this is the first work where probabilistic decision trees are extended using this reference approach. However, under a possible worlds semantics the result of a tree induction process is not a single tree but *a set* of alternative trees. Therefore, the other main contribution of this paper is to introduce a new data structure, called mu-tree, which is shown to be able to represent any probabilistic system of alternative decision trees.

Before introducing our data structure we very briefly recall the main concepts related to decision trees and set some working assumptions.

3 Decision Trees: Basics

In this section we briefly review the basic concepts about decision tree classifiers for tabular data, limited to the ones needed to follow the remaining of the paper.

Consider Figure 1, where we have represented the table used as our working example, taken from [9]. The last column indicates the *class* attribute, and there are two possible class values: `no` and `yes`. For instance, record 5 has class

TID	Home Owner	Marital status	Income	*Defaulted*
1	yes	Single	125K	no
2	no	Married	100K	no
3	no	Single	70K	no
4	yes	Married	120K	no
5	no	Divorced	95K	yes
6	no	Married	60K	no
7	yes	Divorced	220K	no
8	no	Single	85K	yes
9	no	Married	75K	no
10	no	Single	90K	yes

Fig. 1. A table T1 used in the following examples

yes. When there is no uncertainty this dataset can be used to build a classifier (decision tree) which can then take other records as input and assign them to one of the two pre-defined classes[1].

A decision tree classifier has two kinds of node, as illustrated in Figure 2. Internal nodes (on the left in the figure) correspond to an attribute of the input table used to partition the records associated to that node according to their values on the attribute. In the example, all records with value Yes on attribute Home Owner would belong to the left sub-tree. On the right we have illustrated a leaf, indicating the percentage of records belonging to each of the classes. In addition, we will indicate the number of records corresponding to each node: 5 in these examples. As we will see, the internal nodes of the uncertain data structure proposed in this paper will be a mixture of these two kinds of node.

Fig. 2. Nodes of a decision tree: internal and leaf

The basic algorithm used to build a decision tree is very simple and efficient, though it may fail in providing the best solution (which is instead a problem in NP):

1. Start with a single node representing all records in the dataset.
2. Choose one attribute and split the records according to their values on that attribute.
3. Repeat the splitting on all new nodes, until a stop criterion is satisfied.

This is known as Hunt's algorithm, and evidently there are many ways of implementing it. In the remaining of this section we will define the options adopted in this paper, list the topics not treated because they do not directly concern the management of uncertainty, and provide an example of tree construction to which we will then add uncertainty. The three aspects that should be decided to instantiate the algorithm are:

- How to partition (split) the records after a split attribute has been selected.
- How to choose the split attribute.
- When to stop.

[1] Usually part of the data is used to build the classifier and part of it to test its accuracy, but this is not relevant to our discussion.

Partitioning. In the example we have two categorical attributes (`Home Owner` and `Marital Status`, respectively binary and ternary), and a continuous attribute. In this paper we assume that each attribute can be split in a pre-defined number of children. Therefore, for the continuous attribute we will use a discretization procedure to split the records in a finite number of classes, and in our example we will consider two splits with values ≤ 80 and > 80. We do not provide additional details on how to set this threshold, because this is not relevant to our discussion. However, in the following examples we do assume that the threshold is fixed for a given attribute.

Choice of the split attribute. Once we have decided how to split the records when an attribute is chosen, we must select the split attribute. To do this we will use a very common function (among the many available ones) to measure how good a split is. Intuitively, if all records belong to only one of the classes the node outcome will have a high degree of confidence, which is not the case when records are equally distributed among all classes.

Let $p(c_i|n)$ be the percentage of records belonging to class c_i at node n. Entropy at node n is defined as:

$$E(n) = \sum_{\text{all classes}} -p(c_i|n) \log_2 p(c_i|n) \tag{1}$$

The overall quality of a splitting can be measured by a weighted sum of the entropy on all newly created nodes, defining its level of *impurity* [9]. Let N be the total number of records associated to the parent and N_i be the number of records associated to the ith child after the splitting. The impurity of a split node S with children $n_0, \dots, n_k$ is defined as:

$$I(S) = \sum_{i \in [0,k]} \frac{N_i}{N} E(n_i) \tag{2}$$

As an example consider table T1 where we can split on `Home Owner`, `Marital Status` or `Income`. These three alternatives are illustrated in Figure 3 and their impurities are, from left to right:

$$\frac{3}{10}(-0 \log_2 0 - 1 \log_2 1) + \frac{7}{10}(-.43 \log_2 .43 - .57 \log_2 .57) = .69$$

$$\frac{4}{10}(-.5 \log_2 .5 - .5 \log_2 .5) + \frac{4}{10}(-0 \log_2 0 - 1 \log_2 1) + \\ + \frac{2}{10}(-.5 \log_2 .5 - .5 \log_2 .5) = .6$$

$$\frac{3}{10}(-0 \log_2 0 - 1 \log_2 1) + \frac{7}{10}(-.43 \log_2 .43 - .57 \log_2 .57) = .69$$

From these values it appears that the less impure split is the one made on the `Marital Status` attribute. At this point we can recursively split the three new nodes until we satisfy the stop condition.

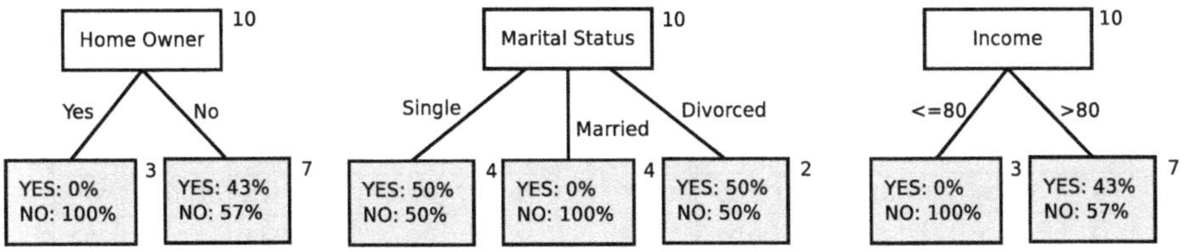

Fig. 3. Alternative split attributes for the dataset T1

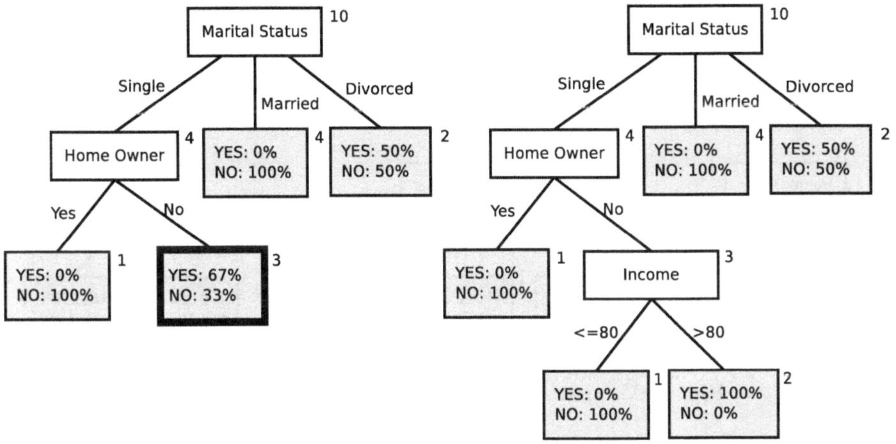

Fig. 4. Decision tree construction for dataset T1

Stop criterion. Stopping the growth of a tree is very important to reduce its size, increase its readability and avoid problems like overfitting, all topics that are not treated here. In the following examples we use the following stop condition:
A node is processed if all the following predicates hold:

1. There is a new attribute to split.
2. The node contains more than two records.
3. There is not any class with at least 75% of the records.

Please notice that the two-record and 75% thresholds are arbitrary and only used as an example.

To conclude our working example, in Figure 4 we have represented the next two steps of the tree building process, omitting the computation of impurity.

4 Decision Trees with Uncertainty: Possible Worlds Semantics

Uncertainty is a state of incomplete knowledge where we have multiple alternatives with regard to our description of the world. In our case each possible

TID	HOwner	MStatus	Income	*Def*
1	yes	Single	125K	no
2	no	Married	100K	no
3	no	Single	70K	no
4	yes	Married	120K	**no**
5	no	Divorced	95K	yes
6	no	Married	60K	no
7	yes	Divorced	220K	no
8	no	Single	85K	yes
9	no	Married	75K	no
10	no	Single	90K	yes

TID	HOwner	MStatus	Income	*Def*
1	yes	Single	125K	no
2	no	Married	100K	no
3	no	Single	70K	no
4	yes	Married	120K	**yes**
5	no	Divorced	95K	yes
6	no	Married	60K	no
7	yes	Divorced	220K	no
8	no	Single	85K	yes
9	no	Married	75K	no
10	no	Single	90K	yes

Fig. 5. Two alternative tables T1 and T2, with probability .8 and .2

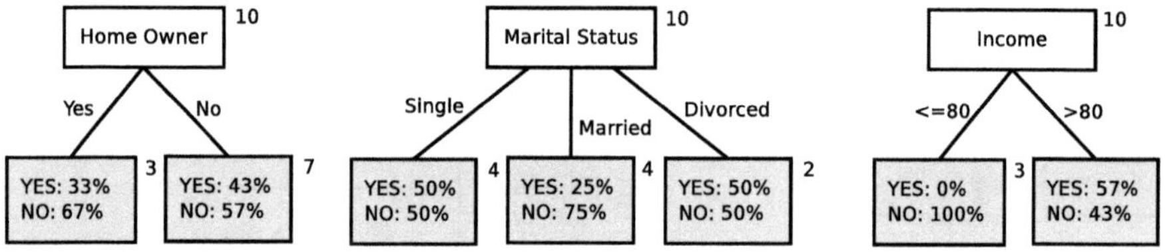

Fig. 6. Alternative split attributes for the dataset T2

world is defined by an alternative table, e.g., in Figure 5 we have two tables T1 and T2 which differ in one class value, represented in bold face (record 4). Each alternative table constitutes a possible world, and numerical values can be associated to each alternative to indicate our degree of belief in it. In this work we will use a probabilistic setting and assign probability .8 to table T1 and .2 to table T2, as indicated in the figure. We do not focus on how these probabilities can be computed, which is an orthogonal problem out of the scope of this work: we assume to have uncertain data and focus on the analysis step.

For each possible world we can build a decision tree using the procedure reported in the previous section. However, this time with regard to table T2 and differently from the other possible world the best split is on the `Income` attribute (we omit calculations for space reasons). Continuing the expansion, we obtain the tree illustrated in Figure 7.

If we want to classify a certain record we can now use each alternative classifier and consider the aggregated output. Let $T_0, \ldots, T_n$ be our alternative classifiers. If $p(\text{class} = c|T_i, r)$ is the probability that record r belongs to class c according to T_i and $p(T_i)$ is the probability that the correct classifier is T_i, then the probability that record r belongs to class c is:

$$p(\text{class} = c|r) = \sum_{i \in [0,n]} p(\text{class} = c|T_i, r)p(T_i) \qquad (3)$$

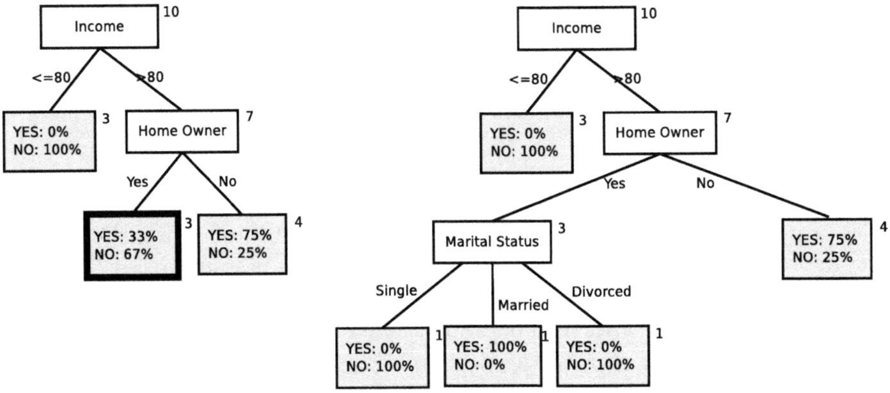

Fig. 7. Decision tree construction for dataset T2

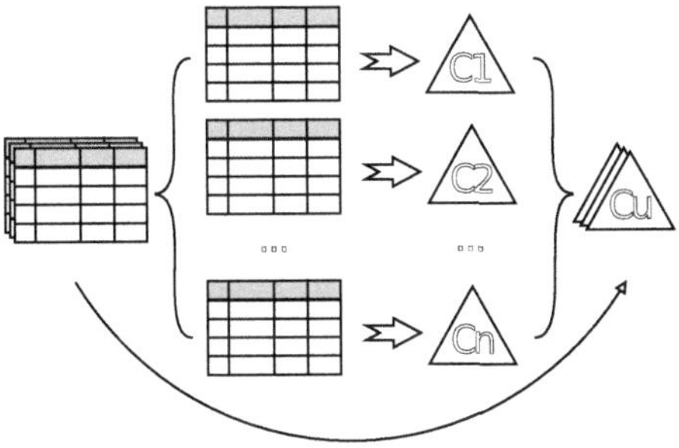

Fig. 8. Possible world semantics for uncertain classification trees

As an example, assume we want to classify the record $\langle$yes, Married, 120$\rangle$. It would be classified as 100% NO by the first classifier, and 100% YES by the second. The probability associated to each class will then be:

YES $.8 \cdot 0 + .2 \cdot 1 = .2$
NO $.8 \cdot 1 + .2 \cdot 0 = .8$

As usual, and especially in presence of many possible worlds, we would like to define compact data structures to represent the multiple input tables and the multiple classifiers, and the result of a classification on the compact uncertain classifier should be the same that we would obtain by performing the classification in each possible world.

This scenario is illustrated in Figure 8, representing a *strong interpretation system*, and it is the standard semantics underlying uncertain relational models. However, when we merge multiple tables into one compact uncertain table we usually have one single schema. On the contrary, when we want to build a compact classifier we may need to merge trees with different structures. This is the case in our working example — Figures 4 and 7, on the right.

In the next section we introduce a compact data structure that can represent multiple heterogeneous possible trees.

5 Mu-Trees

In Figure 9 we have represented the nodes of a mu-tree. Let us start with the leaf, on the right. This is like a leaf in a certain decision tree, but it is annotated with a probability. Internal nodes are instead a mixture of a branching node, like in traditional trees, and a leaf node, which can be present or not (in which case we may think of it as being present with probability 0).

A mu-tree is a tree with the nodes mentioned above where for all paths from the root to a leaf the sum of the probabilities (internal and leaf nodes) is 1. We will also use the notation α-mu-tree when probabilities add to α instead of 1. Finally, we will notate MT.predict(r,c) the class probability computed by a mu-tree MT on record r. As we can see in Figure 9 all nodes in a mu-tree (may) have an associated node-probability and express a probability for each class. We can then define MT.predict(r,c) as follows:

Definition 1. *If a record r to be classified traverses n nodes with node-probability* $(p_1, \ldots, p_n)$ *and class probability* $(c_1, \ldots, c_n)$ *for class c, then MT.predict(r,c)* $= \sum_{i \in [i,n]} p_i c_i$.

The following theorem tells us that our data structure can always correctly represent a set of alternative decision trees with cumulative probability 1.

Theorem 1. *Let T1,...,Tn be decision trees with the same values for each split attribute, P be a probability distribution over {T1,...,Tn}, and let Ti.predict(r,c) be the class probability computed by Ti on record r. There is always a mu-tree MT such that for all records r and classes c: MT.predict(r,c)=* $\sum$ *P(Ti)*Ti.predict(r,c).*

The proof of this theorem is provided constructively through the `compact` algorithm indicated in Figure 10, where we express the input trees as α-mu-trees by annotating their leaves with the probability of their possible world. Let T1 be an α_1-mu-tree and T2 an α_2-mu-tree. For each leaf l in T1 we insert it into T2 using the method `process`(T2.root, l). At the end, this produces an $(\alpha_1 + \alpha_2)$-mu-tree, and it follows that a mu-tree can be constructed by applying iteratively this algorithm until all alternative trees have been processed.

It is worth noticing that this algorithm is provided as part of the proof, and it is not intended to be used directly to build mu-trees in practice.

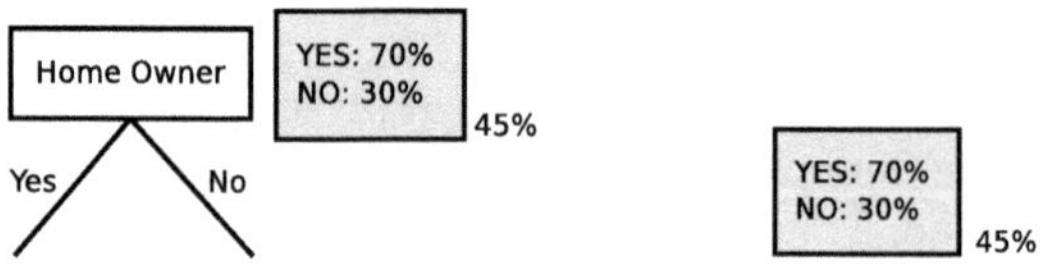

Fig. 9. Nodes of a decision tree: internal and leaf

Proof. We split the proof into three parts: we first show that the algorithm terminates, i.e., it always builds an output tree. Then, we show that all new paths added to the tree have cumulative probability $\alpha_1+\alpha_2$ (soundness). Finally, we show that an attribute can always be classified, i.e., there is a branch for all branches in the input trees (completeness).

- (Termination) Consider the `process` method indicated in Figure 11, which is called a finite number of times. When `node` is a leaf (line 08) the function ends, because both functions `merge` (line 09 and Figure 12) and `expand` (line 10 and Figure 13) contain no infinite loops. Otherwise, the `process` method is called recursively at lines 03 or 07. In both cases, it is called on one or more children of `node`, moving to the lower level of the tree until it reaches a leaf. Therefore the length of each recursion chain is bounded by the height of the tree, and as we are dealing with finite trees the process eventually ends.
- (Soundness) To add or modify a leaf we perform the functions `merge` (line 09) and `expand` (line 10). In the first case, line 04 of function merge (Figure 12) sets the probability of the new leaf to the sum of the two input probabilities. In the second case, line 06 of function expand (Figure 13) sets the probability of the new leaf created *under* the existing leaf of T2 to the probability of T1. Therefore, the path to this new leaf will have cumulative probability summing the two input probabilities.
- (Completeness) With regard to this point, each combination of split values in T2 is obviously present in the resulting α-mu-tree, because it is built starting from T2. Then, each combination of split attributes and values in T1 corresponds to a specific leaf, and as we insert all leaves from T1 into T2 we will also replicate all the original paths — notice that if T2 contains split attributes not present in T1, the new leaf is inserted into *all* alternative sub-trees (line 03 of Figure 11).

```
FUNCTION compact(T1, T2):
    for each leaf l in T1
        process(T2.root, l)
```

Fig. 10. Constructive proof of existence (1)

5.1 An Example of Building a Mu-Tree from a Set of Decision Trees

In this section we show how to build a mu-tree merging the two trees built in our working example. We start by converting the tree built from table T2 into an α-mu-tree, annotating its leaves with the probability of the corresponding possible world, .2 in our example.

We then start the insertion of leaves from tree T1 from its left-most leaf (Figure 4), corresponding to the records with: Marital Status: Single, Home Owner: yes, Income: * (meaning: all possible values), and insert it into the root of T2,

```
FUNCTION process(node, inserted):
    if node is not a leaf then:
        if (inserted does not contain node.splitAtt) then
            for each child n of node process(n, inserted)
        else
            val = inserted.value(node.splitAtt)
            n = node.child(val)
            process(n, inserted.remove(splitAtt))
    else
        if (inserted is empty) then node = merge(node,inserted)
        else expand(node,inserted)
```

Fig. 11. Constructive proof of existence (2): `inserted` corresponds to a leaf in T1, and contains several split attributes (splitAtt) with a specific value (`inserted.value(splitAtt)`). `node` is a node of T2, represents a split on attribute `node.splitAtt` when it is not a leaf, and has one child `node.child(val)` for each split value `val`, e.g., > 80.

```
FUNCTION merge(leaf1, leaf2): leaf
    declare newLeaf of type leaf;
    for each class c:
        newLeaf.c = leaf1.c * leaf1.prob + leaf2.c * leaf2.prob
    newLeaf.prob = leaf1.prob + leaf2.prob
    return newLeaf
```

Fig. 12. Constructive proof of existence (3): `l.c` is the probability associated to class c at leaf l, while `l.prob` is the probability of the leaf — that is the probability of the corresponding possible world

```
FUNCTION expand(leaf1, leaf2):
    declare newNode of type node;
    for each splitAtt in leaf2:
        tmpNode = newNode.addChild(splitAtt,leaf2.value(splitAtt))
        newNode = tmpNode
        leaf2.remove(splitAtt)
    newNode.prob = leaf2.prob
```

Fig. 13. Constructive proof of existence (4). This method creates a new path in the tree with one node for each split attribute/value in `leaf2` (example in Figure 15).

which separates records with income less or greater than 80. This corresponds to the first high level call of the `process` method, and is illustrated starting from Figure 14. The leaf under consideration applies to both branches, because it does not specify any constraints on the Income attribute. Therefore, we continue inside both. Now, let us focus on the right branch: this time the split regards Home Owners, and our leaf contains only records with Home Owners: yes, therefore we continue only inside the left branch and remove the Home Owner attribute from the case under consideration. At this point, we continue by considering the Marital Status attribute and thus following the left branch, ending in the highlighted leaf of the mu-tree under construction (thick border, in the figure). As there are no more constraints to be processed we can compute the weighted probability for each class using the `merge` function. You can see that the overall probability associated to this leaf is now 1.

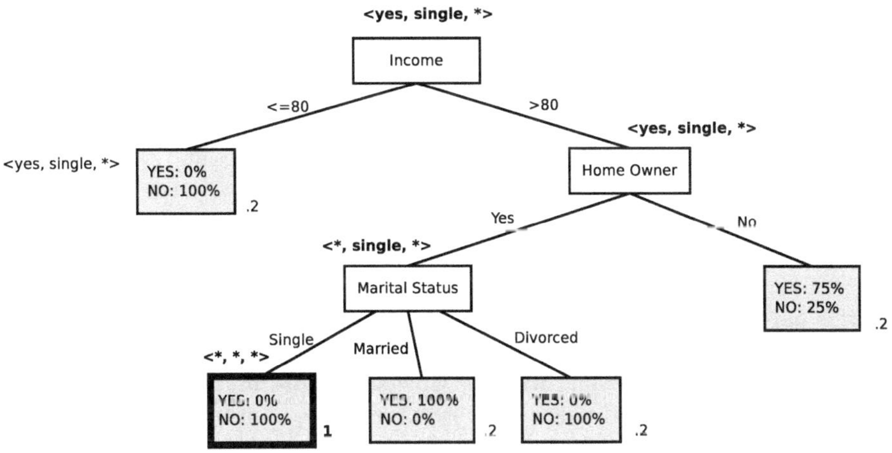

Fig. 14. Merging two alternative trees: common leaf (merge)

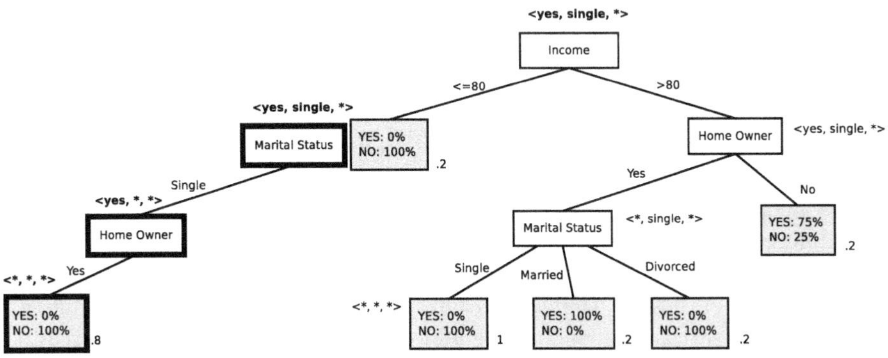

Fig. 15. Merging two alternative trees: expansion

To complete the insertion of this leaf into the mu-tree we must still process the left branch at the root node. Here we reach a leaf of the temporary mu-tree corresponding to records with Income less then 80, but the node we are inserting still contains the split conditions Home Owner: yes and Marital Status: Single. As a consequence, we must expand the corresponding sub-tree, producing the branch highlighted in Figure 15.

In Figure 16 we have represented the final result of the algorithm after the insertion of all leaves from tree T1 into T2.

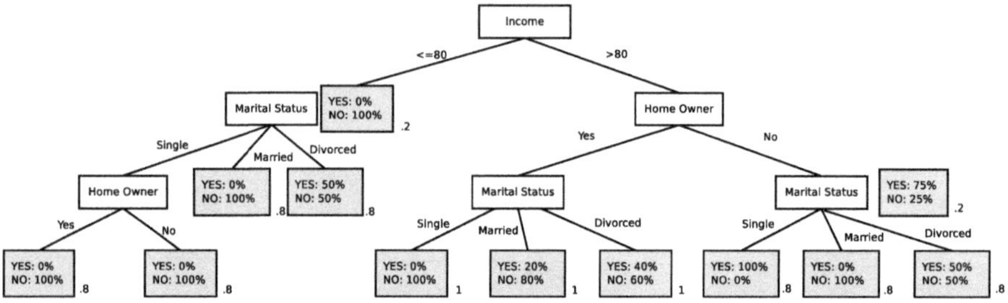

Fig. 16. Merging two alternative trees: final result

6 Concluding Remarks and Future Works

In this paper we have defined a possible worlds semantics for classification tasks using decision trees. The compact representation of multiple alternative trees poses a new problem: dealing with multiple structures inside a single tree. For this reason we introduced a new data structure, called mu-tree, showing that it can be used to this aim. In particular, we provided a constructive proof showing how to merge several probabilistic alternative trees.

While this work can be seen as a theoretical foundation of uncertain decision tree classification, it opens some interesting and challenging practical questions that will be addressed in future works. What is the complexity of building a mu-tree directly from a data set (this corresponds to the lower arrow in Figure 8)? From the point of view of space complexity, in the best case all the alternative trees will have the same leaves (and therefore the same structure), and the corresponding mu-tree will have no additional nodes. On the contrary, if the attributes used to induce two alternative trees are disjoint our algorithm would create a copy of the first tree on each leaf of the second. However, in this specific case different copies of the same tree could be referenced without being explicitly created. The working example presented in this paper shows an intermediate case. Apart from these final considerations, mu-tree induction and its complexity need further investigation, and to this aim we may use ideas developed in related works more focused on tree induction.

As a final remark, notice that given a mu-tree it is possible to convert it into a traditional tree with probabilistic results by pushing probabilities from internal nodes to the leaves. This can be useful for example to compare it with other approaches and verify if they comply with possible worlds semantics. However, keeping intermediate nodes with probabilities can be useful because while classifying a new record we would know that at some point it would have already been classified in some possible worlds, even before reaching a leaf of the mu-tree, and if the accumulated probability is over some user-defined threshold this information can be used to stop the process or to prune the sub-tree. Also this intuition deserves further investigation.

References

1. Aggarwal, C.C.: Managing and Mining Uncertain Data. Springer, Heidelberg (2009)
2. Clark, D.E.: Computational methods for probabilistic decision trees. Computers and Biomedical Research 30 (1997)
3. Quinlan, J.R.: Probabilistic decision trees. Machine Learning: An Artificial Intelligence Approach 3 (1990)
4. Quinlan, J.R.: C4.5: Programs for Machine Learning. Morgan Kaufmann, San Francisco (1993)
5. Hawarah, L., Simonet, A., Simonet, M.: Dealing with missing values in a probabilistic decision tree during classification. In: Sixth International Conference on Data Mining – Workshops (2006)
6. Yuan, Y., Shaw, M.J.: Induction of fuzzy decision trees. Fuzzy Sets and Systems 2 (1995)
7. Tsang, S., Kao, B., Yip, K.Y., Ho, W.S., Lee, S.D.: Decision trees for uncertain data. In: IEEE International conference on Data Engineering (2009)
8. Qin, B., Xia, Y., Li, F.: Dtu: A decision tree for uncertain data. In: Theeramunkong, T., Kijsirikul, B., Cercone, N., Ho, T.-B. (eds.) PAKDD 2009. LNCS, vol. 5476, pp. 4–15. Springer, Heidelberg (2009)
9. Tan, P.N., Steinbach, M., Kumar, V.: Introduction to Data Mining. Addison-Wesley, Reading (2005)

Efficient Policy-Based Inconsistency Management in Relational Knowledge Bases

Maria Vanina Martinez[1], Francesco Parisi[2], Andrea Pugliese[2], Gerardo I. Simari[1], and V.S. Subrahmanian[1]

[1] Department of Computer Science and UMIACS
University of Maryland College Park
College Park, MD 20742, USA
{mvm,gisimari,vs}@cs.umd.edu
[2] Università della Calabria
Via Bucci – 87036 Rende (CS), Italy
{fparisi,apugliese}@deis.unical.it

Abstract. Real-world databases are frequently inconsistent. Even though the users who work with a body of data are far more familiar not only with that data, but also their own job and the risks they are willing to take and the inferences they are willing to make from inconsistent data, most DBMSs force them to use the policy embedded in the DBMS. Inconsistency management policies (IMPs) were introduced so that users can apply policies that they deem are appropriate for data they know and understand better than anyone else. In this paper, we develop an efficient "cluster table" method to implement IMPs and show that using cluster tables instead of a standard DBMS index is far more efficient when less than about 3% of a table is involved in an inconsistency (which is hopefully the case in most real world DBs), while standard DBMS indexes perform better when the amount of inconsistency in a database is over 3%.

1 Introduction

Reasoning about inconsistent knowledge bases (KB) has led to a huge amount of research in AI for over 30 years. Paraconsistent logics were introduced in the 60s, and logics of inconsistency were later developed [4,7,14]. [4] introduced a four valued logic that was used for handling inconsistency in logic programming [7] and extended to the case of bilattices [16]. Later, frameworks such as default logic [27], maximal consistent subsets [3], inheritance networks [30], and others [6,15] were used to generate multiple plausible consistent scenarios (or "extensions"), and methods to draw inferences were developed that looked at truth in all (or some) extensions. Kifer and Lozinskii [24] extended annotated logics of inconsistency developed by Blair and Subrahmanian [7] to handle a full first order case. In the last few years, the problem of managing inconsistent data has deserved much interest in the database community as well [1,9]. These methods clean data and/or provide consistent query answers in the presence of inconsistent data [8,11,23]. For instance, [11] addresses the basic concepts and results of the area of consistent query answering (in the standard model-theoretic sense). They consider universal and binary integrity constraints, denial constraints, functional dependencies, and referential integrity constraints. [8] presents a cost-based framework that

A. Deshpande and A. Hunter (Eds.): SUM 2010, LNAI 6379, pp. 264–277, 2010.

allows finding "good" repairs for databases that exhibit inconsistencies in the form of violations to either functional or inclusion dependencies. They propose heuristic approaches to constructing repairs based on equivalence classes of attribute values; the algorithms presented are based on greedy selection of least repair cost, and a number of performance optimizations are also explored. Furthermore, for conjunctive queries and primary key FDs, the problem of consistent query answering is intractable [11]. Efficient query-rewriting methods such as the one introduced in [1] for quantifier-free conjunctive queries and binary universal constraints, and later extended in [19] to work for a subclass of conjunctive queries in the presence of key constraints, have been proposed. The *conflict hypergraph* proposed in [12] is able to compactly represent the set of repairs of a given database instance and enables tractable computations of consistent query answers of quantifier-free query in the presence of denial constraints [13]. An extended version of the conflict hypergraph, which allows us to capture all repairs w.r.t. a set of universal constraints, and a polynomial-time algorithm for computing consistent answers to ground quantifier-free queries was proposed in [28]. Even though recent work such as [10] shows encouraging results for consistent query answering in the presence of functional dependencies, the knowledge bases they deal with can be considered rather small (approximately 10,000 tuples). The above-cited approaches and the logic-based frameworks in [2,21] assume that tuple insertions and deletions are the basic primitives for repairing inconsistent data. Repairs also consisting of value-update operations were considered in [5,8,17,18,31]. Several important works have also tried to quantify the amount of inconsistency in a database [20,22,25].

In most such past efforts, the "owner" or "administrator" of the KB decides what inconsistency management policy should be used, even though he may have no understanding of the data the user is working with, the user's mission, and/or the risks that the user takes when making decisions on the basis of the data that is present. In recent years, [26] and [29] have tried to get rid of this assumption. [29] develops a method to remove inconsistencies in accordance with an arbitrary objective function (basically by selecting consistent subsets of a KB that maximize a given numeric objective function) – the user gets to specify the objective function. In contrast, [26] develops a formal definition of an *inconsistency management policy* (IMP for short) – IMPs are functions that transform an inconsistent KB into a less (intuitively) inconsistent one. All IMPs were required in [26] to satisfy various reasonable axioms, and were showed to be capable of expressing many very intuitive and useful methods to handle inconsistency that can be combined with classical relational algebra operators. For instance, in a standard employee database with inconsistency about John's salary in 2005 (*e.g.*, via two records saying his salary in 2005 is both 60K and 70K), a tax auditor auditing John might want to assume John's salary is 70K (and issue a letter to John asking why he reported it as 60K), while another auditor auditing the company and questioning its claimed expenses may assume that expense is 60K (and issue a letter to the company demanding why they reported it as 70K). Even from this simple example, we see that users should be able to apply their own IMPs to handle inconsistency.

For settings in which computing repairs and certain answers is not computationally feasible, the IMPs framework proposes a user-based inconsistency management for RKBs where the execution time is polynomial as long as the policy defined by the user is polynomial as well. IMPs can be used to express many different known methods of removing inconsistency such as maximal consistent subsets, resolving inconsistency based on various preference criteria (based on time, reliability of sources, etc.). Moreover, unlike past inconsistency management methods which focused mainly on inconsistency removal, IMPs also allow users to specify that some or all inconsistencies should persist (wholly or partially). As a consequence, IMPs are a powerful tool for end users to express what they wish to do with their data, rather than have a system manager or a DB engine that does not understand their domain problem to dictate how they should handle inconsistencies.

In this paper, we ask: *how can we implement IMPs efficiently*? We focus on knowledge bases that consist of a relational database, together with a very specific set of integrity constraints called *functional dependencies* (FDs for short); these knowledge bases are called *relational KBs* (or RKBs for short). Though the expressive power of RKBs is less general than many other frameworks for inconsistency management, the space of deployed applications in the real world is huge. This paper shows promising empirical results on inconsistency management efforts capable of handling millions of tuples in a reasonable amount of time for RKBs.

Section 2 provides a quick overview of IMPs from [26]. Then, in Section 3, we detail our proposed data structure to index the content of an RKB and support efficient implementation of inconsistency management policies when the relational database part of an RKB is changing rapidly while the set of associated functional dependencies are relatively static. This is the case for most real world applications of the type mentioned above where the structure of the relations (and hence the prevailing set of FDs) does not change much, but the flow of transactions changes very frequently. For instance, the FDs associated with employee or stock databases do not change often, but the rate of change of the employee data or the stock market data is very high. In Section 4 we report an experimental evaluation of our approach. Finally, Section 5 outlines conclusions and possible future work.

2 Background on IMPs

In this section, we provide basic definitions and background on IMPs from [26]. The content of this section is not new.

We assume the existence of relational schemas of the form $S(A_1, \ldots, A_n)$ where the A_i's are attribute names. Each attribute A_i has an associated domain, $dom(A_i)$. A *tuple over* S is a member of $dom(A_1) \times \cdots \times dom(A_n)$, and a finite set R of such tuples is called a *relation*; we will also use $schema(R)$ to denote S. We use $Attr(S)$ to denote the set of all attributes in S. Moreover, we use (i) $t[A_i]$ to denote the value of the A_i attribute of tuple t, and (ii) $t[\mathcal{A}]$, with $\mathcal{A}$ being the ordered set $\{ A'_1, \ldots, A'_p \} \subseteq Attr(S)$, to denote the tuple $(t[A'_1], \ldots, t[A'_p])$.

Given the relational schema $S(A_1, \ldots, A_n)$, a functional dependency *fd* over S is an expression of the form $A'_1 \cdots A'_k \rightarrow A'_{k+1} \cdots A'_m$ where $\{A'_1, \ldots, A'_m\} \subseteq Attr(S)$. A relation R over the schema S *satisfies* the above functional dependency iff $\forall t_1, t_2 \in R$, $t_1[\{ A'_1, \ldots, A'_k \}] = t_2[\{ A'_1, \ldots, A'_k \}] \Rightarrow t_1[\{ A'_{k+1}, \ldots, A'_m \}] = t_2[\{ A'_{k+1}, \ldots, A'_m \}]$. We denote the set of attributes on the left-hand side of a functional dependency *fd* as $LHS(fd)$. Moreover, without loss of generality, we assume

	Aline	FNo	Orig	Dest	Deptime	Arrtime	Source
t_1	BA	299	LHR	IAD	0900	1300	s_1
t_2	BA	299	LGW	IAD	0900	1300	s_2
t_3	BA	299	LGW	WAS	0900	1315	s_3
t_4	AF	100	TLS	LHR	1100	1200	s_1
t_5	AF	100	TLS	LHR	1100	1300	s_5
t_6	AF	117	CDG	LHR	1400	1500	s_6

Fig. 1. Example relation

that every functional dependency *fd* has exactly one attribute on the right-hand side (*i.e.*, $k + 1 = m$) and denote this attribute as $RHS(fd)$. Finally, with a little abuse of notation, we say that *fd* is defined over R. A *relational knowledge base* (RKB for short) is a pair $(R, \mathcal{F})$ where R is a relation and $\mathcal{F}$ is a finite set of FDs over R.

Example 1. Throughout this paper, we use the *Flight* relation in Fig. 1 to illustrate our definitions and results.

This relation has the schema $Flight(Aline, FNo, Orig, Dest, Deptime, Arrtime)$ where $dom(Aline)$ is a finite set of airline codes, $dom(FNo)$ is the set of all flight numbers, $dom(Orig)$ and $dom(Dest)$ are the airport codes of all airports in the world, and $dom(Deptime)$ and $dom(Arrtime)$ is the set of all times expressed in military time (*e.g.*, 1425 hrs or 1700 hours and so forth).[1] In this case, $fd = Aline, FNo \rightarrow Orig$ might be an FD that says that each (Aline,FNo) pair uniquely determines an origin.

The notion of *culprits*, *clusters*, and *inconsistency management policies* from [26] are recapitulated below.

Definition 1 (Culprits and clusters). *Let R be a relation and $\mathcal{F}$ a set of functional dependencies. A* culprit *is a minimal set $c \subseteq R$ that does not satisfy $\mathcal{F}$. Moreover, let $culprits(R, \mathcal{F})$ be the set of all culprits in R w.r.t. $\mathcal{F}$. Given two culprits $c, c' \in culprits(R, \mathcal{F})$, we write $c \bigtriangleup c'$ iff $c \cap c' \neq \emptyset$. Let $\bigtriangleup^*$ be the reflexive transitive closure of relation $\bigtriangleup$; a* cluster *is a set $cl = \bigcup_{c \in e} c$ where e is an equivalence class of $\bigtriangleup^*$. We use $clusters(R, \mathcal{F})$ to denote the set of all clusters in R w.r.t. $\mathcal{F}$.*

With a little abuse of notation, we write $clusters(R, fd)$ and $culprits(R, fd)$ to denote the sets $clusters(R, \{fd\})$ and $culprits(R, \{fd\})$, respectively.

Example 2. It is easy to see that $\{t_1, t_2\}$ and $\{t_1, t_3\}$ from Example 1 are all culprits w.r.t. the RKB $(R, \{fd\})$. The only cluster is $\{t_1, t_2, t_3\}$.

Definition 2. *An* inconsistency management policy *(IMP for short) for a relation R w.r.t. a functional dependency fd over R is a function γ_{fd} that takes a relation R and returns another relation $R' = \gamma_{fd}(R)$ that satisfies the following axioms:*

[1] For the sake of simplicity, we are not considering cases where flights arrive on the day after departure, etc. – these can be accommodated through an expanded schema.

– *If* $t \in R - \bigcup_{c \in culprits(R,fd)} c$, *then* $t \in R'$ *(tuples that do not belong to any culprit cannot be eliminated or changed).*
– *If* $t' \in R' - R$, *then there exists a cluster* cl *and a tuple* $t \in cl$ *such that for each attribute* A *not appearing in fd,* $t[A] = t'[A]$ *(every tuple in* R' *must somehow be linked to a tuple in* R*).*
– $|culprits(R,fd)| \geq |culprits(R', fd)|$ *(the* **IMP** *cannot increase the number of culprits).*
– $|R| \geq |R'|$ *(the* **IMP** *cannot increase the cardinality of the relation).*

[26] shows that IMPs can be used to express many different types of policies. For instance, in our *flight* example, IMPs can be used to express policies such as ignoring anything said by s_2 when an inconsistency occurs involving a tuple whose source is s_2, or always choosing the latest arrival time and the earliest departure time when an inconsistency arises. For our complexity results, we will assume that all IMPs can be computed in polynomial time w.r.t. the size of the clusters on which they are applied.

Suppose each $fd \in \mathcal{F}$ has an associated IMP which specifies how to manage the inconsistencies in the relation with respect to that dependency. We assume that the user or system manager also specifies a partial ordering $\leq_{\mathcal{F}}$ on the FDs, specifying their relative importance. Let $Tot_{\leq_{\mathcal{F}}}(\mathcal{F})$ be the set of all possible total orderings of FDs w.r.t. $\leq_{\mathcal{F}}$: this can be obtained by topological sorting. Given a relation R, a set of functional dependencies $\mathcal{F}$, a partial ordering $\leq_{\mathcal{F}}$, and an total order $o = \langle fd_1, \ldots, fd_k \rangle \in Tot_{\leq_{\mathcal{F}}}(\mathcal{F})$, a *multi-dependency* IMP for R w.r.t. o and $\mathcal{F}$ is a function $\mu^o_{\mathcal{F}}$ from a relation R to a relation $\gamma_{fd_k}(\ldots \gamma_{fd_2}(\gamma_{fd_1}(R))\ldots)$, where $\gamma_{fd_1}, \ldots, \gamma_{fd_k}$ are the inconsistency management policies associated with $fd_1, \ldots, fd_k$, respectively.

In [26], three semantics are defined for inconsistency management policies on an RKB with multiple FDs. Suppose $(R, \mathcal{F})$ is an RKB and suppose $\leq_{\mathcal{F}}$ is a partial ordering on $\mathcal{F}$. In the *deterministic* semantics, the user chooses a specific total ordering o in $Tot(\leq_{\mathcal{F}})$ and a tuple t is in $\mu^o_{\mathcal{F}}$ iff $t \in \gamma_{fd_k}(\ldots \gamma_{fd_2}(\gamma_{fd_1}(R))\ldots)$. Alternatively, in the *brave* semantics (cautious semantics, resp.), we must check if there exists at least one ordering (all ordering, resp.) that satisfies the previous condition.

[26] also shows the intractability of applying multi-dependency IMPs in the general case under the cautious semantics. Thus, in this paper we focus on the management of RKBs where all IMPs are defined w.r.t. one functional dependency.

3 Applying IMPs

In this section, we will address the following question: *how can we implement* IMPs *efficiently*? The heart of the problem of applying an IMP lies in the fact that the clusters must be identified. In the following, we start by discussing how classical DBMS indexes can be used to carry out these operations, and then we present a new data structure that can be used to identify the set of clusters more efficiently: the *cluster table*.

3.1 Using DBMS-Based Indexes

A basic approach to the problem of identifying clusters is to directly define one DBMS index (DBMSs in general provide hash indexes, B-trees, etc.) for each functional dependency's left-hand side. Assuming that the DBMS index used allows $O(1)$ access to individual tuples, this approach has several advantages:

– Takes advantage of the highly optimized implementation of operations which is provided by the DBMS. Insertion, deletion, lookup, and update are therefore all inexpensive operations in this case.

– Identifying a single cluster (for given values for the left hand side of a certain functional dependency) can be done by issuing a simple query to the DBMS, which can be executed in $O(\max_{cl \in clusters(R,fd)} |cl|)$ time, in the (optimistic) assumption of $O(1)$ time for accessing a single tuple. However, the exact cost depends on the particular DBMS implementation, especially that of the query planner.

– Identifying all clusters can be done in two steps, each in time in $O(|R|)$: *(i)* Issue a query with a GROUP BY on the left hand side of the functional dependency of interest and count the number of tuples associated with each one; and *(ii)* Take those LHS values with a count ≥ 1 and obtain the cluster. This can be easily done in a single nested query.

There is, however, one important disadvantage to this approach: clusters must be identified time and time again and are not explicitly maintained. This means that, in situations where a large portion of the table constitutes clean tuples (and we therefore have few clusters), the $O(|R|)$ operations associated with obtaining all clusters become quite costly because they may entail actually going through the entire table.

3.2 Cluster Table

We now introduce a novel data structure that we call *cluster table*. For each $fd \in \mathcal{F}$, we maintain a cluster table focused on that one dependency. When relation R gets updated, each FD's associated cluster table must be updated. This section defines the cluster table associated with one FD, how that cluster table gets updated, and how it can be employed to efficiently apply an IMP. Note that even though we do not cover the application of multiple policies, we assume that for each relation a set of cluster tables associated with $\mathcal{F}$ must be maintained. Therefore, when a policy w.r.t. an FD is applied to a relation, the cluster tables corresponding to other FDs in $\mathcal{F}$ might need to be updated as well.

Definition 3 (Tuple group). *Given a relation R and a set of attributes $\mathcal{A} \subseteq$ Attr$(schema(R))$, a* tuple group *w.r.t. $\mathcal{A}$ is a maximal set $g \subseteq R$ such that $\forall t, t' \in g, t[\mathcal{A}] = t'[\mathcal{A}]$.*

For instance, in the case of Example 1, suppose $\mathcal{A} = \{Aline, FNo\}$. Then $\{t_1, t_2, t_3\}$ is a group, as are $\{t_4, t_5\}$ and $\{t_6\}$ – but $\{t_4, t_5\}$ and $\{t_6\}$ are not clusters. We use *groups*$(R, \mathcal{A})$ to denote the set of all tuple groups in R w.r.t. $\mathcal{A}$, and M to denote the maximum size of a group, i.e., $M = \max_{g \in groups(R,\mathcal{A})} |g|$. The following result shows that all clusters are groups, but not viceversa.

Proposition 1. *Given a relation R and a functional dependency fd defined over R, clusters$(R,fd) \subseteq$ groups$(R, LHS(fd))$.*

The reason a group may not be a cluster is because the FD may be satisfied by the tuples in the group. In the cluster table approach, we store all groups associated with a table together with an indication of whether the group is a cluster or not. When tuples are inserted into the RKB, or when they are deleted or modified, the cluster table can be easily updated using procedures we will present shortly.

Definition 4 (Cluster table). *Given an RKB* (R,fd)*, a* cluster table *w.r.t.* (R,fd)*, denoted as* $\mathsf{CT}(R,fd)$ *is a pair* (G, D) *where:*
– G *is a set containing, for each tuple group* $g \in groups(R, LHS(fd))$ *s.t.* $|g| > 1$*, a tuple of the form* $(v, \overrightarrow{g}, flag)$*, where:*

- $v = t[LHS(fd)]$ *where* $t \in g$*;*
- $\overrightarrow{g}$ *is a set of pointers to the tuples in* g*;*
- $flag$ *is* $true$ *iff* $g \in clusters(R,fd)$*,* $false$ *otherwise.*

– D *is a set of pointers to the tuples in* $R \setminus \bigcup_{g \in groups(R,LHS(fd)),|g|>1} g$*;*
– *both* G *and* D *are sorted by* $LHS(fd)$*.*

The following example shows the contents of the cluster table for our running example.

Example 3. The cluster table associated with our flight example has the following form: $G = \{((AF, 100), \{\overrightarrow{t_4}, \overrightarrow{t_5}\}, false), ((BA, 299), \{\overrightarrow{t_1}, \overrightarrow{t_2}, \overrightarrow{t_3}\}, true)\}$, $D = \{\overrightarrow{t_6}\}$. A graphical representation of the table is reported in Fig. 2.

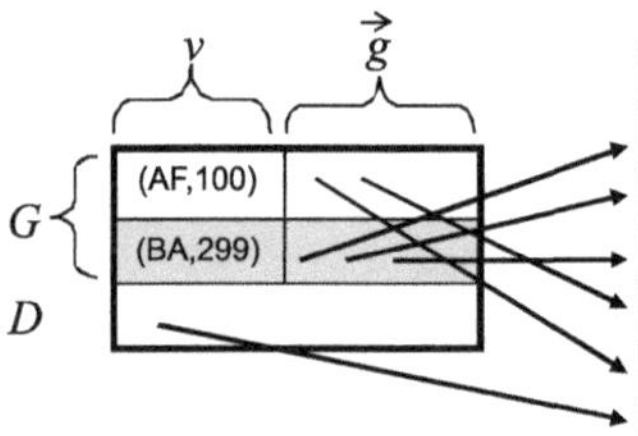

Aline	FNo	Orig	Dest	Deptime	Arrtime
BA	299	LHR	IAD	0900	1300
BA	299	LGW	IAD	0900	1300
BA	299	LGW	WAS	0900	1315
AF	100	TLS	LHR	1100	1200
AF	100	TLS	LHR	1100	1300
AF	117	CDG	LHR	1400	1500

Fig. 2. Cluster table for the relation of Example 1. The shaded row in G is flagged as a cluster.

A cluster table can be built through a simple procedure that, given an FD *fd*, identifies the clusters w.r.t. *fd* in a relation R by first sorting the tuples in R according to the left-hand side of *fd*, then performing a linear scan of the ordered list of tuples.

Proposition 2. *Given a relation* R *and a functional dependency fd defined over* R*, the worst-case running time for building* $\mathsf{CT}(R,fd)$ *is* $O(|R| \cdot log|R|)$*.*

Maintaining cluster tables. We now study how to update a cluster table for the RKB (R,fd) under three kinds of updates: (i) when a tuple is inserted into R, (ii) when a tuple is deleted from R, and (iii) when a tuple in R is modified.

Insertion. Fig. 3 describes an algorithm for updating $\mathsf{CT}(R,fd)$ after inserting a new tuple t in R. The algorithm starts by checking whether t belongs to a tuple group already present in R (line 1) and, if this is the case, it (*i*) adds $\overrightarrow{t}$ to the corresponding entry in G and (*ii*) checks if the group is a cluster (lines 3–6). If t does not already belong to a tuple group, the algorithm checks whether it forms a new group when paired to a tuple pointed to by D (line 8). If this is the case, it adds the new group to G (lines 9–11); otherwise, it just adds $\overrightarrow{t}$ to D (line 13).

The following example shows how this algorithm works.

```
Algorithm CT-insert
Input: Relation R, functional dependency fd, cluster table (G, D) = CT(R, fd), new tuple t
if ∃(t[LHS(fd)], →g, flag) ∈ G then
  add →t to →g
  if flag = false then
    pick the first →t' from →g
    if t[RHS(fd)] ≠ t'[RHS(fd)] then
      flag ← true
  end-algorithm
if ∃→t' ∈ D s.t. t[LHS(fd)] = t'[LHS(fd)] then
  remove →t' from D
  add (t[LHS(fd)], {→t, →t'}, flag) to G where
      flag = true iff t[RHS(fd)] ≠ t'[RHS(fd)]
  end-algorithm
add →t to D
```

Fig. 3. Updating a cluster table after inserting a tuple

Example 4. Consider the cluster tables for our flight example (Example 3). Suppose a new tuple $t = (AF, 100, CDG, LHR, 1100, 1200)$ is inserted into the relation *Flight* of Example 1. Algorithm CT-insert first adds $\overrightarrow{t}$ to set $\overrightarrow{g}$ in the first row of the cluster table. Then, it picks $\overrightarrow{t_4}$ and, since $t[RHS(fd)] \neq t_4[RHS(fd)]$, it assigns $true$ to the $flag$ of the second row. Now suppose that the new tuple is $t = (AF, 117, CDG, LHR, 1400, 1500)$. Here, the algorithm removes $\overrightarrow{t_6}$ from D and, since $t[RHS(fd)] = t_6[RHS(fd)]$, adds a new row $((AF, 117), \{\overrightarrow{t}, \overrightarrow{t_6}\}, false)$ to G.

The following results ensure the correctness and complexity of CT-insert.

Proposition 3. *Algorithm CT-insert terminates and correctly computes* $\mathrm{CT}(R \cup \{t\}, fd)$.

Proposition 4. *The worst-case running time of CT-insert is* $O(\log(|G|) + \log(|D|))$.

Deletion. Fig. 4 presents an algorithm for updating a cluster table $\mathrm{CT}(R, fd)$ after deleting a tuple t from R. The algorithm checks whether t belongs to a tuple group (line 1) and, if this is the case, it removes $\overrightarrow{t}$ from the corresponding entry in G and checks if the group is a cluster (lines 3–4). Otherwise, it just removes $\overrightarrow{t}$ from D (line 6).

```
  Algorithm CT-delete
  Input: Relation R, functional dependency fd, cluster table (G, D) = CT(R, fd), deleted tuple t
1 if ∃(t[LHS(fd)], →g, flag) ∈ G then
2   remove →t from →g
3   if |→g| = 1 then
4     remove (t[LHS(fd)], →g, flag) from G
5     add →g to D
6   else if flag = true and ∄→t1, →t2 ∈ →g s.t.
        t1[RHS(fd)] ≠ t2[RHS(fd)] then
7     flag ← false
8   end-algorithm
9 remove →t from D
```

Fig. 4. Updating a cluster table after deleting a tuple

Example 5. Consider the cluster table in our flight example (Example 3). Suppose tuple t_5 is removed from the relation *Flight* of Example 1. Algorithm CT-delete first removes $\vec{t_5}$ from set $\vec{g}$ in the first row of the cluster table. Then, since the group has been reduced to a singleton, it moves $\vec{t}$ to set D and removes the first row from G. Now suppose that tuple t_1 is removed from the relation *Flight*. In this case, the algorithm removes $\vec{t_1}$ from set $\vec{g}$ in the second row of the cluster table. As the group is no longer a cluster (t_2 and t_3 agree on $Orig$), the algorithm sets the corresponding $flag$ to $false$.

The following results specify the correctness and complexity of the CT-delete algorithm.

Proposition 5. *CT-delete terminates and correctly computes* CT$(R \setminus \{t\}, fd)$.

Proposition 6. *The worst-case running time of* CT-delete *is* $O(\log|G| + \log|D| + M)$.

Update. Fig. 5 shows an algorithm for updating a cluster table CT(R, fd) after updating a tuple t to t' in R (clearly, $\vec{t} = \vec{t'}$).

```
   Algorithm CT-update
   Input: Relation R, functional dependency fd, cluster table (G, D) = CT(R, fd), tuples t, t'
1  if t[LHS(fd)] = t'[LHS(fd)] and t[RHS(fd)] = t'[RHS(fd)] then
2     end-algorithm
3  if t[LHS(fd)] = t'[LHS(fd)] and ∃(t[LHS(fd)], g⃗, flag) ∈ G then
4     if flag = true and ∄t⃗'' ∈ g⃗ s.t. t'[RHS(fd)] ≠ t''[RHS(fd)] then
5        flag ← false
6        end-algorithm
7     if flag = false then
8        pick the first t⃗'' from g⃗
9        if t'[RHS(fd)] ≠ t''[RHS(fd)] then
10          flag ← true
11    end-algorithm
12 if t[LHS(fd)] = t'[LHS(fd)] then
13    end-algorithm
14 execute CT-delete with t
15 execute CT-insert with t'
```

Fig. 5. Updating a cluster table after updating a tuple

The algorithm first checks whether anything regarding *fd* has changed in the update (lines 1–2). If this is the case and t belongs to a group (line 3), the algorithm checks if the group was a cluster whose inconsistency has been removed by the update (lines 4–6) or the other way around (lines 7-11). At this point, as t does not belong to any group and the values of the attributes in the LHS of *fd* did not change, the algorithm ends (lines 12–13) because this means that the updated tuple simply remains in D. If none of the above conditions apply, the algorithm simply calls CT-delete and then CT-insert.

Example 6. Consider the cluster table for the flight example (Example 3). Suppose the value of the $Orig$ attribute of tuple t_1 is changed to LGW. Tuple t_1 belongs to the group represented by the second row in G, which is a cluster. However, after the update to t, no two tuples in the group have different values of $Orig$, and thus Algorithm CT-update changes the corresponding $flag$ to $false$. Now suppose the value of the $Orig$ attribute of tuple t_5 is changed to LGW. In this case, the algorithm picks $\vec{t_4}$ and, since $t_4[RHS(fd)] \neq t_5[RHS(fd)]$, it assigns $true$ to the $flag$ of the second row.

The following results specify the correctness and complexity of CT-update.

Proposition 7. *CT-update terminates and correctly computes* $CT(R \setminus \{t\} \cup \{t'\}, fd)$.

Proposition 8. *The worst-case running time of CT-update is* $O(\log |G| + \log |D| + M)$.

Applying cluster tables to compute IMPs. We now show how to use the cluster table to compute an IMP over an RKB with an FD. Fig. 6 shows the proposed algorithm. For each cluster in G, procedure $apply(\gamma_{fd}, \overrightarrow{g})$ applies policy γ_{fd} to the set of tuples $\overrightarrow{g}$. As a consequence of this, dependending on the nature of the policy, some tuples in $\overrightarrow{g}$ might be deleted or updates according to what the policy determines. Therefore, $changes$ keeps the list on changes performed by the policy in a cluster. After applying the IMP, the algorithm updates the cluster table in order to preserve its integrity; it checks whether all the inconsistencies have been removed (lines 3–4) and whether the cluster has been reduced to a single tuple (lines 5–7). The first check is necessary so the flag can be updated and future applications of a policy do not need to consider that group if it is no longer a cluster; in the latter case, the pointer is moved from G to D since that tuple is no longer in conflict with any other tuple w.r.t. fd. Finally, the changes performed by the policy are propagated to the rest of the cluster tables for relation R. This is, for every other functional dependency in $\mathcal{F}$ either *CT-delete* or *CT-update* are called on the corresponding cluster table depending on the natura of the change.

```
Algorithm CT-applyIMP
Input: Relation R, functional dependency fd, cluster table (G, D) = CT(R, fd), IMP γ_fd
1  for all (v, g⃗, true) ∈ G
2    changes ← apply(γ_fd, g⃗)
3    if ∄ t⃗1, t⃗2 ∈ g⃗ s.t. t1[RHS(fd)] ≠ t2[RHS(fd)] then
4      flag ← false
5    if |g⃗| = 1 then
6      remove (v, g⃗, true) from G
7      add g⃗ to D
8    for all fd' ∈ F and fd' ≠ fd
9      let (G', D') be the cluster table associated with fd'
10     for all change ch ∈ changes
11       if ch = delete(t, R) then CT-delete(R, fd', (G', D'), t)
12       if ch = update(t, t', R) then CT-update(R, fd', (G', D'), t, t')
```

Fig. 6. Applying an IMP using a cluster table

The following results show the correctness and complexity of CT-applyIMP.

Proposition 9. *CT-applyIMP terminates and correctly computes* $\gamma_{fd}(R)$ *and* $CT(\gamma_{fd}(R), fd)$.

Proposition 10. *The worst-case time complexity of CT-applyIMP is* $O(|G| \cdot (poly(M) + log|D| + |\mathcal{F}| \cdot |M| \cdot (log|G'| + log|D'| + M')))$*, where* G' *(resp.* D'*) is the largest set* G *(resp.* D*) among all cluster tables, and* M' *is the maximum* M *among all cluster tables.*

It is important to note that algorithm CT-applyIMP assumes that the application of a policy can be done on a cluster-by-cluster basis, *i.e.*, applying a policy to a relation has the same effect as applying the policy to every cluster independently – a wide variety of policies satisfy this property. Moreover, in cases where the actions taken on one cluster depend on other clusters, it is possible the prove that there always exists and equivalent policy that can be applied on a cluster-by-cluster basis. A formal treatment of this, including the definition of a language to express complex policies, will be developed in future work.

In the next section we will present the results of our preliminary experimental evaluation of cluster tables vs. the DBMS-based approach discussed above.

4 Experimental Evaluation

Our experiments measure the running time performance of applying IMPs using cluster tables; moreover, we analyzed the required storage space on disk. We compared these measures with those obtained through the use of a heavily optimized DBMS-based index. The parameters varied were the size of the database and the amount of inconsistency present.

Our prototype JAVA implementation consists of roughly 9,000 lines of code, relying on Berkeley DB Java Edition[2] database for implementation of our disk-based index structures. The DBMS-based index was implemented on top of PostgreSQL version 7.4.16; a B-Tree index (PostgreSQL does not currently allow hash indexes on more than one attribute) was defined for the LHS of each functional dependency. All experiments were run on multiple multi-core Intel Xeon E5345 processors at 2.33GHz, 8GB of memory, running the Scientific Linux distribution of the GNU/Linux operating system (our implementation makes use of only 1 processor and 1 core at a time, the cluster is used for multiple runs). The numbers reported are the result of averaging between 5 and 50 runs to minimize experimental error. All tables had 15 attributes and 5 functional dependencies associated with them. Tables were randomly generated with a certain percentage of inconsistent tuples[3] divided in clusters of 5 tuples each. The cluster tables were implemented on top of BerkeleyDB; for each table, both G and D were kept in the hash structures provided by BerkeleyDB.

Fig. 7 shows comparisons of running times when varying the size of the database and the percentage of inconsistent tuples. The operation carried out was the application of a value-based policy that replaces the RHS of tuples in a cluster with the median value in the cluster; this policy was applied to all clusters in the table.

We can see that the amount of inconsistency clearly affected the cluster table-based approach more than it did the DBMS-based index. For the runs with 0.1% and 1% inconsistent tuples, the cluster table clearly outperformed the DBMS-based approach – in the case of a database with 2 million tuples and 0.1% inconsistency, applying the policy took 2.12 seconds with the cluster table and 27.56 seconds with the DBMS index. This is due to the fact that relatively few clusters are present and thus many tuples can be ignored, while the DBMS index must process all of them. Further experiments with 0.1% inconsistency showed that the cluster table approach remains quite scalable over

[2] `http://www.oracle.com/database/berkeley-db/je/index.html`

[3] Though of course tuples themselves are not inconsistent, we use this term to refer to tuples that are involved in some inconsistency, *i.e.*, belong to a cluster.

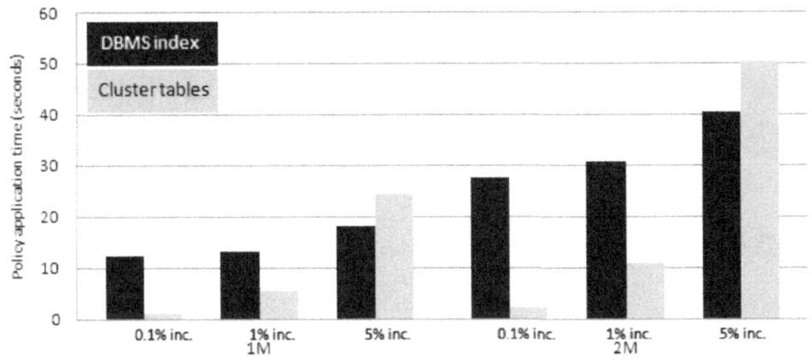

Fig. 7. Average policy application times for *(i)* 1M and 2M tuples and *(ii)* varying percentage of inconsistency

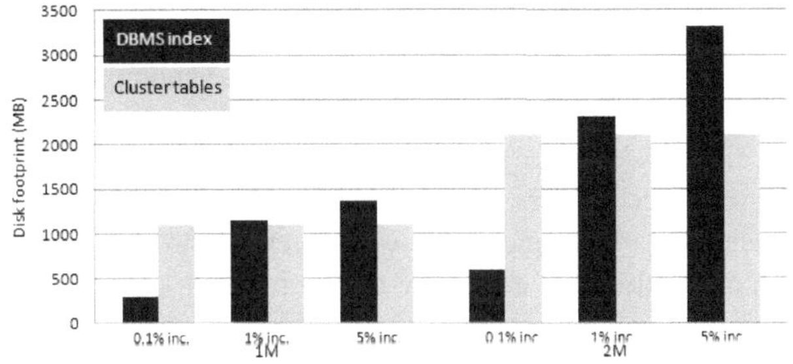

Fig. 8. Disk footprint for *(i)* 1M and 2M tuples and *(ii)* varying percentage of inconsistency

much larger databases, while the performance of the DBMS index degrades quickly – for a database with 5 millon tuples, applying the policy took 3.7 seconds with the cluster table and 82.9 seconds with the DBMS index. Overall, our experiments suggest that under about 3% inconsistency the cluster table approach is able to provide much better performance in the application of IMPs.

Fig. 8 shows comparisons of disk footprints when varying the size of the database and the percentage of inconsistent tuples – note that the numbers reported include the sizes of the structures needed to index all of the functional dependencies used in the experiments. In this case, the cluster table approach provides a smaller footprint with respect to the DMBS index in the case of higher inconsistency. For instance, in the case of a database with 2 million tuples and 5% inconsistency, the cluster tables size was 63% of that of the DBMS index.[4]

In performing update operations, the cluster table approach performed at most 1 order of magnitude worse than the DBMS index. This result is not surprising since these kinds of operations are the specific target of DBMS indexes, which are thus able to provide extremely good performance in these cases (*e.g.*, 2 seconds for 1,000 update operations over a database containing 2 million tuples).

[4] In addition, we point out that our current implementation is not yet optimized for an intelligent use of disk space, as the DBMS is.

Overall, our evaluation showed that the cluster table approach is capable of providing very good performance in scenarios where an 1%-3% inconsistency is present, which are fearly common [8]. For lower inconsistency, the rationale behind this approach becomes even more relevant and makes the application of IMPs much faster.

5 Conclusions and Future Work

In this paper we have extended previous work on policy-based inconsistency management by proposing a new approach that can be used to support this process. We focused on indexing the content of RKBs assuming that their content is dynamic while the constraints defined over them are relatively static. We developed a theoretical analysis of our approach as well as a preliminary empirical study of its performance on synthetic data. Our study showed that the amount of inconsistency present in the table plays an important role in the performance of the approach being adopted. This work is the first to report on inconsistency management efforts capable of handling millions of tuples in a reasonable amount of time for RKBs. Future work will involve carrying out a more extensive empirical analysis varying other parameters that may have an important impact, such as the size of the clusters, the type of policy being applied, etc. Moreover, we will study the translation of complex policies (expressed through a specific language) into ones that can be applied on a cluster-by-cluster basis.

Acknowledgments. The first, fourth, and fifth authors were funded in part by AFOSR grant FA95500610405, ARO grant W911NF0910206 and ONR grant N000140910685. The second and third authors were supported by PRIN grant "EASE", funded by the Italian Ministry for Education, University and Research.

References

1. Arenas, M., Bertossi, L.E., Chomicki, J.: Consistent query answers in inconsistent databases. In: PODS, pp. 68–79 (1999)
2. Arenas, M., Bertossi, L.E., Chomicki, J.: Answer sets for consistent query answering in inconsistent databases. TPLP 3(4-5), 393–424 (2003)
3. Baral, C., Kraus, S., Minker, J.: Combining multiple knowledge bases. TKDE 3(2), 208–220 (1991)
4. Belnap, N.: A useful four valued logic. Modern Uses of Many Valued Logic, 8–37 (1977)
5. Bertossi, L.E., Bravo, L., Franconi, E., Lopatenko, A.: The complexity and approximation of fixing numerical attributes in databases under integrity constraints. Information Systems 33(4-5), 407–434 (2008)
6. Besnard, P.: Remedying inconsistent sets of premises. Int. J. Approx. Reasoning 45(2), 308–320 (2007)
7. Blair, H.A., Subrahmanian, V.S.: Paraconsistent logic programming. Theor. Comp. Sci. 68(2), 135–154 (1989)
8. Bohannon, P., Fan, W., Flaster, M., Rastogi, R.: A cost-based model and effective heuristic for repairing constraints by value modification. In: SIGMOD 2005, pp. 143–154 (2005)
9. Calì, A., Lembo, D., Rosati, R.: On the decidability and complexity of query answering over inconsistent and incomplete databases. In: PODS, pp. 260–271 (2003)

10. Caniupán Marileo, M., Bertossi, L.E.: The consistency extractor system: Answer set programs for consistent query answering in databases. Data Knowl. Eng. 69(6), 545–572 (2010)
11. Chomicki, J.: Consistent query answering: Five easy pieces. In: Schwentick, T., Suciu, D. (eds.) ICDT 2007. LNCS, vol. 4353, pp. 1–17. Springer, Heidelberg (2006)
12. Chomicki, J., Marcinkowski, J.: Minimal-change integrity maintenance using tuple deletions. Inf. Comp. 197(1-2), 90–121 (2005)
13. Chomicki, J., Marcinkowski, J., Staworko, S.: Computing consistent query answers using conflict hypergraphs. In: Proc. 13th ACM Conf. on Information and Knowledge Management (CIKM), pp. 417–426 (2004)
14. da Costa, N.: On the theory of inconsistent formal systems. N. Dame J. of Formal Logic 15(4), 497–510 (1974)
15. de Saint-Cyr, F.D., Prade, H.: Handling uncertainty and defeasibility in a possibilistic logic setting. Int. J. Approx. Reasoning 49(1), 67–82 (2008)
16. Fitting, M.: Bilattices and the semantics of logic programming. J. of Log. Prog. 11(1-2), 91–116 (1991)
17. Flesca, S., Furfaro, F., Parisi, F.: Querying and repairing inconsistent numerical databases. ACM Trans. Database Syst. 35 (2) (2010)
18. Franconi, E., Palma, A.L., Leone, N., Perri, S., Scarcello, F.: Census data repair: a challenging application of disjunctive logic programming. In: LPAR, pp. 561–578 (2001)
19. Fuxman, A., Miller, R.J.: First-order query rewriting for inconsistent databases. J. Comput. Syst. Sci. 73(4), 610–635 (2007)
20. Grant, J., Hunter, A.: Measuring inconsistency in knowledgebases. J. of Intel. Inf. Syst. 27(2), 159–184 (2006)
21. Greco, G., Greco, S., Zumpano, E.: A logical framework for querying and repairing inconsistent databases. IEEE TKDE 15(6), 1389–1408 (2003)
22. Hunter, A., Konieczny, S.: Approaches to measuring inconsistent information. In: Bertossi, L., Hunter, A., Schaub, T. (eds.) Inconsistency Tolerance. LNCS, vol. 3300, pp. 191–236. Springer, Heidelberg (2005)
23. Jermyn, P., Dixon, M., Read, B.J.: Preparing clean views of data for data mining. In: ERCIM Work. on Database Res., pp. 1–15 (1999)
24. Kifer, M., Lozinskii, E.L.: A logic for reasoning with inconsistency. J. of Autom. Reas. 9(2), 179–215 (1992)
25. Lozinskii, E.L.: Resolving contradictions: A plausible semantics for inconsistent systems. J. of Autom. Reas. 12(1), 1–31 (1994)
26. Martinez, M.V., Parisi, F., Pugliese, A., Simari, G.I., Subrahmanian, V.S.: Inconsistency management policies. In: KR, pp. 367–377 (2008)
27. Reiter, R.: A logic for default reasoning. Artif. Intel. 13(1-2), 81–132 (1980)
28. Staworko, S., Chomicki, J.: Consistent query answers in the presence of universal constraints. Inf. Syst. 35(1), 1–22 (2010)
29. Subrahmanian, V.S., Amgoud, L.: A general framework for reasoning about inconsistency. In: IJCAI, pp. 599–504 (2007)
30. Touretzky, D.: The mathematics of inheritance systems. Morgan Kaufmann, San Francisco (1986)
31. Wijsen, J.: Database repairing using updates. ACM TODS 30(3), 722–768 (2005)

Modelling Probabilistic Inference Networks and Classification in Probabilistic Datalog

Miguel Martinez-Alvarez and Thomas Roelleke

Queen Mary, University of London
{miguel,thor}@eecs.qmul.ac.uk

Abstract. Probabilistic Graphical Models (PGM) are a well-established approach for modelling uncertain knowledge and reasoning. Since we focus on inference, this paper explores Probabilistic Inference Networks (PIN's) which are a special case of PGM. PIN's, commonly referred as Bayesian Networks, are used in Information Retrieval to model tasks such as classification and ad-hoc retrieval.

Intuitively, a probabilistic logical framework such as Probabilistic Datalog (PDatalog) should provide the expressiveness required to model PIN's. However, this modelling turned out to be more challenging than expected, requiring to extend the expressiveness of PDatalog. Also, for IR and when modelling more general tasks, it turned out that 1st generation PDatalog has expressiveness and scalability bottlenecks. Therefore, this paper makes a case for 2nd generation PDatalog which supports the modelling of PIN's. In addition, the paper reports the implementation of a particular PIN application: Bayesian Classifiers to investigate and demonstrate the feasibility of the proposed approach.

1 Introduction

1.1 Motivation and Background

Nowadays, there is a big productivity challenge when designing customisable IR systems which are usually developed for specific cases, having to rewrite a high portion of the original code for other purposes. This problem in IR is comparable with that happened in the Software Industry, when Software Engineering evolved from programs focused in one specific context to the developing of frameworks for general tasks that could be adapted for specific ones. We propose a generic module for PIN's based on probabilistic logic. This concept would provide a generic framework for any task that requires its use. Furthermore, thanks to the logical implementation, the functionality would be defined in a high-level, making it more understandable. In addition, we show the implementation of some classifiers and proof that the module could be adapted for specific cases. PIN's are a mechanism that allows modelling knowledge and reasoning.

Figure 1 presents the famous example [10] about burglary, earthquake, and alarm. It shows the network expressing that a burglary implies an alarm, and so does an earthquake. Indeed, there is also an implication from earthquake to burglary (since burglaries tend to happen during and after earthquakes). This arc increases the complexity of the network, since the event earthquake implies alarm via two different paths.

A. Deshpande and A. Hunter (Eds.): SUM 2010, LNAI 6379, pp. 278–291, 2010.

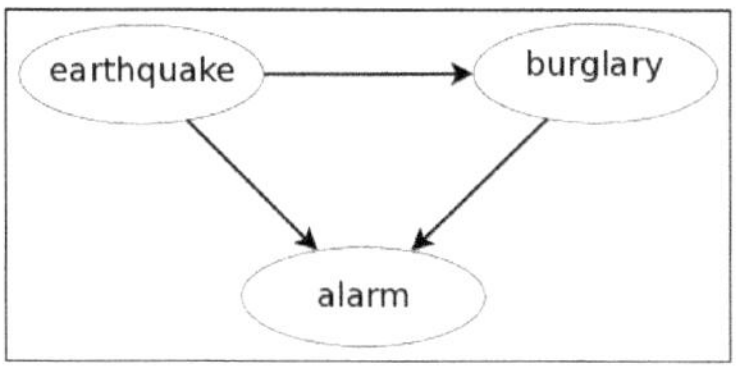

Fig. 1. PIN representing the hypothesis of an alarm being triggered in case of burglary/earthquake

The theory around PIN's has been influential in many domains, including artificial intelligence and information retrieval. However, the latter tend to use more complex representations. This is because in IR the application is large-scale in the sense that there is a PIN for each document to retrieve. [17] describe the dominant approach of how IR is modelled in a PIN.

Probabilistic inference also impacted the logical approach to IR [18,19,6], and the approach to utilise a probabilistic version of Datalog to model IR [5].

Intuitively, PDatalog should allow to model a PIN and we report in this paper that this intuition is true to a certain degree, but in "real-world" applications, the modelling is usually much more complex than initially expected.

Moreover, since the PIN theory is a natural candidate to model classification, we investigate the modelling of classifiers in 2nd generation PDatalog, reporting results on expressiveness, processing issues, and quality measures.

1.2 Structure and Contributions

The remainder of this paper is structured as follows: Section 2 reviews PIN's and their application in IR. Section 3 reviews PDatalog, reflecting on the 1st generation [5], and the 2nd generation (which incorporates the relational Bayes [13]). Section 4 binds the sections on PIN's and PDatalog: Modelling PIN's in PDatalog. The main contributions in this first part of the paper are the introduction and discussion of the 2nd generation PDatalog and the modelling of PIN's. Then, Section 5 presents a specific application of PIN's: Bayesian classifiers. Section 6 focuses on the modelling of Bayesian classifiers in PDatalog. The contribution of the sections on classification is to study how PDatalog copes with this concrete task. Finally, section 7 presents study of feasibility and experiments, showing that the implementation achieves quality levels that could be expected for other approaches.

2 Probabilistic Inference Networks

Probabilistic Inference Networks (PIN's), also referred as Bayesian Networks, are one of the most established technique for different IR and AI tasks. PIN's are used for representing conditional probabilities between different events. The definition of a PIN can be formulate as:

Definition 1. *A PIN is a directed acyclic graph (DAG). Let* (N, V) *denote a PIN where* N *is the set of nodes and* V *is a set of arcs, where an arc is a pair* (n_i, n_j)*, and* n_i *and*

n_j are nodes. For each node, there is a so-called conditional dependence probability (CDP) matrix. This matrix represents the probability $P(n_i|parents_i)$.

This technique allows to represent and use for reasoning conditional probabilities between different events.

2.1 PIN-Based Modelling of IR

[17,16] utilised the PIN framework to investigate how the ranking provided by IR models (e.g. TF-IDF) can be explained via a PIN and its interpretation. The PIN model is a formal framework which infers the probability that each document in the collection satisfies the user's information need.

It uses document representations as sources of evidence about its content and multiple query representations as source of information need. In addition, it provides representation nodes that reflect different concepts in the model.

An inference network model (taken from [16]) is presented in figure 2. It contains four different type of nodes and the conections (with a weight assigned) between them. The links between nodes d and r are not shown in the diagram for clarity reasons:

- Document nodes (d): Representing documents in the corpus
- Representation nodes (r): Modelling the concepts considered (terms, phrases,...)
- Query nodes (q): They are related to parts of the information needed by the user
- Information need (I): It models the complete information needed by the user

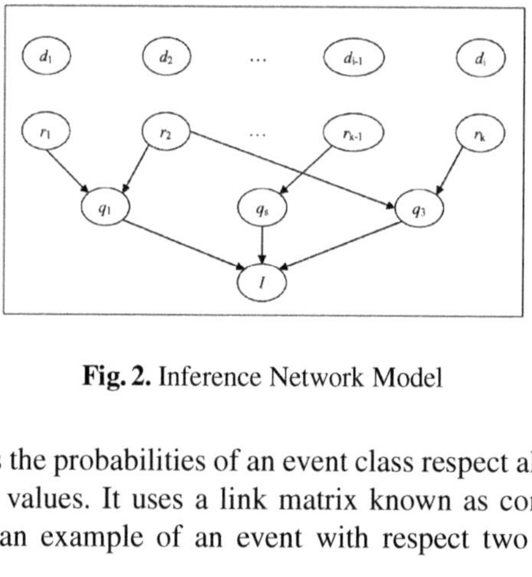

Fig. 2. Inference Network Model

This model represents the probabilities of an event class respect all the possible combinations in its parents values. It uses a link matrix known as conditional dependence probabilities (CDP), an example of an event with respect two parents is shown in equation 1.

$$L = \left[\begin{array}{c|cccc} & x_3 & x_2 & x_1 & x_0 \\ & 11 & 10 & 01 & 00 \\ \hline P(q|x) & a_{11} & a_{12} & a_{13} & a_{14} \\ P(\bar{q}|x) & a_{21} & a_{22} & a_{23} & a_{24} \end{array} \right] \quad (1)$$

One of the problems of this model is that the number of combinations exponentially rises making its processing extremely expensive in terms of computational power.

Therefore, this model can not be directly applied to problems with a large quantity of features such as *ad hoc* retrieval or text classification.

As a solution for the problems of the original PIN model, a modification was designed by Turtle and Croft [16]. They assumed independence between terms and defined a special setting of link matrix with a normalization over the total weight of the features. These modifications lead to the equation 2. Using this approach, the model only needs the probability of being and not being in a class respect the existence of each term.

$$P(q|d) = \sum_{t} \frac{P(q|r)}{\sum_{t'} P(q|r')} \cdot P(r|d) \tag{2}$$

The reason for the normalization in equation 2 is that the sum over $P(q|r)$ for the terms/representations in a query could be greater than one. Therefore, $P(q|d)$ could be it as well.

3 Probabilistic Datalog

PDatalog is a probabilistic logical retrieval framework that combines deterministic Datalog (a query language used in deductive databases) and probability theory ([4,12]). It was extended in [13,20] to improve its expressiveness and scalability for modelling IR models (ranking functions). In addition, it is a flexible platform for modelling and prototyping different IR tasks. We utilise PDatalog here for implementing an abstraction layer for modelling PIN's. Moreover, it is also used for the implementation, as a proof of concept, of Bayesian classifiers.

3.1 1st Generation PDatalog

The 1st generation PDatalog uses free Horn clauses with a probability attached to rules and facts. It was introduced for IR in [5]. The main idea is to allow for probabilities in facts and rules. Figure 3 describes the syntax utilized in traditional datalog and PDatalog.

A PDatalog rule consists of a head and a body. A head is a goal, and a body is a subgoal list. A rule is evaluated such that the head is true if and only if the body is true. So far, the syntax is the one of ordinary Datalog.

3.2 2nd Generation PDatalog

The 2nd generation PDatalog includes a more complex syntax allowing assumptions and probability estimation. These modifications include, among others, score agregation (SUM, PROD) for the facts following certain patterns. For example, given "grade(Mike, A, DCS225); grade(Mike, B, DCS115); grade(Mike, A, DCS111);", a rule can be defined for computing $P(grade|student)$:

"$p\_grade\_studentSUM(Grade, Student)\text{:}-grade(Student, Grade, Module)|(Student);$"

This line uses probability estimation and an aggregation assumption (SUM) which is needed for the score agregation. A simplified version (for improving readability) of

<table>
<tr><th colspan="2">Traditional Datalog</th></tr>
<tr><td>fact</td><td>::= NAME '(' constants ')'</td></tr>
<tr><td>rule</td><td>::= head ':-' body</td></tr>
<tr><td>head</td><td>::= goal</td></tr>
<tr><td>body</td><td>::= subgoals</td></tr>
<tr><td>goal</td><td>::= NAME '(' args ')'</td></tr>
<tr><td>subgoal</td><td>::= pos_subgoal |
neg_subgoal</td></tr>
<tr><td>pos_subgoal</td><td>::= atom</td></tr>
<tr><td>neg_subgoal</td><td>::= '!' atom</td></tr>
<tr><td>atom</td><td>::= NAME '(' args ')'</td></tr>
<tr><td>arg</td><td>::= constant | variable</td></tr>
<tr><td>constant</td><td>::= NAME | STRING |
NUMBER</td></tr>
<tr><td>variable</td><td>::= VAR_NAME</td></tr>
<tr><td>args</td><td>::= | arg ',' args</td></tr>
<tr><td>constants</td><td>::= | constant ',' constants</td></tr>
<tr><td>subgoals</td><td>::= | subgoal ',' subgoals</td></tr>
</table>

<table>
<tr><th colspan="2">1st Generation Probabilistic Datalog</th></tr>
<tr><td>prob_fact</td><td>::= prob fact</td></tr>
<tr><td>prob_rule</td><td>::= prob rule</td></tr>
</table>

<table>
<tr><th colspan="2">2nd Generation Probabilistic Datalog</th></tr>
<tr><td>goal</td><td>::= tradGoal | bayesGoal | aggGoal</td></tr>
<tr><td>subgoal</td><td>::= tradSubgoal |
bayesSubgoal | aggGoal</td></tr>
<tr><td>tradGoal</td><td>::= see 1st Generation</td></tr>
<tr><td>tradSubgoal</td><td>::= see 1st Generation</td></tr>
<tr><td>bayesGoal</td><td>::= tradGoal '|'
{estAssump} evidenceKey</td></tr>
<tr><td>bayesSubgoal</td><td>::= tradSubgoal '|'
{estAssump} evidenceKey</td></tr>
<tr><td>evidenceKey</td><td>::= '(' variables ')'</td></tr>
<tr><td>aggGoal</td><td>::= NAME {aggAssump} '(' args ')'</td></tr>
<tr><td>aggSuboal[1]</td><td>::= NAME {aggAssump} '(' args ')'</td></tr>
<tr><td>tradAssump</td><td>::= 'DISJOINT' | 'INDEPENDENT' |
'SUBSUMED'</td></tr>
<tr><td>irAssump</td><td>::= 'DF' | 'TF' |
'MAX_IDF' | 'MAX_ITF' | ...</td></tr>
<tr><td>probAssump</td><td>::= tradAssump | irAssump</td></tr>
<tr><td>algAssump</td><td>::= 'SUM' | 'PROD'</td></tr>
<tr><td>aggAssump</td><td>::= probAssump
probAssump | complexAssump</td></tr>
</table>

Fig. 3. PDatalog syntax

the syntax specification for the 2nd generation PDatalog is outlined in Figure 3. The assumption between predicate name and argument list is the so-called *aggregation* assumption (aggAssump). For example, for disjoint events, the sum of probabilities is the resulting tuple probability. In this case, the assumptions 'DISJOINT' and 'SUM' are synonyms, and so are 'INDEPENDENT' and 'PROD'. The assumption in a conditional is the so-called *estimation* assumption (estAssump). For example, for disjoint events, the subgoal "index(Term, Doc) | DISJOINT(Doc)" expresses the conditional probability $P(Term|Doc)$ derived from the statistics in the relation called "index".

Complex assumptions such as DF (for document frequency) and MAX_IDF (max inverse document frequency) can be specified to describe in a convenient way probabilistic parameters commonly used in IR. Expressions with complex assumptions can be decomposed in PDatalog programs with traditional assumptions only. However, for improving the readability and processing (optimisation), complex assumptions can be specified. The decomposition of complex assumptions is shown in [13].

4 Modelling PIN in PDatalog

In this section we explain the modelling of PIN's in PDatalog, showing one example for each of its generations (Figure 4 for the 1st generation and Figure 5 for the 2nd). The former illustrates how can we represent prior and conditional probabilities while the latter represents all the input information about earthquakes, burglaries and alarms being triggered. The main difference is that probabilistic rules are manually specified in the 1st generation whereas probability estimation is being used in 2nd gen. In the second case we model $P(region|burglary)$ and $P(alarm \wedge burglary)$ for each region using the relational bayes as en example of probability estimation.

According to the first example, the probability of an alarm being triggered if there is a burglary is 90% while it is 40% in case of an earthquake. In addition, the example represent the prior probability of a burglary (0.1%) and an earthquake (0.001%). Due to the limitations of 1st generation the representation of all possible combinations between any parents and their son is needed.

```
0.001 hypo(burglary);
0.00001 hypo(earquake);
0.1 hypo(burglary) :- hypo(earthquake);
0.9 evidence(alarm) :- hypo(burglary);
0.4 evidence(alarm) :- hypo(earthquake);
0.35 evidence(alarm) :- hypo2(burglary,earthquake);
0.55 evidence(alarm) :- hypo2(burglary,n_earthquake);
0.05 evidence(alarm) :- hypo2(n_burglary,earthquake);
0.05 evidence(alarm) :- hypo2(n_burglary,n_earthquake);
```

Fig. 4. Modelling PIN in 1st Generation PDatalog

```
event(burglary); #... 200 more facts similar this one representing 20 different crimes.
event(earthquake); #... 3 facts
event(alarm); #... 150 more facts representing 10 different alarms being triggered
event2(burglary, earthquake); #... 10 facts
event2(n_burglary, earthquake); #... 30 facts
event2(alarm, burglary); #... 35 facts
event2(n_alarm, burglary); #... 5 facts
event2(alarm, earthquake); #... 10 facts
event2(n_alarm, earthquake); #... 30 facts
...
p_event1_event2 SUM(Event1, Event2) :- event2(Event1, Event2)|(Event2);
...
```

Fig. 5. Modelling PIN in 2nd Generation PDatalog

5 Bayesian Classifiers

Bayesian classifiers are a set of different classifiers that uses the Bayes Theorem for inference knowledge (Equation 3). However, different models use a different event space for representation, different techniques for calculate certain probabilities or other assumptions. Applying the Bayes theorem we can calculate the probability of a class given a document, being d a document for classify and c one of the classes. This equation could be extended by referring $P(d|c)$ and P(d) to the terms inside document d (Equation 3). Finally, its numerator can be rewritten, applying equation 4.

$$P(c|d) = \frac{P(d|c) \cdot P(c)}{P(d)} = \frac{P(t_1, t_2, ..., t_n|c) \cdot P(c)}{P(t_1, t_2, ..., t_n)} \tag{3}$$

$$P(t_1, t_2, ..., t_n|c) = P(t_1|c) \cdot P(t_2|c, t_1) \cdot ...P(t_n|c, t_1, ...t_{n-1}) \tag{4}$$

5.1 Independence Assumption

The computational power required for Bayes inference exponentially grows with the number of features making this method difficult to apply in large scale environments.

One of the most common solutions for this problem, known as "Independence Assumption", is assuming independence between features given the context of a class. Applying this assumption, the join probability from the general equation for Bayesian classifiers (equation 4) is modified, leading to:

$$P(t_1, t_2, ..., t_n|c) = P(t_1|c) \cdot P(t_2|c)... \cdot P(t_n|c) \tag{5}$$

Assuming independence between features we can define the probability of a document being labelled in a class as follows, where n(t,d) is the number of times that word t appears in document d,

$$P(c|d) = \frac{P(c)}{P(d)} \cdot \prod_{t \in d} P(t|c)^{n(t,d)} \tag{6}$$

Classifiers that make this assumption are usually referred as Naive-Bayes, even if there are differences between them [8]. This is a common assumption that allows the application of this algorithms to larger collections. However, it could be not correct.

5.2 Uniform Prior Distribution Assumption

We can erase $P(d)$ and $P(c)$ from equation 6 if we assume that they are uniformly distributed. In that case the prior probabilities of a document and a class are constants, not having any effect in the ranking. Therefore, they can be erased.

5.3 Multi-variate Bernoulli

In this model [8] we represent different features (i.e. terms) using a binary vector that indicates which features are present in which elements (i.e. documents). We can apply this model for classification using the general Bayes formula (equation 3) substituting the equations shown in this section. This model computes the class prior probability by the maximum likelihood estimate (Equation 7), assuming $P(d) = \frac{1}{|D|}$ for all documents and the document prior (Equation 8). In addition, $P(d|c)$ and $P(t|c)$ are specified in equations 9 and 10 respectively where B_t is the binary value indicating if term t appears in document d.

$$P(c) = \frac{\sum_{d \in D} P(c|d)}{|D|} \tag{7}$$

$$P(d) = \sum_{c \in C} P(c) \cdot P(d|c) \tag{8}$$

$$P(d|c) = \prod_{t \in V} (B_t \cdot P(t|c) + (1 - B_t) \cdot (1 - P(t|c))) \tag{9}$$

$$P(t|c) = \frac{\sum_{d \in D} B_{dt} \cdot P(c|d)}{\sum_{d \in D} P(c|d)} \tag{10}$$

This model applies the naive independence assumption explained in section 5.1 and it explicitly takes into account the non-occurrence probability of features that are not in the element.

5.4 Multinomial

This model, explained in [8], uses a non-binary vector for representing different features. It uses the frequency of each parameter (i.e. term) for each element (i.e. document).

We can apply this model for classification using the general Bayes formula (equation 3) substituting the equations shown in this section. It computes the probability of a document given a class using Equation 11, where $n(t, d)$ represents the number of times term t occurs in document d and V is the set of terms contained in the document. The definition of $P(t|c)$ in Multinomial-Bayes is illustrated in equation 12.

$$P(d|c) = P(|d|) \cdot |d|! \prod_{t \in V} \frac{P(t|c)^{n(t,d)}}{n(t,d)!} \tag{11}$$

$$P(t|c) = \frac{\sum_{d \in D} n(t,d) \cdot P(c|d)}{|V| + \sum_{t \in V} \sum_{d \in D} n(t,d) \cdot P(c|d)} \tag{12}$$

Class and document priors are computed as they were in the Bernoulli model, applying equations 7 and 8 respectively.

6 Modelling Bayesian Classifiers in PDatalog

The modelling of Bayesian classifiers in PDatalog is a proof of concept regarding the expressiveness of PDatalog. This section outlines the case for Naive-Bayes, and then underlines that PDatalog programs are the result of a translation process, namely the translation of a PIN/BN specification to PDatalog. This translation frees the developer from actually "writing" PDatalog.

6.1 Naive-Bayes Classifier

Figure 6 shows a PDatalog program for modelling Naive-Bayes. This program uses as input facts tuples representing terms in training documents (termDoc_sample), terms contained in documents to classify (termDoc_classify) and the class labelled for each training document (part_of) (i.e. *0.2 termDoc_sample(car, d1); 1.0 part_of(d1, earn);*)

First of all, we define termClass(Term, Class), the representation of classes as it is derived from the members of the class and the prior probability of classes prior(Class). Secondly, It specifies the rule for predicate p_t_c, to model the feature likelihood $P(t|c)$. The expressiveness of 2nd-generation PDatalog supports the description of this step, namely the estimation of the feature probability. For IR, termClass(Term, Class) will be term-based representation of the classes derived from the documents that are part of the class. Then, there is a rule describing conditional probabilities of a document given a class, $P(d|c)$. Finally, it shows the probability of a class given a document using the expression $P(c|d)$.

```
prior(Class) :- part_of(Doc, Class) | ();
termClass(Term, Class) :- termDoc_sample(Term, Doc) & part_of(Doc, Class);
p_t_c SUM(Term,Class) :- termClass(Term, Class) | (Class);
p_d_c PROD(Doc, Class) :- termDoc_classify(Term, Doc) & p_t_c(Term, Class);
p_c_d(Class, Doc) :- p_d_c(Doc, Class) & prior(Class);
```

Fig. 6. Naive-Bayes Classifier in PD

6.2 Turtle-Croft-PIN-Based Classifier

Figure 7 shows a PDatalog program for using the PIN model as a classifier. It follows the same notation and input data explained for the Naive-Bayes classifier in section 6.1.

```
termClass(Term, Class) :- termDoc_sample(Term, Doc) & part_of(Doc, Class);
p_t_c SUM(Term, Class) :- termClass(Term, Class) | (Class);
p_d_t SUM(Doc, Term) :- termDoc_classify(Term, Doc) | (Term);
p_d_c SUM(Class, Doc) :- p_d_t(Doc, Term) & p_t_c(Term, Class);
```

Fig. 7. Turtle-Croft-PIN-based Classifier in PD

6.3 Generation of PDatalog Classifier Programs

A PDatalog program representing a classifier can be viewed as the result of translating a PIN specification into a PDatalog program. This is currently a manual process.

The future idea is to automatically generate PDatalog programs for specific problems. We can use generic definitions of Bayesian classifiers that could be adapted to a single problem using a mapping layer. This layer would be a connection between the abstract implementation of classifiers and some specific facts representing the problem to be solved. In addition, this strategy would be extended to other classification algorithms, creating an abstraction layer for classification in PDatalog.

7 Feasibility and Experimental Study

The main focus at this stage of research is the feasibility of modelling PINs using PD. However, we also present, as a proof of concept, the quality evaluation of our approach for the Enron collection.

7.1 Collection

The Enron collection [1] contains emails from many of the senior management of Enron Corporation. This data was made public by SRI after clean-up and attachments removal. This corpus has a large number of emails, although there are many users with folders almost empty. The final corpus used in this paper is a subset of this collection. It contains the emails of seven employees with a large number of folders and mails. In addition, non topical folders such as "all_documents", "calendar" and "contacts" were removed. After this, folder hierarchies were flatten and folders with less than three messages were deleted. Finally, the "X-folder" field in mail headers were removed as it contains the class label. This collection was obtained from http://www.cs.umass.edu/~ronb/enron_dataset.html.

7.2 Training Splits

The most common strategy to divide a collection between train an test documents for classification is using random splits. However, this method could create unnatural dependencies of earlier documents in latter documents for email foldering because this collections could have strong chronological dependences. Some authors have recommended splits based on the chronological order of the emails, using incremental splits [1] or using only one big split using half of the collections as a training set [7]. The former strategy can report non-realistic high rates for quality measures whereas the latter has the problem that some classes could have documents only in one of the splits.

We propose a modification of the second strategy applying a chronological split for each of the classes. By doing this we represent chronological dependencies and we guarantee that all the classes have documents in both splits. The focus of this paper is not a comparison between different split algorithms. However, we have decide to do experiments using four different methods:

Global chronological split. The train set is formed by the first $n/2$ emails, where n is the size of the collection.

Class chronological split. The train set is formed by the first $n_i/2$ emails of each class, where n_i is the size of the class i.

Global random split. The documents in the train set are randomly chosen from the collection until its size is $n/2$, where n is the size of the collection.

Class random split. The documents in the train set are randomly chosen from each class until the number of emails selected is $n_i/2$, where n_i is the size of class i.

7.3 Document Representation

For our experiments we have used a "bag of words" representation. In addition, we use the terms that appear in more than 75% of the classes as stopwords.

7.4 Measures

The measures used for evaluation are the macro and micro-averaged F_1. They measure the classifier's quality based on its precision and recall values for each class. Precision is defined as the ratio of correctly classified documents respect to the number of documents classified by the system, while recall calculates the ratio between correctly classified documents and total number of documents truly belonging to the class [15].

Macro-averaged is usually computed in two different ways. The "correct" method, according to [21], is calculate the F_1 values for each category and then compute the average. On the other hand, F_1 could be computed using the macro-averaged recall and precision. Both methods give different results and the "correct" one is often significantly lower than the "incorrect" one [21]. We are using in the experiments the first method described.

7.5 Results and Discussion

Figure 8 shows the macro-average F_1 values for Naive-Bayes (implementation is shown in figure 6) and PIN implementation in PDatalog. The NB code follows the definition in equation 6 assuming uniform document frequency. $P(t|d)$ is represented by the number of times t appear in d divided by the total number of terms in it. Figure 9 presents the micro-averageF_1 (best value for each user in bold). They illustrate this quality measures using four different split algorithms for the collection. The values shown for random splits represent the average of three different executions for each configuration.

	Global Random		Class Random		Global Chronological		Class Chronological	
Enron user	NB	PIN	NB	PIN	NB	PIN	NB	PIN
beck-s	0.4642	0.3546	**0.5099**	0.3546	0.2035	0.1346	0.4108	0.3170
farmer-d	0.6166	0.3949	**0.6169**	0.3615	0.4541	0.2167	0.3937	0.2630
kaminski-v	0.5893	0.3295	**0.6009**	0.3357	0.3619	0.2223	0.4086	0.2540
kitchen-l	0.4655	0.2823	**0.4799**	0.2522	0.2116	0.0534	0.3204	0.2462
lokay-m	0.7072	0.5188	**0.7276**	0.5262	0.5481	0.4994	0.6115	0.4132
sanders-r	0.6466	0.6010	**0.6580**	0.6206	0.4900	0.4508	0.4904	0.6362
williams-w3	0.6465	0.4055	0.6363	0.3631	0.8989	**0.8998**	0.5935	0.2612
Average	0.5908	0.4124	**0.6042**	0.4020	0.4526	0.3539	0.4613	0.3415

Fig. 8. Macro-Average F1 for different split strategies

As we can see, the choice of splitting algorithm significantly changes the quality measure assigned to it. The best performance for the Naive-Bayes implementation has been obtained, both for macro and micro-averaged, using the random split per class. The explanation for these results is that a random split per class uses different periods of time as an information source, representing the "meaning" of the folder in all the time considered. In addition, this split algorithm guarantees that all the classes have at least one representative in the test model. We consider that it could be beneficial doing

	Global Random		Class Random		Global Chronological		Class Chronological	
Enron user	NB	PIN	NB	PIN	NB	PIN	NB	PIN
beck-s	0.5019	0.3656	**0.5377**	0.3672	0.2315	0.1076	0.4086	0.2978
farmer-d	0.7932	0.4700	**0.7983**	0.4261	0.5708	0.1531	0.6284	0.2612
kaminski-v	**0.6534**	0.3044	0.6514	0.3050	0.4176	0.2242	0.4955	0.2327
kitchen-l	0.5119	0.2164	**0.5333**	0.2016	0.1574	0.0279	0.3786	0.2051
lokay-m	0.8095	0.5677	**0.8163**	0.6546	0.6996	0.4683	0.7454	0.2981
sanders-r	0.6953	0.5387	**0.7156**	0.5830	0.5286	0.3384	0.5751	0.5392
williams-w3	0.9275	0.5798	0.9360	0.5926	0.9949	**0.9993**	0.9160	0.5757
Average	0.6990	0.4347	**0.7127**	0.4472	0.5143	0.3313	0.5925	0.3443

Fig. 9. Micro-Average F1 for different split strategies

more experiments for an exhaustive comparison between different splitting methods. However, such research is beyond the scope if this paper.

The F_1 value for the user "williams-w3" with a global chronological split is much higher than the ones obtained applying other strategies because using that configuration the test collection only has documents from two different classes.

Given the assumption that "macro-averaged is higher for classifiers that behave well for few positive training documents while micro-averaged is better in the opposite case" [15] and the figures shown in this paper, we can say that the classifiers considered perform slightly better for common classes in this specific case.

Results obtained by the PDatalog-based Naive-Bayes classifier are similar to the ones expected for other implementations, showing its feasibility. On the other hand, Turtle-Croft-PIN based classifier has obtained lower results than Naive-Bayes in almost all the executions.

8 Summary and Conclusions

This paper has presented the modelling of PIN's in 1st and 2nd generation PDatalog. In 1st generation PDatalog, probabilistic rules have to be used to model the conditional probabilities in PIN's. In addition, it has no means to express the probability estimation that is required to derive/learn a PIN from sample data, i.e. the "learning probabilities" process is external to PDatalog. In the 2nd generation PDatalog, Bayesian goals and subgoals support, on one hand the modelling of probability estimation, and, on the other hand, the modelling of conditional probabilities.

The main contribution of this paper is to present and discuss the issues when modelling PIN's in PDatalog. Having used PDatalog for a decade, it was surprising to realise that the modelling of PIN's turned out to be much more challenging than could be expected, since, intuitively, a probabilistic logical framework could be expected to naturally provide what the modelling of PIN's requires. In addition to the conceptual and theoretical aspects of modelling PIN's, this paper contributes the modelling of classifiers in PDatalog, and an experimental study to confirm feasibility and investigate the quality of a PDatalog-based implementation. We have demonstrated that it is possible to model PIN's in PDatalog. In addition, we adapted these models to the concrete task of text classification using a mapping layer, that would be automatically generated in the future. Furthermore, we have shown that this implementation achieves quality measures than could be expected by other implementations of Bayesian classifiers.

The potentially high impact of this research lies in the fact that PDatalog is an abstraction layer that gives access to more than just the modelling of PIN's. It has been used as an intermediate processing layer for semantic/terminological logics in different IR tasks such as *ad-hoc* retrieval [9], annotated document retrieval [3] and summarization [2].

Furthermore, probabilistic versions of Datalog are regarded for the semantic web as a platform layer on which other modelling paradigms (ontology-based logic) can rest and rely upon [11,14]. The 2nd generation of PDatalog provides extended expressiveness using probability estimation and conditional probabilities. It also improved scalability because probabilistic rules are not required and extensional relations and assumptions can be used in order to achieve efficient and scalable programs. Therefore, it can be expected to have an impact beyond the flexible modelling of classification.

The ultimate goal is to achieve a framework of logical building blocks that offers classifiers, retrieval models, information extractors, and other functions, and those functional blocks can be composed in a possibly web-based service infrastructure. Thereby, high-level languages can be used that are translated to PDatalog for the purpose of composition and execution.

The next objective in the creation of a logical framework for Information Retrieval is developing an abstraction layer for PIN's and classification in PDatalog. This module should include not only Naive-Bayes but also non-probabilistic classifiers such as Support Vector Machines (SVM) or k-NN. In addition, it should allow single and multilabel strategies applying flat or hierarchical classification. Furthermore, the module would be able to deal with large-scale classification tasks (in terms of documents and classes).

Moreover, we have started to work on design techniques (design patterns, UML-like) to assist the design, composition and test of PDatalog programs. This framework supports teams of knowledge engineers to efficiently "plug-and-play" components they need for solving complex scenarios as they occur in information management tasks.

References

1. Bekkerman, R., et al.: Automatic categorization of email into folders: Benchmark experiments on enron and sri corpora. Tech. Rep. Center for Intelligent Information Retrieval (2004)
2. Forst, J.F., Tombros, A., Roelleke, T.: Polis: A probabilistic logic for document summarisation. In: Studies in Theory of Information Retrieval, pp. 201–212 (2007)
3. Frommholz, I.: Annotation-based document retrieval with probabilistic logics. In: Kovács, L., Fuhr, N., Meghini, C. (eds.) ECDL 2007. LNCS, vol. 4675, pp. 321–332. Springer, Heidelberg (2007)
4. Fuhr, N.: Probabilistic datalog - a logic for powerful retrieval methods. In: ACM SIGIR, pp. 282–290 (1995)
5. Fuhr, N.: Optimum database selection in networked ir. In: NIR 1996, SIGIR (1996)
6. Kheirbeck, A., Chiaramella, Y.: Integrating hypermedia and information retrieval with conceptual graphs formalism. In: Hypertext - Information Retrieval - Multimedia, Synergieeffekte elektronischer Informationssysteme, pp. 47–60 (1995)

7. Klimt, B., Yang, Y.: The Enron corpus: A new dataset for email classification research. In: Boulicaut, J.-F., Esposito, F., Giannotti, F., Pedreschi, D. (eds.) ECML 2004. LNCS (LNAI), vol. 3201, pp. 217–226. Springer, Heidelberg (2004)
8. McCallum, A., Nigam, K.: A comparison of event models for naive bayes text classification. In: AAAI/ICML-1998 Workshop on Learning for Text Categorization, p. 41 (1998)
9. Meghini, C., Sebastiani, F., Straccia, U., Thanos, C.: A model of information retrieval based on a terminological logic. In: ACM SIGIR, pp. 298–308 (1993)
10. Pearl, J.: Probabilistic Reasoning in Intelligent Systems: Networks of Plausible Inference. Morgan Kaufman, San Mateo (1988)
11. Polleres, A.: From SPARQL to rules (and back). In: 16th international conference on World Wide Web (WWW), pp. 787–796. ACM, New York (2007)
12. Roelleke, T., Fuhr, N.: Information retrieval with probabilistic datalog. In: Uncertainty and Logics - Advanced Models for the Representation and Retrieval of Information (1998)
13. Roelleke, T., Wu, H., Wang, J., Azzam, H.: Modelling retrieval models in a probabilistic relational algebra with a new operator: The relational Bayes. VLDB Journal (2009)
14. Schenk, S.: A SPARQL semantics based on Datalog. In: Hertzberg, J., Beetz, M., Englert, R. (eds.) KI 2007. LNCS (LNAI), vol. 4667, pp. 160–174. Springer, Heidelberg (2007)
15. Sebastiani, F.: Machine learning in automated text categorization. ACM Comput. Surv. 34(1), 1–47 (2002)
16. Turtle, H., Croft, W.: Efficient probabilistic inference for text retrieval. In: Proceedings RIAO 1991, pp. 644–661 (1991)
17. Turtle, H., Croft, W.B.: Inference networks for document retrieval. In: ACM SIGIR, New York, pp. 1–24 (1990)
18. van Rijsbergen, C.J.: Towards an information logic. In: ACM SIGIR, pp. 77–86 (1989)
19. Wong, S., Yao, Y.: On modeling information retrieval with probabilistic inference. ACM Transactions on Information Systems 13(1), 38–68 (1995)
20. Wu, H., Kazai, G., Roelleke, T.: Modelling anchor text retrieval in book search based on back-of-book index. In: SIGIR Workshop on Focused Retrieval, pp. 51–58 (2008)
21. Yang, Y.: A study on thresholding strategies for text categorization. In: ACM SIGIR, pp. 137–145 (2001) (press)

Handling Dirty Databases: From User Warning to Data Cleaning — Towards an Interactive Approach

Olivier Pivert[1] and Henri Prade[2]

[1] Irisa – Enssat, University of Rennes 1
Technopole Anticipa 22305 Lannion Cedex France
[2] IRIT, CNRS and University of Toulouse, 31062 Toulouse Cedex 9, France
pivert@enssat.fr, prade@irit.fr

Abstract. One can conceive many reasonable ways of characterizing how dirty a database is with respect to a set of integrity constraints (e.g., functional dependencies). However, dirtiness measures, as good as they can be, are difficult to interpret for an end-user and do not give the database administrator much hint about how to clean the base. This paper discusses these aspects and proposes some methods aimed at either helping the user or the administrator overcome the limitations of dirtiness measures when it comes to handling dirty databases.

1 Introduction

In the database world, there is a strong concern about the semantic consistency of stored data and the avoidance of contradictory information. However, as noted in [1], the support for specifying and enforcing semantic integrity in commercial products and practical systems remains scant.

Theoretically speaking, dirty databases should not exist, since declarative solutions — that have been widely proclaimed and advertised in the scientific and technical literature of the field — are available for specifying and enforcing integrity constraints [2]. However, in practice, dirty databases are quite common, for various reasons: i) the database may be ill-designed and/or some integrity constraints (in particular functional dependencies) are not properly enforced [1], ii) when a database results from the integration of multiple data sources, or when the data comes from unverified sources or is uncertain, the resulting database generally contains inconsistencies [3].

Assessing the dirtiness of a database and devising methods to clean it or to overcome it are important problems which have been addressed from different points of view, see e.g. [4,5,6,7]. More particularly, the measures of dirtiness that can be found in the literature are somewhat debatable from a semantic point of view. Moreover, it is not quite clear how these measures can be used in practice and exploited in relation with cleaning methods. In this paper, we consider the case where dirtiness corresponds to the violation of one or several functional dependencies (FDs) — we do not deal with other forms of "dirtiness"

A. Deshpande and A. Hunter (Eds.): SUM 2010, LNAI 6379, pp. 292–305, 2010.

such as null values, uncertain values, or the violation of other types of integrity constraints, even though this would of course make sense. We both discuss the limitations of dirtiness measures and propose new tools for helping users and administrators handle dirty data.

The remainder of the paper is structured as follows. Section 2 deals with the concept of FD violations, defines new dirtiness measures based on the graded notion of proximity, and discusses the shortcomings of dirtiness measures in general. In Section 3, we take a pragmatic viewpoint and define some methods for detecting suspect answers in the result of a query (which can be a valuable piece of information to provide the user with). Section 4 deals with data cleaning. Section 5 discusses related works about database repair approaches, whereas Section 6 concludes the paper and outlines perspectives for further research.

2 About Dirty Databases and Dirtiness Measures

In this section, we first recall different ways to define an FD violation, as well as the related notions of culprits and clusters first introduced in [6]. We then give a general definition of a dirtiness measure, propose some new measures that take into account the graded notion of proximity between attribute values, and discuss the limitations of such measures.

2.1 Preamble

We assume the existence of a relation r of schema $R = (A_1, \ldots, A_n)$ where the A_i's are attributes. Each attribute A_i has an associated domain, $dom(A_i)$. A tuple of r is a member of $dom(A_1) \times \ldots \times dom(A_n)$.

Definition 1. The regular functional dependency $X \to Y$ is defined as:

$$\forall t_1,\ t_2 \in r,\ t_1.X = t_2.X \Rightarrow t_1.Y = t_2.Y \tag{1}$$

where X and Y denote sets of attributes. For instance, in Table 1, Name $\to$ Age is an FD saying that two tuples about the same person should agree on age. One may notice that in this table, tuples t_3 and t_5 are identical. In the following, we assume that relations are bags (rather than sets) of tuples. Indeed, a typical situation dealt with is when a database results from the fusion of different data sources, which implies that the primary key constraints have been disabled.

Remark. Since any FD $X \to \{Y_1 \ldots Y_q\}$ can be decomposed into a set of FDs $\{X \to Y_1, \ldots, X \to Y_q\}$, it will be assumed in the following with no loss of generality that the right-hand side of an FD is made of a single attribute.

As noted in [8], there are two basic approaches to define exceptions to (or violations of) an FD $X \to Y$:

- exceptions as pairs of tuples. An exception is a pair of tuples $(t_1, t_2) \in r$ verifying $t_1.X = t_2.X$ and $t1.Y \neq t_2.Y$. The corresponding measure is denoted by g_1 in [9]:

Table 1. An example database

	Name	*Age*	*Height*
t_1	Mary	28	170
t_2	Mary	28	176
t_3	John	30	163
t_4	John	30	160
t_5	John	30	163
t_6	Matthew	32	170
t_7	Matthew	32	170
t_8	Paul	37	172
t_9	Paul	37	171
t_{10}	Paul	37	176
t_{11}	James	45	177

$$g_1(X \to Y,\, r) = \frac{|\{(t,\, t') \mid t,\, t' \in r \,\wedge\, t.X = t'.X \,\wedge\, t.Y \neq t'.Y\}|}{|r|^2}$$

- exceptions as individual tuples. The two most straightforward definitions corresponding to this view are [9]:

$$g_2(X \to Y,\, r) = \frac{|\{t \mid t \in r \,\wedge\, \exists t' \in r,\, t.X = t'.X \,\wedge\, t.Y \neq t'.Y\}|}{|r|}$$

$$g_3(X \to Y,\, r) = \frac{m_e}{|r|}$$

where m_e is the minimal size of of a set r_e of exceptions such that the dependency holds in $r - r_e$.

In a database cleaning perspective, one may prefer g_3 that assesses how large the maximal consistent part of the database is. It is proven in [9] that

$$\frac{g_2}{|r|} \leq g_1 \leq g_2^2 - \frac{g_2}{|r|},$$
$$\frac{2 \cdot g_3}{|r|} \leq g_1 \leq 2 \cdot g_3 - g_3^2 - \frac{g_3}{|r|},$$
$$g_3 + \frac{1}{|r|} \leq g_2 \leq 1,$$

which shows that the three measures are pairwise related through inequalities. Besides, it is also possible to consider the proximity between the Y-values so as to take into account the intensity of an exception. For instance, if an exception is seen as a pair of tuples, one can say that $(t,\, t')$ is all the more intense as $t.Y$ is far away from $t'.Y$. Clearly, the notion of proximity would depend on the attribute domain. This idea is developed in Subsection 2.4.

2.2 Culprits and Clusters

Martinez et al. [6] introduce the notions of *culprits* and *clusters*. Culprits are the duals of the maximal consistent subsets (studied, e.g., in [10]). The following three definitions are drawn from [6].

Definition 2. Let r be a relation and F a set of functional dependencies. A *culprit* is a set $c \subseteq r$ such that $c \cup F$ is inconsistent and $\forall c' \subset c$, $c' \cup F$ is consistent.

Thus, culprits are minimal sets of database tuples that cause a functional dependency violation. In the case where a single FD $X \rightarrow Y$ is considered, every culprit is a pair of tuples that agree on X and disagree on Y. Let $culprits(r, fd)$ denote the set of culprits in relation r w.r.t. the functional dependency fd.

Example 1. Consider a functional dependency *fd* stating that

$$\forall t,\ t' \in r,\ t.Name = t'.Name \Rightarrow t.Height = t'.Height.$$

The relation in Table 1 has six culprits, namely $c_1 = \{t_1, t_2\}$, $c_2 = \{t_3, t_4\}$, $c_3 = \{t_4, t_5\}$, $c_4 = \{t_8, t_9\}$, $c_5 = \{t_8, t_{10}\}$, and $c_6 = \{t_9, t_{10}\}$.$\diamond$

Definition 3. Given two culprits $c, c' \in culprits(r, fd)$, we say that c and c' overlap, denoted by $c\, \Delta\, c'$, iff $c \cap c' \neq \emptyset$.

Definition 4. Let Δ^* be the reflexive transitive closure of relation Δ. A *cluster* is a set $cl = \bigcup_{c \in e} c$ where e is an equivalence class of Δ^*.

Example 2. In Table 1, the clusters are the sets $cl_1 = \{t_1, t_2\}$, $cl_2 = \{t_3, t_4, t_5\}$, and $cl_3 = \{t_8, t_9, t_{10}\}$.$\diamond$

2.3 About Dirtiness Measures

The dirtiness of a relation r w.r.t. an FD can be measured in different ways. Some measures only take into account the number of exceptions, cf. [9,8,11,12,13]. On the other hand, some approaches (in particular [6]) also take into account the *intensity* of the exceptions, e.g., by using the statistical notions of standard deviation $\sigma(X)$ and variance $var(X)$ (which is the square of standard deviation). Let us recall that:

$$\sigma(X)^2 = E[(X - E[X])^2] = E[X^2] - E[X]^2$$

where $E[X]$ is the average of the random variable X. A single-dependency dirtiness function δ takes a relation r and a functional dependency *fd*, and returns a real number in $[0, 1]$ — or in $[0, \infty[$ as in [6]. Martinez *et al.* [6] consider three axioms related to a dirtiness measure δ. The first one — which is rather self-evident — says that consistent databases have a dirtiness level of zero.

Axiom S1. If $culprits(r, fd) = \emptyset$, then $\delta(r, fd) = 0$.

The two other axioms take into account the intensity of the exceptions expressed by a variance and depend on the statistical point of view chosen by the authors.

In [6], the authors assume that the attributes in a table are totally ordered by a reliability ordering. Here, we consider a simplified version where all the attributes are assumed equally reliable. Let r be a relation, *fd* a functional dependency over r, $\{A_1, \ldots, A_m\}$ the set of attributes involved in *fd*, and $var_{max}(A_i)$, with $i \in [1, m]$, be the maximal possible variance for attribute A_i assumed to be given. Hereafter, $cl.A_i$ denotes the restriction of cluster cl to attribute A_i. The definition of dirtiness measure δ_{var} is as follows:

$$\delta_{var(r,\, fd)} = \sum_{cl \in clusters(r,\, fd)} wtVar(cl,\, fd)$$

where $wtVar(cl,\, fd) = \sum_{i=1}^{m} \frac{var_{A_i}(cl.A_i)}{var_{max}(A_i)}$.

Example 3. With the data from Table 1 and $var_{max}(Height) = 5$, one gets:
$wtVar(cl_1,\, fd) = \frac{(170-173)^2+(176-173)^2}{2} = 9$
$wtVar(cl_2,\, fd) = \frac{(163-162)^2+(160-162)^2+(163-162)^2}{3} = 2$
$wtVar(cl_3,\, fd) = \frac{(172-173)^2+(171-173)^2+(176-173)^2}{3} \approx 4.67$
$\delta_{var(r,\, Name \rightarrow Height)} \approx \frac{9+2+4.67}{5} \approx 3.13.\diamond$

This provides a cumulative amount of dirtiness, but cannot distinguish between the existence of one very dirty cluster with a large violation of the FD, and the existence of several clusters with light violations of the FD.

2.4 Proximity-Based Dirtiness Measures

Even though the authors of [6] adopt a statistical point of view for measuring the discrepancy between two attribute values that should be equal, it can be argued that such an approach is not always suitable. Intuitively, the discrepancy between two values depends on the attribute domain in different respects:

- if we consider person's heights and salaries, values 160 and 163 representing heights are not as close as 1600 and 1630 representing salaries, which the variance-based approach does not reflect;
- if the difference between two attribute values exceeds some threshold (which depends on the attribute domain), they can be considered as definitively different (in the sense that there is no hope to "reconciliate" them, e.g., in assuming some kind of rounding error), whatever the difference is. This view is more in the spirit of a *fuzzy* (dis)similarity point of view, than a statistical point of view and may lead to a different way of assessing dirtiness, as shown hereafter.

We consider a proximity function on $domain(A_i)$ of the form:

$$\mu_{prox_{A_i}}(t.A_i,\, t'.A_i) = max(0,\, \frac{M_i - dist_{A_i}(t.A_i,\, t'.A_i)}{M_i})$$

where $dist_{A_i}$ is a distance defined on $domain(A_i)$ and M_i is the maximal distance inside which a pair of values can be considered somewhat similar. Notice that only when the distance is zero, the proximity degree equals 1. The following statements (expressing the extent to which a database is "clean" w.r.t. to an FD) may serve as a basis for the definition of a dirtiness function. We assume that $\forall$ is interpreted by min and consider scalar cardinalities of fuzzy sets [14].

- the pairs (t, t') from r such that $t.X = t'.X$ verify $prox(t.Y, t'.Y)$ (in the spirit of measure g_1):

$$\delta_1(r, fd) = 0 \text{ if } \forall(t, t') \in r \times r,\ t.X \neq t'.X,$$
$$= 1 - \frac{\sum_{(t,\, t') \in r \times r \,|\, t \neq t' \wedge t.X = t'.X} \mu_{prox_Y}(t.Y,\, t'.Y)}{|\{(t,\, t') \in r \times r \,|\, t \neq t' \wedge t.X = t'.X\}|} \text{ otherwise.}$$

- $\forall t' \in r$, the t's from r such that $t.X = t'.X$ verify $prox(t.Y, t'.Y)$ (in the spirit of measure g_2):

$$\delta_2(r, fd) = 0 \text{ if } \forall(t, t') \in r \times r,\ t.X \neq t'.X,$$
$$= 1 - min_{t' \in r} \frac{\sum_{t \in r \,|\, t' \neq t \wedge t.X = t'.X} \mu_{prox_Y}(t'.Y,\, t.Y)}{|\{t \in r \,|\, t \neq t' \wedge t.X = t'.X\}|} \text{ otherwise.}$$

Note: here, we choose not to give "true generalizations" of g_1 and g_2 because these measures consider the whole set of tuples (or pairs of tuples) from r instead of the pairs of tuples which agree on X — which, in our opinion, makes more sense. However, it would be straightforward to give true generalizations of g_1 and g_2 involving a proximity relation.

- Let us define:

$$\psi(r, fd, \alpha) = \frac{|r| - max\{|s| \,|\, s \subseteq r \wedge min_{t,\, t' \in s \,|\, t.X = t'.X}\, \mu_{prox_Y}(t.Y,\, t'.Y) \geq \alpha\}}{|r|}.$$

which can also be expressed as

$$\psi(r, fd, \alpha) = \frac{|r| - m_e(\alpha)}{|r|}$$

where $m_e(\alpha)$ denotes the minimal size of a set $r_e(\alpha)$ of exceptions such that $r - r_e(\alpha)$ does not contain any exception (i.e., any pair (t, t') such that $t.X = t'.X$ and $t.Y \neq t'.Y$) whose "intensity" $1 - \mu_{prox_Y}(t.Y, t'.Y)$ is greater than or equal to $1 - \alpha$. A possible extension of measure g_3, which relies on a Choquet integral for aggregating the "ratios of dirtiness" (ψ) associated with the different levels of violation, is:

$$\delta_3(r, fd) = \sum_{i=1}^{k} \psi(r, fd, \alpha_{i-1}) \cdot (\alpha_i - \alpha_{i-1})$$

where $\alpha_0 = 0$ and $\alpha_1 < \ldots < \alpha_k = 1$ denote the proximity degrees such that $\exists(t, t') \in r \times r$, $\mu_{prox_Y}(t.Y, t'.Y) = \alpha_i$. Notice that one always has $\psi(r, fd, 0) = 0$.

- for every cluster in $clusters(r, fd)$, the minimal proximity between the pairs of Y-values is high. The associated dirtiness function is defined as:

$$\delta_{prox}(r, fd) = 1 - min_{cl_i \in clusters(r, fd)} min_{(t, t') \in cl_i} \mu_{prox_Y}(t.Y, t'.Y).$$

  Contrary to δ_{var}, measure δ_{prox} reflects the fact that two values which are too distant from each other cannot be "reconcilated".

It is trivial to prove that functions δ_1, δ_2, δ_3, and δ_{prox} satisfy Axiom S1.

Example 4. Let us consider the proximity relation on the domain of attribute *Height* defined as $\mu_{prox}(x, y) = max(0, \frac{5-|x-y|}{5})$. With relation r from Table 1 and the FD $Name \rightarrow Height$, the dirtiness functions defined above yield:

- $\delta_1(r, fd) = 1 - \frac{0+\frac{2}{5}+1+\frac{2}{5}+1+\frac{4}{5}+\frac{1}{5}+0}{8} \approx 0.52$,
- $\delta_2(r, fd) = 1 - min(0, 0, \frac{7}{10}, \frac{4}{10}, \frac{7}{10}, 1, 1, \frac{1}{2}, \frac{4}{10}, \frac{1}{10}, 1) = 1$,
- $\delta_3(r, fd) = \frac{14}{55} \approx 0.25$
  since $\psi(0) = 0$, $\psi(\frac{1}{5}) = \psi(\frac{2}{5}) = \frac{2}{11}$, $\psi(\frac{3}{5}) = \psi(\frac{4}{5}) = \frac{3}{11}$, and $\psi(1) = \frac{4}{11}$,
- $\delta_{prox}(r, fd) = 1 - min(0, min(0.4, 0, 0.4), min(0.8, 0.2, 0)) = 1.\diamond$

Note that both δ_2 and δ_{prox} are sensitive to the existence of large violations of the FD. Still, this effect is softened in δ_2 when there exists both large and slight violations in the same cluster (a feature of δ_2 which may be considered as not very desirable). Note also that the global information provided by δ_{prox} can be completed by the number of clusters for which there exists at least one full violation, i.e., leading to a zero proximity between Y-values.

2.5 Discussion

First, we discuss the respective merits of variance and proximity for characterizing dirtiness, then we point out the limitations of dirtiness measures in general.

Proximity vs. variance. First, one may observe that variance is *computed* inside the clusters for each attribute whereas proximity rather expresses a tolerance which is defined *a priori*. Besides, it appears that variance and proximity are respectively associated with two ways of characterizing dirtiness:

- based on the probability of wrongly retrieving a tuple for a threshold query $A \geq \theta$ or $A \leq \theta$ taken at random on attribute Y. For instance, if we both have $\langle John, 160\rangle$ and $\langle John, 180\rangle$, *John* will be (wrongly) returned for more queries of the form "*height* $\geq \theta$" than if we have $\langle John, 160\rangle$ and $\langle John, 161\rangle$, assuming that $\langle John, 160\rangle$ is the right piece of data. Therefore, one may consider that a tuple is all the more dirty as it wrongly belongs to the result of such a threshold query where θ is far from the correct A-value for the tuple. This viewpoint is not consistent with the use of proximity since proximity-based measures do not make it possible to discriminate between databases in which all of the exceptions (t, t') are such that $\mu_{prox_Y}(t, t') = 0$.

- based on the maximal error that may occur in a query result. One may say that if every cluster satisfies the proximity-based tolerance to a degree $\geq \rho$, then any answer to a threshold query will be acceptable to a degree $\geq \rho$. This view corresponds to the definition of measure δ_{prox}. It implies that, beyond a certain tolerance level, all the wrong answers are considered equally wrong.

Limitations of dirtiness measures. One may notice that:

- it is not easy to compare two databases with these measures. Indeed, dirtiness is a problem with many facets (e.g., number of clusters, cardinality of each cluster, proximity between the Y-values in a given cluster), and too many aggregation steps between the partial measures related to these aspects can make the result difficult to interpret. In Subsection 4.1, we rather propose some indicators which assess these aspects separately.
- at best, the measure makes it possible to compare the dirtiness of a database before and after a cleaning operation has been performed, but it does not give any precise hint about the way to clean it.

3 User Warning

First, let us emphasize the difference between the type of approach considered here and the problematic known as *consistent query answering* [4,15,5,16]. The starting point of CQA is to consider that in many cases cleaning the database from inconsistencies may not be an option, e.g. in virtual data integration, or doing it may be costly, non-deterministic, and may lead to loss of potentially useful data [15]. In consequence, CQA constitutes an alternative approach to data cleaning which consists in basically living with the inconsistent data, but making sure that the consistent data can be identified when queries are answered. In this approach, the first problem that has to be confronted is the one of characterizing in precise terms the notion of consistent data in a possibly inconsistent database. The basic intuition is that a piece of data in the database D is consistent if it is invariant under minimal forms of restoring the consistency of the database, i.e. it remains in every database instance D' that shares the schema with D, is consistent wrt the given ICs, and "minimally differs" from D [4]. Note the similarity of the notion of consistent query answers to that of sure or certain answers studied in the context of incomplete databases [17].

Hereafter, rather than identifying the certain answers to a given (conjunctive) query, we aim at warning the user about the presence of suspect elements in the query result. Roughly speaking, the idea is that such elements can be identified inasmuch they can be found in the answers to contradictory queries.

3.1 Crisp View

Let us consider a conjunctive query Q of the form:

$$\textbf{select } XZ \textbf{ from } r \textbf{ where } cond_1(Y_1) \textbf{ and} \ldots \textbf{and } cond_n(Y_n)$$

where $cond_i$ denotes an atomic condition on attribute Y_i of the form $(Y_i\,\theta_i\,v_i)$, θ_i being a comparator and v_i a constant. Query Q may produce suspicious answers iff there exists an FD $X \to Y_i$ over relation r such that $\delta(r,\, X \to Y_i) > 0$, whatever the measure of dirtiness δ satisfying Axiom S1.

Definition 5. An answer $\langle x,\, z\rangle$ to a query Q of the form described above is suspicious iff $\exists i \in \{1,\, \ldots,\, n\}$ such that

- the FD $X \to Y_i$ over relation r should hold,
- $\exists (t,\, t') \in r$ s.t. $t.X = t'.X = x \wedge truth(cond_i(t.Y_i)) \neq truth(cond_i(t'.Y_i))$.

Note that a value $x = t.X$ such that t is a "suspicious tuple" w.r.t. the FD $X \to Y$ — i.e., $\exists t' \in r$ such that $t.X = t'.X \wedge t.Y \neq t'.Y$ — is not necessarily a suspicious answer to Q. A counter-example is $r = \{\langle John,\, 175\rangle,\, \langle John,\, 190\rangle\}$ where both tuples are suspicious w.r.t. the FD $name \to height$ but *John* is *not* a suspicious answer to the query:

$$\textbf{select } name \textbf{ from } r \textbf{ where } height \geq 170,$$

although the information about *John* is inconsistent, since the disjunction of inconsistent pieces is here implicitly held as true. Let us denote by D_X the subset of $\{Y_1,\, \ldots,\, Y_n\}$ such that the FD $X \to Y_i$ should hold. An efficient way to detect the suspicious answers to a conjunctive query Q of the form

$$\textbf{select } XZ \textbf{ from } r \textbf{ where } cond_1(Y_1) \textbf{ and } \ldots \textbf{and } cond_n(Y_n)$$

consists of the following steps:

- for every $Y_i \in D_X$, compute the result pos_i of

$$\textbf{select } X \textbf{ from } r \textbf{ where } cond_i(Y_i),$$

- for every $Y_i \in D_X$, compute the result neg_i of

$$\textbf{select } X \textbf{ from } r \textbf{ where not } cond_i(Y_i),$$

- Let us denote by res the result of Q. The suspicious answers consist of the set $sus = \{t \in res \,|\, \exists Y_i \in D_X \text{ s.t. } t.X \in pos_i \cap neg_i\}$. Note that the answers which can be considered as certain in the sense that they would be obtained from any maximal consistent subpart of the database, are in $res - sus$. Thus we retrieve both the answers that are "certain", but also suspicious answers, which may be of interest especially when there is no answer of the first type.

Using this procedure, it is possible to order the answers according to their level of "suspiciousness" in the case of conjunctive queries: an answer t is all the more suspicious as $t.X$ belongs to a high number of $(pos_i \cap neg_i)$, i.e., as t is connected with an important number of suspicious attribute values. Let us emphasize that an answer is suspicious *with respect to a query*. For instance, in the example above, *John* is suspicious w.r.t. a query involving the condition $height \geq 180$.

3.2 Graded View

We may also take into account the proximity relation over the domain of Y_i in order to quantify the "suspiciousness" of an answer $u = \langle x, z \rangle$. Let us denote:

$$\mu_{disc_{Y_i}}(u) = 1 - min_{(t,\, t') \in r \mid t.X = t'.X = x \wedge x \in (pos_i \cap neg_i)}\ \mu_{prox_{Y_i}}(t.Y_i,\ t'.Y_i),$$

which corresponds to the highest degree of discrepancy on attribute Y_i among the tuples t such that $t.X = x$.

Now let us build a vector $V(u) = \langle \alpha_1, \ldots, \alpha_n \rangle$ such that α_i is the ith highest degree among the $\mu_{disc_{Y_i}}(u)$'s. The answers u_j can be ranked according to the lexicographic order applied to the vectors $V(u_j)$'s in order to display the most suspicious answers first. One may as well present the less suspicious answers first (using $1 - \mu_{disc_{Y_i}}(u)$ instead).

4 Towards Data Cleaning

Data cleaning deals with detecting and removing errors and inconsistencies from data [18]. Since there are several ways of restoring consistency even in simple cases, it cannot be an objective *per se*. As a matter of fact, restoring consistency by means of a fully automated process implies taking the chance of introducing false data. The approach we advocate hereafter rather aims at helping the administrator clean the database in an intelligent way by i) giving him/her some information about the violated FDs, ii) detecting the "dirty attributes", i.e., the attributes which are the cause of the violation of one or several FDs.

4.1 Assessing the Violated FDs

A first step is to compute, for each FD and for each cluster, some synthesis of the violations of the FD. For instance, one could provide the administrator with the following type of information related to the FD $X \rightarrow Y$:

$$\text{cluster 1 } (X\text{-value } x_1)\text{: } 0/n_{1,\,0}, \sigma_1/n_{1,\,\sigma_1},\ \ldots,\ \sigma_m/n_{1,\,\sigma_m},\ 1/n_{1,\,1},$$

$$\ldots$$

$$\text{cluster } p \text{ } (X\text{-value } x_p)\text{: } 0/n_{p,\,0}, \sigma_1/n_{p,\,\sigma_1},\ \ldots,\ \sigma_m/n_{p,\,\sigma_m},\ 1/n_{p,\,1}.$$

where $\sigma_i/n_{k,\,\sigma_i}$ means that in cluster k, there are $n_{k,\,\sigma_i}$ pairs of tuples which have a proximity degree on Y equal to σ_i.

An additional interesting piece of information is the percentage of tuples from the relation which belong to at least one cluster.

These indicators should help the administrator focus on the attributes which are involved in highly violated FDs (where "highly" is meant qualitatively or quantitatively or both). In particular, the administrator may choose to focus on clusters containing many violations (including small ones), or rather on clusters including severe violations (even if not many of them).

4.2 Suspicious Attributes

A second idea is to consider that an attribute is suspicious if it is involved, as a right-part or a left-part attribute or both, in at least one FD that is violated. Both cases, i.e., right part and left part, must be considered since the inconsistency of (t, t') with respect to an FD $X \rightarrow Y$ either means that $t.Y$ or $t'.Y$ are wrong, or that we have $t.X = t'.Y$ by error (i.e., that $t.X$ or $t'.X$ are wrong). Let us denote by F^+ the transitive closure of the set F of FDs considered.

Definition 6. An attribute A is said to be suspicious if

- there exists an FD $X \rightarrow A$ in F^+ s.t. $\exists (t, t') \in r \times r,\ t.X = t'.X \wedge t.A \neq t'.A$ or
- there exists an FD $AZ \rightarrow Y$ in F^+ (where Z may be empty) such that $\exists (t, t') \in r \times r,\ t.(AZ) = t'.(AZ) \wedge t.Y \neq t'.Y$.

Crisp View. For each attribute A involved in the right part (resp. left part) of a functional dependency from F^+, one measures the ratio $\gamma_R(A)$ (resp. $\gamma_L(A)$) of A-values in the database which are involved in the violation of such an FD. Let us denote by $R(A) = \{f_1, \ldots, f_n\}$ (resp. $L(A) = \{f'_1, \ldots, f'_m\}$) the set of FDs from F^+ whose right part is A (resp. whose left part X contains A, i.e., $X = AZ$) and by $f_1.X, \ldots, f_n.X$ (resp. $f'_1.Y, \ldots, f'_m.Y$) their respective left parts (resp. right parts). Let us define the predicates $except_R(a, r, f_i)$ and $except_L(a, r, f_i)$ as follows:

$$except_R(a, r, f_i) \equiv \exists (t, t') \in r,\ t.(f_i.X) = t'.(f_i.X) \wedge t.A = a \wedge t'.A \neq a.$$

$$except_L(a, r, f'_i) \equiv \\ \exists (t, t') \in r,\ t.(f'_i.X) = t'.(f'_i.X) = \langle a, z \rangle \wedge t.(f'_i.Y) \neq t'.(f'_i.Y).$$

Ratios $\gamma_R(Y)$ and $\gamma_L(Y)$ may be defined as follows:

$$\gamma_R(A) = \frac{|\{a \in r[A] \mid \exists f_i \in R(A),\ except_R(a, r, f_i)\}|}{|r[A]|},$$
$$\gamma_L(A) = \frac{|\{a \in r[A] \mid \exists f'_i \in L(A),\ except_L(a, r, f'_i)\}|}{|r[A]|}.$$

Of course, an attribute A may be such that both $\gamma_R(A)$ and $\gamma_L(A)$ are nonzero. This examplifies the idea that dirtiness-like measures may be useful when used in a focused way.

Graded View. Taking into account the proximity over the domain of attribute A, we can now compute the ratios $\gamma_R(A, \alpha)$ and $\gamma_L(A, \alpha)$ of A-values in the database which are involved in a violation of an FD whose intensity is at least equal to α. The predicates $except_R(a, r, f_i, \alpha)$ and $except_L(a, r, f'_i, \alpha)$ may be defined as follows:

$$except_R(a, r, f_i, \alpha) \equiv \exists (t, t') \in r,\ t.(f_i.X) = t'.(f_i.X) \wedge t.A = a \wedge \\ \mu_{prox_A}(t.A, t'.A) \leq 1 - \alpha.$$

$$except_L(a, r, f'_i, \alpha) \equiv \exists(t, t') \in r, t.(f'_i.X) = t'.(f'_i.X) = \langle a, z\rangle \wedge \mu_{prox_{f'_i.Y}}(t.(f'_i.Y), t'.(f'_i.Y)) \leq 1 - \alpha.$$

Then $\gamma_R(A, \alpha)$ and $\gamma_L(A, \alpha)$ are defined as $\gamma_R(A)$ and $\gamma_L(A)$, except that the predicates $except_R(a, r, f_i)$ and $except_L(a, r, f_i)$ are replaced respectively by $except_R(a, r, f_i, \alpha)$ and $except_L(a, r, f_i, \alpha)$.

Notice that one might also take advantage of the joint existence of several FDs in order to identify those attributes which are particularly suspicious and should be cleaned first. For instance, intuitively, if two FDs $X \rightarrow Y$ and $Y \rightarrow Z$ are violated, while the FD $X \rightarrow Z$ is not, it is likely that attribute Y is taking wrong values in some tuples.

5 Related Work

The literature about data cleaning is quite rich, and in the following we only consider works which use data dependencies. These works are mainly oriented towards an *automated repair process*, which fundamentally distinguishes them from the type of approach outlined in the present paper.

In [19], the author deals with the *constraint repair problem* which attempts to bring a database in accordance with a given set of integrity constraints by applying modifications that are as small as possible. Unlike previous works on this topic, the approach proposed in [19] allows tuple *updates* as a repair primitive and shows that for conjunctive queries and a certain type of dependencies (called *full dependencies*), there exists a condensed representation of all repairs that permits computing trustable query answers.

In [20], Bohannon *et al.* define a database repair as a set of *value modifications* and introduce a cost framework that allows for the application of techniques from record-linkage to the search for good repairs. After proving that finding minimal cost repairs in this model is NP-complete in the size of the database, they introduce an approach to heuristic repair-construction based on equivalence classes of attribute values and define two greedy algorithms.

In [7,21], the authors consider extensions of functional dependencies and inclusion dependencies, referred to as conditional functional dependencies (CFDs) and conditional inclusion dependencies (CINDs), respectively, by additionally specifying patterns of semantically related values; these patterns impose conditions on what part of the relation(s) the dependencies are to hold and which combinations of values should occur together. An example CFD is customer([cc = 44; zip] $\rightarrow$ [$street$]), which asserts that for customers in the UK (cc = 44), zip code determines street. It is an "FD" that is to hold on the subset of tuples that satisfies the pattern $cc = 44$. These constraints are then extended with similarity. In [22], the same authors present Semandaq, a research prototype system for data repairing which supports (a) specifications of CFDs, (b) automatic detections of CFD violations, based on SQL-based techniques, and (c) repairing, i.e., given a set of CFDs and a dirty database, it finds a candidate repair that minimally differs from the original data and satisfies the CFDs [23].

6 Conclusion

In this paper, we have provided a detailed and systematic discussion of different facets of the problem raised by the presence of FD violations in a database. The contributions of this paper are as follows:

- we have discussed two possible approaches to dirtiness measures: one from the literature [6], based on statistical notions (e.g., variance), and a novel one, based on the graded concept of proximity over attribute domains;
- after acknowledging the limitations of dirtiness measures as practical tools for data cleaning, we have proposed an approach *which does not aim at automatically cleaning a database* (which could lead to a loss of useful information), but rather at (i) warning the user in case a query result contains suspicious answers (i.e., elements related to inconsistencies), (ii) helping the administrator detect the attributes which constitute the causes of the inconsistencies, so that he/she can "intelligently" clean the database.

Several perspectives can be thought of, among which:

- generalization to other kinds of integrity constraints than functional dependencies, in particular association constraints, whose basic form is:

  $$\forall t \in r,\ t.A_1\, \theta_1\, v_1\ \wedge\ \ldots\ \wedge\ t.A_m\, \theta_m\, v_m \Rightarrow t.A_p\, \theta_p\, v_p$$

  where A_i is an attribute, θ_i a comparator, and v_i a constant. An example is "a person with age below 2 years should have a weight below 50 pounds". A particular case is when the left part of the implication is replaced by *true* and we get an existence constraint, of the form:

  $$t \in r \Rightarrow t.A_1\, \theta_p\, v_1\ \wedge\ \ldots\ \wedge t.A_n \theta_n v_n.$$

  A extension of the basic patterns above consists in allowing comparisons between attributes (i.e., conditions of the form $A_{i,1}\, \theta_i\, A_{i,2}$) in the left and/or right parts of the implication.
- implementation and user study aimed at assessing the practical interest of the methods proposed here.

References

1. Decker, H., Martinenghi, D.: Getting rid of straitjackets for flexible integrity checking. In: DEXA Workshops, pp. 360–364. IEEE Computer Society, Los Alamitos (2007)
2. Martinenghi, D., Christiansen, H., Decker, H.: Integrity checking and maintenance in relational and deductive databases and beyond. In: Ma, Z. (ed.) Intelligent Databases: Technologies and Applications, pp. 238–285. Idea Group, USA (2006)
3. Decker, H., Martinenghi, D.: Avenues to flexible data integrity checking. In: DEXA Workshops, pp. 425–429. IEEE Computer Society, Los Alamitos (2006)

4. Arenas, M., Bertossi, L.E., Chomicki, J.: Answer sets for consistent query answering in inconsistent databases. TPLP 3(4-5), 393–424 (2003)
5. Wijsen, J.: Project-join-repair: An approach to consistent query answering under functional dependencies. In: Larsen, H.L., Pasi, G., Ortiz-Arroyo, D., Andreasen, T., Christiansen, H. (eds.) FQAS 2006. LNCS (LNAI), vol. 4027, pp. 1–12. Springer, Heidelberg (2006)
6. Martinez, M.V., Pugliese, A., Simari, G.I., Subrahmanian, V.S., Prade, H.: How dirty is your relational database? An axiomatic approach. In: Mellouli, K. (ed.) ECSQARU 2007. LNCS (LNAI), vol. 4724, pp. 103–114. Springer, Heidelberg (2007)
7. Bohannon, P., Fan, W., Geerts, F., Jia, X., Kementsietsidis, A.: Conditional functional dependencies for data cleaning. In: Proc. of ICDE 2007, pp. 746–755 (2007)
8. Delgado, M., Martin-Bautista, M.-J., Sanchez, D., Vila, M.-A.: Mining strong approximate dependencies from relational databases. In: Proc. of IPMU 2000, pp. 1123–1130 (2000)
9. Kivinen, J., Mannila, H.: Approximate inference of functional dependencies from relations. Theor. Comput. Sci. 149(1), 129–149 (1995)
10. Baral, C., Kraus, S., Minker, J., Subrahmanian, V.S.: Combining knowledge bases consisting of first-order analysis. Computational Intelligence 8, 45–71 (1992)
11. Lozinskii, E.L.: Resolving contradictions: A plausible semantics for inconsistent systems. J. Autom. Reasoning 12(1), 1–32 (1994)
12. Hunter, A., Konieczny, S.: Approaches to measuring inconsistent information. In: Bertossi, L., Hunter, A., Schaub, T. (eds.) Inconsistency Tolerance. LNCS, vol. 3300, pp. 191–236. Springer, Heidelberg (2005)
13. Grant, J., Hunter, A.: Measuring inconsistency in knowledgebases. J. Intell. Inf. Syst. 27(2), 159–184 (2006)
14. De Luca, A., Termini, S.: A definition of a nonprobabilistic entropy in the setting of fuzzy sets theory. Information and Control 20(4), 301–312 (1972)
15. Bertossi, L.E.: Consistent query answering in databases. SIGMOD Record 35(2), 68–76 (2006)
16. Chomicki, J.: Consistent query answering: Five easy pieces. In: Schwentick, T., Suciu, D. (eds.) ICDT 2007. LNCS, vol. 4353, pp. 1–17. Springer, Heidelberg (2006)
17. Lipski, W.: On semantic issues connected with incomplete information databases. ACM Transactions on Database Systems 4(3), 262–296 (1979)
18. Rahm, E., Do, H.H.: Data cleaning: Problems and current approaches. IEEE Data Eng. Bull. 23(4), 3–13 (2000)
19. Wijsen, J.: Condensed representation of database repairs for consistent query answering. In: Calvanese, D., Lenzerini, M., Motwani, R. (eds.) ICDT 2003. LNCS, vol. 2572, pp. 375–390. Springer, Heidelberg (2002)
20. Bohannon, P., Flaster, M., Fan, W., Rastogi, R.: A cost-based model and effective heuristic for repairing constraints by value modification. In: SIGMOD Conference, pp. 143–154 (2005)
21. Fan, W., Geerts, F., Jia, X.: Conditional dependencies: A principled approach to improving data quality. In: Proc. of BNCOD 2009, pp. 8–20 (2009)
22. Fan, W., Geerts, F., Jia, X.: Semandaq: a data quality system based on conditional functional dependencies. PVLDB 1(2), 1460–1463 (2008)
23. Cong, G., Fan, W., Geerts, F., Jia, X., Ma, S.: Improving data quality: Consistency and accuracy. In: Proc. of VLDB 2007 07, pp. 315–326 (2007)

Disjunctive Fuzzy Logic Programs with Fuzzy Answer Set Semantics

Emad Saad

Department of Computer Science
Gulf University for Science and Technology
Mishref, Kuwait
saad.e@gust.edu.kw

Abstract. Reasoning under fuzzy uncertainty arises in many applications including planning and scheduling in fuzzy environments. In many real-world applications, it is necessary to define fuzzy uncertainty over qualitative uncertainty, where fuzzy values are assigned over the possible outcomes of qualitative uncertainty. However, current fuzzy logic programming frameworks support only reasoning under fuzzy uncertainty. Moreover, disjunctive logic programs, although used for reasoning under qualitative uncertainty it cannot be used for reasoning with fuzzy uncertainty. In this paper we combine extended and normal fuzzy logic programs [30, 23], for reasoning under fuzzy uncertainty, with disjunctive logic programs [7, 4], for reasoning under qualitative uncertainty, in a unified logic programming framework, namely extended and normal disjunctive fuzzy logic programs. This is to allow directly and intuitively to represent and reason in the presence of both fuzzy uncertainty and qualitative uncertainty. The syntax and semantics of extended and normal disjunctive fuzzy logic programs naturally extends and subsumes the syntax and semantics of extended and normal fuzzy logic programs [30, 23] and disjunctive logic programs [7, 4]. Moreover, we show that extended and normal disjunctive fuzzy logic programs can be intuitively used for representing and reasoning about scheduling with fuzzy preferences.

1 Introduction

Reasoning under fuzzy uncertainty arises in many applications including planning and scheduling in fuzzy environments as in robotics planning in real-world environments. Among the approaches for reasoning in the presence of fuzzy uncertainty is *fuzzy logic programming* [18, 9, 29, 15, 28, 31, 10, 30, 23]. Disjunctive logic programs with classical answer set semantics [7, 4] present an alternative approach for reasoning under uncertainty; namely qualitative uncertainty reasoning. It often be the case that $p \vee q \vee r$ occurs while we are uncertain which of these propositions is true [7, 4]. There might be states of the world where p is true or q is true or r is true.

One of the earliest fuzzy logic programming frameworks has been introduced in [31] that generalizes definite logic programs (logic programs without either

A. Deshpande and A. Hunter (Eds.): SUM 2010, LNAI 6379, pp. 306–318, 2010.

classical or non-monotonic negation). Considering fuzzy uncertainty, a more general fuzzy logic programs (without negation) framework has been introduced in [10], whose semantics subsumes the fuzzy logic programming framework of [31]. A model theoretic semantics has been defined for fuzzy logic programs of [10]. Normal fuzzy logic programs with stable fuzzy model semantics and alternating fixed point well-founded fuzzy model semantics have been presented in [30]. Normal fuzzy logic programs extend fuzzy logic programs of [10] to allow non-monotonic negation. In addition, normal fuzzy logic programs of [30] have been extended in [23] to extended fuzzy logic programs with fuzzy answer set semantics that allow both classical negation and non-monotonic negation. Moreover, in [23] fixpoint semantics for extended fuzzy logic programs with and without non-monotonic negation has been defined and their relationship to the fuzzy answer set semantics is studied. Furthermore, it has been shown in [23] that fuzzy answer set semantics of extended fuzzy logic programs is a natural extension to the classical answer set semantics of classical extended logic programs [7] and fuzzy answer set semantics of extended fuzzy logic programs is reduced to stable fuzzy model semantics of normal fuzzy logic programs of [30].

The importance of the fuzzy logic programming frameworks of [10, 30, 23] lies in the fact that [10, 30, 23] are strictly more expressive than the fuzzy logic programming frameworks of [18, 9, 29, 15, 28, 31], in addition, the way a rule is fired in [10, 30, 23] is close to the way it fires in classical logic programming. This is an important feature in these frameworks because it makes any possible extension to [10, 30, 23] to more expressive forms of logic programming with uncertainty, in general, and with fuzzy uncertainty, in particular, including the addition of negation as failure, classical negation, and disjunctions, is more intuitive and more flexible.

Therefore, normal and extended fuzzy logic programs [30, 23] are used for reasoning under fuzzy uncertainty, in addition, disjunctive logic programs [7, 4] are used for representing and reasoning under qualitative uncertainty. However, in many real-world applications, representing and reasoning with both forms of uncertainty is necessary. In real-world applications, it can be the case that fuzzy uncertainty need to be defined over qualitative uncertainty, where fuzzy values are assigned over the possible outcomes of qualitative uncertainty.

For example, consider a simple course assignment problem where one of two courses, c_1, c_2, need to be assigned to an instructor i such that instructor i is assigned exactly one course. If the instructor is neutral regarding teaching either course, then disjunctive logic program can be used to model this problem as a disjunctive logic program of the form $teaches(i, c_1)$ *or* $teaches(i, c_2)$ with $\{teaches(i, c_1)\}$ and $\{teaches(i, c_2)\}$ are the possible answer sets, according to the answer set semantics of disjunctive logic programs [7, 4]. Consider instructor i prefers to teach c_1 over c_2, where this preference relation is specified as a fuzzy set over the courses c_1, c_2. Consider instructor i preference in teaching c_1 is characterized by the grade membership value 0.8 and instructor i preference in teaching c_2 is characterized by the grade membership value 0.3. In this case, disjunctive logic programs cannot represent instructor preferences over courses, as disjunctive logic programs are incapable of reasoning in the presence of fuzzy uncertainty. On the other hand, this course assignment problem cannot be represented intuitively and directly in normal fuzzy logic program or extended fuzzy logic program either, since disjunctions are not allowed in these kind of fuzzy logic programs.

In this paper we integrate disjunctive logic programs [7, 4] with normal fuzzy logic programs [30] and extended fuzzy logic programs [23] in a unified logic programming framework to allow directly and intuitively to represent and reason in the presence of both fuzzy uncertainty and qualitative uncertainty.

This is achieved by defining the notions of extended and normal disjunctive fuzzy logic programs, which generalize extended and normal disjunctive logic programs of classical logic programming [7, 4], respectively. In addition, extended and normal disjunctive fuzzy logic programs generalize extended fuzzy logic programs [23] and normal fuzzy logic programs [30], respectively. The semantics of extended and normal disjunctive fuzzy logic programs are based on the answer sets semantics and stable model semantics of extended and normal disjunctive logic programs [7, 4]. We show that extended disjunctive fuzzy logic programs naturally subsumes extended disjunctive logic programs [7] and extended fuzzy logic programs [23], and normal disjunctive fuzzy logic programs naturally subsumes normal disjunctive logic programs [4] and normal fuzzy logic programs [30]. Moreover, we show that the fuzzy answer set semantics of extended disjunctive fuzzy logic programs is reduced to stable fuzzy model semantics of normal disjunctive fuzzy logic programs. The importance of that is computational methods developed for normal disjunctive fuzzy logic programs can be applied to extended disjunctive fuzzy logic programs.

2 Fuzzy Sets

In this section we review the basic notions of fuzzy sets as presented in [32]. Let U be a set of objects. A fuzzy set, F, in U is defined by the grade membership function $\mu_F : U \to [0,1]$, where for each element $x \in U$, μ_F assigns to x a value $\mu_F(x)$ in $[0,1]$. The support for F denotes the set of all objects x in U for which the grade membership of x in F is a non-zero value. Formally, $support(F) = \{x \in U \mid \mu_F(x) > 0\}$. The intersection (conjunction) of two fuzzy sets F and F' in U, denoted by $F \wedge_f F'$ is a fuzzy set G in U where the grade membership function of G is $\mu_G(x) = \min(\mu_F(x), \mu_{F'}(x))$ for all $x \in U$. However, the union (disjunction) of two fuzzy sets F and F' in U, denoted by $F \vee_f F'$ is a fuzzy set G in U where the grade membership function of G is $\mu_G(x) = \max(\mu_F(x), \mu_{F'}(x))$ for all $x \in U$. The complement (negation) of a fuzzy set F in U is a fuzzy set in U denoted by $\overline{F}$ where the grade membership function of $\overline{F}$ is $\mu_{\overline{F}}(x) = 1 - \mu_F(x)$ for all $x \in U$. A fuzzy set F in U is said to be contained in another fuzzy set G in U if and only if $\mu_F(x) \leq \mu_G(x)$ for all $x \in U$. Notice that we use the notations $\wedge_f$ and $\vee_f$ to denote fuzzy conjunction and fuzzy disjunction respectively to distinguish them from $\wedge$ and $\vee$ for propositional conjunction and disjunction respectively. Furthermore, other function characterizations for the fuzzy conjunction and fuzzy disjunction operators can be used. However, we will stick with the min and max function characterizations for the fuzzy conjunction and fuzzy disjunction as originally proposed in [32].

3 Extended and Normal Disjunctive Fuzzy Logic Programs

Extended and normal disjunctive fuzzy logic programs syntax is presented in this section, which are fuzzy logic programs with classical negation, non-monotonic negation, and disjunction in head of rules, whose underlying semantics is the fuzzy set theory.

We consider first-order language L with finitely many predicate symbols, constants, and infinitely many variables. A literal is either an atom a in B_L or the negation of a ($\neg a$), where B_L is the Herbrand base of L and $\neg$ is the classical negation. Non-monotonic negation or the negation as failure is denoted by *not*. Let Lit be the set of all literals in L, where $Lit = \{a | a \in B_L\} \cup \{\neg a | a \in B_L\}$. The grade membership are assigned to literals in L as values from $[0, 1]$. Let $\alpha_1, \alpha_2 \in [0, 1]$. The set $[0, 1]$ and the relation $\leq$ form a complete lattice, where the join ($\oplus$) operation is defined as $\alpha_1 \oplus \alpha_2 = \max(\alpha_1, \alpha_2)$ and the meet ($\otimes$) is defined as $\alpha_1 \otimes \alpha_2 = \min(\alpha_1, \alpha_2)$. An *annotation*, α, is either a constant in $[0, 1]$, a variable (*annotation variable*) ranging over $[0, 1]$, or $f(\alpha_1, \ldots, \alpha_n)$ (called *annotation function*) where f is a representation of a computable total function $f : ([0, 1])^n \to [0, 1]$ and $\alpha_1, \ldots, \alpha_n$ are annotations.

Definition 1 (Rules). *An extended disjunctive fuzzy rule (ed-rule) is an expression of the form*

$$l_1 : \mu_1 \ or \ldots or \ l_k : \mu_k \leftarrow l_{k+1} : \mu_{k+1}, \ldots, l_m : \mu_m, not \ l_{m+1} : \mu_{m+1}, \ldots, not \ l_n : \mu_n,$$

whereas a normal disjunctive fuzzy rule (nd-rule) is an expression of the form

$$a_1 : \mu_1 \ or \ldots or \ a_k : \mu_k \leftarrow a_{k+1} : \mu_{k+1}, \ldots, a_m : \mu_m, not \ a_{m+1} : \mu_{m+1}, \ldots, not \ a_n : \mu_n,$$

where $\forall (1 \leq i \leq n)$, l_i *is a literal,* a_i *is an atom, and* μ_i *is an annotation. An ed-rule*^{not} *and nd-rule*^{not} *are ed-rule and nd-rule without non-monotonic negation respectively—i.e.,* $n = m$.

Intuitively, for any ed-rule, if $\forall (k+1 \leq i \leq m)$ $l_i : \mu_i$ it is *known* that the grade membership of l_i is at least μ_i and $\forall (m+1 \leq j \leq n)$ *not* $l_j : \mu_j$ it is *not known (undecidable)* that the grade membership of l_j is at least μ_j, then there exist at least $(1 \leq i \leq k)$ l_i such that the grade membership of l_i is at least μ_i. However, for any nd-rule, if $\forall (k+1 \leq i \leq m)$ $a_i : \mu_i$ it is *believable* that the grade membership of a_i is at least μ_i and $\forall (m+1 \leq j \leq n)$ *not* $a_j : \mu_j$ it is *not believable* that the grade membership of a_j is at least μ_j, then there exist at least $(1 \leq i \leq k)$ a_i such that the grade membership of a_i is at least μ_i.

Definition 2 (Programs). *An extended (normal) disjunctive fuzzy logic program, ed-program (nd-program), is a finite set of ed-rules (nd-rules). An ed-program*^{not} *(nd-program*^{not}*) is an ed-program (nd-program) whose rules are ed-rules*^{not} *(nd-rules*^{not}*).*

An ed-program^{not} (nd-program^{not}) is an ed-program (nd-program) without non-monotonic negation. An extended (normal) disjunctive fuzzy logic program is ground if no variables appear in any of its rules. The following is a typical extended (normal) disjunctive fuzzy logic program inspired from [2].

Example 1. Assume that we have n instructors (denoted by $l_1, \ldots, l_n$) that are assigned n different courses (denoted by $c_1, \ldots, c_n$) in m rooms (denoted by $r_1, \ldots, r_m$) at k different time slots (denoted by $s_1, \ldots, s_k$) such that; each instructor is assigned exactly one course; no two different courses can be taught in the same room at the same time; each instructor prefers to teach the courses (s)he likes, where instructors preferences over courses is a fuzzy set over courses; each instructor likes to teach at some time slots over others, where instructors preferences over time slots is a fuzzy set over time slots; and each instructor likes to teach in some rooms over others, where instructors preferences over rooms is a fuzzy set over rooms. This course scheduling problem can be represented as an ed-program as follows:

$$teaches(l_i, c_1) : \mu_{i,1} \ or \ teaches(l_i, c_2) : \mu_{i,2} \ or \ \ldots or \ teaches(l_i, c_n) : \mu_{i,n} \leftarrow \qquad \forall i \in n$$

$$in(r_1, C) : \nu_{i,1} \ or \ in(r_2, C) : \nu_{i,2} \ldots or \ in(r_m, C) : \nu_{i,m} \leftarrow teaches(l_i, C) : V, course(C) : 1 \qquad \forall i \in n$$

$$at(s_1, C) : v_{i,1} \ or \ at(s_2, C) : v_{i,2} \ or \ldots or \ at(s_k, C) : v_{i,k} \leftarrow teaches(l_i, C) : V, course(C) : 1 \qquad \forall i \in n$$

$$inconsistent : 1 \leftarrow not \ inconsistent : 1, teaches(I, C) : V_1, teaches(I, C') : V_2, C \neq C'$$

$$inconsistent : 1 \leftarrow not \ inconsistent : 1, in(R, C) : V_1, in(R, C') : V_2, at(S, C) : V_3, at(S, C') : V_4, C \neq C'$$

where $V, V_1, \ldots, V_n$ are annotation variables act as place holders, $teaches(l_i, c_j) : \mu_{i,j}$ represents that instructor l_i likes to teach course c_j with grade membership $\mu_{i,j}$ (which is the instructor preference in teaching the course), $in(r_j, C) : \nu_{i,j}$, for any $j \in m$, represents that instructor l_i likes to teach course C in room r_j with grade membership $\nu_{i,j}$, $at(s_j, C) : v_{i,j}$, for any $j \in k$, represents that instructor l_i likes to teach course C in time slot s_j with grade membership $v_{i,j}$.

The first three ed-rules encode the instructors preference over courses, rooms, and time slots. However, the last two ed-rules encode the constraints that an instructor is assigned exactly one course and different courses cannot be taught in the same room at the same time.

3.1 Satisfaction and Models

Interpretations, models, satisfaction, and the semantics of extended and normal disjunctive fuzzy logic programs are defined in this section.

Definition 3. *A fuzzy interpretation, I, of an ed-program is a fuzzy set in the set of all literals Lit where the grade membership function of I is a mapping $\mu_I : Lit \to [0, 1]$. We say that a fuzzy interpretation I is a partial fuzzy interpretation* *iff the grade membership function of I is a partial mapping from Lit to $[0,1]$. A fuzzy interpretation of an nd-program is a fuzzy set in the set of all atoms B_L, whose grade membership function is a mapping $B_L \rightarrow [0,1]$.*

For the ease of the presentation, we use $I : Lit \rightarrow [0,1]$ to refer to a fuzzy interpretation of an ed-program, where the grade membership of a literal l in the fuzzy interpretation I is $I(l)$. Similarly, for nd-programs, a fuzzy interpretation, I, is viewed as a mapping $I : B_L \rightarrow [0,1]$, where the grade membership of an atom, a in I is given by $I(a)$. If the grade membership of a literal, l, in the fuzzy interpretation, I, of an ed-program is $I(l)$, then the grade membership of the negation of l $(\neg\, l)$ in I is $I(\neg\, l) = 1 - I(l)$. As a literal and its negation are allowed in fuzzy interpretations of ed-programs, more conditions are required to ensure their consistency, which are specified by in the following definitions. Let $dom(I)$ denotes the domain of I.

Definition 4. *Let I be a total or partial fuzzy interpretation of an ed-program. We say I is inconsistent if there exists $l, \neg l \in Lit$ $(l, \neg l \in dom(I))$ such that $I(\neg l) \neq 1 - I(l)$.*

Definition 5. *Let S be a subset of literals from Lit. We say that S is a set of consistent literals if there is no pair of complementary literals l and $\neg l$ belonging to S.*

Definition 6. *Let I be fuzzy interpretation of an ed-program. Then, I is consistent if it is either not inconsistent or maps a consistent set of literals S to $[0,1]$.*

Intuitively, a consistent fuzzy interpretation of an ed-program, I, is a fuzzy interpretation such that for every $l, \neg l \in dom(I)$, $I(\neg l) = 1 - I(l)$ or it maps a consistent set of literals into $[0,1]$. Let I_1 and I_2 be two (partial or total) fuzzy interpretations in Lit. If I is total fuzzy interpretation then $dom(I) \subseteq Lit$, however, if I is partial fuzzy interpretation then $dom(I) \subsetneqq Lit$. We say $I_1 \leq I_2$ iff $dom(I_1) \subseteq dom(I_2)$ and $\forall l \in dom(I_1)$ we have $I_1(l) \leq I_2(l)$. The set of all fuzzy interpretations in Lit (denoted by $\mathcal{F}$) and the relation $\leq$ form a complete lattice. The meet $\otimes$ and the join $\oplus$ operations on $\mathcal{F}$ are defined as follows.

Definition 7. *Let I_1 and I_2 be two partial fuzzy interpretations of an ed-program. The meet $\otimes$ and join $\oplus$ operations corresponding to the partial order $\leq$ are defined respectively as:*

- $(I_1 \otimes I_2)(l) = I_1(l) \otimes I_2(l) = \min(I_1(l), I_2(l))$ *for all l defined in both I_1 and I_2, otherwise, undefined.*
- $(I_1 \oplus I_2)(l)$ *is equal to*
  - $I_1(l) \oplus I_2(l) = \max(I_1(l), I_2(l))$ *for all l defined in both I_1 and I_2.*
  - $(I_1 \oplus I_2)(l) = I_1(l)$ *for all l defined in I_1 but not defined in I_2*
  - $(I_1 \oplus I_2)(l) = I_2(l)$ *for all l defined in I_2 but not defined in I_1*
  - *otherwise, undefined.*

Definition 8 (Fuzzy Satisfaction). *Let P be a ground ed-program, I be a fuzzy interpretation of P, and r be*

$$l_1 : \mu_1 \ or \ldots or \ l_k : \mu_k \leftarrow l_{k+1} : \mu_{k+1}, \ldots, l_m : \mu_m, not \ l_{m+1} : \mu_{m+1}, \ldots, not \ l_n : \mu_n,$$

Then

- *I satisfies $l_i : \mu_i$ (denoted by $I \models l_i : \mu_i$) iff $\mu_i \leq I(l_i)$ and $l_i \in dom(I)$.*
- *I satisfies not $l_i : \mu_i$ (denoted by $I \models not \ l_i : \mu_i$) iff $\mu_i \nleq I(l_i)$ and $l_i \in dom(I)$ or $l_i \notin dom(I)$.*
- *I satisfies $Body \equiv l_{k+1} : \mu_{k+1}, \ldots, l_m : \mu_m, not \ l_{m+1} : \mu_{m+1}, \ldots, not \ l_n : \mu_n$ (denoted by $I \models Body$) iff $\forall (k+1 \leq i \leq m), I \models l_i : \mu_i$ and $\forall (m+1 \leq i \leq n), I \models not \ l_i : \mu_i$.*
- *I satisfies $Head \equiv l_1 : \mu_1 \ or \ldots or \ l_k : \mu_k$ (denoted by $I \models Head$) iff there exists at least i $(1 \leq i \leq k)$ such that $I \models l_i : \mu_i$.*
- *I satisfies $Head \leftarrow Body$ iff $I \models Head$ whenever $I \models Body$ or I does not satisfy $Body$.*
- *I satisfies P iff I satisfies every ed-rule in P and for every literal $l_i \in dom(I)$, we have $\max\{\!\{\mu_i \ (1 \leq i \leq k) \mid l_1 : \mu_1 \ or \ldots or \ l_k : \mu_k \leftarrow Body \in P, I \models Body, \text{ and } I \models l_i : \mu_i\}\!\} \leq I(l_i)$.*

The definition of fuzzy satisfaction of nd-programs is similar to the definition of fuzzy satisfaction of ed-programs as described in Definition 8. The only difference is that nd-programs disallow classical negation and contain only atoms. Moreover, a fuzzy interpretation of an nd-program is a total mapping from B_L to $[0, 1]$.

Definition 9 (Models). *A fuzzy model for an ed-program (nd-program), P, with or without non-monotoinc negation is a fuzzy interpretation of P that satisfies P.*

Let I be a fuzzy model of an ed-program (nd-program), P, with or without non-monotonic negation. Then I is a minimal fuzzy model of P if there is no fuzzy model I' of P such that $I' < I$ w.r.t. $\leq$.

Example 2. Consider the following ed-program P, without non-monotonic negation, where P includes

$$\begin{array}{ll} p : 0.93 & or \ \neg q : 0.8 \\ \neg r : 0.78 & \leftarrow p : 0.86 \\ s : 0.9 & \leftarrow \neg q : 0.7 \\ \neg s : 0.1 & \leftarrow p : 0.65, \neg q : 0.7, \neg r : 0.94 \end{array}$$

It can be easily seen that P has two minimal fuzzy models I_1 and I_2, where

$$\begin{array}{ll} I_1(p) & = 0.93 \\ I_1(\neg r) & = 0.78 \\ I_2(\neg q) & = 0.8 \\ I_2(s) & = 0.9 \end{array}$$

4 Fuzzy Answer Sets and Fuzzy Stable Models Semantics

In this section, we define the *fuzzy answer set semantics* of ed-programs and the *stable fuzzy model semantics* of nd-programs, which generalize the classical answer set semantics and stable model semantics of classical disjunctive logic programs. The proposed semantics are defined by guess and verify as follows; guessing a fuzzy answer set (stable fuzzy model), I, of an ed-program (nd-program), P, then verifying whether I is a fuzzy answer set (stable fuzzy model) of P. I is a fuzzy answer set (stable fuzzy model) of P if I is a minimal fuzzy model of the fuzzy reduct of P w.r.t. I.

Definition 10 (Fuzzy Reduct). *Let P be a ground ed-program (nd-program) and I be a fuzzy interpretation. The fuzzy reduct P^I of P w.r.t. I is P^I where:*

$$l_1 : \mu_1 \; or \ldots \; or \; l_k : \mu_k \leftarrow l_{k+1} : \mu_{k+1}, \ldots, l_m : \mu_m \in P^I$$

iff

$$l_1 : \mu_1 \; or \ldots \; or \; l_k : \mu_k \leftarrow l_{k+1} : \mu_{k+1}, \ldots, l_m : \mu_m, not \; l_{m+1} : \mu_{m+1}, \ldots, not \; l_n : \mu_n \in P$$

such that $\forall (m+1 \leq j \leq n)$, $\mu_j \nleq_t I(l_j)$ *or* $l_j \notin dom(I)$

Observe that the fuzzy reduct definition of nd-programs is similar to the definition of fuzzy reduct of ed-programs. However, contrary to ed-programs, nd-programs do not allow classical negation, furthermore, fuzzy interpretations in nd-programs are total mappings from B_L to $[0,1]$, therefore, the condition $l_j \notin dom(I)$ is not applicable for nd-programs.

The fuzzy reduct, P^I, of P w.r.t. I is an ed-program (nd-program) without non-monotonic negation whose intuitive meaning is that if $\mu_j \nleq I(l_j)$ for *not* $l_j : \mu_j$ in the body of a rule $r \in P$, then it is *not known (not believable for nd-program)* that the grade membership of l_j is at least μ_j given the available knowledge in P. This means *not* $l_j : \mu_j$ is satisfied and hence removed from the body of r. Nevertheless, if $l_j \notin dom(I)$ (for ed-program only), i.e., l_j is undefined in I, then it is entirely *not known* that the grade membership of l_j is at least μ_j. This implies that *not* $l_j : \mu_j$ is satisfied and thus removed from the body of r. However, if $\mu_j \leq I(l_j)$ (for both ed-programs and nd-programs), implies that the grade membership of l_j is at least μ_j hence the body of r is not satisfied and r is trivially ignored.

Definition 11. *A fuzzy interpretation I of an ed-program (nd-program) P is a fuzzy answer set (stable fuzzy model) of P if I is a minimal fuzzy model of P^I.*

Since, ed-programs allow classical negation, it is possible to have an ed-program that is inconsistent. We say an ed-program, P, with or without non-monotonic negation is inconsistent if it has an inconsistent fuzzy answer set. In this case we say, LIT, where $LIT : Lit \rightarrow \{1\}$, is the fuzzy answer set of P. This implies that every literal with the grade membership 1 follows from P. This extends the definition of inconsistent classical extended disjunctive logic programs [7].

Observe that ed-programs without classical negation are nd-programs whose fuzzy answer sets have domains that consists of only atoms. Furthermore, the definition of fuzzy answer sets semantics is equivalent to the definition of stable fuzzy models semantics for the class of nd-programs. This means that the application of the fuzzy answer set semantics to nd-programs boils down to the stable fuzzy model semantics for nd-programs. However, there are two main differences between the two semantics. A fuzzy answer set of an nd-program may be a partial fuzzy model, but a stable fuzzy model for an nd-program is a total fuzzy model. In addition, if an atom a is undefined in a fuzzy answer set, I, of an nd-program P, implies that a has a grade membership equal to 0 in a stable fuzzy model I' of P that is equivalent to I. This means a is undefined and hence unknown in the fuzzy answer set of P but a is false in the stable fuzzy model P.

Proposition 1. *Let P be an nd-program. Then I is a fuzzy answer set for P iff I' is a stable fuzzy model of P, where $I(a) = I'(a)$ for each $I'(a) \neq 0$ and $I(a)$ is undefined for each $I'(a) = 0$.*

Proposition 1 shows that ed-programs can be reduced to nd-programs via simple reduction. The importance of that is computational methods developed for nd-programs can be applied to ed-programs under the consistency condition.

Example 3. Assume that we want to schedule two different courses, denoted by c_1, c_2, to two different instructors, named i_1, i_2, given that we have only one room, denoted by r_1, and two different time slots, denoted by s_1, s_2. Thus, the ed-program in Example 1 can be written as follows given that the last two ed-rules of Example 1 are kept unchanged and the instructors preferences over courses, rooms, and time slots are given as described below.

$$
\begin{array}{l}
teaches(i_1, c_1) : 0.9 \text{ or } teaches(i_1, c_2) : 0.5 \leftarrow \\
teaches(i_2, c_1) : 0.4 \text{ or } teaches(i_2, c_2) : 0.7 \leftarrow \\
in(r_1, C) : 0.8 \leftarrow teaches(i_1, C) : V, course(C) : 1 \\
in(r_1, C) : 0.3 \leftarrow teaches(i_2, C) : V, course(C) : 1 \\
at(s_1, C) : 0.5 \text{ or } at(s_2, C) : 0.5 \leftarrow teaches(i_1, C) : V, course(C) : 1 \\
at(s_1, C) : 0.9 \text{ or } at(s_2, C) : 0.2 \leftarrow teaches(i_2, C) : V, course(C) : 1 \\
course(c_1) : 1 \leftarrow \\
course(c_2) : 1 \leftarrow
\end{array}
$$

The above ed-program has four fuzzy answer sets as follows. For the ease of presentation, we present these fuzzy answer sets as sets of annotated literals as:

$$
\begin{array}{ll}
I_1 = \{ & teaches(i_1, c_1) : 0.9,\ teaches(i_2, c_2) : 0.7,\ at(s_1, c_1) : 0.5,\ at(s_2, c_2) : 0.2, \\
& in(r_1, c_1) : 0.8,\ in(r_1, c_2) : 0.3,\ course(c_1) : 1,\ course(c_2) : 1\}
\end{array}
$$

$$
\begin{array}{ll}
I_2 = \{ & teaches(i_1, c_1) : 0.9,\ teaches(i_2, c_2) : 0.7,\ at(s_2, c_1) : 0.5,\ at(s_1, c_2) : 0.9, \\
& in(r_1, c_1) : 0.8,\ in(r_1, c_2) : 0.3,\ course(c_1) : 1,\ course(c_2) : 1\}
\end{array}
$$

$$I_3 = \{teaches(i_1, c_2) : 0.5,\ teaches(i_2, c_1) : 0.4,\ at(s_1, c_1) : 0.9,\ at(s_2, c_2) : 0.5,\ in(r_1, c_1) : 0.3,\ in(r_1, c_2) : 0.8,\ course(c_1) : 1,\ course(c_2) : 1\}$$

$$I_4 = \{teaches(i_1, c_2) : 0.5,\ teaches(i_2, c_1) : 0.4,\ at(s_2, c_1) : 0.2,\ at(s_1, c_2) : 0.5,\ in(r_1, c_1) : 0.3,\ in(r_1, c_2) : 0.8,\ course(c_1) : 1,\ course(c_2) : 1\}$$

It can be seen that the fuzzy answer sets I_2 and I_3 are more preferable over the answer sets I_1 and I_4 and I_2 is more preferable over I_3.

Now we show that ed-programs and nd-programs naturally extend extended fuzzy logic programs [23] and normal fuzzy logic programs respectively [30].

Proposition 2. *The fuzzy answer set semantics of ed-programs is equivalent to the fuzzy answer set semantics of extended fuzzy logic programs [23] for all ed-programs P such that $\forall\ r \in P, k = 1$. In addition, the stable fuzzy model semantics of nd-programs is equivalent to the stable fuzzy model semantics of normal fuzzy logic programs [30] for all nd-programs P such that $\forall\ r \in P, k = 1$.*

In the rest of this section we show that the fuzzy answer set semantics of ed-programs and the stable fuzzy model semantics of nd-programs naturally extend the answer set semantics and the stable model semantics of classical extended and normal disjunctive logic programs respectively [7, 4]. A classical extended disjunctive logic program P can be represented as an ed-program P' where each classical extended disjunctive rule

$$l_1 \ or \ldots or\ l_k \leftarrow l_{k+1}, \ldots, l_m, not\ l_{m+1}, \ldots, not\ l_n \in P$$

can be represented, in P', as an ed-rule of the form

$$l_1 : 1\ or \ldots or\ l_k : 1 \leftarrow l_{k+1} : 1, \ldots, l_m : 1, not\ l_{m+1} : 1, \ldots, not\ l_n : 1 \in P'$$

where $l_1, \ldots, l_n$ are literals and 1 represents the truth value *true*. We denote the class of ed-programs that consist of only ed-rules of the above form as ed-programs^{one}. We say that nd-programs^{one} are the same as ed-programs^{one}, except that, only atoms (positive literals) are allowed to appear in rules of ed-programs^{one}. The following result shows that ed-programs^{one} and nd-programs^{one} are equivalent to classical extended and normal disjunctive logic programs [4, 7], hence ed-programs and nd-programs subsume classical extended and normal disjunctive logic programs, respectively.

Proposition 3. *Let P be an extended disjunctive logic program. Then S is an answer set of P iff I is a fuzzy answer of $P' \in$ ed-programs^{one} that corresponds to P where $I(l) = 1$ iff $l \in S$ and $I(l')$ is undefined iff $l' \notin S$. Let Q be a normal disjunctive logic program. Then S' is a stable model of Q iff I' is a stable fuzzy model of $Q' \in$ nd-programs^{one} that corresponds to Q where $I'(a) = 1$ iff $a \in S'$ and $I'(b) = 0$ iff $b \in B_L \setminus S'$.*

5 Conclusions and Related Work

We defined the notions of extended and normal disjunctive fuzzy logic programs that generalize disjunctive logic programs [7, 4], extended fuzzy logic programs [23], and normal fuzzy logic programs [30] in a unified logic programming framework to allow disjunctions, classical negation, and non-monotonic negation under fuzzy uncertainty. The proposed framework is necessary to provide the ability to assign fuzzy uncertainly over the possible outcomes of qualitative uncertainty, which is required in real-world applications. We developed the fuzzy answer set semantics and stable fuzzy model semantics for extended and normal disjunctive fuzzy logic programs respectively. It has been shown that the stable fuzzy model semantics of normal disjunctive fuzzy logic programs subsumes the stable fuzzy model semantics of normal fuzzy logic programs [30] and stable model semantics of normal disjunctive logic programs [7, 4]. Furthermore, the fuzzy answer set semantics of extended disjunctive fuzzy logic programs subsumes the fuzzy answer set semantics of extended fuzzy logic programs [7, 4] and the answer set semantics of extended disjunctive logic programs [7, 4]. Moreover, it has been shown that fuzzy answer set semantics of extended disjunctive fuzzy logic programs is reduced to stable fuzzy model semantics of normal disjunctive fuzzy logic programs.

Various approaches to logic programming with uncertainty have been proposed for reasoning under different forms of uncertainty including fuzzy, probabilistic, possibilistic, and multi-valued uncertainty [18, 9, 29, 15, 28, 31, 23, 5, 10, 3, 11, 19, 20, 26, 24, 21, 25, 22, 12, 14]. A detailed survey and comparisons among these approaches to uncertain logic programming can be found in [23]. Fuzzy logic programming approaches for reasoning under fuzzy uncertainty include [18, 9, 29, 15, 28, 31, 10, 30, 23]. Recall, disjunctive logic programs [7, 4] are used for reasoning under qualitative uncertainty, where extended and normal disjunctive fuzzy logic programs subsume. Although, representing and reasoning with both fuzzy and qualitative uncertainty is necessary as evident by some applications, this issue has not been addressed by the current work in fuzzy and qualitative uncertainty in logic programming.

The main difference in this work is that we generalize reasoning with fuzzy uncertainty and qualitative uncertainty in a unified logic programming framework represented by extended and normal disjunctive fuzzy logic programs, that allow the assignment of fuzzy uncertainty over the possible outcomes of qualitative uncertainty. The current work in the literature supports either reasoning under fuzzy uncertainty [18, 9, 29, 15, 28, 31, 10, 30, 23] or reasoning under qualitative uncertainty [7, 4]. The closest to the work presented in this paper is a probabilistic logic programming framework described in [22]. However, the probabilistic logic programming framework presented in [22] allows reasoning in the presence of both probabilistic uncertainty and qualitative uncertainty, where problems with fuzzy uncertainly over qualitative uncertainty cannot to neither presented nor reasoned about. A detailed survey on the work related to reasoning in the presence of both qualitative uncertainty (in general) and qualitative uncertainty can be found in [22].

References

1. Apt, K.R., Bol, R.N.: Logic programming and negation:a survey. Journal of Logic Programming 19(20), 9–71 (1994)
2. Brewwka, G.: Complex preferences for answer set optimization. In: Ninth International Conference on Principles of Knowledge Representation and Reasoning (2004)
3. Dekhtyar, A., Subrahmanian, V.S.: Hybrid probabilistic program. Journal of Logic Programming 43(3), 187–250 (2000)
4. Brewka, G., Dix, J.: Knowledge representation with logic programs. In: Dix, J., Moniz Pereira, L., Przymusinski, T.C. (eds.) LPKR 1997. LNCS (LNAI), vol. 1471, p. 1. Springer, Heidelberg (1998)
5. Dubois, D., et al.: Towards possibilistic logic programming. In: ICLP. MIT Press, Cambridge (1991)
6. Gelfond, M., Lifschitz, V.: The stable model semantics for logic programming. In: ICSLP. MIT Press, Cambridge (1988)
7. Gelfond, M., Lifschitz, V.: Classical negation in logic programs and disjunctive databases. New Generation Computing 9(3-4), 363–385 (1991)
8. Van Gelder, A.: The alternating fixpoint of logic programs with negation. Journal of Computer and System Sciences 47(1), 185–221 (1993)
9. Janssen, J., Schockaert, S., Vermeir, D., De Cock, M.: General fuzzy answer set programs. In: International Workshop on Fuzzy Logic and Applications (2009)
10. Kifer, M., Subrahmanian, V.S.: Theory of generalized annotated logic programming and its applications. Journal of Logic Programming 12, 335–367 (1992)
11. Lakshmanan, V.S.L., Shiri, N.: A parametric approach to deductive databases with uncertainty. IEEE TKDE 13(4), 554–570 (2001)
12. Loyer, Y., Straccia, U.: The approximate well-founded semantics for logic programs with uncertainty. In: Rovan, B., Vojtáš, P. (eds.) MFCS 2003. LNCS, vol. 2747, pp. 541–550. Springer, Heidelberg (2003)
13. Lukasiewicz, T.: Fuzzy description logic programs under the answer set semantics for the semantic Web. Fundamenta Informaticae 82(3), 289–310 (2008)
14. Lukasiewicz, T.: Many-valued disjunctive logic programs with probabilistic semantics. In: Gelfond, M., Leone, N., Pfeifer, G. (eds.) LPNMR 1999. LNCS (LNAI), vol. 1730, p. 277. Springer, Heidelberg (1999)
15. Madrid, N., Ojeda-Aciego, M.: Towards a fuzzy answer set semantics for residuated logic programs. In: Proceedings of the IEEE/WIC/ACM International Conference on Web Intelligence and Intelligent Agent Technology (2008)
16. Nerode, A., Remmel, J., Subrahmanian, V.S.: Annotated nonmonotone rule systems. Theoretical Computer Science 171(1-2), 77–109 (1997)
17. Niemela, I., Simons, P.: Efficient implementation of the well-founded and stable model semantics. In: Joint International Conference and Symposium on Logic Programming, pp. 289–303 (1996)
18. Nieuwenborgh, D., Cock, M., Vermeir, D.: An introduction to fuzzy answer set programming. Annals of Mathematics and Artificial Intelligence 50(3-4), 363–388 (2007)
19. Ng, R.T., Subrahmanian, V.S.: Probabilistic logic programming. Information & Computation, 101(2) (1992)
20. Ng, R.T., Subrahmanian, V.S.: Stable semantics for probabilistic deductive databases. Information & Computation, 110(1) (1994)

21. Saad, E.: Incomplete knowlege in hybrid probabilistic logic programs. In: Fisher, M., van der Hoek, W., Konev, B., Lisitsa, A. (eds.) JELIA 2006. LNCS (LNAI), vol. 4160, pp. 399–412. Springer, Heidelberg (2006)
22. Saad, E.: A logical approach to qualitative and quantitative reasoning. In: Mellouli, K. (ed.) ECSQARU 2007. LNCS (LNAI), vol. 4724, pp. 173–186. Springer, Heidelberg (2007)
23. Saad, E.: Extended fuzzy logic programs with fuzzy answer set semantics. In: Godo, L., Pugliese, A. (eds.) SUM 2009. LNCS, vol. 5785, pp. 223–239. Springer, Heidelberg (2009)
24. Saad, E., Pontelli, E.: Towards a more practical hybrid probabilistic logic programming framework. In: Hermenegildo, M.V., Cabeza, D. (eds.) PADL 2004. LNCS, vol. 3350, pp. 67–82. Springer, Heidelberg (2005)
25. Saad, E., Pontelli, E.: Hybrid probabilistic logic programs with non-monotonic negation. In: Gabbrielli, M., Gupta, G. (eds.) ICLP 2005. LNCS, vol. 3668, pp. 204–220. Springer, Heidelberg (2005)
26. Saad, E., Pontelli, E.: A new approach to hybrid probabilistic logic programs. Annals of Mathematics and Artificial Intelligence Journal 48(3-4), 187–243 (2006)
27. Saad, E., Elmorsy, S., Gabr, M., Hassan, Y.: Reasoning about actions in fuzzy environment. In: the World Congress of the International Fuzzy Systems Association/European society for Fuzzy Logic and Technology, IFSA/EUSFLAT 2009 (2009)
28. Shapiro, E.: Logic programs with uncertainties: A tool for implementing expert systems. In: Proc. of IJCAI, pp. 529–532 (1983)
29. Straccia, U., Ojeda-Aciego, M., Damasio, C.V.: On fixed-points of multivalued functions on complete lattices and their application to generalized logic programs. SIAM Journal on Computing 38(5), 1881–1911 (2009)
30. Subrahmanian, V.S.: Amalgamating knowledge bases. ACM TDS 19(2), 291–331 (1994)
31. van Emden, M.H.: Quantitative deduction and its fixpoint theory. Journal of Logic Programming 4(1), 37–53 (1986)
32. Zadeh, L.: Fuzzy Sets. Information and Control 8(3), 338–353 (1965)
33. Zadeh, L.: Outline of a new approach to the analysis of complex systems and decision processes. IEEE Trans. on Systems, Man, and Cybernetics SMC-3, 28–44 (1973)

Cost-Based Query Answering in Action Probabilistic Logic Programs

Gerardo I. Simari, John P. Dickerson, and V.S. Subrahmanian

Department of Computer Science and UMIACS
University of Maryland College Park
College Park, MD 20742, USA
{gisimari,jdicker1,vs}@cs.umd.edu

Abstract. Action-probabilistic logic programs (*ap*-programs), a class of probabilistic logic programs, have been applied during the last few years for modeling behaviors of entities. Rules in *ap*-programs have the form "If the environment in which entity E operates satisfies certain conditions, then the probability that E will take some action A is between L and U". Given an *ap*-program, we have addressed the problem of deciding if there is a way to change the environment (subject to some constraints) so that the probability that entity E takes some action (or combination of actions) is maximized. In this work we tackle a related problem, in which we are interested in reasoning about the expected reactions of the entity being modeled when the environment is changed. Therefore, rather than merely deciding if there is a way to obtain the desired outcome, we wish to find the *best* way to do so, given costs of possible outcomes. This is called the Cost-based Query Answering Problem (CBQA). We first formally define and study an exact (intractable) approach to CBQA, and then go on to propose a more efficient algorithm for a specific subclass of *ap*-programs that builds on past work in a basic version of this problem.

1 Introduction

Action probabilistic logic programs (*ap*-programs for short) [10] are a class of probabilistic logic programs (PLPs) [14,15,9]. *ap*-programs have been used extensively to model and reason about the behavior of groups; for instance – an application for reasoning about terror groups based on *ap*-programs has users from over 12 US government entities [7]. *ap*-programs use a two sorted logic where there are "state" predicate symbols and "action" predicate symbols, where action atoms only represent the fact that an action is taken, and not the action itself; they are therefore quite different from actions in domains such as AI planning or reasoning about actions, in which effects, preconditions, and postconditions are part of the specification. We assume that *effects and preconditions are generally not known*. These programs can be used to represent behaviors of arbitrary entities (ranging from users of web sites to institutional investors in the finance sector to corporate behavior) because they consist of rules of the form "if a conjunction C of atoms is true in a given state S, then entity E (the entity whose behavior is being modeled) will take action A with a probability in the interval $[L, U]$."

In such applications, it is essential to avoid making probabilistic independence assumptions as the goal is to *discover* probabilistic dependencies and then exploit these

A. Deshpande and A. Hunter (Eds.): SUM 2010, LNAI 6379, pp. 319–332, 2010.

findings for forecasting. For instance, Figure 1 shows a small set of rules automatically extracted from data [1] about Hezbollah's past. Rule 1 says that Hezbollah uses kidnappings as an organizational strategy with probability between 0.5 and 0.56 whenever no political support was provided to it by a foreign state (`forstpolsup`), and the severity of inter-organizational conflict involving it (`intersev1`) is at level "*c*". Rules 2 and 3 state that kidnappings will be used as a strategy with 80-86% probability when no external support is solicited by the organization (`extsup`) and either the organization does not advocate democratic practices (`demorg`) or electoral politics is not used as a strategy (`elecpol`). Similarly, Rules 4 and 5 refer to the action "civilian targets chosen based on ethnicity" (`tlethciv`). Rule 4 states that this action will be taken with probability 0.49 to 0.55 whenever the organization advocates democratic practices, while Rule 5 states that the probability rises to between 0.71 and 0.77 when electoral politics are used as a strategy and the severity of inter-organizational conflict (with the organization with which the second highest level of conflict occurred) was not negligible (`intersev2`). *ap*-programs have been used extensively by terrorism analysts to make predictions about terror group actions [7,13].

r_1. $\mathtt{kidnap}(1) : [0.50, 0.56] \leftarrow \mathtt{forstpolsup}(0) \wedge \mathtt{intersev1}(c)$.
r_2. $\mathtt{kidnap}(1) : [0.80, 0.86] \leftarrow \mathtt{extsup}(1) \wedge \mathtt{demorg}(0)$.
r_3. $\mathtt{kidnap}(1) : [0.80, 0.86] \leftarrow \mathtt{extsup}(1) \wedge \mathtt{elecpol}(0)$.
r_4. $\mathtt{tlethciv}(1) : [0.49, 0.55] \leftarrow \mathtt{demorg}(1)$.
r_5. $\mathtt{tlethciv}(1) : [0.71, 0.77] \leftarrow \mathtt{elecpol}(1) \wedge \mathtt{intersev2}(c)$.

Fig. 1. A small set of rules modeling Hezbollah

In [21], we explored the problem of determining what *we can do* in order to induce a given behavior by the group. For example, a policy maker might want to understand what we can do so that a given goal (*e.g.*, the probability of Hezbollah using kidnappings as a strategy is below some percentage) is achieved, given some constraints on what is feasible. In this paper, we take the problem one step further by adding the desire to reason about how the entity being modeled reacts to our efforts. We are therefore interested in finding what the *best* course of action on our part is given some additional input regarding how desirable certain outcomes are; this is called the *cost-based query answering problem* (CBQA). In the following, we first briefly recall the basics of *ap*-programs and then present CBQA formally. We then investigate an approach to solving this problem exactly based on Markov Decision Processes, showing that this approach quickly becomes infeasible in practice. Afterwards, we describe a novel heuristic algorithm based on probability density estimation techniques that can be used to tackle CBQA with much larger instances. Finally, we describe a prototype implementation and experimental results showing that our heuristic algorithm scales well in practice.

A brief note on related work: almost all past work on abduction in such settings has been devised under various independence assumptions [19,18,5]. Apart from our first proposal [21], we are aware of no work to date on abduction in possible worlds-based probabilistic logic systems such as those of [8], [16], and [6] where independence assumptions are not made.

2 Preliminaries

2.1 Syntax

We assume the existence of a logical alphabet consisting of a finite set $\mathcal{L}_{cons}$ of constant symbols, a finite set $\mathcal{L}_{pred}$ of predicate symbols (each with an associated arity), and an infinite set $\mathcal{L}_{var}$ of variable symbols; function symbols are not allowed. Terms, atoms, and literals are defined in the usual way [12]. We assume $\mathcal{L}_{pred}$ is partitioned into disjoint sets: $\mathcal{L}_{act}$ of *action symbols* and $\mathcal{L}_{sta}$ of *state* symbols. If $t_1, \ldots, t_n$ are terms, and p is an n-ary action (resp. state) symbol, then $p(t_1, \ldots, t_n)$, is an *action (resp. state) atom*.

Definition 1 (Action formula). *(i) A (ground) action atom is a (ground) action formula; (ii) if F and G are (ground) action formulas, then $\neg F$, $F \wedge G$, and $F \vee G$ are also (ground) action formulas.*

The set of all possible action formulas is denoted by *formulas*$(B_{\mathcal{L}_{act}})$, where $B_{\mathcal{L}_{act}}$ is the Herbrand base associated with $\mathcal{L}_{act}$, $\mathcal{L}_{cons}$, and $\mathcal{L}_{var}$.

Definition 2 (*ap*-formula). *If F is an action formula and $\mu = [\alpha, \beta] \subseteq [0, 1]$, then $F : \mu$ is called an* annotated action formula *(or* ap-*formula), and μ is called the* ap-*annotation of F.*

We will use $\mathcal{APF}$ to denote the (infinite) set of all possible *ap*-formulas.

Definition 3 (World/State). *A* world *is any finite set of ground action atoms. A* state *is any finite set of ground state atoms.*

It is assumed that all actions in the world are carried out more or less in parallel and at once, given the temporal granularity adopted along with the model. Contrary to (related but essentially different) approaches such as stochastic planning, we are not concerned here with reasoning about the effects of actions. We now define *ap*-rules.

Definition 4 (*ap*-rule). *If F is an action formula, $B_1, \ldots, B_n$ are state atoms, and μ is an* ap-*annotation, then $F : \mu \leftarrow B_1 \wedge \ldots \wedge B_m$ is called an* ap-rule. *If this rule is named r, then Head(r) denotes $F : \mu$ and Body(r) denotes $B_1 \wedge \ldots \wedge B_n$.*

Intuitively, the rule specified above says that if $B_1, \ldots, B_m$ are all true in a given state, then there is a probability in the interval μ that the action combination F is performed by the entity modeled by the *ap*-rule.

Definition 5 (*ap*-program). *An* action probabilistic logic program *(*ap-*program for short) is a finite set of* ap-*rules. An* ap-*program Π' s.t. $\Pi' \subseteq \Pi$ is called a* subprogram *of Π.*

Figure 1 shows a small part of an *ap*-program derived automatically from data about Hezbollah. Henceforth, we use *Heads*(Π) to denote the set of all annotated formulas appearing in the head of some rule in Π. Given a ground *ap*-program Π, *sta*(Π) (resp., *act*(Π)) denotes the set of all state (resp., action) atoms in Π.

Example 1 (Worlds and states). Coming back to the *ap*-program in Figure 1, the following are examples of worlds: $\{\mathtt{kidnap}(1)\}, \{\mathtt{kidnap}(1), \mathtt{tlethciv}(1)\}, \{\}$. The following are examples of states: $\{\mathtt{forstpolsup}(0), \mathtt{elecpol}(0)\}$, $\{\mathtt{demorg}(1)\}$, and $\{\mathtt{extsup}(1), \mathtt{elecpol}(1)\}$.

2.2 Semantics of *ap*-Programs

We use $\mathcal{W}$ to denote the set of all possible worlds, and $\mathcal{S}$ to denote the set of all possible states. It is clear what it means for a state to satisfy the body of a rule [12].

Definition 6 (Satisfaction of a rule body by a state). *Let Π be an* ap-*program and s a state. We say that s* satisfies *the body of a rule $F : \mu \leftarrow B_1 \wedge \ldots \wedge B_m$ if and only if $\{B_1, \ldots, B_M\} \subseteq s$.*

Similarly, we define what it means for a world to satisfy a ground action formula:

Definition 7 (Satisfaction of an action formula by a world). *Let F be a ground action formula and w a world. We say that w* satisfies *F if and only if: (i) if $F \equiv a$, for some atom $a \in B_{\mathcal{L}_{act}}$, then $a \in w$; (ii) if $F \equiv F_1 \wedge F_2$, for action formulas $F_1, F_2 \in formulas(B_{\mathcal{L}_{act}})$, then w satisfies F_1 and w satisfies F_2; (iii) if $F \equiv F_1 \vee F_2$, for action formulas $F_1, F_2 \in formulas(B_{\mathcal{L}_{act}})$, then w satisfies F_1 or w satisfies F_2; (iv) if $F \equiv \neg F'$, for action formula $F' \in formulas(B_{\mathcal{L}_{act}})$, then w does not satisfy F'.*

Finally, we will use the concept of *reduction* of an *ap*-program w.r.t. a state:

Definition 8 (Reduction of an *ap*-program w.r.t. a state). *Let Π be an* ap-*program and s a state. The* reduction of Π w.r.t. s*, denoted Π_s, is the set $\{F : \mu \mid s$ satisfies Body and $F : \mu \leftarrow$ Body is a ground instance of a rule in $\Pi\}$. Rules in this set are said to be* relevant *in state s.*

The semantics of *ap*-programs uses possible worlds in the spirit of [6,8,16]. Given an *ap*-program Π and a state s, we can define a set $LC(\Pi, s)$ of linear constraints associated with s. Each world w_i expressible in the language $\mathcal{L}_{act}$ has an associated variable v_i denoting the probability that it will actually occur. $LC(\Pi, s)$ consists of the following constraints.

1. For each $Head(r) \in \Pi_s$ of the form $F : [\ell, u]$, $LC(\Pi, s)$ contains the constraint $\ell \leq \sum_{w_i \in \mathcal{W} \wedge w_i \models F} v_i \leq u$.
2. $LC(\Pi, s)$ contains the constraint $\sum_{w_i \in \mathcal{W}} v_i = 1$.
3. All variables are non-negative.
4. $LC(\Pi, s)$ contains only the constraints described in $1 - 3$.

While [10] provides a more formal model theory for *ap*-programs, we merely provide the definition below. Π_s is *consistent* iff $LC(\Pi, s)$ is solvable over the reals, $\mathbb{R}$.

Definition 9 (Entailment of an *ap*-formula by an *ap*-program). *Let Π be an* ap-*program, s a state, and $F : [\ell, u]$ a ground action formula. Π_s* entails *$F : [\ell, u]$, denoted $\Pi_s \models F : [\ell, u]$ iff $[\ell', u'] \subseteq [\ell, u]$ where:*
*$\ell' =$ **minimize** $\sum_{w_i \in \mathcal{W} \wedge w_i \models F} v_i$ **subject to** $LC(\Pi, s)$.*
*$u' =$ **maximize** $\sum_{w_i \in \mathcal{W} \wedge w_i \models F} v_i$ **subject to** $LC(\Pi, s)$.*

The following is an example of both $LC(\Pi, s)$ and entailment of an *ap*-formula.

Example 2 (Multiple probability distributions given $LC(\Pi, s)$ and entailment). Consider *ap*-program Π from Figure 1 and state s_2 from Figure 2. The set of all possible worlds is: $w_0 = \{\}$, $w_1 = \{\texttt{kidnap(1)}\}$, $w_2 = \{\texttt{tlethciv(1)}\}$, and $w_3 = \{\texttt{kidnap(1)}, \texttt{tlethciv(1)}\}$. Suppose p_i denotes the probability of world w_i. $LC(\Pi, s_2)$ then consists of the following constraints:

$s_1 = \{\texttt{forstpolsup}(0), \texttt{intersev1}(c), \texttt{intersev2}(0), \texttt{elecpol}(1), \texttt{extsup}(0), \texttt{demorg}(0)\}$
$s_2 = \{\texttt{forstpolsup}(0), \texttt{intersev1}(c), \texttt{intersev2}(0), \texttt{elecpol}(0), \texttt{extsup}(0), \texttt{demorg}(1)\}$
$s_3 = \{\texttt{forstpolsup}(0), \texttt{intersev1}(c), \texttt{intersev2}(0), \texttt{elecpol}(0), \texttt{extsup}(0), \texttt{demorg}(0)\}$
$s_4 = \{\texttt{forstpolsup}(1), \texttt{intersev1}(c), \texttt{intersev2}(c), \texttt{elecpol}(1), \texttt{extsup}(1), \texttt{demorg}(0)\}$
$s_5 = \{\texttt{forstpolsup}(0), \texttt{intersev1}(c), \texttt{intersev2}(c), \texttt{elecpol}(0), \texttt{extsup}(1), \texttt{demorg}(0)\}$

Fig. 2. A small set of possible states

$$0.5 \le p_1 + p_3 \le 0.56$$
$$0.49 \le p_2 + p_3 \le 0.55$$
$$p_0 + p_1 + p_2 + p_3 = 1$$

One possible solution to this set of constraints is $p_0 = 0$, $p_1 = 0.51$, $p_2 = 0.05$, and $p_3 = 0.44$; in this case, there are other possible distributions that are also solutions. Consider formula $\texttt{kidnap}(1) \wedge \texttt{tlethciv}(1)$, which is satisfied only by world w_3. This formula is entailed with probability in $[0, 0.55]$, meaning that one cannot assign a probability greater than 0.55 to this formula[1].

3 The Cost-Bounded Query Answering Problem

Suppose s is a state (the current state), G is a goal (an action formula), and $[\ell, u] \subseteq [0, 1]$ is a probability interval. We are interested in finding a new state s' such that $\Pi_{s'}$ entails $G : [\ell, u]$. However, s' must be *reachable* from s. In this paper, we assume that there are costs associated with *transforming the current state* into another state, and also an associated probability of success of this transformation; *e.g.*, the fact that we may try to reduce foreign state political support for Hezbollah may only succeed with some probability. To model this, we will make use of three functions:

Definition 10. *A* transition function *is any function* $T : \mathcal{S} \times \mathcal{S} \to [0, 1]$, *and a* cost function *is any function* $cost : \mathcal{S} \to [0, \infty)$. *A* transition cost function, *defined w.r.t. a transition function T and some cost function cost, is a function* $cost_T : \mathcal{S} \times \mathcal{S} \to [0, \infty)$, *with* $cost_T(s, s') = \frac{cost(s')}{T(s,s')}$ *whenever* $T(s, s') \neq 0$, *and* ∞ *otherwise*[2].

Example 3. Suppose that the only state predicate symbols are those that appear in the rules of Figure 1, and consider the set of states in Figure 2. Then, an example of a transition function is: $T(s_1, s_2) = 0.93$, $T(s_1, s_3) = 0.68$, $T(s_2, s_1) = 0.31$, $T(s_4, s_1) = 1$, $T(s_2, s_5) = 0$, $T(s_3, s_5) = 0$, and $T(s_i, s_j) = 0$ for any pair s_i, s_j other than the ones considered above. Note that, if state s_5 is reachable, then the *ap*-program is inconsistent, since both rules 1 and 2 are relevant in that state.

Function $cost_T$ describes *reachability* between any pair of states – a cost of ∞ represents an impossible transition. The cost of transforming a state s_0 into state s_n by

[1] Note that, contrary to what one might think, the interval $[0, 1]$ is not necessarily a solution.
[2] We assume that ∞ represents a value for which, in finite-precision arithmetic, $\frac{1}{\infty} = 0$ and $x^\infty = \infty$ when $x > 1$. The IEEE 754 floating point standard satisfies these rules.

intermediate transformations through state the sequence of states $seq = \langle s_0, s_1, \ldots, s_n \rangle$ is defined:

$$cost^*_{seq}(s_0, s_n) = e^{\sum_{0 \le i < n, s_i \in seq} cost_T(s_i, s_{i+1})} \quad (1)$$

One way in which cost functions can be specified is in terms of *reward functions*.

Definition 11 (Reward functions). *An* action reward function *is a partial function* $R : \mathcal{APF} \rightarrow [0,1]$. *An action reward function is* finite *if* $dom(R)$ *is finite.*

Let R *be a finite reward function and* Π *be an* ap-*program. An* entailment-based reward function *for* Π *and* R *is a function* $E_{\Pi,R} : \mathcal{S} \rightarrow [0, \infty)$, *defined as:*

$$E_{\Pi,R}(s) = \sum_{F:[\ell,u] \in dom(R) \wedge \Pi_s \models F:[\ell,u]} R(F : [\ell, u]) \quad (2)$$

Reward functions are used to represent how desirable it is, from the reasoning agent's point of view, for a given annotated action formula to be entailed in a given state by the model being used (this is the intuition behind Equation 2; other ways of defining this are also possible). *In this paper, we will assume that all reward functions are finite.* We use this notion of reward to define a natural *canonical cost function* as $cost^{\circ}(s) = \frac{1}{E_{\Pi,R}(s)}$ when $E_{\Pi,R}(s) \neq 0$, and ∞ otherwise, for each state s. In the rest of this paper, we assume that all transition cost functions are defined in terms of a canonical cost function.

Example 4. An example of an entailment-based reward function is as follows. Consider state s_2 from Figure 2, and annotated formulas $F_1 = \texttt{kidnap}(1) \wedge \texttt{tlethciv}(1) : [0, 0.60]$, $F_2 = \texttt{kidnap}(1) : [0, 0.05]$, and $F_3 = \texttt{tlethciv}(1) : [0, 0.5]$. Suppose we have action reward function R such that $R(F_1) = 0.2$, $R(F_2) = 0.54$, and $R(F_3) = 0.14$. Now, considering that $\Pi_{s_2} \models F_1$, $\Pi_{s_2} \not\models F_2$, and $\Pi_{s_2} \models F_3$, we have that, according to Equation 2 in Definition 11, $E_{\Pi,R}(s_2) = 0.2 + 0.14 = 0.34$. Assuming $T(s_1, s_2) = 0.93$ as in Example 3, we have $cost_T(s_1, s_2) = \frac{1}{0.34} * \frac{1}{0.93} \approx 3.162$.

Definition 12. *A* cost based query *is a 4-tuple* $\langle G : [\ell, u], s, cost_T, k \rangle$, *where* $G : [\ell, u]$ *is an* ap-*formula,* $s \in \mathcal{S}$, $cost_T$ *is a cost function, and* $k \in \mathbb{R}^+ \cup \{0\}$.

CBQA Problem. Given an *ap*-program Π and a cost-based query $\langle G : [\ell, u], s, cost_T, k \rangle$, return "Yes" if and only if there exists a state s' and sequence of states $seq = \langle s, s_1, \ldots, s' \rangle$ such that $cost^*_{seq}(s, s') \le k$, and $\Pi_{s'} \models G : [\ell, u]$; the answer is "No" otherwise.

In [21], a related problem called the *Basic Probabilistic Logic Abduction Problem* (Basic PLAP) is proposed; the main difference is that in Basic PLAP there is no notion of cost, and we are only interested in the *existence of some sequence* of states leading to a state that entails the *ap*-formula.

Example 5. Consider once again the program in the running example and the set of states from Figure 2. Suppose the goal is *kidnap*$(1) : [0, 0.6]$ (we want the probability of Hezbollah using kidnappings to be at most 0.6) and the current state is s_4, $k = 3$. Suppose we have a reward function $E_{\Pi,R}$ such that $E_{\Pi,R}(s_1) = 0.5$, $E_{\Pi,R}(s_2) = 0.15$, $E_{\Pi,R}(s_3) = 0.5$, $E_{\Pi,R}(s_4) = 0.1$, $E_{\Pi,R}(s_5) = 0$, and $E_{\Pi,R}(s_i) = 0$ for all other $s_i \in \mathcal{S}$. Finally, for the sake of simplicity, suppose transition function T states that all transitions have probability 1.

The states that make relevant a subprogram that entails the goal are: s_1, s_2, s_3, and s_5. The objective is then to find a finite sequence of states starting at s_4 and finishing in any other state such that the total cost of the sequence is less than 3 (recall that cost is defined $cost_T(s, s') = cost^\circ(s')/T(s, s')$). We can easily see that directly moving to either state s_1 or s_3 satisfies these conditions, with a cost of 2; moving to s_2 or s_5 does not, since the cost would be ≈ 6.67 and ∞, respectively.

The following proposition is a direct consequence of Proposition 1 in [21], which states that the Basic PLAP problem is EXPTIME-complete. Unfortunately, due to space constraints proofs will be omitted.

Proposition 1. *CBQA is EXPTIME-complete.*

Furthermore, we can show that CBQA is NP-hard whenever one of two simplifying assumptions hold: (1) the cardinality of the set of ground action atoms is bounded by a constant, and (2) the cardinality of the set of ground state atoms is bounded by a constant. These results are interesting because they show that the complexity of CBQA is caused by having to solve two independent problems: (P1) Finding a subprogram $\Pi' \subseteq \Pi$ such that when the body of all rules in Π' is deleted, the resulting subprogram entails the goal, and (P2) Decide if there exists a state s' such that $\Pi' = \Pi_s$ and s is reachable from the initial state within the cost budget allowed.

In the next sections we will investigate algorithms for CBQA when the cost function is defined in terms of entailment-based reward functions. We will begin by presenting an exact algorithm, and then go on to investigate a more tractable approach to finding solutions, albeit not optimal ones.

4 CBQA Algorithms for Threshold Queries

A *threshold goal* is an annotated action formula of the form $F : [0, u]$ or $F : [\ell, 1]$; this kind of goals can be used to express the desire that certain formulas (actions) should only be entailed with a certain maximum probability (upper bound) or should be entailed with at least a certain minimum probability (lower bound). In this paper, we only give algorithms for such queries.

4.1 An Exact Algorithm for CBQA

We show that any CBQA problem can be mapped to a *Markov Decision Process* [2,20] problem. An instance of an MDP consists of: a finite set S of environment *states*; a finite set A of *actions*; a *transition function* $T : S \times A \rightarrow \Pi(S)$ specifying the probability of arriving at every possible state given that a certain action is taken in a given state; and a *reward function* $R : S \times A \rightarrow \mathbb{R}$ specifying the expected immediate reward gained by taking an action in a state. The objective is to compute a policy $\pi : S \rightarrow A$ specifying what action should be taken in each state – the policy should be optimal w.r.t. the expected utility obtained from executing it.

Obtaining an MDP from the Specification of a CBQA Instance. We show how any instance of a CBQA problem can be mapped to an MDP in such a way that an optimal policy for this MDP corresponds to solutions to the original CBQA problem.

State Space: The set S_{MDP} of MDP states corresponds directly to the set $\mathcal{S}$.
Actions: The set A_{MDP} of possible actions in the MDP domain corresponds to the set of all possible attempts at changing the current state. We can think of the set of actions as containing one action per state in $s \in \mathcal{S}$, which represents the change from the current state to s. We will therefore say that action a specifying that the state will be changed to s is *congruent* with s, denoted $a \cong s$.
Transition Function: The transition function T_{MDP} for the MDP can be directly obtained from the transition function T in the CBQA instance. Formally, let $s, s' \in S_{MDP}$ and $a \in A_{MDP}$; we define:

$$T_{MDP}(s,a,s') = \begin{cases} 0 & \text{if } a \ncong s', \\ T(s,s') & \text{otherwise}; \end{cases} \tag{3}$$

$$T_{MDP}(s,a,s) = 1 - T(s,a,s') \text{for } a \cong s'; \tag{4}$$

the last case represents the fact that, when actions fail to have the desired effect, the current state is unchanged.
Reward Function: The reward function of the MDP, which describes the reward directly obtained from performing action $a \in A$ in state $s \in S$, can also be directly obtained from the CBQA instance. Let $s \in S_{MDP}$, $a \in A_{MDP}$, Π be an *ap*-program, $G : [\ell, u]$ be the goal, and $E_{\Pi,R}$ be an entailment-based reward function:

$$R(s,a) = \begin{cases} -1 * cost_T(s,s') & \text{for state } s' \in \mathcal{S} \text{ such that } a \cong s', \\ 1 & \text{for states } s' \in \mathcal{S} \text{ such that } \Pi_{s'} \models G : [\ell, u]. \end{cases} \tag{5}$$

To conclude, we present the following results. The first states that given an instance of CBQA, our proposed translation into an MDP is such that an optimal policy under Maximum Expected Utility (MEU) for such an MDP expresses a solution for the original instance. In the following, we say that a sequence of states $\langle s_0, s_1, \ldots, s_k \rangle$ is the result of *following* a policy π if $\pi(s_i) = a_{i+1}$, where $0 \le i < k$ and $a_{i+1} \cong s_{i+1}$.

Proposition 2. *Let* $O = (\Pi, \mathcal{S}, s_0, G : [\ell, u], cost, T, E_{\Pi,R}, k)$ *be an instance of a CBQA problem that has a solution (output "Yes"), and* $M = (S_{MDP}, A_{MDP}, T_{MDP}, R_{MDP})$ *be its corresponding translation into an MDP. If* π *is a policy for* M *that is optimal w.r.t. the MEU criterion, then following* π *starting at state* $s_0 \in S_{MDP}$ *yields a sequence of states that satisfies the conditions for a solution to* O.

Second, we analyze the computational cost of taking this approach. As there are numerous algorithms to solve MDPs, we only analyze the size of the MDP resulting from the translation of an instance of CBQA. The well-known Value Iteration algorithm [2] iterates over the entire state space a number of times that is polynomial in $|S|$, $|A|$, β, and B, where B is an upper bound on the number of bits that are needed to represent any numerator or denominator of β [11]. Now, each iteration takes time in $O(|A| \cdot |\mathcal{S}|^2)$, which is equivalent to $O(|\mathcal{S}|^3)$ since $|A| = |\mathcal{S}|$; this means that only for very small instances will solving the corresponding MDP be feasible.

As can be seen from the above mapping, the *key point in which our problem differs* from approaches like planning under uncertainty is that finding a sequence of states that is a solution to CBQA involves executing actions in parallel which, among other things, means that the number of possible actions that can be considered in a given state is *very large*. This makes planning approaches infeasible since their computational cost is intimately tied to the number of possible actions in the domain (generally assumed to be fixed at a relatively small number). In the case of MDPs, even though state aggregation techniques have been investigated to keep the number of states being considered manageable [4,23], similar techniques for *action aggregation* have not been developed.

4.2 A Heuristic Algorithm Based on Iterative Sampling

Given the exponential search space, we would like to find a tractable heuristic approach. We now show how this can be done by developing an algorithm in the class of *iterated density estimation* algorithms (IDEAs) [3,17]. The main idea behind these algorithms is to improve on other approaches such as Hill Climbing, Simulated Annealing, and Genetic Algorithms by maintaining a *probabilistic model* characterizing the best solutions found so far. An iteration then proceeds by (1) generating new candidate solutions using the current model, (2) singling out the best out of the new samples, and (3) updating the model with the samples from Step 2. One of the main advantages of these algorithms over classical approaches is that the probabilistic model, a "byproduct" of the effort to find an optimum, contains a wealth of information about the problem at hand.

Algorithm DE_CBQA (Figure 3) follows this approach to finding a solution to our problem. The algorithm begins by identifying certain *goal states*, which are states s' such that $\Pi_{s'} \models G : [\ell, u]$; these states are pivotal, since any sequence of states from s_0 to a goal state is a candidate solution. The algorithms in [21] can be used to compute a set of goal states. Continuing with the preparation phase, the algorithm then tests how good the direct transitions from the initial state s_0 to each of the goal states is; ϕ^* now represents the current best sequence (though it might not actually be a solution). The final step before the sampling begins occurs in Line 5, where we initialize a probability distribution over all states[3], which is initialized with the uniform distribution.

The `while` loop in Lines 6-13 then performs the main search; $giveUp$ is a predicate given by parameter which simply tells us when the algorithm should stop (it can be based on total number of samples, time elapsed, etc). The value j represents the length of the sequence of states currently considered, and *numIter* is a parameter indicating how many iterations we wish to perform for each length. Line 9 performs the sampling of sequences, while Line 10 assigns a score to each based on the transition cost function. After updating the score of the best solution found up to now, Line 13 updates the probabilistic model P being used by keeping only the best solutions found during the last sampling phase. The algorithm finally returns the best solution it found (if any). An attractive feature of DE_CBQA is that it is an anytime algorithm, *i.e.*, once it finds a solution, given more time it may be able to refine it into a better one while always being able to return the best so far. We now show an example of this algorithm at work.

Example 6. Consider once again the *ap*-program from Figure 1, and the states from Figure 2. Suppose that we have the following inputs. The goal is $\texttt{kidnap}(1) : [0, 0.6]$; the

[3] In an actual implementation, the probability distribution should be represented implicitly, as storing a probability for an exponential number of states would be intractable.

Algorithm DE_CBQA($\Pi, G : [\ell, u], s_0, T, h, k, \mathit{numIter}, \mathit{giveUp}$)
1. $\mathcal{S}_G$:= $\mathit{getGoalStates}(\Pi, G : [\ell, u])$;
2. test all transitions (s_0, s_G), for $s_G \in \mathcal{S}_G$; calculate $\mathit{cost}^*_{seq}(s_0, s_G)$ for each;
3. let ϕ_{best} be the two-state sequence that has the lowest cost, denoted c_{best};
4. let $\mathcal{S}' = \mathcal{S} - \mathcal{S}_G - \{s_0\}$; set $j := 2$;
5. initialize probability distribution P over $\mathcal{S}'$ s.t. $P(s) = \frac{1}{|\mathcal{S}'|}$ for each $s \in \mathcal{S}'$;
6. while $!\mathit{giveUp}$ do
7. $\quad j := j + 1$;
8. $\quad$ for $i = 1$ to $\mathit{numIter}$ do
9. $\quad\quad$ randomly sample (using P) a set H of h sequences of states of length j starting at s_0 and ending at some $s_G \in \mathcal{S}_G$;
10. $\quad\quad$ rank each sequence ϕ with $\mathit{cost}^*_{seq}(s_0, \phi(j))$;
11. $\quad\quad$ pick the sequence in H with the lowest cost c^*, call it ϕ^*;
12. $\quad\quad$ if $c^* < c_{best}$ then ϕ_{best}:= ϕ^*; c_{best}:= c^*;
13. $\quad\quad$ P:= generate new distribution based on H;
14. return ϕ_{best};

Fig. 3. An algorithm for CBQA based on probability density estimation

transition probabilities are as follows: $T(s_4, s_1) = 0.1$, $T(s_4, s_2) = 0.1$, $T(s_4, s_3) = 0.1$, $T(s_2, s_1) = 0.9$, $T(s_3, s_2) = 0.8$, $T(s_5, s_2) = 0.9$, $T(s_5, s_3) = 0.2$, $T(s_5, s_1) = 0.3$, $T(s_1, s_3) = 0.01$, and $T(s_i, s_j) = 1$ for any pair of states s_i, s_j not previously mentioned; the initial state is s_4; the reward function $E_{\Pi,R}$ is defined as follows: $E_{\Pi,R}(s_1) = 0.5$, $E_{\Pi,R}(s_2) = 0.15$, $E_{\Pi,R}(s_3) = 0.5$, $E_{\Pi,R}(s_4) = 0.1$, and $E_{\Pi,R}(s_5) = 0.7$; *giveUp* is a predicate that simply checks if we've sampled a total of 5 or more sequences; *numIter* $= 2$; $h = 3$; and $k = 1{,}000$.

The three states that make relevant a subprogram that entails the goal are s_1, s_2, and s_3. The costs of the two-state direct sequences are the following: $\mathit{cost}_{seq}(s_4, s_1) \approx 10^{8.68}$, $\mathit{cost}_{seq}(s_4, s_2) \approx 10^{28.9}$, and $\mathit{cost}_{seq}(s_4, s_3) \approx 10^{8.68}$; therefore, $c_{best} = 10^{8.68}$ and $\phi_{best} = \langle s_4, s_3 \rangle$. Next, since we are assuming that s_1-s_5 are the only states for the sake of brevity, the algorithm sets up a probability distribution P that starts out as $(0.2, 0.2, 0.2, 0.2, 0.2)$. Suppose we sample $H = \{\langle s_4, s_5, s_3 \rangle, \langle s_4, s_5, s_2 \rangle, \langle s_4, s_1, s_3 \rangle\}$. These sequences have respective costs of $10^{9.23}$, $10^{3.21}$, and $10^{21.71}$. The update step in line 13 of the algorithm will then look at the two best sequences in H and, depending on how it is implemented, might update P to $(0.1, 0.1, 0.1, 0, 0.7)$. Thus, the algorithm has learned that s_4, s_5 seems to be a good way to start. For brevity, suppose that the next iteration of samples (the last one according to *giveUp*) contains $\langle s_4, s_5, s_1 \rangle$, whose cost is $\approx 10^{2.89}$; it is the best seen so far, and since $10^{2.89} < k$, it is a valid answer.

Next, we present the results of our experimental evaluation of this algorithm.

5 Empirical Evaluation

We carried out all experiments on an Intel Core2 Q6600 processor running at 2.4GHz with 8GB of memory available, using code written in Java 1.6; all runs were preformed on Windows 7 Ultimate 64-bit OS, and made use of a single core.

First, we compare the run time and accuracy of the MDP formulation against that of the DE_CBQA algorithm. Recall that DE_CBQA randomly selects states with respect to a probability distribution that is updated from one iteration to the next. The simplest way to represent this probability distribution is with a vector of size $|S|$, where the element at position i represents the proportion of "good" samples that contained state i. This representation does not scale as $|S|$ increases; our implementation thus only keeps track of the states we have visited, implicitly assigning proportion 0 to all nonvisited states. Second, we explore instances of CBQA that are beyond the scope of the exact MDP implementation, but within reach of the DE_CBQA heuristic algorithm.

For all experiments, we assume an instance of the CBQA problem with *ap*-program Π and cost-based query $Q = \langle G : [\ell, u], s, cost_T, k\rangle$. The required cost, transition, and reward values for both algorithms are assigned randomly in accordance with their definitions. We assume an infinite budget for our experiments, choosing instead to compare the numeric costs associated with the sequences returned by the algorithms.

Exact MDP versus Heuristic DE_CBQA. Let S_{MDP} and A_{MDP} be the state and action spaces of the MDP corresponding to a given CBQA – each iteration of the Value Iteration algorithm requires $O(|S_{MDP}|^2 \cdot |A_{MDP}|)$ time. From the transformation discussed in Section 4.1, we see that $|A_{MDP}| = |S_{MDP}|$; furthermore, since $|S_{MDP}|$ is exponentially larger than the number of state atoms found in Π, we expect running the multiple iterations of Value Iteration required to obtain an optimal policy to be intractable for all but very small instances of our problem. Our experimental results support this intuition.

For this set of experiments, we varied the number of state atoms, action atoms, and *ap*-rules in an *ap*-program Π; 10 unique *ap*-programs were created per combination of these inputs. We tested 10 randomly generated cost, transition, and reward assignments for each unique *ap*-program. Then, for each of these generations, we tested multiple runs of the MDP and DE_CBQA algorithms. We varied the discount factor γ and maximum error ϵ for the MDP[4], while exploring different completion predicates, maximum and minimum sequence lengths, and number of iterations per sequence length for DE_CBQA. We provide a space-constrained overview of the results here.

Figure 4 compares the running time (log-scale) of both algorithms. Immediately clear is the fact that, although increasing state and rule space size slows down both algorithms, DE_CBQA consistently outperforms the standard MDP implementation. More subtle is the observation that the difference in run times between the two algorithms increases with the number of states, with DE_CBQA maintaining nearly constant run time across small numbers of states as the MDP implementation increases noticeably. This disparity is explained at least in part by the MDP's optimality requirement; it requires an exhaustive list of *all* goal states while DE_CBQA can rely on faster heuristic search methods (see [21]). As the state space increases, so too does the list of states that must be tested for entailment of the goal *ap*-formula.

We now compare the costs of sequences returned by MDP and DE_CBQA, as given by Equation 1. Typically, the recommended sequences' costs are close[5]; however, in

[4] Given γ and ϵ, one can calculate an error threshold that gaurantees an optimal policy [24].

[5] In terms of relative error, $\eta = \frac{|v-v'|}{|v|}$, for true cost v (MDP) and approx. cost v' (DE_CBQA).

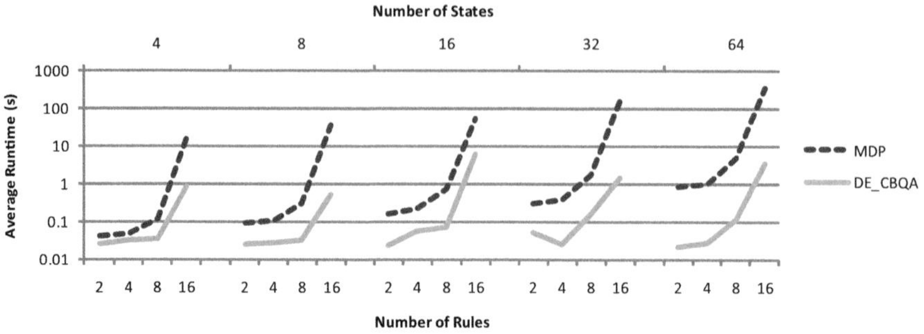

Fig. 4. Log-scale run time comparison of MDP and DE_CBQA, shown with increasing state size (top axis) for each of 2, 4, 8, and 16 rules (bottom axis). Note the sharp jump in run time as the number of rules increases compared to the gradual upward trend as the number of states rises.

rare cases, DE_CBQA performs poorly. We believe this is due to the initial probability distribution assigning mass uniformly to all states – meaning that "good" and "bad" states are equally likely to be selected, at least initially. When DE_CBQA randomly selects bad states at the start, its ability to find better, lower-cost states in future iterations is hampered. Given its low run time, one strategy for dealing with these fringe cases is executing DE_CBQA multiple times, selecting and returning the overall lowest-cost sequence over all runs. In general, increasing the number of iterations (Line 8) did not affect sequence cost; however, increasing the number of samples per iteration (Line 9) often resulted in a better sequence. This hints that allowing the probability mass to converge to a small number of states too quickly is not desirable, as low-cost candidates that are not immediately evident can be ignored. Furthermore, increasing the minimum and maximum sequence lengths (Lines 4 and 6) did not benefit the final result.

Finally, we tried using Policy Iteration [22] instead of Value Iteration to solve the MDP; however, this method was either slower than Value Iteration or, if faster, forced to use such a low discount factor γ and error limit ϵ that following the resultant policy often yielded a *worse* sequence than DE_CBQA's recommendation – at a slower speed!

Scaling the Heuristic DE_CBQA Algorithm. The MDP formulation of CBQA quickly becomes intractable as Π becomes more complex. In this section, we discuss how DE_CBQA scales beyond the reach of MDP as the number of states, actions, and rules increase. In order to avoid a direct exponential blowup when increasing the number of rules, we made one small change to the algorithm: whenever no goal states are found with the fast heuristics (line 1), it fails to return an answer; *i.e.*, it takes a pessimistic approach. Figure 5 compares an increase in number of states to a similar increase in number of rules; observe that the number of rules seems to have a larger effect on overall run time, with an increase in state space being barely noticeable. This is due to two characteristics of our algorithm. First, the heuristic sampling strategy to find states that entail the goal formula visits every rule, but not every state. Second, once entailing states are found, the run time of the DE_CBQA algorithm is not related to the size of the state space at all. Following these intuitions, we see that the algorithm scales gracefully

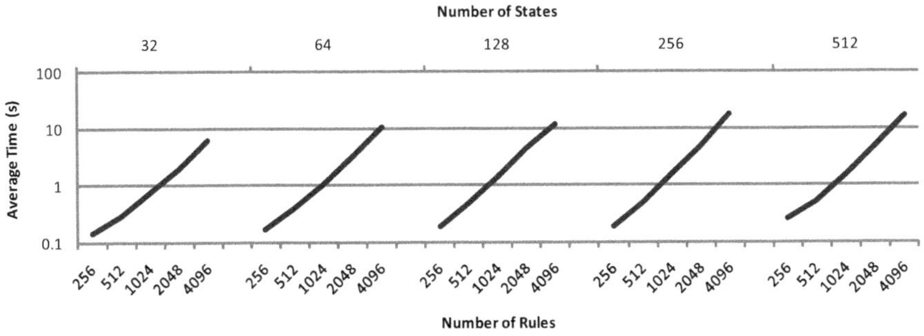

Fig. 5. Log-scale run time as DE_CBQA scales with respect to number of states (top axis) and number of rules (bottom axis). Note the addition of extra rules slows down algorithm execution time much more significantly than a similar increase in state space size.

Fig. 6. *Towards the limits of our current implementation.* Timing results taken by maximizing an individual parameter. The size of the state space was limited by system memory in this implementation.

States	*Actions*	*Rules*	*Time (s)*
4,096	2^{20}	4,096	35.817
64	$\mathbf{2^{25,600}}$	1,024	6.881
64	2^{20}	**16,384**	213.511

to larger state/action spaces. In our experience, real-world instances of CBQA tend to contain significantly *fewer* rules than states and actions [10]; as such, in these cases DE_CBQA scales quite well.

6 Conclusions and Future Work

In this paper, we introduce the Cost-based Query Answering Problem (CBQA), and show that computing an optimal solution to this problem is computationally intractable, both in theory and in practice. We then propose a heuristic algorithm (DE_CBQA) based on iterative random sampling and show experimentally that it provides comparably accurate solutions in significantly less time. Finally, we show that DE_CBQA scales to very large problem sizes.

In the future, we will explore different formalisms used to learn the probability distribution in Line 13 of DE_CBQA. Currently, we use a simple probability vector to update weights for different states; however, such a representation assumes complete independence between any pair of states. A model that takes into account relationships between states (*e.g.*, Bayesian or neural nets) could provide a more intelligent sampling strategy. It is likely that we will see both a higher computational cost and higher quality solutions as the complexity of the formalism increases.

Acknolwedgements. The authors were funded in part by AFOSR grant FA95500610405 and ARO grant W911NF0910206.

References

1. Asal, V., Carter, J., Wilkenfeld, J.: Ethnopolitical violence and terrorism in the middle east. In: Hewitt, J., Wilkenfeld, J., Gurr, T. (eds.) Peace and Conflict 2008, Paradigm (2008)
2. Bellman, R. A markovian decision process. J. Mathematics and Mechanics 6 (1957)
3. Bonet, J.S.D., Isbell Jr., C.L., Viola, P.A.: MIMIC: Finding optima by estimating probability densities. In: Proceedings of NIPS 1996, pp. 424–430. MIT Press, Cambridge (1996)
4. Boutilier, C., Dearden, R., Goldszmidt, M.: Stochastic dynamic programming with factored representations. Artificial Intelligence 121(1-2), 49–107 (2000)
5. Christiansen, H.: Implementing probabilistic abductive logic programming with constraint handling rules. In: Schrijvers, T., Frühwirth, T. (eds.) Constraint Handling Rules. LNCS (LNAI), vol. 5388, pp. 85–118. Springer, Heidelberg (2008)
6. Fagin, R., Halpern, J.Y., Megiddo, N.: A logic for reasoning about probabilities. Information and Computation 87(1/2), 78–128 (1990)
7. Giles, J.: Can conflict forecasts predict violence hotspots? New Scientist, 2647 (March 2008)
8. Hailperin, T.: Probability logic. Notre Dame J. Formal Logic 25(3), 198–212 (1984)
9. Kern-Isberner, G., Lukasiewicz, T.: Combining probabilistic logic programming with the power of maximum entropy. Artif. Intell. 157(1-2), 139–202 (2004)
10. Khuller, S., Martinez, M.V., Nau, D.S., Sliva, A., Simari, G.I., Subrahmanian, V.S.: Computing most probable worlds of action probabilistic logic programs: scalable estimation for 10^30,000 worlds. AMAI 52(2-4), 295–331 (2007)
11. Littman, M.L.: Algorithms for Sequential Decision Making. PhD thesis, Department of Computer Science, Brown University, Providence, RI (February 1996)
12. Lloyd, J.W.: Foundations of Logic Programming, 2nd edn. Springer, Heidelberg (1987)
13. Mannes, A., Michael, M., Pate, A., Sliva, A., Subrahmanian, V.S., Wilkenfeld, J.: Stochastic opponent modelling agents: A case study with Hezbollah. In: Liu, H., Salerno, J. (eds.) Proceedings of IWSCBMP (2008)
14. Ng, R.T., Subrahmanian, V.S.: Probabilistic logic programming. Information and Computation 101(2), 150–201 (1992)
15. Ng, R.T., Subrahmanian, V.S.: A semantical framework for supporting subjective and conditional probabilities in deductive databases. J. Autom. Reas. 10(2), 191–235 (1993)
16. Nilsson, N.: Probabilistic logic. Artificial Intelligence 28, 71–87 (1986)
17. Pelikan, M., Goldberg, D.E., Lobo, F.G.: A survey of optimization by building and using probabilistic models. Comput. Optim. Appl. 21(1), 5–20 (2002)
18. Poole, D.: Probabilistic horn abduction and bayesian networks. Artif. Intell. 64(1), 81–129 (1993)
19. Poole, D.: The independent choice logic for modelling multiple agents under uncertainty. Artif. Intell. 94(1-2), 7–56 (1997)
20. Puterman, M.L.: Markov decision processes: Discrete stochastic dynamic programming. John Wiley & Sons, Inc., Chichester (1994)
21. Simari, G.I., Subrahmanian, V.S.: Abductive inference in probabilistic logic programs. In: Proceedings of ICLP 2010 (Tech. Comm.) (to appear, 2010)
22. Tseng, P.: Solving H-horizon, stationary Markov decision problems in time proportional to log(H). Operations Research Letters 9(5), 287–297 (1990)
23. Tsitsiklis, J., van Roy, B.: Feature-based methods for large scale dynamic programming. Machine Learning 22(1/2/3), 59–94 (1996)
24. Williams, R., Baird, L.: Tight performance bounds on greedy policies based on imperfect value functions. In: 10th Yale Workshop on Adaptive and Learning Systems (1994)

Clustering Fuzzy Data Using the Fuzzy EM Algorithm

Benjamin Quost and Thierry Denœux

laboratoire HeuDiaSyC, Université de Technologie de Compiègne
Centre de Recherches de Royallieu, B.P. 20529
F-60205 Compiègne Cedex
{quostben,tdenoeux}@hds.utc.fr

Abstract. In this article, we address the problem of clustering imprecise data using finite mixtures of Gaussians. We propose to estimate the parameters of the mixture model using the fuzzy EM algorithm. This extension of the EM algorithm allows us to handle imprecise data represented by fuzzy numbers. First, we briefly recall the principle of the fuzzy EM algorithm. Then, we provide the update equations for the parameters of a Gaussian mixture model for fuzzy data. Experiments carried out on synthetic and real data demonstrate the interest of our approach for clustering data that are only imprecisely known.

1 Introduction

Gaussian mixture modelling is a very powerful tool for estimating a multivariate distribution [19]. This model assumes the data to arise from a random sample, whose distribution is a finite mixture of Gaussians. The major difficulty is to estimate the parameters of the model. Generally, these estimates are computed using the maximum-likelihood (ML) approach, through an iterative procedure known as the EM algorithm. Once the parameter values are known, the posterior probabilities of each data point may be computed. Then, classifying each point into the class with highest posterior probability gives a partition of the data. The choice of Gaussian mixtures, rather than geometrical models, is motivated by several arguments. Additional assumptions, for example regarding the shape or the volume of the classes, may be easily taken into account, giving birth to parsimonious variants of the general model. This approach also provides a theoretical framework in which solutions to complex problems, such that determining the number of classes or validating the structure of the partition obtained, may be proposed.

When estimating the parameters using the EM algorithm, the observed data are assumed to be precisely known. However, in some applications, the precise value taken by the variables may be difficult or even impossible to know. For example, acoustic emission control may be used to detect flaws on pressure equipments. This technique provides locations of acoustic events associated with imprecision degrees [8]. The interest of taking into account uncertainty measurements has been demonstrated [16]. Many works advocate the use of fuzzy sets

A. Deshpande and A. Hunter (Eds.): SUM 2010, LNAI 6379, pp. 333–346, 2010.

theory for dealing with imprecise data [12,13,15,23]. Some of them consider that the data at hand are intrinsically fuzzy, a position that has been known as the *physical interpretation* of fuzziness. Here, we rather adopt an *epistemic interpretation*, in which fuzzy numbers "imperfectly specify a value that is existing and precise, but not measurable with exactitude under the given observation conditions" [12]. In this setting, a data sample is a collection of possibility distributions. Each one represents the partial knowledge of the precise value taken by the random variable of interest. The problem of clustering fuzzy data has been addressed in a number of recent papers [2,6,10,11,20,21,22,27,28]. These approaches differ from the type of data considered and from the clustering approach used. However, to our knowledge, clustering imprecise data using mixtures of distributions has only been addressed in [9], when data are intervals.

In this paper, we propose to fit a Gaussian mixture model to the fuzzy data at hand. The likelihood of the sample may be computed using Zadeh's extension principle [29]; then, an EM-like procedure may be used to estimate the parameters maximizing this likelihood. Denœux [4] recently proposed an extension of the EM algorithm for imprecise data in the framework of belief functions. As a possibility distribution may be identified with the plausibility function of a consonant belief mass, this extension is also valid for fuzzy data [5].

The paper is organized as follows. In Section 2, the Gaussian mixture model for clustering data is briefly recalled, along with the procedure for estimating the parameters using the EM algorithm. We focus on the particular case where the covariance matrices are diagonal. In Section 3, we present the fuzzy EM algorithm for estimating the parameters of a Gaussian mixture model with diagonal covariance matrices, when the data are fuzzy numbers. Section 4 presents the experiments on synthetic and real data, and we conclude in Section 5.

2 Gaussian Mixtures Models for Crisp Data

Here, we recall the main results of Gaussian mixture modeling using the EM algorithm. More information on Gaussian mixture models may be found in [19,14]. For a thorough study of the EM algorithm, the reader may refer to [17].

2.1 Model

We suppose that $(\mathbf{x}_1, \ldots, \mathbf{x}_n)$ is the realization of a random sample $(X_1, \ldots, X_n)$. Each $\mathbf{x}_i$ is a p-dimensional vector: $\mathbf{x}_i = (x_{i1}, \ldots, x_{ip})$, supposed to be drawn from a mixture of g Gaussians of probability density function (pdf):

$$g(\mathbf{x}; \Psi) = \sum_{k=1}^{g} \pi_k g_k(\mathbf{x}; \Psi_k), \tag{1}$$

where $\Psi \in \Omega$ is the vector of parameters of the model, and $g_k(\mathbf{x}; \Psi_k)$ denotes the kth Gaussian component with parameters $\Psi_k = (\mathbf{m}_k, \Sigma_k, \pi_k)^\top$:

$$g_k(\mathbf{x}; \Psi_k) = \frac{1}{(2\pi)^{p/2}|\Sigma_k|^{1/2}} \exp\left(-\frac{1}{2}(\mathbf{x} - \mathbf{m}_k)^\top \Sigma_k^{-1}(\mathbf{x} - \mathbf{m}_k)\right). \tag{2}$$

Let the g-dimensional vector $\mathbf{z}_i = (z_{i1}, \ldots, z_{ig})$ indicate the membership of $\mathbf{x}_i$: $z_{ik} = 1$ if $\mathbf{x}_i$ was generated by the kth component, and 0 otherwise. Now, let us introduce the notations $\mathbf{y} = \{\mathbf{x}_1, \ldots, \mathbf{x}_n\}$ and $\mathbf{z} = \{\mathbf{z}_1, \ldots, \mathbf{z}_n\}$, and let $(\mathbf{y}, \mathbf{z})$ be the complete data sample, with pdf $g_c(\mathbf{y}, \mathbf{z}; \Psi)$. The EM algorithm aims at maximizing the observed data log-likelihood $L(\Psi) = \sum_{\mathbf{z}} g(\mathbf{y}|\mathbf{z}; \Psi)\, \mathbb{P}(\mathbf{z}; \Psi)$. The algorithm solves this problem by proceeding iteratively with the complete data log-likelihood $\log L_c$. In the case of Gaussian mixtures, we have:

$$\log L_c(\Psi) = \log g_c(\mathbf{y}, \mathbf{z}; \Psi) = \sum_{i=1}^{n} \sum_{k=1}^{g} z_{ik} \log \pi_k + \sum_{i=1}^{n} \sum_{k=1}^{g} z_{ik} \log g_k(\mathbf{x}_i; \Psi_k),$$

$$= \sum_{k=1}^{g} \log \pi_k \sum_{i=1}^{n} z_{ik} - \frac{np}{2} \log(2\pi) - \sum_{k=1}^{g} \sum_{i=1}^{n} \frac{z_{ik}}{2} \log |\Sigma_k|$$

$$- \frac{1}{2} \sum_{i=1}^{n} \sum_{k=1}^{g} z_{ik} (\mathbf{x}_i - \mathbf{m}_k)^\top \Sigma_k^{-1} (\mathbf{x}_i - \mathbf{m}_k). \quad (3)$$

In this article, we restrict ourselves to the particular case where the variables are independent conditionally to each class: by definition,

$$g_k(\mathbf{x}; \Psi_k^{(q)}) = \prod_{j=1}^{p} g_k(x_j; \Psi_k^{(q)}); \quad (4)$$

this is equivalent to requiring that the covariance matrices be diagonal: for each $k = 1, \ldots, g$, we have $\Sigma_k = \text{diag}(\sigma_{1k}^2, \ldots, \sigma_{pk}^2)$, where $\text{diag}(\mathbf{u})$ denotes the matrix whose diagonal is the vector $\mathbf{u}$. The complete log-likelihood thus becomes:

$$\log L_c(\Psi) = \sum_{k=1}^{g} \log \pi_k \sum_{i=1}^{n} z_{ik} - \frac{np}{2} \log(2\pi) - \sum_{k=1}^{g} \sum_{i=1}^{n} z_{ik} \sum_{j=1}^{p} \log(\sigma_{jk})$$

$$- \frac{1}{2} \sum_{i=1}^{n} \sum_{k=1}^{g} \sum_{j=1}^{p} \frac{z_{ik}}{\sigma_{jk}^2} (x_{ij} - m_{jk})^2. \quad (5)$$

2.2 Estimating the Parameters Using the EM Algorithm

The EM algorithm estimates the parameters so as to maximize the likelihood of the observed data. For this purpose, it proceeds iteratively with the complete log-likelihood $\log L_c(\Psi)$, alternating between two steps that we briefly recall here.

E-step of the EM algorithm. The E-step consists in computing

$$Q(\Psi, \Psi^{(q)}) = \mathbb{E}_{\Psi^{(q)}} [\log L_c(\Psi) | \mathbf{y}, \mathbf{z}]; \quad (6)$$

here, $\Psi^{(q)}$ denotes the current fit of Ψ at iteration q, and $\mathbb{E}_{\Psi^{(q)}}$ represents the expectation computed using parameters $\Psi^{(q)}$. Let

$$t_{ik} = \mathbb{E}_{\Psi^{(q)}} [Z_{ik} | \mathbf{x}_i, \Psi_k] = \mathbb{P}_{\Psi^{(q)}} [Z_{ik} = 1 | \mathbf{x}_i] = \frac{\pi_k\, g_k(\mathbf{x}_i; \Psi_k)}{\sum_{k=1}^{g} \pi_k\, g_k(\mathbf{x}_i; \Psi_k)}; \quad (7)$$

then,

$$Q(\Psi,\Psi^{(q)}) = \sum_{k=1}^{g} \log \pi_k \sum_{i=1}^{n} t_{ik} - \frac{np}{2}\log(2\pi) - \sum_{k=1}^{g}\sum_{i=1}^{n} t_{ik} \sum_{j=1}^{p} \log(\sigma_{jk}) - \frac{1}{2}\sum_{i=1}^{n}\sum_{k=1}^{g}\sum_{j=1}^{p} \frac{t_{ik}}{\sigma_{jk}^2}(x_{ij}-m_{jk})^2. \quad (8)$$

M-step of the EM algorithm. The M-step then consists in maximizing the expectation $Q(\Psi,\Psi^{(q)})$ with respect to $\Psi^{(q)}$; that is, in computing Ψ_{q+1} such that $Q(\Psi_{q+1},\Psi^{(q)}) \geq Q(\Psi,\Psi^{(q)})$, for all $\Psi \in \Omega$. In practice, the update equations are given by setting the derivatives of $Q(\Psi,\Psi^{(q)})$ with respect to each component of Ψ to zero. Assume that the covariance matrices are diagonal (4); then:

$$\pi_k^{(q+1)} = \frac{1}{n}\sum_{i=1}^{n} t_{ik},$$

$$m_{jk}^{(q+1)} = \frac{\sum_{i=1}^{n} t_{ik}x_{ij}}{\sum_{i=1}^{n} t_{ik}},$$

$$\sigma_{jk}^{(q+1)} = \sqrt{\frac{\sum_{i=1}^{n} t_{ik}\left(x_{ij}-m_{jk}\right)^2}{\sum_{i=1}^{n} t_{ik}}}.$$

Remark 1 (Spherical model). Assume that in each class, the variances of all the variables are equal: $\Sigma_k = \sigma_k^2 \mathrm{Id}_\mathrm{p}$, for $k = 1,\ldots,g$. (Here, Id_p is the $p \times p$ identity matrix.) In this case, the level curves of the density are hyper-spheres, and the classes are said to be spherical. Then, the update equations for the proportions and the means are unchanged; the standard deviations are updated using:

$$\sigma_k^{(q+1)} = \sqrt{\frac{\sum_{i=1}^{n} t_{ik}\sum_{j=1}^{p}\left(x_{ij}-m_{jk}\right)^2}{p\sum_{i=1}^{n} t_{ik}}}. \quad (9)$$

Convergence of the EM algorithm. The convergence of the EM algorithm to a local maximum for the observed log-likelihood L has been proved in [3,24]. Under some conditions on the initial values of the parameters, L is bounded from above. As the observed log-likelihood increases at each iteration of the algorithm [3], the convergence is ensured. In practice, the algorithm is stopped when the difference between two successive values of $L(\Psi)$ is less than a given threshold ϵ:

$$\log L(\Psi^{(q+1)}) - \log L(\Psi^{(q)}) \leq \epsilon. \quad (10)$$

As noted in [17, page 85], in many practical applications, the EM algorithm converges to a local maximizer of the observed log-likelihood. However, it is underlined that this convergence towards nontrivial solutions relies on the compactness of the parameter space. This assumption may not hold in certain cases.

For example, when computing ML estimators of the parameters in a mixture of Gaussians, setting the mean of a class to be one of the data points and letting its variance tend to zero will let $L(\Psi)$ tend to infinity.

To avoid such degenerate solutions, prior knowledge on the actual value of the parameters may be integrated in the estimation process, using an adequate distribution $p(\Psi)$. Then, the maximum *a posteriori* (MAP) estimate of the vector parameter Ψ may be computed so as to maximize the log (incomplete) posterior density $\log p(\Psi|\mathbf{y},\mathbf{z}) = \log L_c(\Psi) + \log p(\Psi)$. The analytic formulation for the update equations of the parameter estimates is simpler if $p(\Psi)$ is a *conjugate prior* for the distribution of the model. In the case of a mixture of Gaussians, the conjugate prior for a covariance matrix Σ is the inverse-Wishart distribution:

$$f(\Sigma) = \frac{|\Lambda|^{m/2}|\Sigma|^{-(m+p+1)/2}\exp\left(-\operatorname{trace}\left(\Lambda\Sigma^{-1}\right)/2\right)}{2^{mp/2}\Gamma_p(m/2)}, \tag{11}$$

where Γ_p stands for the (p-dimensional) multivariate Gamma distribution, $m \geq p$ is the number of degrees of freedom, and Λ is a positive definite matrix. The mean and the mode of this pdf are $\Lambda/(m-p-1)$ and $\Lambda/(m+p+1)$, respectively.

3 Gaussian Mixture Models for Fuzzy Data

3.1 The Fuzzy EM Algorithm Applied to Gaussian Mixtures

Here, we briefly present the fuzzy EM (FEM) algorithm [5] that may be derived from Denœux's EM algorithm for credal data [4]. Assume that the available data are imprecise and represented using fuzzy numbers: instead of a crisp value $\mathbf{x}_i$, we have a sample $\tilde{\mathbf{y}}$ of fuzzy numbers, of which each element $\tilde{\mathbf{x}}_i$ has a membership function $\mu_{\tilde{\mathbf{x}}_i}$. The value $\mu_{\tilde{\mathbf{x}}_i}(\mathbf{x})$ may be interpreted as the degree of possibility that the actual value taken by the random variable X_i is $\mathbf{x}$. Thus, the complete-data sample is now $(\tilde{\mathbf{y}},\mathbf{z})$. Then, Zadeh's definition of the probability of a fuzzy event [30] may be used to compute the observed data log-likelihood:

$$L(\Psi) = \sum_{\mathbf{z}} \mathbb{P}(\mathbf{z};\Psi)\int g(\mathbf{y}|\mathbf{z};\Psi)d\mathbf{y}. \tag{12}$$

Thus, the E-step now consists in computing

$$Q(\Psi,\Psi^{(q)}) = \mathbb{E}_{\Psi^{(q)}}\left[\log L_c(\Psi)|\tilde{\mathbf{y}},\mathbf{z}\right]. \tag{13}$$

Note that the expectation is now taken with respect to the fuzzy sample $\tilde{\mathbf{y}}$. We remind here that the conditional density of a continuous random variable X with pdf g_X, with respect to a fuzzy event $\tilde{\mathbf{x}}$ with fuzzy membership function $\mu_{\tilde{\mathbf{x}}}$, is:

$$g_X(\mathbf{x}|\tilde{\mathbf{x}}) = \frac{\mu_{\tilde{\mathbf{x}}}(\mathbf{x})g_X(\mathbf{x})}{\int \mu_{\tilde{\mathbf{x}}}(\mathbf{x})g_X(\mathbf{x})d\mathbf{x}}. \tag{14}$$

The M-step still consists, at iteration q, in maximizing $Q(\Psi,\Psi^{(q)})$ with respect to Ψ. The FEM algorithm iterates alternatedly between steps E and M, until the difference between two successive values is small. Its convergence has been proved [4,5], using similar arguments to those proposed in [3].

3.2 Update Equations of the Parameters

We describe here how mixtures of Gaussians may be fit to fuzzy data using the FEM algorithm. In addition to the conditional independence of the variables, we assume that the membership function of a multidimensional fuzzy number may be expressed as the product of the membership functions of its components:

$$\mu_{\tilde{\mathbf{x}}_i}(\mathbf{x}) = \prod_{j=1}^{p} \mu_{\tilde{x}_{ij}}(x_j). \tag{15}$$

E-step of the FEM algorithm. Let us compute $Q(\Psi, \Psi^{(q)}) = \mathbb{E}[\log L_c(\Psi)|\tilde{\mathbf{y}}, \mathbf{z}]$:

$$Q(\Psi, \Psi^{(q)}) = \sum_{k=1}^{g} \log \pi_k \sum_{i=1}^{n} \mathbb{E}_{\Psi^{(q)}}[Z_{ik}|\tilde{\mathbf{x}}_i] - \frac{1}{2} \sum_{k=1}^{g} \log|\Sigma_k| \sum_{i=1}^{n} \mathbb{E}_{\Psi^{(q)}}[Z_{ik}|\tilde{\mathbf{x}}_i]$$
$$- \frac{1}{2} \sum_{i=1}^{n} \sum_{k=1}^{g} \mathbb{E}_{\Psi^{(q)}} \left[\sum_{j=1}^{p} \frac{Z_{ik}}{\sigma_{jk}^2}(x_{ij} - m_{jk})^2 |\tilde{\mathbf{x}}_i \right] - \frac{np}{2}\log(2\pi). \tag{16}$$

Let us introduce the following notations:

$$\gamma_{ik}^{(q)} = \mathbb{P}_{\Psi^{(q)}}(\tilde{\mathbf{x}}_i|Z_{ik} = 1) = \int \mu_{\tilde{\mathbf{x}}_i}(\mathbf{x}) g_k(\mathbf{x}; \Psi_k^{(q)}) d\mathbf{x}, \tag{17}$$

$$\gamma_{ijk}^{(q)} = \mathbb{P}_{\Psi^{(q)}}(\tilde{x}_{ij}|Z_{ik} = 1) = \int \mu_{\tilde{x}_{ij}}(w_j) g_k(w_j; \Psi_k^{(q)}) dw_j; \tag{18}$$

$$p_i^{(q)} = \mathbb{P}_{\Psi^{(q)}}(\tilde{\mathbf{x}}_i) = \sum_{k=1}^{p} \pi_k^{(q)} \int \mu_{\tilde{\mathbf{x}}_i}(\mathbf{x}) g_k(\mathbf{x}; \Psi_k^{(q)}) d\mathbf{x}; \tag{19}$$

$$\eta_{ijk}^{(q)} = \mathbb{E}_{\Psi^{(q)}}[x_{ij}|\tilde{x}_{ij}, Z_{ik} = 1] = \frac{\int x_j \mu_{\tilde{x}_{ij}}(x_j) g_{jk}(x_j, \Psi_k^{(q)}) dx_j}{\gamma_{ijk}^{(q)}}; \tag{20}$$

$$\xi_{ijk}^{(q)} = \mathbb{E}_{\Psi^{(q)}}[x_{ij}^2|\tilde{\mathbf{x}}_i, Z_{ik} = 1] = \frac{\int x_j^2 \mu_{\tilde{x}_{ij}}(x_j) g_{jk}(x_j, \Psi_k^{(q)}) dx_j}{\gamma_{ijk}^{(q)}}. \tag{21}$$

With these notations, using Bayes' theorem, we have:

$$t_{ik}^{(q)} = \mathbb{E}_{\Psi^{(q)}}[Z_{ik}|\tilde{\mathbf{x}}_i] = \frac{\mathbb{P}_{\Psi^{(q)}}(\tilde{\mathbf{x}}_i|Z_{ik=1})\mathbb{P}_{\Psi^{(q)}}(Z_{ik} = 1)}{\mathbb{P}_{\Psi^{(q)}}(\tilde{\mathbf{x}}_i)} = \frac{\gamma_{ik}^{(q)} \pi_k^{(q)}}{p_i^{(q)}}. \tag{22}$$

Now, using assumptions (4) and (15), we get:

$$\mathbb{E}_{\Psi^{(q)}} \left[\sum_{j=1}^{p} \frac{Z_{ik}}{\sigma_{jk}^2}(x_{ij} - m_{jk})^2 |\tilde{\mathbf{x}}_i \right] = \sum_{j=1}^{p} \frac{1}{\sigma_{jk}^{(q)2}} \Big(\mathbb{E}_{\Psi^{(q)}} \left[Z_{ik} x_{ij}^2 |\tilde{\mathbf{x}}_i \right]$$
$$- 2\, m_{jk}^{(q)}\, \mathbb{E}_{\Psi^{(q)}} \left[Z_{ik} x_{ij} |\tilde{\mathbf{x}}_i \right] + m_{jk}^{(q)2}\, \mathbb{E}_{\Psi^{(q)}} \left[Z_{ik} |\tilde{\mathbf{x}}_i \right] \Big). \tag{23}$$

Furthermore,

$$\mathbb{E}_{\Psi^{(q)}}\left[Z_{ik}x_{ij}^2|\tilde{\mathbf{x}}_i\right] = \mathbb{E}_{\Psi^{(q)}}\left[x_{ij}^2|\tilde{x}_{ij}, Z_{ik}=1\right]\mathbb{P}\left(Z_{ik}=1|\tilde{\mathbf{x}}_i\right) = \xi_{ijk}^{(q)}t_{ik}^{(q)}, \quad (24)$$

$$\mathbb{E}_{\Psi^{(q)}}\left[Z_{ik}x_{ij}|\tilde{\mathbf{x}}_i\right] = \mathbb{E}_{\Psi^{(q)}}\left[x_{ij}|\tilde{x}_{ij}, Z_{ik}=1\right]\mathbb{P}\left(Z_{ik}=1|\tilde{\mathbf{x}}_i\right) = \eta_{ijk}^{(q)}t_{ik}^{(q)}. \quad (25)$$

Hence, finally, Equation (16) becomes:

$$Q(\Psi,\Psi^{(q)}) = \sum_{k=1}^{g}\log\pi_k\sum_{i=1}^{n}t_{ik}^{(q)} - \frac{np}{2}\log(2\,\pi) - \sum_{k=1}^{g}\sum_{j=1}^{p}\log\sigma_{jk}^{(q)}\sum_{i=1}^{n}t_{ik}^{(q)}$$
$$-\frac{1}{2}\sum_{i=1}^{n}\sum_{k=1}^{g}t_{ik}^{(q)}\left(\sum_{j=1}^{p}\frac{1}{\sigma_{jk}^{(q)2}}\xi_{ijk}^{(q)} - 2\sum_{j=1}^{p}\frac{m_{jk}^{(q)}}{\sigma_{jk}^{(q)2}}\eta_{ijk}^{(q)} + \sum_{j=1}^{p}\frac{m_{jk}^{(q)2}}{\sigma_{jk}^{(q)2}}\right). \quad (26)$$

M-step of the FEM algorithm. In order to maximize $Q(\Psi,\Psi^{(q)})$ defined by Equation (26), its partial derivatives with respect to the various parameters have to be set to zero. The partial derivatives with respect to the proportions π_k are:

$$\frac{\partial Q(\Psi,\Psi^{(q)})}{\partial\pi_k} = \frac{1}{\pi_k}\sum_{i=1}^{n}t_{ik}^{(q)};$$

equating these derivatives to zero, under the constraint $\sum_{k=1}^{g}\pi_k = 1$, give similar results to the EM algorithm for crisp data:

$$\pi_k^{(q+1)} = \frac{1}{n}\sum_{i=1}^{n}t_{ik}^{(q)}. \quad (27)$$

Computing derivatives with respect to each element m_{jk} of the means gives:

$$\frac{\partial Q(\Psi,\Psi^{(q)})}{\partial m_{jk}} = \frac{1}{\sigma_{jk}^{(q)2}}\sum_{i=1}^{n}t_{ik}^{(q)}\eta_{ijk}^{(q)} - m_{jk}\sum_{i=1}^{n}t_{ik}^{(q)}; \quad (28)$$

setting this partial derivative to zero, we get the following update equations:

$$m_{jk}^{(q+1)} = \frac{\sum_{i=1}^{n}t_{ik}^{(q)}\eta_{ijk}^{(q)}}{\sum_{i=1}^{n}t_{ik}^{(q)}}. \quad (29)$$

Eventually, the first-order derivative of $Q(\Psi,\Psi^{(q)})$ with respect to σ_{jk} is:

$$\frac{\partial Q(\Psi,\Psi^{(q)})}{\partial\sigma_{jk}} = -\frac{1}{\sigma_{jk}}\sum_{i=1}^{n}t_{ik}^{(q)} + \frac{1}{\sigma_{jk}^3}\sum_{i=1}^{n}t_{ik}^{(q)}\left(\xi_{ijk}^{(q)} - 2\,m_{jk}^{(q)}\eta_{ijk}^{(q)} + m_{jk}^{(q)2}\right).$$

Setting this partial derivative to zero gives:

$$\sigma_{jk}^{(q+1)} = \sqrt{\frac{\sum_{i=1}^{n}t_{ik}^{(q)}\left(\xi_{ijk}^{(q)} - 2\,m_{jk}^{(q)}\eta_{ijk}^{(q)} + m_{jk}^{(q)2}\right)}{\sum_{i=1}^{n}t_{ik}^{(q)}}}. \quad (30)$$

Remark 2 (Spherical model). Assume, as in Remark 1, that the classes are spherical: $\Sigma_k = \sigma_k^2 \mathrm{Id_p}$. Then, the update equations for the proportions and the means are unchanged, and the update equations for the standard deviations become:

$$\sigma_k^{(q+1)} = \sqrt{\frac{\sum_{i=1}^n t_{ik}^{(q)} \sum_{j=1}^p \left(\xi_{ijk}^{(q)} - 2m_{jk}^{(q)} \eta_{ijk}^{(q)} + m_{jk}^{(q)2} \right)}{p \sum_{i=1}^n t_{ik}^{(q)}}}. \quad (31)$$

Remark 3 (Relationship between the update equations for crisp and fuzzy data). We may notice the similarity with the update equations obtained for crisp data. The difference is that the crisp quantities x_{ij} and x_{ij}^2 are replaced with the conditional expectations η_{ijk} and ξ_{ijk} of the fuzzy variables $\tilde{x_{ij}}$ and $\tilde{x_{ij}}^2$, respectively.

Remark 4 (Prior on the covariance matrices). Suppose that a prior is set on each covariance matrix Σ_k using the inverse-Wishart distribution with parameters m_0 and $\Lambda_{k\,0} = \mathrm{diag}\,(\lambda_{k1}, \ldots, \lambda_{kp})$. Then, the estimates for the standard deviations for the conditional independence case becomes:

$$\sigma_{jk}^{(q+1)} = \sqrt{\frac{\sum_{i=1}^n t_{ik}^{(q)} \left(\xi_{ijk}^{(q)} - 2\, m_{jk}^{(q)} \eta_{ijk}^{(q)} + m_{jk}^{(q)2} \right) + \lambda_{jk}}{\sum_{i=1}^n t_{ik}^{(q)} + (m_0 + p + 1)}}. \quad (32)$$

In the spherical case, using $\Lambda_{k\,0} = \lambda_k \mathrm{Id_p}$, we get:

$$\sigma_k^{(q+1)} = \sqrt{\frac{\sum_{i=1}^n t_{ik}^{(q)} \sum_{j=1}^p \left(\xi_{ijk}^{(q)} - 2m_{jk}^{(q)} \eta_{ijk}^{(q)} + m_{jk}^{(q)2} \right) + \lambda_k}{p \sum_{i=1}^n t_{ik}^{(q)} + p\,(m_0 + p + 1)}}. \quad (33)$$

4 Experiments

4.1 Synthetic Data

First, we ran experiments over synthetic two-dimensional data. We placed ourselves in the experimental setting considered in [8]: here, a fuzzy datum represents the imprecise knowledge of the actual (precise) value of a variable. We generated data as follows. First, we drew a sample of $n = 300$ realizations $\mathbf{x}_1, \ldots, \mathbf{x}_n$ of a Gaussian mixture of $g = 3$ components with the parameters given in Table 1. Level curves of the corresponding density are represented in Figure 1. The curves correspond to levels 0.01, 0.02, 0.03, 0.04 and 0.05 of the density.

Each data point was classified according to the Bayes' rule. Then, each $\mathbf{x}_i$ was transformed into a fuzzy number. Let a (monovariate) trapezoidal fuzzy number $\tilde{w}$ be defined by four scalars a, b, c and d, such that:

$$\mu_{\tilde{w}}(w) = \begin{cases} (w-b)/(b-a), & \text{if } a \leq w \leq b, \\ 1, & \text{if } b \leq w \leq c, \\ (d-w)/(d-c), & \text{if } c \leq w \leq d, \\ 0 & \text{otherwise.} \end{cases} \quad (34)$$

Table 1. Parameters of the components of the Gaussian mixture

	comp. 1	comp. 2	comp. 3
π_k	0.3	0.4	0.3
m_k	$(-2,-2)^\top$	$(-1,+1)^\top$	$(+2,-2)^\top$
Σ_k	$\begin{pmatrix} 2.5 & 0.25 \\ 0.25 & 0.75 \end{pmatrix}$	$\begin{pmatrix} 2 & 0 \\ 0 & 1.5 \end{pmatrix}$	$\begin{pmatrix} 1.25 & -0.25 \\ -0.25 & 1 \end{pmatrix}$

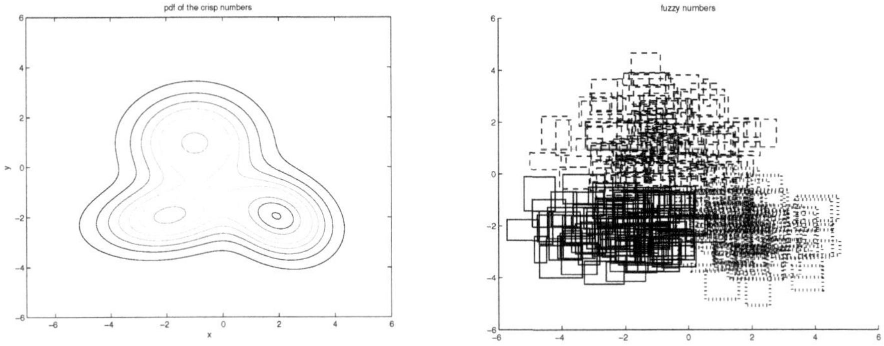

Fig. 1. Pdf of the Gaussian mixture (left); fuzzy numbers obtained with $r = 0.5$ and $s = 2$ (right)

Here, each coordinate x_{ij} of a crisp data point $\mathbf{x}_i$ was transformed into a trapezoidal fuzzy number $\tilde{x_{ij}}$ as follows. Four iid realizations u_1, u_2, u_3, and u_4 were drawn from an uniform distribution $\mathcal{U}_{[0;1]}$; then, r and s being user-defined:

$$\begin{aligned} b_{ij} &= u_1(x_{ij} - r), \quad a_{ij} = u_2(b_{ij} - s); \\ c_{ij} &= u_3(x_{ij} + r), \quad d_{ij} = u_4(x_{ij} + s). \end{aligned} \tag{35}$$

Figure 1 displays the fuzzy numbers thus obtained using $r = 0.5$ and $s = 2$: each rectangle corresponds to the alpha-cut of a membership function, with $\alpha = 0.75$. The line style of each rectangle (plain, dashed, or dotted) represents the class with highest probability, determined using the true distribution of the data.

Parameters were estimated using the FEM algorithm, for the two models studied in Section 3.2. From now on, the model with diagonal covariance matrices will be referred to as diagonal model. The means were initialized at random, according to a centered and scaled normal distribution; the initial covariance matrices were set to the identity. The quality of each partition was evaluated using the pairwise Rand index (PRI). For each model, we performed $N = 20$ runs of the algorithm; we retained the best result (for which the log-likelihood was maximal). Parameter estimates are given in Table 2, as well as the corresponding value of the fuzzy log-likelihood, the number q of iterations, and the PRI. Figure 2 displays the densities estimated by both models, at the same levels as previously. The fuzzy numbers are also represented; now, the line style of each rectangle indicates the class into which the corresponding data point was classified.

Table 2. Parameter estimates, number q of iterations, log-likelihood, and PRI obtained with $r = 0.5$ and $s = 2$

	π_k	m_k	Σ_k	q	$\log L$	PRI
comp. 1	0.41	$(-2, -1.58)^\top$	diag $(1.37, 1.37)$			
spherical comp. 2	0.28	$(-0.75, 1.49)^\top$	diag $(1.06, 1.06)$	18	-969.1	0.7984
comp. 3	0.31	$(1.74, -2.04)^\top$	diag $(0.75, 0.75)$			
comp. 1	0.31	$(-1.94, -2.15)^\top$	diag $(1.78, 0.39)$			
diagonal comp. 2	0.41	$(-1, 0.94)^\top$	diag $(1.59, 1.53)$	68	-963.8	0.9136
comp. 3	0.28	$(1.82, -2.11)^\top$	diag $(0.79, 0.60)$			

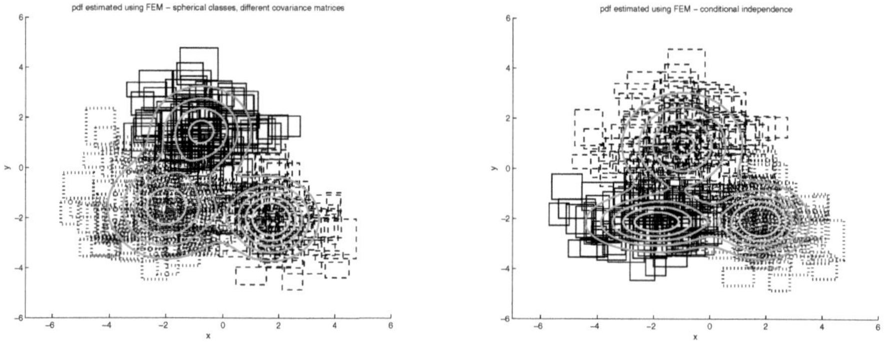

Fig. 2. Pdf estimated using the spherical model (left) and the diagonal model (right)

Table 3. Parameter estimates, number q of iterations, log-likelihood, and PRI obtained with $r = 2$ and $s = 2$

	π_k	m_k	Σ_k	q	$\log L$	PRI
comp. 1	0.49	$(-1.71, -1.26)^\top$	diag $(1.29, 1.29)$			
spherical comp. 2	0.21	$(-0.63, 1.73)^\top$	diag $(0.59, 0.59)$	165	-587.7	0.6937
comp. 3	0.30	$(1.70, -1.90)^\top$	diag $(0.63, 0.63)$			
comp. 1	0.53	$(-1.27, -1.50)^\top$	diag $(1.99, 0.73)$			
diagonal comp. 2	0.25	$(-0.86, 1.63)^\top$	diag $(1.05, 0.48)$	209	-585	0.6536
comp. 3	0.22	$(2.02, -2.06)^\top$	diag $(0.27, 0.73)$			

Then, we modified the synthetic dataset as follows. We fuzzified the same crisp data using values $r = 2$ and $s = 2$. Figure 3 represents the alpha-cuts (with $\alpha = 0.75$) of the fuzzy numbers thus obtained. As previously, the algorithms were run $N = 20$ times using the same initialization procedure, and the best result was retained for each model. The results are given in Table 3. Figure 3 also displays the density estimated using the diagonal model. The estimates of the means are very similar to the previous; those of the variances, however, are much lower. Indeed, the imprecision on the actual realizations of the random variables is higher. In other terms, the intervals in which these realizations may fall with

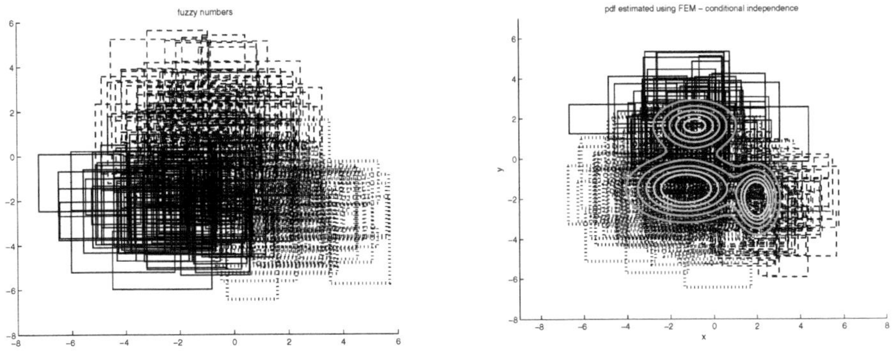

Fig. 3. Fuzzy numbers obtained with $r = 2$ and $s = 2$ (left); density estimated using the diagonal model (right)

Table 4. Parameter estimates, number q of iterations, log-likelihood, and PRI obtained with $r = 2$ and $s = 2$, with an inverse-Wishart prior on the covariance matrices

		π_k	m_k	Σ_k	q	$\log L$	PRI
	comp. 1	0.37	$(-2.16, -1.54)^\top$	diag $(1.14, 1.14)$			
spherical	comp. 2	0.31	$(0.78, 1.16)^\top$	diag $(1.18, 1.18)$	95	-622.1	0.7913
	comp. 3	0.32	$(1.47, -1.97)^\top$	diag $(0.83, 0.83)$			
	comp. 1	0.34	$(-2.19, -1.62)^\top$	diag $(1.27, 0.95)$			
diagonal	comp. 2	0.33	$(-0.82, 1.09)^\top$	diag $(1.23, 1.19)$	331	-620.9	0.8427
	comp. 3	0.33	$(1.38, -2)^\top$	diag $(1.11, 0.64)$			

the same degree of possibility as previously are larger. Then, the algorithm obviously favours a solution with as small variances as possible, as it maximizes the likelihood. Finally, we set an inverse Wishart prior on the covariance matrices. Table 4 presents the results obtained with $m_0 = 2$, and $\Lambda_0 = \text{diag}\,(2\ 2)$ for both models.

4.2 Real Data

We ran the FEM algorithm on the blood data used in [6]. This dataset presents statistics on daily measurements of systolic and diastolic pressures on $n = 108$ patients. Remark that here, each measurement is precise; however, for each patient, center and spread values only were stored. Thus, the fuzziness of a datum stems from the variability of the measurements performed on each patient and the choice to summarize these measurements using their center and their spread only. Although this interpretation differs from the point of view adopted in this paper, we used these data to compare our results with those presented in [6].

We interpreted these data as triangular fuzzy numbers, which are a special case for trapezoidal fuzzy numbers. In addition, we assumed that each center was equidistant to the minimal and maximal values. First, the data were centered

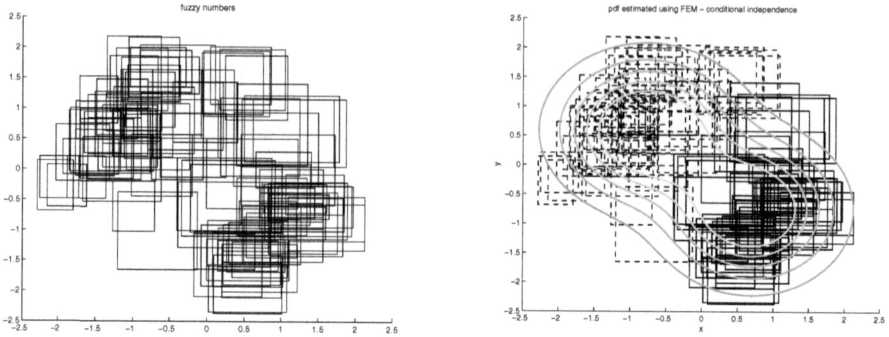

Fig. 4. Fuzzy numbers obtained from the blood data (left); density estimated using the diagonal model with an inverse-Wishart prior on the covariance matrices (right)

Table 5. Parameter estimates, number q of iterations, and log-likelihood obtained for the blood data, with an inverse-Wishart prior on the covariance matrices

	π_k	m_k	Σ_k	q	$\log L$
spherical comp. 1	0.523	$(-0.768, 0.599)^\top$	diag (0.558, 0.558)	66	-220.4
comp. 2	0.477	$(0.690, -0.757)^\top$	diag (0.532, 0.532)		
diagonal comp. 1	0.523	$(-0.767, 0.598)^\top$	diag (0.567, 0.549)	32	-220.4
comp. 2	0.477	$(0.691, -0.757)^\top$	diag (0.520, 0.543)		

and scaled with respect to the mean and standard deviation of the center values. Then, the density of a two-component mixture of Gaussians was estimated using the FEM algorithm. The parameters were initialized as previously.

The alpha-cuts of the fuzzy numbers are represented in Figure 4 (again, with $\alpha = 0.75$). Table 5 presents the results obtained with both models using an inverse-Wishart prior on Σ_1 and Σ_2, with $m_0 = 2$ and $\Lambda_{01} = \Lambda_{02} = \text{diag}\,(2, 2)$. Figure 4 displays the density estimated using the diagonal model, at levels 0.02, 0.04, 0.06, 0.08 and 0.1. The results are quite similar to those obtained in [6]. Here, 55 patients were assigned to the first class, and 53 to the second one.

5 Conclusion

In this paper, we addressed the problem of clustering fuzzy data using mixture models. Our approach is based on an extention of the EM algorithm for fuzzy data, proposed by Denœux [4,5]. The likelihood of a mixture of Gaussians may be computed, given a sample of fuzzy numbers, using Zadeh's extention principle. Then, the estimates maximizing this likelihood may be estimated using an iterative procedure. At each iteration, the expectation $Q(\Psi, \Psi^{(q)}) = \mathbb{E}[\log L_c(\Psi)|\tilde{\mathbf{y}}]$ of the log-likelihood is first computed. Then, the parameters of the model may be updated so as to maximize this expectation. In this paper, we detailed the computation of the update equations under the assumption that the covariance matrices considered are diagonal, in the case of a finite mixture of Gaussians.

We conducted experiments on synthetic and real data. Experiments show that our algorithm estimates accurately the distribution of imprecisely known data. Our approach may be sensitive to the amount of imprecision in the available information. In particular, the covariance matrices may be under-estimated if the degree of fuzziness is high. However, the algorithm may be guided towards a desired solution by setting a prior distribution on these parameters. Thus, our algorithm constitutes a generic approach for clustering imprecise data.

The extension of the fuzzy EM algorithm to finite mixture of Gaussians with full covariance matrices is straightforward. However, in such cases, it may be necessary to rely on Monte Carlo processes in order to perform the E-step of the EM algorithm. Therefore, this work is left for further research.

References

1. Coppi, R., D'Urso, P.: Fuzzy K-means clustering models for triangular fuzzy time trajectories. Statistical Methods and Applications 11(1), 21–40 (2002)
2. Coppi, R., D'Urso, P.: Three-way fuzzy clustering models for LR fuzzy time trajectories. Computational Statistics and Data Analysis 43, 149–177 (2003)
3. Dempster, A.P., Laird, N.M., Rubin, D.B.: Maximum likelihood from incomplete data via the EM algorithm. Journal of the Royal Statistical Society B 39, 1–38 (1977)
4. Denœux, T.: Maximum likelihood estimation from evidential data. In: Proceedings of the first workshop on the theory of belief functions and their applications, Brest, France (2010) (manuscript), `http://www.ensieta.fr/belief2010/`
5. Denœux, T.: Maximum likelihood estimation from fuzzy data using the Fuzzy EM algorithm (working paper)
6. D'Urso, P., Giordani, P.: A weighted fuzzy *c*-means clustering model for fuzzy data. Computational Statistics and Data Analysis 50, 1496–1523 (2006)
7. Auephanwiriyakul, S., Keller, J.M.: Analysis and efficient implementation of a linguistic fuzzy c-means. IEEE Trans. on Fuzzy Systems 10(5), 563–582 (2002)
8. Hamdan, H., Govaert, G.: CEM algorithm for imprecise data. Application to flaw diagnosis using acoustic emission. In: Proc. of the IEEE International Conference on Systems, Man and Cybernetics, vol. (5), pp. 4774–4779. The Hague, Netherlands (2004)
9. Hamdan, H., Govaert, G.: Mixture model clustering of uncertain data. In: Proc. of the IEEE International Conference on Fuzzy Systems, Reno, Nevada, USA, pp. 879–884 (2005)
10. Hathaway, R.J., Bezdek, J.C., Pedrycz, W.: A parametric model for fusing heterogeneous fuzzy data. IEEE Trans. on Fuzzy Systems 4(3), 1277–1282 (1996)
11. Hung, W.-L., Yang, M.-S.: Fuzzy clustering on LR-type fuzzy numbers with an application in Taiwanese tea evaluation. Fuzzy Sets and Systems 150(3), 561–577 (2005)
12. Gebhardt, J., Gil, M.A., Kruse, R.: Fuzzy set-theoretic methods in statistics. In: Slowinski, R. (ed.) Fuzzy Sets in Decision Analysis, Operations Research and Statistics, pp. 311–347. Kluwer Academic Publishers, Boston (1998)

13. Gil, M.A., López-Díaz, M., Ralescu, D.A.: Overview on the development of fuzzy random variables. Fuzzy Sets and Systems 157(19), 2546–2557 (2006)
14. Jordan, M., Jacobs, R.: Hierarchical mixtures of experts and the EM algorithm. Neural Computation 6, 181–214 (1994)
15. Kruse, R., Meyer, K.D.: Statistics with vague data. Kluwer, Dordrecht (1987)
16. Mauris, G.: Expression of measurement uncertainty in a very limited knowledge context: a possibility theory-based approach. IEEE Trans. on Instrumentation and Measurement 56(3), 731–735 (2007)
17. McLachlan, G.J., Krishnan, T.: The EM Algorithm and Extensions. Wiley, New York (1997)
18. Pelekis, N., Iakovidis, D., Kotsifakos, E., Kopanakis, I.: Fuzzy Clustering of Intuitionistic Fuzzy Data. International Journal of Business Intelligence and Data Mining 3(1), 45–65 (2008)
19. Redner, R., Walker, H.: Mixture densities, maximum likelihood and the EM algorithm. SIAM review 26(2), 195–239 (1984)
20. Sato, M., Sato, Y.: Fuzzy clustering model for fuzzy data. In: Proc. of the 4th IEEE Conf. on Fuzzy Systems, Yokohama, Japan, pp. 2123–2128 (1995)
21. Takata, O., Miyamoto, S., Umayahara, K.: Clustering of data with uncertainties using Hausdorff distance. In: Proc. of the 2nd IEEE International Conference on Intelligence Processing Systems, Gold Coast, Australia, pp. 67–71 (1998)
22. Takata, O., Miyamoto, S., Umayahara, K.: Fuzzy clustering of data with uncertainties using minimum and maximum distances based on L1 metric. In: Proc. of the Joint 9th IFSA World Congress and 20th NAFIPS International Conference, Vancouver, British Columbia, Canada, pp. 2511–2516 (2001)
23. Viertl, R.: Univariate statistical analysis with fuzzy data. Computational Statistics & Data Analysis 51(1), 133–147 (2006)
24. Wu, C.F.J.: On the convergence properties of the EM algorithm. Annals of Statistics 11, 95–103 (1983)
25. Xu, Z., Chen, J.: WU, J. Clustering algorithm for intuitionistic fuzzy sets. Information Sciences 178, 3775–3790 (2008)
26. Yang, M.S., Ko, C.H.: On a class of fuzzy c-numbers clustering procedures for fuzzy data. Fuzzy Sets and Systems 84, 49–60 (1996)
27. Yang, M.S., Liu, H.H.: Fuzzy clustering procedures for conical fuzzy vector data. Fuzzy Sets and Systems 106, 189–200 (1999)
28. Yang, M.S., Hwang, P.Y., Chen, D.H.: Fuzzy clustering algorithms for mixed feature variables. Fuzzy Sets and Systems 141, 301–317 (2004)
29. Zadeh, L.A.: Fuzzy sets. Information and Control 8, 338–353 (1965)
30. Zadeh, L.A.: Probability measures of fuzzy events. Journal of Mathematical Analysis and Applications 10, 421–427 (1968)

Combining Multi-resolution Evidence for Georeferencing Flickr Images

Olivier Van Laere[1], Steven Schockaert[2], and Bart Dhoedt[1]

[1] Department of Information Technology, Ghent University, IBBT, Belgium
{olivier.vanlaere,bart.dhoedt}@intec.ugent.be
[2] Dept. of Applied Mathematics and Computer Science, Ghent University, Belgium
steven.schockaert@ugent.be

Abstract. We explore the task of determining the geographic location of photos on Flickr, using combined evidence from Naive Bayes classifiers that are trained at different spatial resolutions. In particular, we estimate the location of Flickr photos, based on their tags, at four different scales, ranging from a city-level granularity to fine-grained intra-city areas. Using Dempster-Shafer's evidence theory, we combine the output of the different classifiers into a single mass assignment. We demonstrate experimentally that the induced belief and plausibility measures are useful to determine whether there is sufficient evidence to classify the photo at a given granularity. Thus an adaptive method is obtained, by which photos are georeferenced at the most appropriate resolution.

1 Introduction

An increasing number of web systems allow users to organize and share resources, such as photos, videos, or scientific papers. The predominant way of organizing such resources is by the use of short textual descriptions called tags. These tags are added by users in an uncontrolled way, without the need for any semantic resources. Nonetheless, due to the wide availability of such tags, statistically analyzing tag distributions has proven to be a successful way of obtaining (shallow) semantic information in an automated way [13].

Considering photo sharing websites such as Flickr[1] or Panoramio[2], the most important kind of metadata is arguably the location where a photo was taken. Accordingly, these websites typically allow to attach explicit geographical coordinates to a photo, in addition to tag-based descriptions of its content. This is important for at least two reasons. First, it allows to put photos on a map, providing an interesting addition to e.g. Google Maps[3], and allowing users to quickly retrieve photos that were taken in a particular region [1,4,11]. Second, by looking at correlations between tag occurrences and locations, it becomes possible to find approximate boundaries of geographic regions [8]. This is particularly important for vernacular (i.e. informal) place names, which have no

[1] http://www.flickr.com/
[2] http://www.panoramio.com/
[3] http://maps.google.com

A. Deshpande and A. Hunter (Eds.): SUM 2010, LNAI 6379, pp. 347–360, 2010.

official boundaries that could be retrieved from gazetteers or other geographic resources. As a consequence, a thorough analysis of Flickr tags may result in rich geographic models that can be used to support geographically informed web search engines.

The question remains of how the location of a photo could be acquired. Certain cameras have a built-in GPS, in which case the exact coordinates are obtained automatically. In most cases, however, users need to manually specify where a photo was taken. Because this puts an extra burden on the user, without any immediate benefit, only a small minority of the users go through this step. Another approach is taken by Suggestify[4], a web application which allows users to suggest the location of a photo of another Flickr user, which she can then choose to accept or refuse. Nonetheless, for the vast majority of all photos on Flickr and Panoramio, the location is not known.

To solve this problem, we may attempt to derive the approximate location of a photo automatically, by comparing its tags with the tags of photos whose exact location is known [4,11,14]. These approximate locations may be sufficient to put the photo on a map (at a certain resolution), or to help determine the approximate boundaries of a vernacular region. Alternatively, by establishing the area in which the photo was taken, we may also assist users that are willing to manually specify the exact location, by centering a map on the area that was found, and zooming in at an appropriate level. In each case, it is important not only to find a location that is approximately correct, but also to provide a reliable estimate of how accurate that location is (e.g. street-level, neighborhood-level, city-level, regional level, ...). For some photos, we can easily find a very precise location (e.g. a photo tagged "Eiffel tower"), while for other photos we cannot even indicate in which country it was taken, by only looking at its tags (e.g. a photo tagged "birthday party"). Thus it is of interest to study adaptive techniques that provide reliable estimates at an appropriate resolution, or admit that no reliable location could be established.

To georeference Flickr photos (i.e. to assign a location), in previous work [14] we have proposed to discretize space by clustering the photos from some training set, and then train a Naive Bayes classifier to find the most appropriate cluster for previously unseen photos. Different resolutions can then be considered by repeating the whole process for more or less fine-grained clusterings, i.e. by adopting a larger or smaller number of clusters. Thus we obtain a series of different classifiers, operating at different levels of resolution. In this way, for a given photo, the most appropriate resolution can be chosen by looking at the confidence each of the classifiers has in its respective outcome.

In this paper, we look at how the results of these different classifiers can be combined to find the most appropriate location and resolution. In particular, we experimentally investigate the use of Dempster-Shafer theory, which naturally allows to combine evidence from sources that operate at different levels of granularity. Our hypothesis in using Dempster-Shafer theory is that agreement between the classifiers is a strong indicator of the correctness of the location that

[4] http://suggestify.appspot.com/

was found. For instance, if classifier C_1 finds locations at the neighborhood-level and C_2 at the city level, and the neighborhood that was found by C_1 is not from the city that was found by C_2, our confidence that the neighborhood is correct should be low, regardless of the confidence of classifier C_1 in its choice.

The remainder of this paper is structured as follows. Section 2 summarizes our methodology in obtaining training and test data from Flickr. We also explain how we have clustered the images in the training set, and which preprocessing techniques were applied. Then in Section 3 we discuss the details of our proposed method. We briefly recall how a Naive Bayes classifier can be trained to find plausible areas where a photo might be located, at a fixed resolution. Subsequently we provide details on how Dempster-Shafer theory is applied to combine the classifiers that were trained at different resolutions. In Section 4, we present our experimental results, demonstrating substantial improvements over a baseline system. Finally, related work is discussed in Section 5.

2 Methodology

To obtain suitable training and test data, we composed a list of 55 large European cities. These cities were selected by intersecting the set of the 100 most densely populated European cities[5] with the set of the 160 most important European cities for tourism[6]. This choice was motivated by the intuition that a high population should ensure that allocating photos to locations is non-trivial (as opposed to villages where all activity is centered around a small area), while tourist activity should ensure that a sufficient number of photos is available on Flickr. For each georeferenced photo in these cities, we collected the corresponding tags and coordinates using the Flickr API, leading to a total of 3738072 photos. In addition to the coordinates themselves, Flickr provides information about the accuracy of coordinates as a number between 1 (world level) and 16 (street level). From our initial set of photos, we removed those photos whose coordinates had an accuracy of 13 or less, to ensure that all coordinates were meaningful w.r.t. within-city location. Furthermore, we removed photos whose tag set and user name was identical to a photo that is already in our collection (to reduce the impact of bulk uploads [11]). After these two filtering steps, a set of 1029761 photos remained from 54 cities (no photos from Bremen had coordinates whose accuracy was above 13), which was split into 686193 photos for training ($\approx 66\%$) and 343568 photos for testing ($\approx 33\%$). In separating training data from test data, we ensured that all photos from the same user were either in the training set, or in the test set (to avoid an unfair exploitation of user-specific tags).

We then divided the 54 remaining cities into a set of disjoint areas that will serve as classification labels. The areas themselves were obtained by clustering the locations of the photos in the training set using the k-medoids algorithm with geodesic distance. Below, we consider four different resolutions, corresponding to

[5] `http://www.nga.mil`

[6] `http://www.visiteuropeancities.info`

the city level (in which case there are 54 areas, each corresponding to an entire city), as well as the result of clustering all photos in 250, 500, or 1000 clusters. In each case, we chose the number of clusters per city proportional to the number of georeferenced photos we had available for that city (in the training set), with the exception that every city should contain at least one cluster centre. As a result, cities for which we had only few georeferenced photos were divided in areas of a larger scale. This conforms to our intuition that we should try to be precise in estimating the location of a photo only when sufficient information is available for making that decision. In addition, whenever the number of photos in a given cluster dropped below 50, after an iteration of the k-medoids algorithm, that cluster was eliminated and the associated photos added to the nearest remaining cluster. The actual number of areas after the clustering algorithm had converged was respectively 54, 217, 401 and 677.

For efficiency, and to increase the robustness of the approach, we removed all tags that were used by 2 users or less. Next, we applied χ^2 feature selection to eliminate tags that are not indicative of a particular area. In particular, the vocabulary V that was used for classification was obtained by taking for each area a those 25 tags whose χ^2 value was highest. This led to a total number of 1269, 4701, 8452 and 13727 distinct tags, respectively in the case where the initial number of clusters k was 54, 250, 500 and 1000.

3 Georeferencing Images

3.1 Naive Bayes Classification

Let $\mathcal{A}$ be a set of (disjoint) areas, obtained by clustering the locations of the images in our training set. For each area $a \in \mathcal{A}$, we write X_a to denote the set of images from our training set that were taken in area a. Given a previously unseen image x, we try to determine in which area x was most likely taken by comparing its tags with those of the images in the training set. In [14], we proposed a (multinomial) Naive Bayes classifier to this end, which has the advantage of being simple, efficient, and robust. An additional advantage, which will be crucial for combining classifiers that operate at different resolutions, is the fact that Naive Bayes produces probabilities, in contrast to e.g. support vector machines. Specifically, we assume that an image x is represented as its set of tags. Using Bayes' rule, and assuming that occurrences of different tags are independent, the probability $P(a|x)$ that image x was taken in area a is proportional to

$$P(a|x) \propto P(a) \cdot \prod_{t \in x} P(t|a) \tag{1}$$

Using a multinomial language model with Laplace smoothing [18], the probability $P(t|a)$ is estimated as

$$P(t|a) = \frac{N_t + 1}{\left(\sum_{y \in X_a} |y|\right) + |V|}$$

where N_t is the number of images in area a containing tag t, $\sum_{y \in X_a} |y|$ is the total number of tag occurrences over all images in area a, and V is the vocabulary, as before. Note that this technique of estimating $P(t|a)$ originates from Laplace's rule of succession. The maximum likelihood estimation $\frac{N_t}{\sum_{y \in X_a} |y|}$ would not be useful here, as it would imply $P(a|x) = 0$ as soon as x has one tag which does not occur with any image of the training set that is located in area a. The prior probability $P(a)$ of area a, on the other hand, can reliably be estimated using the maximum likelihood method:

$$P(a) = \frac{|X_a|}{\sum_{b \in \mathcal{A}} |X_b|}$$

Finally note that the actual value of $P(a|x)$, for all $a \in \mathcal{A}$, is found from (1) after normalization.

3.2 Combining Classifiers Using Dempster-Shafer Theory

Motivation. The fact that areas are spatially distributed should intuitively help to assign photos to areas more accurately. For example, assume that $\mathcal{A} = \{a, b, c, d\}$ and that the Naive Bayes classifier finds for a given photo x that that $P(a|x) = 0.3$, $P(b|x) = 0.25$, $P(c|x) = 0.25$ and $P(d|x) = 0.2$. Now assume furthermore that b, c, and d are adjacent neighborhoods, while a is located in a different city. Then in fact, the correct location is more likely to be near areas $\{b, c, d\}$ than near a. Naive Bayes in its basic form ignores this information and simply treats areas as abstract classes. To make Naive Bayes more spatially-aware, we propose to apply the approach outlined in Section 3.1 at multiple resolutions and combine the results. A classifier working at a higher resolution will then hopefully find the region containing regions b, c, d to be more likely than the region containing a. Based on the agreement between fine-grained classifiers and coarse-grained classifiers, we may then try to find the most appropriate resolution for a given photo: in cases of disagreement, coarser results are preferred, while in cases of strong agreement, fine-grained results may be better suited.

Specifically, let $\{\mathcal{A}_1, ..., \mathcal{A}_k\}$ be different clusterings of the cities of interest into disjoint areas, where $\mathcal{A}_1$ corresponds to the finest clustering and $\mathcal{A}_k$ corresponds to the coarsest clustering, i.e. $|\mathcal{A}_1| > |\mathcal{A}_2| > ... > |\mathcal{A}_k|$. Furthermore let C_i be a classifier that was trained to find the area from $\mathcal{A}_i$ in which a given photo was taken. With each area in $\mathcal{A}_i$, we can now associate a set of areas from the finest level $\mathcal{A}_1$. In particular, for $a \in \mathcal{A}_i$, we let $areas(a)$ denote the set of areas from $\mathcal{A}_1$ that overlap with area a. In this way, classifications at coarser resolutions can be seen as incomplete classifications at the finest resolution. For instance, if classifier $\mathcal{A}_k$ suggests that a is the most plausible area, we can take this as evidence that the correct area, at the finest level, is among those of the set $area(a)$. Such incomplete conclusions are naturally represented in the theory of evidence that was proposed by Dempster and Shafer [5,12]. In Dempster-Shafer theory, evidence is encoded by a probability distribution on the power set of the universe. This probability distribution is called a belief function, or mass assignment, to distinguish it from probability distributions on the universe itself.

Obtaining mass assignments. In Dempster-Shafer theory, a mass assignment m in the universe U maps any subset of U to a value in $[0,1]$ such that $\sum_{X \subseteq U} m(X) = 1$ and $m(\emptyset) = 0$. Intuitively, $m(X)$ represents the amount of evidence that the correct value is among those in X. Subsets X such that $m(X) > 0$ are called focal elements. If all focal elements are disjoint, then $m(X)$ can be interpreted as the probability that the correct area is among those in X. In general, two measures of uncertainty are typically defined in Dempster-Shafer theory, for any $X \subseteq U$:

$$Bel(X) = \sum_{Y \subseteq X} m(Y) \qquad\qquad Pl(X) = \sum_{Y \cap X \neq \emptyset} m(Y)$$

The degree of belief $Bel(X)$ can be interpreted as a lower bound on the probability that X contains the correct value, while the degree of plausibility $Pl(X)$ is an upper bound for this probability.

In the context of this paper, the universe will always be the set of areas (clusters) in the most fine-grained clustering, viz. the set $\mathcal{A}_1$. Let $p_i(a)$ be the probability that classifier C_i has assigned to area $a \in \mathcal{A}_i$ for the photo under consideration. Intuitively, we can take this information as evidence that the correct area, among the fine-grained areas in $\mathcal{A}_1$, is among those that overlap with a, i.e. among those in $areas(a)$. This idea leads to the following mass assignment corresponding to classifier C_i ($X \subseteq \mathcal{A}_1$):

$$m_i(X) = \begin{cases} p_i(a) & \text{if } X = areas(a) \text{ for some } a \in A_i \\ \sum_{a \in (\mathcal{A}_i \setminus A_i)} p_i(a) & \text{if } X = \mathcal{A}_1 \\ 0 & \text{otherwise} \end{cases} \tag{2}$$

where $A_i \subseteq \mathcal{A}_i$ is the set of areas that are most likely according to classifier C_i. In principle, we may take $A_i = \mathcal{A}_i$ but there are at least two reasons for taking A_i to be a much smaller set of areas. The mass assigned to the universe $\mathcal{A}_1$ corresponds to a degree of ignorance, i.e. we only put belief in the most plausible areas of each classification, and admit that we are ignorant about the correct area when it turns out that none of the most plausible areas is correct. The underlying motivation is that Naive Bayes can be useful to find which are the most likely areas, but that the probability estimates for the remaining areas are not meaningful. Moreover, restricting attention to a relatively small subset of areas A_i is a prerequisite for obtaining a sufficiently scalable method. In our experiments, the set A_i was constructed by adding areas in decreasing order of likelihood (according to C_i), until $\sum_{a \in A} p_i(a) \geq 0.95$.

Note that alternatively, we could also assign the mass $\sum_{a \in \mathcal{A}_i \setminus A_i} p_i(a)$ to $\mathcal{A}_i \setminus A_i$ instead of $\mathcal{A}_i$; we do not consider this possibility, however, in the remainder of this paper.

Combining mass assignments. An important advantage of using Dempster-Shafer theory in this context is that it allows to combine evidence from different sources. In particular, for two mass assignments m and m' in the universe $\mathcal{A}_1$, the joint mass assignment $m \oplus m'$ is defined using Dempster's rule of combination as

$$(m \oplus m')(\emptyset) = 0 \tag{3}$$

$$(m \oplus m')(X) = \frac{\sum_{Y \cap Z = X} m(Y) \cdot m'(Z)}{1 - \sum_{Y \cap Z = \emptyset} m(Y) \cdot m'(Z)} \tag{4}$$

for any subset $\emptyset \subset X \subseteq \mathcal{A}_1$, and provided that $\sum_{Y \cap Z = \emptyset} m(Y) \cdot m'(Z) < 1$. It can be shown that this combination rule is associative. By treating the classifiers $C_1, ..., C_k$ as independent sources, we obtain the following mass assignment:

$$m = m_1 \oplus m_2 \oplus ... \oplus m_k \tag{5}$$

Note that the assumption that classifiers $C_1, ..., C_k$ are independent sources is a simplification, as they have essentially been trained on the same data. However, as different classifiers operate at different resolutions, implying among others that different tags have been retained by the χ^2 method in each case, this simplification appears to be reasonable.

The combination rule (3)–(4) is the combination rule proposed by Dempster. It is not entirely uncontroversial, however, and in particular when the degree of conflict $\sum_{Y \cap Z = \emptyset} m(Y) \cdot m'(Z)$ is close to 1, it is reputed to provide counterintuitive results [17]. As an alternative, Yager [16] proposed the following rule for combining k mass assignments in a universe U ($X \subset U$).

$$m(X) = \sum_{\bigcap_i Y_i = X} m_1(Y_1) \cdot ... \cdot m_k(Y_k) \tag{6}$$

$$m(U) = m_1(U) \cdot ... \cdot m_k(U) + \sum_{\bigcap_i Y_i = \emptyset} m_1(Y_1) \cdot ... \cdot m_k(Y_k) \tag{7}$$

$$m(\emptyset) = 0 \tag{8}$$

Clearly, Yager's combination rule only differs from the one proposed by Dempster in what happens with the mass $\sum_{Y \cap Z = \emptyset} m(Y) \cdot m'(Z)$ that would normally be assigned to the empty set. While Dempster's rule distributes this mass over all focal elements, leading to an associative operator, in Yager's rule this mass is assigned to the universe U. As such, Yager's rule can be considered more cautious as the degree of ignorance increases when different sources are in conflict with each other.

4 Experimental Results

In this section, we present the results of a number of experiments which we have carried out to compare the performance of the Dempster-Shafer based approach with a baseline that uses the probabilities from Naive Bayes in a more straightforward way.

In a first experiment, we have verified whether we could improve the accuracy of Naive Bayes by using the combined mass assignment defined by (3)–(4). In particular, the task consists of choosing one area from the clustering $\mathcal{A}_i$ at a given resolution ($k = 54, 250, 500, 1000$), and we have compared the following three methods:

Probability. Choose the cluster for which the highest probability was found using Naive Bayes.

Plausibility. Choose the cluster a for which the value of $Pl(areas(a))$ is maximal.

Belief. Choose the cluster a for which the value of $Bel(areas(a))$ is maximal.

The result is summarized in Table 1. The evaluation metric that was used is accuracy, i.e. the percentage of photos in the test set for which the correct area was found. Clearly, for higher values of k a lower accuracy is generally obtained, as there are more areas to choose from, and less information is available for each area. In approximately 87% of the cases, a photo can be assigned to the correct city, while the correct area at the finest level can only be found in about 40% of the cases. Comparing the different methods, we find that except for $k = 500$, using *belief* leads to slightly better performance than using *probability*. *Plausibility*, on the other hand, leads to worse performance than *probability*, except for the case $k = 54$. This latter fact is not surprising as for any area a which represents an entire city, the only focal elements that overlap with this city will actually be contained in the city, hence $Bel(areas(a)) = Pl(areas(a))$. Overall, in this task, the Dempster-Shafer approach does not allow to substantially improve over the standard Naive Bayes approach.

Table 1. Comparing the use of probability, plausibility and belief for finding the area in which a photo was taken (accuracy)

	54	250	500	1000
Probability	0.8694	0.5137	0.4622	0.4126
Plausibility	0.8729	0.4756	0.3838	0.4134
Belief	0.8729	0.5211	0.4457	0.4151

A second experiment was targeted at evaluating the behavior of the Dempster-Shafer approach when it comes to finding the right resolution for a given photo. Here the task is as follows. For a given photo, choose the most appropriate value of k (54, 250, 500 or 1000) and choose an area from the corresponding clustering. Accuracy is defined as the percentage of cases in which the true location of the photo was within the area that was chosen. However, the idea is that as often as possible, areas should be chosen from the more fine-grained clusterings, while accuracy will clearly be higher when selecting areas from the more coarse-grained clusterings. In addition to accuracy, it is therefore important to compare the average size of the areas that are returned by different methods.

As clusters are simply defined as sets of photos (as opposed to e.g. polygons) we have measured the size of an area (cluster) in terms of the distance between the centroid of that cluster and the remaining photos of the cluster. In particular, for a given area a, represented by the set of photos X_a, the centroid c_a of a is the most central photo, i.e.:

$$c_a = \arg\min_{x \in X_a} \sum_{y \in X_a} d(x, y)$$

where $d(x, y)$ is the geodesic distance between the locations of photos x and y. To measure the size $size(a)$ of area a, we have used the median value of the set $\{d(x, c_a) | x \in X_a\}$. The size intuitively corresponds to the radius of area a, if we think of this area as a circle. Note that the median is used, rather than the maximum or average, because several areas contain outliers, i.e. photos that are not close to any other photos, and that are added to the cluster centre that happens to be closest. The median is more robust against such outliers than the maximum or average, and thus appears to be better suited as an evaluation measure. As an evaluation criterium, in addition to accuracy, we consider the average of $size(a)$ over all areas a that were chosen by a particular method. Ideally, methods should exhibit a high accuracy and a small average size.

The methods that have been compared all follow the same basic strategy. First, the areas from the clustering corresponding to $k = 1000$ are ranked. For the top-ranked area a_1, it is checked whether sufficient support is available. If this is the case, area a_1 is returned as the chosen area. If not, the process is repeated for the clustering at level $k = 500$ and, if necessary for $k = 250$. If there is insufficient support for the top-ranked area a_3 at level $k = 250$, the best area at level $k = 54$ is always chosen. Thus our method is parametrized by a ranking function, a way of measuring support, and a threshold value. The threshold value will be used to control the trade-off between accuracy and average size. As for the remaining two parameters, we have compared the following configurations:

Probability. Areas are ranked according to the probability that was assigned to them by the Naive Bayes classifier. This probability value also serves as a measure of support.

Plausibility. Areas a are ranked according to the plausibility degree $Pl(areas(a))$. This degree also serves as a measure of support.

Belief. Areas a are ranked according to the belief degree $Bel(areas(a))$. This degree also serves as a measure of support.

Hybrid. Areas a are ranked according to the plausibility degree $Pl(areas(a))$. Support is measured as $Bel(areas(a))$.

The result is depicted in Figure 1. This figure was obtained by varying the value of the threshold from 0.01 to 0.99 in steps of 0.01. This led, for each of the four methods, to 99 data points, each of which corresponds to an *(accuracy,average size)* pair. After interpolation of these 99 data points, the graphs in Figure 1 were obtained. Clearly, the three methods based on the combined mass assignment perform substantially better than the method based on the probabilities of the Naive Bayes classifier. For instance, to obtain an accuracy of 75%, using the method *probability* we need to accept an average cluster size of about 2.15

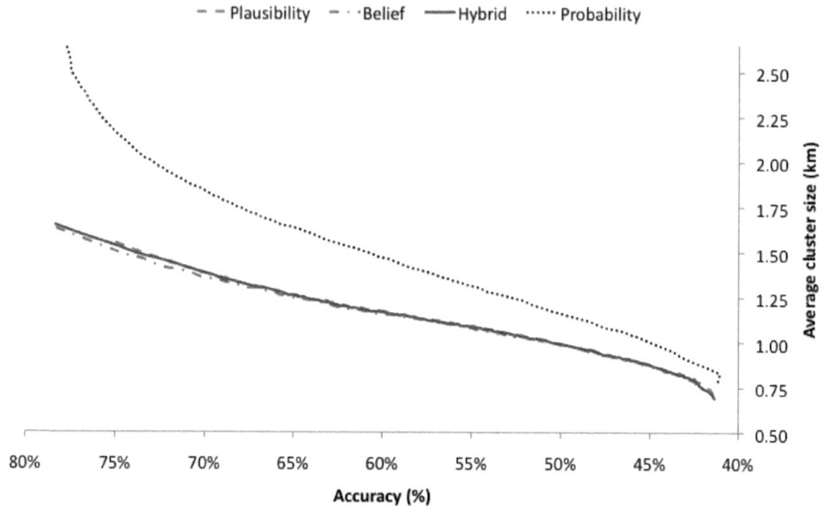

Fig. 1. Comparison of the trade-off between accuracy and average cluster size for four different methods

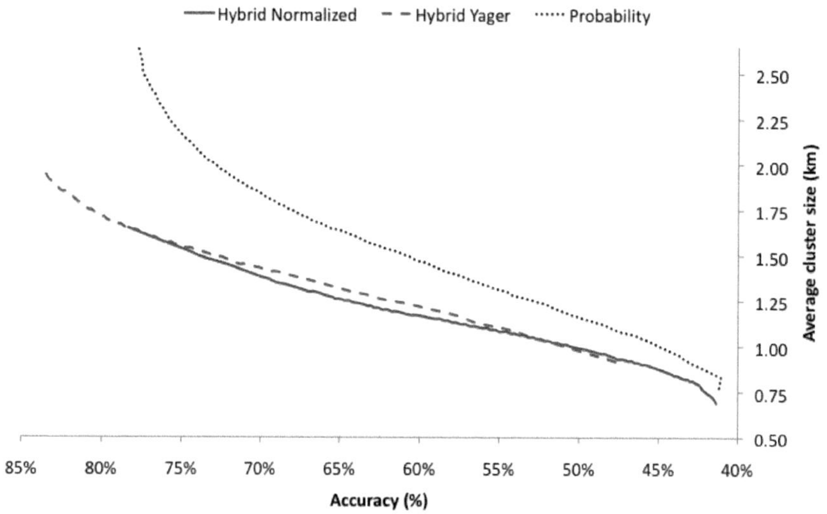

Fig. 2. Comparison the performance of Dempster's and Yager's rule of combination

km, whereas this is around 1.5 for the other methods. At lower accuracy levels, the difference becomes somewhat less pronounced, e.g. an accuracy of 50% corresponds to an average cluster size of about 1.15 using the *probability* method and a size of about 1 km using the other methods.

Figure 1 is based on the combined mass assignment (5) obtained using Dempster's rule. As this task is essentially about deciding whether there is enough support to assign a photo to a cluster at a particular level, Yager's rule may be more suitable. Indeed, Dempster's rule ignores any conflict among the different levels by normalizing the masses. When using Yager's rule, on the other hand, whenever there is conflict, the degrees of belief and plausibility will have lower values. This will lead to photos being assigned to areas of coarser clusterings. In Figure 2, the result of the *hybrid* method is depicted when using either the combined mass assignment (5) based on Dempster's rule or the combined mass assignment (6)–(8) based on Yager's rule. The most important conclusion is that the graph corresponding to Yager's rule is to the left of the graph corresponding to Dempster's rule. This is indeed in accordance with the cautious nature of the method: photos tend to be assigned to coarser levels, leading to higher accuracy at the cost of a higher average size. For the accuracies attained by both methods, i.e. the accuracies in the interval $[0.47, 0.78]$, Yager's rule and Dempster's rule perform comparably, with Dempster's rule performing slightly better.

5 Related Work

Some authors have already studied the task of georeferencing photos based on clustering. One such approach is presented in [4], where target locations are determined using mean shift clustering, a non-parametric clustering technique from the field of image segmentation. The advantage of this clustering method is that an optimal number of clusters is determined automatically, requiring only an estimate of the scale of interest. Specifically, to find good locations, the difference is calculated between the density of photos at a given location and a weighted mean of the densities in the area surrounding that location. To assign locations to new images, both visual (keypoints) and textual (tags) features were used. Experiments were carried out on a sample of over 30 million images, using both Bayesian classifiers and linear support vector machines, with slightly better results for the latter. Two different resolutions were considered corresponding to approximately 100 km (finding the correct metropolitan area) and 100 m (finding the correct landmark). It was found that visual features, when combined with textual features, substantially improve accuracy in the case of landmarks. In [7], an approach is presented which is based purely on visual features. For each new photo, the 120 most similar photos with known coordinates are determined. This weighted set of 120 locations is then interpreted as an estimate of a probability distribution, whose mode is determined using mean-shift clustering. The resulting value is used as prediction of the image's location. Using k-means to spatially cluster geotagged Flickr images has been proposed in [1], where the clusters are used to find representative textual descriptions of each area. The goal is to visualize these textual descriptions on a map, to assist users in finding images of interest.

The idea that when georeferencing images, the spatial distribution of the classes (areas) could be utilized to improve accuracy has already been suggested in [11]. Their starting point is that typically not only the correct area will receive a high probability, but also the areas surrounding the correct area. Indeed, the expected distribution of tags in these areas will typically be quite similar. Hence, if some area a receives a high score, and all of the areas surrounding a also receive a relatively high score, we can be more confident in a being approximately correct than when all the areas surrounding a receive a low score. Motivated by this intuition, [11] proposes to smooth $P(a|x)$ as follows (using a uniform prior):

$$P^*(a|x) \propto \alpha P(x|a) + (1-\alpha) \cdot \sum_{b \in neigh_d(a)} \frac{P(x|b)}{(2d+1)^2 - 1}$$

where $d > 0$ and $neigh_d(a)$ is the set of all areas that are within distance d of a.

Some Flickr tags are intuitively more important than others in determining the location of a photo. Toponyms in particular are by definition indicative of geographic location. One way of recognizing toponyms is by looking for so-called comma-groups. These are groups of words that are comma-separated, e.g. *San Francisco, California, USA*. In this example, there is a clear relationship between the comma-separated values, as San Francisco is a city, located in the state of California, which is in turn one of the states of the USA. As a result, resolution of the toponyms represented by this group reveals an unambiguous geographical reference. Resolution of such comma-groups has been studied by Lieberman in [9]. In [8], Hollenstein studied the way people tag images in order to discover how people refer to a location. She found that the city toponym was by far the most essential reference to a specific location. This is in accordance with our results, where we have also found classification accuracies to be particularly high for the city level. It was furthermore shown in [8] that the average user has a distinct idea of specific places, their location and extent. Despite this tagging behaviour, Hollenstein concluded that the data available in the Flickr database meets the requirements to generate spatial footprints at a sub-city level.

Various authors have investigated the use of Dempster-Shafer theory for combining the results of different classifiers [2,6,10,15]. However, the aim of using Dempster-Shafer theory in this context is quite different from our aim in this paper. Specifically, these methods mainly use Dempster-Shafer theory for its ability to represent partial ignorance. For instance, if a given classifier assigns a probability p_i to each class c_i, a belief function may be constructed by choosing $m(\{c_i\}) = f_i$ for some $f_i < p_i$, and $m(C) = 1 - \sum_i f_i$, for $C = \{c_1, ..., c_n\}$ the set of all classes. The value $1 - \sum_i f_i$ can then intuitively be interpreted in terms of confidence in the associated classifier. Note also that all focal elements are then either singletons or the universe, which makes Dempster-Shafer theory sufficiently scalable to deal with large numbers of classes, although sometimes focal elements of the form $C \setminus \{c_i\}$ are also used. In [3], Dempster-Shafer theory is used for retrieving images of people, combining evidence from a face recognition module and a classifier based on textual descriptions; again only singletons and the entire universe are considered as focal elements.

6 Conclusions

We have studied the problem of finding the geographic location of a photo, particularly emphasizing the importance of determining the appropriate resolution for any given photo. While the precise location of some photos can easily be established, for other photos we can only hope to find a rough idea of where it was taken. Our basic approach consists of clustering the part of geographic space that is of interest, and use standard machine learning techniques (viz. Naive Bayes) to find the cluster which is most likely to contain the correct location of a photo. By varying the number of clusters, different classifiers are obtained which operate at different resolutions.

While adaptive methods, heuristically choosing the most appropriate resolution, can be obtained by straighforwardly analyzing the outputs of these different classifiers, a significant gain in performance is obtained by first combining these outputs using Dempster-Shafer's evidence theory. Experimental results have indicated that the belief and plausibility measures induced by the resulting mass assignment are particularly suitable for determining whether sufficient support is available to classify a photo at a given resolution.

Acknowledgments

Steven Schockaert was funded as a postdoctoral fellow of the Research Foundation – Flanders (FWO).

References

1. Ahern, S., Naaman, M., Nair, R., Yang, J.H.-I.: World explorer: visualizing aggregate data from unstructured text in geo-referenced collections. In: Proceedings of the 7th ACM/IEEE-CS Joint Conference on Digital Libraries, pp. 1–10 (2007)
2. Al-Ani, A., Deriche, M.: A new technique for combining multiple classifiers using the dempster-shafer theory of evidence. J. Artif. Int. Res. 17(1), 333–361 (2002)
3. Aslandogan, Y., Yu, C.: Multiple evidence combination in image retrieval: Diogenes searches for people on the Web. In: Proceedings of the 23rd Annual International ACM SIGIR Conference on Research and Development in Information Retrieval, pp. 88–95 (2000)
4. Crandall, D.J., Backstrom, L., Huttenlocher, D., Kleinberg, J.: Mapping the world's photos. In: Proceedings of the 18th International Conference on World Wide Web, pp. 761–770 (2009)
5. Dempster, A.: A Generalization of Bayesian Inference. Journal of the Royal Statistical Society. Series B (Methodological) 30(2), 205–247 (1968)
6. Denœux, T.: A k-nearest neighbor classification rule based on Dempster-Shafer theory. IEEE Transactions on Systems, Man, and Cybernetics 25(5), 804–813 (1995)
7. Hays, J.H., Efros, A.A.: Im2gps: estimating geographic information from a single image. In: Proc. Computer Vision and Pattern Recognition, CVPR (2008)
8. Hollenstein, L.: Capturing vernacular geography from georeferenced tags. Master's thesis, University of Zurich (2008)

9. Lieberman, M.D., Samet, H., Sankaranayananan, J.: Geotagging: using proximity, sibling, and prominence clues to understand comma groups. In: Proceedings of the 6th Workshop on Geographic Information Retrieval (2010)
10. Rogova, G.: Combining the results of several neural network classifiers. Neural Networks 7(5), 777–781 (1994)
11. Serdyukov, P., Murdock, V., van Zwol, R.: Placing flickr photos on a map. In: Proceedings of the 32nd international ACM SIGIR conference on Research and development in information retrieval, pp. 484–491 (2009)
12. Shafer, G.: A mathematical theory of evidence. Princeton University Press, Princeton (1976)
13. Tang, J., Leung, H.-f., Luo, Q., Chen, D., Gong, J.: Towards ontology learning from folksonomies. In: Proceedings of the 21st international jont conference on Artifical intelligence, pp. 2089–2094 (2009)
14. Van Laere, O., Schockaert, S., Dhoedt, B.: Towards automated georeferencing of flickr photos. In: GIR 2010: Proceedings of the 6th Workshop on Geographic Information Retrieval (2010)
15. Xu, L., Suen, C.: Methods of combining multiple classifiers and their applications to handwriting recognition. IEEE Transactions on Systems, Man, and Cybernetics 22(3), 418–435 (1992)
16. Yager, R.R.: On the dempster-shafer framework and new combination rules. Information Sciences 41(2), 93–137 (1987)
17. Zadeh, L.A.: A simple view of the dempster-shafer theory of evidence and its implication for the rule of combination. AI Mag. 7(2), 85–90 (1986)
18. Zhai, C., Lafferty, J.: A study of smoothing methods for language models applied to information retrieval. ACM Trans. Inf. Syst. 22(2), 179–214 (2004)

A Structure-Based Similarity Spreading Approach for Ontology Matching

Ying Wang, Weiru Liu, and David A. Bell

School of Electronics, Electrical Engineering and Computer Science,
Queen's University Belfast, Belfast, BT7 1NN, UK
{ywang14,w.liu,da.bell}@qub.ac.uk

Abstract. Most of the frequently used ontology mapping methods to date are based on linguistic information implied in ontologies. However, same concepts in different ontologies can represent different semantics under the context of different ontologies, so relationships on mapping cannot be solely recognized by applying linguistic information. Discovering and utilizing structural information in ontology is also very important. In this paper, we propose a structure-based similarity spreading method for ontology matching which consists of three steps. We first select centroid concepts from both ontologies using similarities between entities based on their linguistic information. Second, we partition each ontology based on the set of centroid concepts recognized in it using clustering method. Third, we utilize a similarity spreading method to update the similarities between entities from two ontologies and apply a greedy matching method to establish the final mapping results. The experimental results demonstrate that our approach is very effective and can obtain much better results comparing to other similarity based and similarity flooding based algorithms.

1 Introduction

Ontology mapping is a solution to the semantic heterogeneity problem in information integration and sharing. It establishes correspondences between semantically related entities in different ontologies [1]. Virtually any application that involves multiple ontologies must establish semantic mappings among them, to ensure interoperability [2]. Different applications arise in myriad domains [3]: ontology engineering, information integration, web services, and multi-agent communication etc.

Most of the ontology mapping approaches use the elementary-level matching techniques [4,5,6] (e.g., string-based methods, linguistic-based methods) which map elements by analyzing entities in isolation, ignoring their relationships with other entities [7]. But the determination of the true semantics of an entity is often difficult without a context, so structural information of an ontology plays an important role for ontology mapping.

When an ontology is viewed as a graph, an entity of an ontology (a node in a graph) inherits its parents semantics, and also passes on its own semantics

A. Deshpande and A. Hunter (Eds.): SUM 2010, LNAI 6379, pp. 361–374, 2010.

to its children. Therefore, considering structural information is a natural way for enhancing ontology mapping as illustrated by the following two examples. Example 1 shows that two apparently different entities from two ontologies are similar when their neighboring concepts are similar. On the other hand, in Example 2, even when two entities of two ontologies are similar, if their neighboring concepts are not sufficiently similar, then these two entities are very likely not matched.

Example 1. Figure 1 is a mapping snippet between two ontologies which describes bibliographic references from OAEI benchmark. In this mapping example, concept **Proceedings** and concept **Proc** have different labels and seemingly not very similar semantically. But if they are embedded into the ontologies they belong, the neighbors of **Proceedings**: **Book, Monograph, Collection** are very similar to the neighbors of **Proc**: **Book, Monograph, Collection**, so **Proceedings** and **Proc** are highly possible to be mapped together.

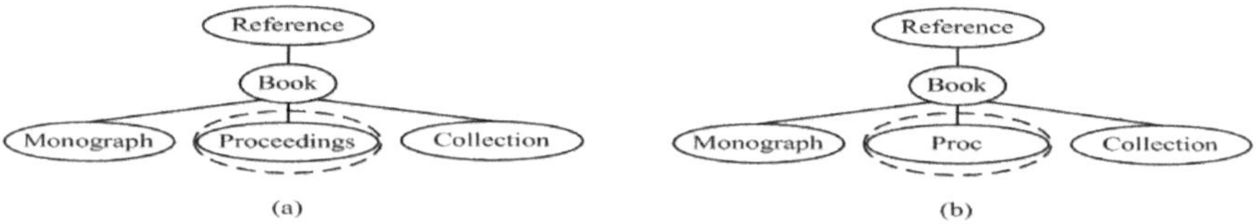

Fig. 1. An example of different concepts with the same meaning

Example 2. Figure 2 shows an example of mapping snippet between two concept hierarchies which are extracted from Google Directory[1] and Yahoo Directory[2] separately. We observe that concept **History** in Figure 2(a) has exactly the same label as the concept **History** in Figure 2(b). When using traditional methods to compare the similarity between these two concepts, such as *edit-distance based or lexical-based methods*, a high similarity value will be obtained which means they should be mapped together. However this is incorrect, because we observe that the neighboring concepts of these two entities are not similar: one talks about the history of music and the other describes the history of architecture. So these two entities should not be mapped.

The idea of semantic propagation was explored in [6] on schema mapping in data integration, called *similarity flooding* which utilize graphs to compute structure similarities between data elements. In this paper, we extend this idea to ontology mapping. The difference between ontology mapping and schema mapping is that ontologies normally have richer semantics and more complex structures than schemas. If we apply the schema mapping similarity flooding algorithm to mapping ontolgies directly, the computational cost for storing graphs will be too high to be practically applicable.

[1] http://www.google.com/dirhp
[2] http://dir.yahoo.com/

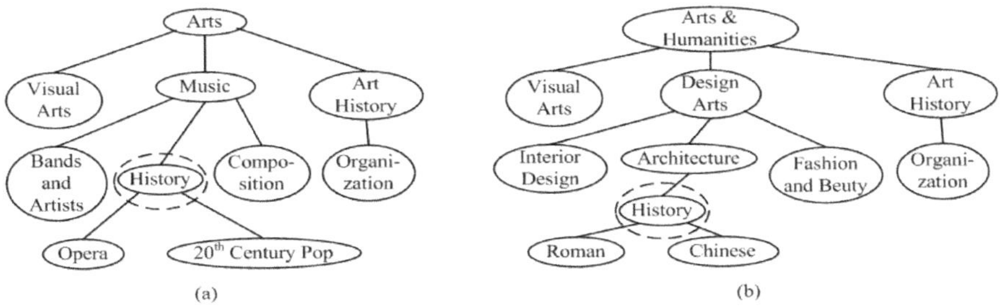

Fig. 2. An example of same concepts with different meaning

Although there have been several extensions of similarity flooding to ontology mapping, such as [8,9], they all suffer from some drawbacks. The main drawback in [8] is that the similarity propagation is done each time for a whole ontology graph. However, similarities can only be preserved to entities that are not too far from a given entity, that is, propagation should be done *locally* in relation to a given entity. The approach proposed in [9] tried to overcome this limitation by adding some restrictions when generating the pairwise connectivity graphs, so that propagations are done through these local graphs. However, the condition for generating the connectivity graphs may eliminate some useful links that should have been used for propagating similarities.

In this paper, we propose another extension to the similarity flooding approach, aiming to explore only locally related neighboring entities through a partition technique. This approach can reduce the computational complexity encountered in [8] on the one hand and avoid the danger of ignoring any useful entities on the other hand as may happen in [9], because only those neighboring concepts which are very close to the given entity structurally and highly related to this concept semantically should be considered. For example, as shown in Figure 2(a), concept **Organization** is not close to concept **History** and their semantics are not closely related, so the interaction (similarity propagation) between them can be ignored.

To propagate similarities locally, we need first to determine how to partition an ontology graph. To facilitate this, a set of centroid concepts (entities) are selected from two given ontologies. From these centroid concepts, some partition sets can be built. Our idea of selecting centroid concepts is built on the recognition that for two ontologies in the same domain, most of the concepts are repeated frequently, for instance, concepts **Reference, Book** of the first ontology also appear in the second ontology. Therefore, it is possible to start with highly matched concepts, selected as centroid concepts. This idea was used in [10] to find *anchors* (a pair of look-alike concepts) for flooding alignments to the neighbors of anchors. The difference between their approach and the idea here is that [10] floods alignments but our approach spreads similarities and revises the similarity value of a nodes when the similarity values of the nodes in its partition are changed.The idea of using *anchors* was also discussed in [11]. The difference between our approach of finding centroid and that in [11] is that the computation of structural similarities between entities are totally different.

Our mapping approach is realized by the following three main steps.

Step one: Establish preliminary mappings between any two entities from two given ontologies based on their descriptive (semantic) information.

Step two: Select centroid concepts from the two ontologies based on similarities between entities where the similarity value is 1. Each centroid is taken as the starting point of a partition set, and every remaining concept in each ontology is assigned to the partition set where the similarity value between the centroid of this partition set and this concept is the highest.

Step three: A structural-based similarity propagation method is performed to update the similarities between entities and a greedy matching method is used to find the final mapping results.

The main contributions of our proposed approach are:

- We propose a structure-based mapping method based on the similarity flooding algorithm.
- We select centroid concepts for determining the similarity propagation scope by deploying several computing similarity methods.
- We utilize a partition method to partition the entities in the same ontology into different similarity propagation scopes.
- Experimental results show that our method performs very well comparing with other similar approaches.

The rest of the paper is organized as follows. Section 2 introduces the basic methods used for computing the similarities between entities from two ontologies and then presents an algorithm to select the centroid concepts for each ontology. Section 3 illustrates our clustering-based method to partition ontologies. Section 4 presents the similarity spreading method. Section 5 describes the experimental datasets and the results of our experiments. Section 6 discusses related work and concludes the paper.

2 Centroid Concepts Selection

In this paper, ontologies are described by OWL and an entity in an ontology is defined as: $e \in C \cup P$ where C and P are the sets of concepts and properties in an ontology respectively.

We first compute the initial similarities between entities and then select the centroid concepts from these two ontologies.

When calculating similarities between entities, we aim to maximize the descriptive (or semantic) information of an entity, such as its ID, its label and its comment to cover diverse situations. The descriptive information of an entity is can be one of the following two forms:

- The descriptive information of a concept:
  $DI(e) = \{ID, label, comment\}$

– The descriptive information of a property:
  $DI(e) = \{ID, label, comment, domain, range\}$

where ID is the name of an entity, *label* provides an optional human readable name of an entity, and *comment* is often expressed in natural language describing an entity. *Domain* and *range* are the basic but important ways to restrict a property. The *domain* of a property limits the individuals to which the property can be applied and the *range* of a property limits the individuals that the property may have as its value[3].

Given two entities e_i from O_1 and e_j from O_2, we first apply the string-based and WordNet-based methods to compute the similarities between words, where these words are from the ID, label, or domain and range of entities. After this, we compute the similarities between entities based on the similarities of words. We also use the Vector Space Model (VSM)-based method to compute the similarities between entities based on their comments. Finally, we combine these similarities obtained from above methods as the final similarities between entities. The details of these methods are given below.

2.1 String-Based Method and WordNet-Based Method for Computing Similarities between Entities

String-based method. It consider strings as sequences of letters in an alphabet. They are typically based on the following intuition: the more similar the strings, the more likely they are to denote the same concepts [7]. In this paper, we apply the method proposed by Stoilos etc in [12] where the similarity measure between words w_i and w_j is defined as:

$$sim_{Str}(w_i, w_j) = comm(w_i, w_j) - diff(w_i, w_j) + winkler(w_i, w_j) \tag{1}$$

where $comm(w_i, w_j)$ stands for the commonality between w_i and w_j, $diff(w_i, w_j)$ for the difference between w_i and w_j, and $winkler(w_i, w_j)$ for the improvement of the result using the method introduced by Winkler in [13].

WordNet-based method. It uses common knowledge or domain specific thesauri to match words. This kind of matcher has been used in many studies [14,15,16]. In this paper, we use an electronic lexicon, WordNet, for calculating the similarity values between words. WordNet is a lexical database developed by Princeton University which is now commonly viewed as an ontology for natural language concepts.

WordNet can be taken as a hierarchical structure and the idea of the path length method [17] is to find the sum of the shortest path from two concepts (words) to their common hypernym. The similarity between two words w_i and w_j is measured by using the inverse of the sum length of the shortest paths:

$$sim_{WN}(w_i, w_j) = \frac{1}{llength + rlength} \tag{2}$$

[3] http://www.w3.org/TR/owl-features/

where *llength* is the shortest path from word node w_i to its common hypernym with word node w_j and *rlength* denotes the shortest path from w_j to its common hypernym with w_i.

Calculating similarities from words to entities. We have computed similarities between pairs of words according to two methods stated above, next we calculate similarities of entities based on the results obtained from the two methods separately. Let us assume that each entity is composed of several words and these individual words are grouped into a set. For example, entity *HistoryBook* is used to generate a set of words {*History, Book*}. Calculating the similarity between two entities is translated into calculating the similarity between two sets of words:

1. First, for each word w_i in one set of words, compute the similarity values between w_i and every word w_j from the other set of words and then pick out the largest similarity value.
2. Second, attach this value to word w_i. Repeat this step until all of the words in both sets have their own values.
3. Third, the final similarity value of a pair of entities is the sum of similarity values of all the words from the two sets divided by the total number of all the words in the two sets.

2.2 VSM-Based Method for Computing Similarities between Entities

In vector space model [18], documents are represented by vectors of words and the similarity between two documents are computed by using the cosine similarity equation. To apply this method, we first regard each comment attached to an entity as a *document* and all of the comments belonging to two ontologies are regarded as the *collection of documents*. Then, we deploy the vector space model based method to compute the similarity between entities.

$$sim_{VSM}(e_i, e_j) = cossim(d_i, d_j) = \frac{\overrightarrow{d_i} \cdot \overrightarrow{d_j}}{|d_i| \cdot |d_j|} \tag{3}$$

where for each document d in a document collection D, a weighted vector can be constructed as $\overrightarrow{d} = (w_1, w_2, ..., w_n)$ where w_i is the weight of word i in document d, and $w_i = tf_i \cdot idf_i = tf_i \cdot \log \frac{|D|}{|d_i|}$ where tf_i is the frequency of word i in document d, $|D|$ is the total number of documents and $|d_i|$ is the number of documents that contains word i.

2.3 Combination of Similarities between Entities

We have obtained the similarities of entities from three different methods by using descriptive information of entities, now we combine these similarities together. If the similarity value of one of three matchers based on descriptive information is 1, then the similarity value between these two entities is set to 1, since there is a method which is very sure about the equivalence between them. Otherwise, the similarity value is set to be the average the three similarities values obtained, as used in [19].

2.4 Centroid Concepts Selection

The selection of centroid concepts ($e \in C$) is based on the similarities between concepts from ontologies. As shown in Algorithm 1, entities (concepts) from two ontologies are selected as centroid concepts if each of them has a perfect match in another ontology.

Algorithm 1. Selecting Centroid Concepts from Ontologies

Input: Ontologies O_1 and O_2
Output: Concept sets C_1 and C_2

1: $C_1 \leftarrow \emptyset, C_2 \leftarrow \emptyset$
2: **for all** $e_i \in O_1, e_j \in O_2$ **do**
3:   **if** $sim(e_i, e_j) = 1$ **then**
4:     $C_1 \leftarrow C_1 \cup \{e_i\}, C_2 \leftarrow C_2 \cup \{e_j\}$
5:   **end if**
6: **end for**

Example 3. After computing the similarities between concepts in Figure 2, we can obtain four centroid concepts for both ontologies: **Visual Arts**, **History**, **Art History**, **Organization**.

3 Ontology Partition

Now, we describe the process of partitioning ontologies based on the centroid concepts selected from ontologies. The intuition of partitioning is that objects in the same partition set should be *close* or related to each other, both semantically and structurally.

3.1 Similarities between Entities in One Ontology

Structure similarity. Wu and Palmer [20] proposed a method to measure the similarity between concepts within one conceptual domain. The conceptual domain is constructed as a hierarchical structure, so this method only considers the structure of the domain and does not use any other extra information.

$$sim_{Stru}(e_i, e_j) = \frac{2 * N_3}{N_1 + N_2 + 2 * N_3} \tag{4}$$

where e_i and e_j are two concepts in the same ontology and we assume that e_p is their most specific, common parent node; N_1 is the number of concepts on the path from e_i to e_p and N_2 is the number of concepts on the path from e_j to e_p; N_3 is the number of nodes on the path from e_p to root.

Semantic similarity. The calculation of semantic similarity between entities in one ontology is the same as for calculating similarities between entities form two different ontologies as seen in Section 2.

3.2 Ontology Partition

The algorithm for partitioning each ontology is illustrated below. We partition an ontology into a partition by considering the sum of the structural based similarity and semantic based similarity outlined in Section 3.1.

Algorithm 2. Partitioning Ontology

Input: Ontology O_1 and centroid concept set C_1 where $C_1 = \{c_1, ...c_k\}$.
Output: A partition $P = \{G_1, ...G_k\}$.

$G_i = \{c_i\}$ for $i = 1, ..., k$.
if $|C_1| > 1$ **then**
  for every entity e_r in O_1 where $e_r \in C$ *and* $e_r \notin G_1 \cup ... \cup G_k$ **do**
    choose a centroid concepts $c_i \in C_1$ which has the maximum similarity with e_r (if there are several such centroid concepts, arbitrarily choose one).
    $G_i = G_i \cup \{e_r\}$.
  end for
end if

In this algorithm, if there is only one centroid concept, there is no need to do the partition, so the partition algorithm is applied when there is more than one centroid concept.

Example 4. Continue Example 3. We can partition the ontologies in Figure 2 as shown below:

- For ontology one demonstrated by Figure 2(a): *P*={{*Visual Arts, Arts, Music, Bands and Artists, Composition*},{*History, Opera, 20th Century Pop*}, {*Art History*}, {*Organization*}}
- For ontology two illustrated by Figure 2(b): *P*={{*Visual Arts, Arts& Humanities, Design Arts, Interior Design, Architecture, Fashion and Beauty*}, {*History, Roman, Chinese*}, {*Art History*},{*Organization*}}

4 Similarity Propagation Method

In order to apply the similarity flooding idea in our paper, we first introduce the similarity flooding (SF) algorithm proposed in [6]. In this method, two schemas *A* and *B* which are described by SQL DDL statements are first translated into graphs by using an import filter SQL2Graph that understands the definitions of relational schemas, in each of these graphs, different components in relational schema which have relationships are connected by edges. For example, there is an edge between table *Personnel* and column *Pno*. Each connection in a graph is represented as a triple (s, p, o) where s and o denote the source node and target node respectively, and p is the label of the edge. Then the three main processes in SF are:

- Construction of pairwise connectivity graph (PCG) between A and B. The PCG is defined as:

$$((x, y), p, (x', y')) \in PCG(A, B) \Leftrightarrow (x, p, x') \in A \ and \ (y, p, y') \in B$$

  Each node in the PCG is an element from $A \times B$, i.e. a possible candidate mapping pairs between two graphs.
- Construction of induced propagation graph (IPG). The induced propagation graph for A and B is constructed from PCG, where for every edge in the PCG, the IPG contains an additional edge with the same source and target nodes but in the opposite direction. The weights placed on the edges of the propagation graph indicate the coefficients of the similarity of a mapping candidate pair to its direct neighbors and back.
- Fixpoint computation. At the very beginning, the algorithm assigns similarities between nodes of two graphs using some traditional similarity calculation methods. Then it runs an propagation of similarities between connected nodes in the IPG iteratively until satisfy some stop conditions.

4.1 Construction of PCG and IPG from Ontologies

Given two ontologies O_1 and O_2, we have partitioned them into partition sets. Here, we need to build a directed label graph for every partition set. In every partition set, the nodes are the concepts and the edges are from the structure information including *SubClassof*, *HasProperty*, *HasRange*, *SubPropertyof*. For each pair of corresponding partition sets from the two ontologies based on matched centroid concepts, we try to build an PCG. In this PCG, any pair of concepts nodes from O_1 and O_2 is merged into a single node when this pair of concepts have the same relationships with their neighbouring concepts, and in turn, their corresponding neighbouring concepts are merged too. For example, in Figure 3, {*history,history*} is merged into one node in Figure 4 because they have similar relationships with their child nodes. The IPG is then constructed based on this PCG as discussed above.

Example 5. Figure 3 shows two connected graphs, each is from one ontology from one partition set.

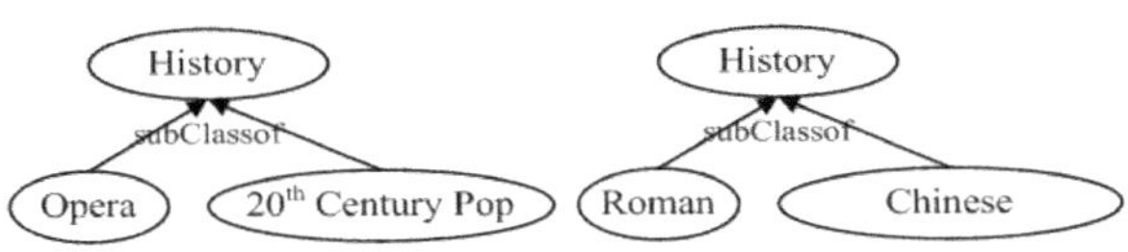

Fig. 3. An example of two connected graphs and each is from one partition set in one ontology

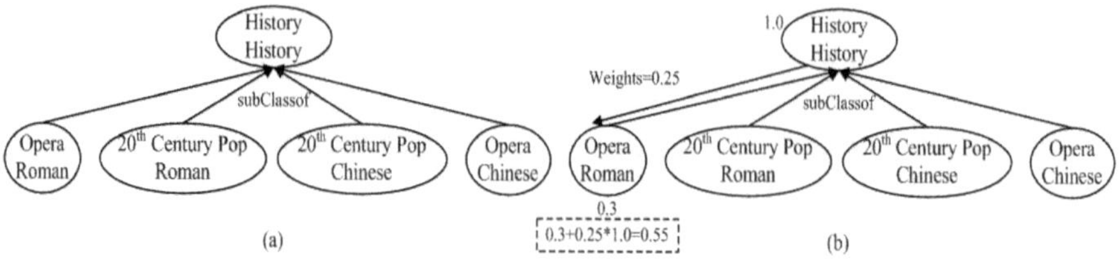

Fig. 4. An example of PCG and IPG

Figure 4 shows the PCG constructed from the two connected graphs in Figure 3, and its corresponding IPG.

4.2 Fixpoint Computation

In SF, $\sigma(x, y)$ denotes the similarity between $x \in O_1$ and $y \in O_2$. The SF is based on the iterative computation of σ-values. In every iteration, the σ-value for a pair (x, y) is incremented by the σ-values of its neighbor pairs in the propagation graph multiplied by the weights on the edges going from the neighbor pairs to (x, y) [6]. The fixpoint formula of iteration for similarity propagation is:

$$\sigma^{i+1} = normalize(\sigma^i + \varphi^i) \quad (5)$$

$$\varphi^i = \Sigma_{j=1}^{m} \sigma_j^i w_j \quad (6)$$

where σ^i and σ^{i+1} are the similarities at the ith and $(i+1)th$ iteration, function φ^i is used to compute the increase of the similarities where m is the total number of neighbouring nodes connected. The value $(\sigma^i + \varphi^i)$ is finally normalized by the maximal σ-value of the current iteration ($(i + 1)th$ iteration). The above computation is repeated until certain conditions are met, that is when there is no changes in similarities produced. To avoid a large number of iterations, a maximum number of propagation can be set to terminate the calculation.

4.3 Greedy Matching

In order to finalize the matching, a greedy matching method in [21] is applied for choosing the best match candidates from the list of ranked matched pairs returned by the similarity spreading approach stated above. Once a pair has the maximum similarity, the entities in the pair are removed from the ontoliges, and the algorithm is applied to match the next pair until no more such pairs can be found.

5 Experiments

5.1 Datasets

We now present the experimental results that demonstrate the performance of different mapping methods on the OAEI 2009 Benchmark Tests. In our experiments, we only focus on classes and properties in ontologies.

Generally, almost all the benchmark tests in OAEI 2009 describe Bibliographic references except Test 102 which is about wine and they can be divided into five groups [22] in terms of their characteristics: Test 101-104, Test 201-210, Test 221-247, Test 248-266 and Test 301-304. A brief description is given below.

- **Test 101-104:** These tests contain classes and properties with either exactly the same or totally different names.
- **Test 201-210:** The tests in this group change some linguistic features compared to Test 101-104. For example, some of the ontologies in this group have no comments or names, names of some ontology have been replaced with synonyms.
- **Test 221-247:** The structures of the ontologies have been changed but the linguistic features have been maintained.
- **Test 248-266:** Both the structures and names of ontologies have been changed and the tests in this group are the most difficult cases in all the benchmark tests.
- **Test 301-304:** Four real-life ontologies about BibTeX.

5.2 Experimental Evaluation Metrics

To evaluate the performance of mapping, like many other papers that use retrieval metrics, *Precision*, *Recall* and *f-measure* to measure a mapping method, we use these measures to evaluate our methods as well. *Precision* describes the number of correctly identified mappings versus the number of all mappings discovered by the three approaches. *Recall* measures the number of correctly identified mappings versus the number of possible existing mappings discovered by hand. *f-measure* is defined as a combination of the *Precision* and *Recall*. Its score is in the range [0, 1]. $precision = \frac{|m_m \cap m_a|}{|m_a|}$, $recall = \frac{|m_m \cap m_a|}{|m_m|}$, $f - measure = \frac{2 * precision * recall}{precision + recall}$ where m_m and m_a represent the mappings results discovered manually and by our method proposed in this paper respectively.

5.3 Comparison of Experimental Results

We now compare the outputs from our system (denoted as A-SP) to the results obtained from the similarity flooding algorithm (denoted as A-SF), and the traditional similarity based methods without using flooding technique (denoted as B-SP). The details are given in Figure 5, which compares the f-measure of three approaches. The benchmark tests are identified as numbers 101 to 304 on the x-axis.

We observe that the overall experimental results of $A - SF$ and $A - SP$ are better than $B - SP$, and the results of $A - SP$ are better than $A - SF$.

From Test 101 to Test 247, the two ontologies to be matched almost contain classes and properties with exactly the same names and comments, so every approach that deploys the computation of similarities of names and comments of entities can get good results. However, there still have three special cases do not obtain good results compared to other tests from Test 101 to Test 247. They are

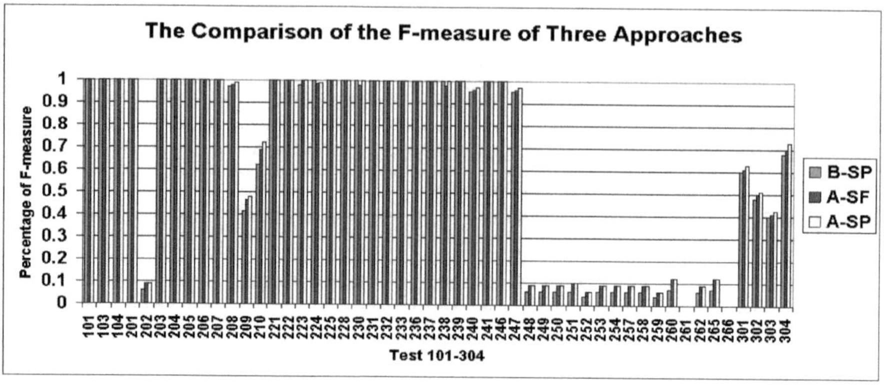

Fig. 5. The comparison of the f-measure of the three approaches

Test 205, 209 and 210. These three tests describe the same kind of information as other ontologies, i.e. publications, however, the class names and comments in them are very different from those in the reference ontology Test 101 so the three approaches does not obtain good results. We think $A-SP$ presents better than $A-SF$ on these tests because in the process of partitioning ontologies and constructing the PCG, $A-SP$ has got the best mapping pairs in each spreading scale. However, $A-SF$ has to construct PCG for the whole ontolgies, and some wrong similarities may generated during this process.

From Test 248 to Test 266, the names of the entities in ontologies are scrambled, meaningless and there are no comments attached to each entity, so string-based similarity method becomes the only useful method for computing similarity, but the similarity results still not very satisfied. These three approaches cannot obtain good results on this group of tests.

Test 301-304 are real-life BibTeX ontologies which also include different words compared to Test 101 describing publications so the results are similar to Test 205, so we do not get quite good similarity results from this data set.

6 Related Work and Conclusion

Related work. Many structure-based ontology mapping approaches have been proposed [11,6,8,9,10,23,24,25]. Anchor-PROMPT [11] takes a set of anchors (pairs of related terms) as input from the source ontologies and traverses the paths between the anchors in the source ontologies. It compares the terms along these paths to identify similar terms and generates a set of new pairs of semantically similar terms. Similarity flooding [6] builds a pairwise connectivity graph and uses the structure features to spread similarity between elements in this graph. RiMOM [8] integrates multiple strategies for ontology alignment. It utilizes a strategy selection module to dynamically determine which strategies should be used in the alignment for different tasks. In its similarity propagation step, it uses the similarity flooding algorithm to generate structure similarities. The similarity propagation method [9] computes the similarity between entities based on the idea of similarity flooding. It first defines some conditions to limit the generation of pairwise connectivity graphs n the process of similarity propagation, it needs to build subgraphs for each element. Anchor-flood [10] starts off with an anchor (a pair of "look-alike") concepts form ontology and collects two blocks of neighboring concepts. The concepts of the pair of blocks are aligned and the process is repeated for finding new alignments. The differences between these approaches and our method proposed here have been discussed in the Introduction and our approach overcomes some weaknesses in these methods.

Conclusion. In this paper, we propose a method for computing similarities based on both the semantic and structural information in ontologies. We particularly integrated the semantic flooding idea into our method in order to reflect structural information. Our methods consists of three steps to find the final mapping results. As a future work, investigation on how to partition ontologies will be carried out to see how different partitions will affect similarity spreading.

References

1. Shvaiko, P., Euzenat, J.: Ten Challenges for Ontology Matching. In: Proceedings of the 7th International Conference on Ontologies, DataBases, and Applications of Semantics (ODBASE 2008), pp. 300–313 (2008)
2. Doan, A., Madhavan, J., Domingos, P., Halevy, A.: Ontology Matching: A Machine Learning Approach. In: Staab, S., Studer, R. (eds.) Handbook on Ontologies in Information Systems, pp. 397–416. Springer, Heidelberg (2004) (invited paper)
3. Euzenat, J., Shvaiko, P.: Ontology Matching. Springer, Heidelberg (2007)
4. Do, H.H., Rahm, E.: COMA - A System for Flexible Combination of Schema Matching Approaches. In: Proceedings of the 27th International Conference on Very Large Data Bases (VLDB 2001), pp. 610–621 (2001)
5. Madhavan, J., Bernstein, P., Rahm, E.: Generic Schema Matching with Cupid. In: Proceedings of the 27th International Conference on Very Large Data Bases (VLDB 2001), pp. 49–58 (2001)
6. Melnik, S., Garcia-Molina, H., Rahm, E.: Similarity Flooding: A Versatile Graph Matching Algorithm and Its Application to Schema Matching. In: Proceedings of the 18th International Conference on Data Engineering (ICDE 2002), pp. 117–128 (2002)
7. Shvaiko, P., Euzenat, J.: A survey of schema-based matching approaches. Journal of Data Semantics 4, 146–171 (2005)
8. Li, J., Tang, J., Li, Y., Luo, Q.: RiMOM: A Dynamic Multistrategy Ontology Alignment Framework. IEEE Transactions on Knowledge and Data Engineering 21(8), 1218–1232 (2009)
9. Wang, P., Xu, B.: An Effective Similarity Propagation Method for Matching Ontologies without Sufficient or Regular Linguistic Information. In: Gómez-Pérez, A., Yu, Y., Ding, Y. (eds.) ASWC 2009. LNCS, vol. 5926, pp. 105–119. Springer, Heidelberg (2009)
10. Hanif, M.S., Aono, M.: An Efficient and Scalable Algorithm for Segmented Alignment of Ontologies of Arbitrary Size. Journal of Web Semantics 7(4), 344–356 (2009)

11. Noy, N.F., Musen, M.A.: Anchor-prompt: Using non-local context for semantic matching. In: Workshop on Ontologies and Information Sharing at the 17th International Joint Conference on Articial Intelligence, IJCAI 2001 (2001)
12. Stoilos, G., Stamou, G.B., Kollias, S.D.: A string metric for ontology alignment. In: Gil, Y., Motta, E., Benjamins, V.R., Musen, M.A. (eds.) ISWC 2005. LNCS, vol. 3729, pp. 624–637. Springer, Heidelberg (2005)
13. Winkler, W.: The state record linkage and current research problems. Technical report, Statistics of Income Division, Internal Revenue Service Publication (1999)
14. Tang, J., Liang, B., Li, Z.: Multiple strategies detection in ontology mapping. In: Proceedings of the 14th international conference on World Wide Web (WWW 2005) (Special interest tracks and posters), pp. 1040–1041 (2005)
15. Madhavan, J., Bernstein, P.A., Rahm, E.: Generic schema matching with cupid. In: Proceedings of the 27th International Conference on Very Large Data Bases (VLDB 2001), pp. 49–58 (2001)
16. Bouquet, P., Serafini, L., Zanobini, S.: Semantic coordination: A new approach and an application. In: Fensel, D., Sycara, K., Mylopoulos, J. (eds.) ISWC 2003. LNCS, vol. 2870, pp. 130–145. Springer, Heidelberg (2003)
17. Resnik, P.: Using information content to evaluate semantic similarity in a taxonomy. In: Proceedings of the 14th International Joint Conference for Artificial Intelligence (IJCAI 1995), pp. 448–453 (1995)
18. Salton, G., Wong, A., Yang, C.S.: A Vector Space Model for Automatic Indexing. Communications of the ACM 18, 613–620 (1975)
19. Zhong, Q., Li, H., Li, J., Xie, G., Tang, J., Zhou, L., Pan, Y.: A Gauss Function based Approach for Unbalanced Ontology Matching. In: Proceeding of the ACM SIGMOD International Conference on Management of Data (SIGMOD 2009), pp. 669–680 (2009)
20. Wu, Z., Palmer, M.S.: Verb Semantics and Lexical Selection. In: Proceedings of the 32nd Annual Meeting of the Association for Computational Linguistics (ACL 1994), pp. 133–138 (1994)
21. Wu, W., Yu, C.T., Doan, A., Meng, W.: An Interactive Clustering-based Approach to Integrating Source Query interfaces on the Deep Web. In: Proceedings of the 30th ACM SIGMOD International Conference on Management of Data (SIGMOD 2004), pp. 95–106 (2004)
22. Qu, Y., Hu, W., Cheng, G.: Constructing virtual documents for ontology matching. In: Proceedings of the 15th international conference on World Wide Web (WWW 2006), pp. 23–31 (2006)
23. Jeh, G., Widom, J.: SimRank: a measure of structural-context similarity. In: Proceedings of the 8th ACM SIGKDD International Conference on Knowledge Discovery and Data Mining (KDD 2002), pp. 538–543 (2002)
24. Hu, W., Jian, N., Qu, Y., Wang, Y.: GMO: A Graph Matching for Ontologies. In: Proceedings of the K-CAP 2005 Workshop on Integrating Ontologies, IO 2005 (2005)
25. Euzenat, J., Guégan, P., Valtchev, P.: OLA in the OAEI 2005 Alignment Contest. In: Proceedings of the K-CAP 2005 Workshop on Integrating Ontologies, IO 2005 (2005)

Risk Modeling for Decision Support

Ronald R. Yager

Machine Intelligence Institute, Iona College
New Rochelle, NY 10801
yager@panix.com

Abstract. Decision-makers require tools to aid in risky situations. Fundamental to this is a need to model uncertainty associated with a course of action, an alternative's uncertainty profile. In addition we need to model the responsible agents decision function, their attitude with respect to different uncertain risky situations. In the real world both these kinds of information are ill defined and imprecise. Here we look at some techniques arising from the modern technologies of computational intelligence and soft computing. The use of fuzzy rule based formulations to model decision functions is investigated. We discuss the role of perception based granular probability distributions as a means of modeling the uncertainty profiles of the alternatives. We suggest a more intuitive and human friendly way of describing uncertainty profiles is in terms of a perception based granular cumulative probability distribution function. We show how these perception based granular cumulative probability distributions can be expressed in terms of a fuzzy rule based model.

Keywords: Fuzzy sets, Granular Probability, Rule Base, Uncertainty Profile.

1 Introduction

The need for risk management arises when we have to make a choice involving a risky alternative. One component of a risky alternative is the uncertainty of the payoff (outcome) resulting from its selection, there are more than one possible outcome. Making decisions in the face of uncertain outcomes requires some of representation of our knowledge of uncertainties associated with the possible outcomes, for example probabilities. Often this information is impossible to obtain precisely and may require an imprecise and fuzzy characterization. Here we shall take advantage of Zadeh's [1-4] work on perception based probability information.

A fundamental difficulty that arises when making decisions involving alternatives with uncertain outcomes is the comparison of the alternatives. This is do to the fact that the multiplicity and complexity of these types of the alternatives makes their direct comparison almost impossible. Here we use rule based valuation functions to circumvent this difficulty.

An additional feature that distinguishes a risky alternative from one that is simply uncertain is that at least one of its possible outcomes is bad, 'undesirable' or 'disturbing.' The concept of undesirable is fuzzy and often involves aspects of human perception. Let us try to provide some intuition. Consider a financial decision in which we

A. Deshpande and A. Hunter (Eds.): SUM 2010, LNAI 6379, pp. 375–388, 2010.

can make a profit ofeither \$50, \$100 or \$200. In this case while we have uncertainty with respect to the outcome and a preference for 200 over 100 over 50, we don't have a risky alternative because none of the payoffs are undesirable. On the other hand, consider an alternative with payoffs {-\$10,000, \$50, \$200}. This can be considered as a risky alternative because in addition to there being an uncertainty with respect to the outcome, it has at least one undesirable outcome. The determination of whether a particular outcome is undesirable is often subjective and context dependent. It is very much dependent on the current state of the decision maker.

A fundamental point that we want to make here is that the construction of decision functions involving these "risky" alternatives often involves some kind of categorization of outcomes with respect to their being undesirable or bad. From a formal point of view decision making with risky alternatives requires that the possible outcomes be expressed on a scale that is richer then an ordinal scale. The scale used must be of a bi-valent nature [5], having positive and negative members, and thereby enabling the capturing of concepts good and bad. An additional feature is that the concepts used to specify "bad" and "good" outcomes are generally fuzzy and imprecise.

2 Modeling the Valuation Function

One approach to addressing the problem of comparing alternatives having uncertain outcomes is to use a valuation function. These functions map the possible payoffs associated with an uncertain alternative into a single scalar value called its valuation. The association of a scalar value with an alternative allows us to easily compare alternatives. Conceptually these valuation functions can be viewed as a mechanism to enable the responsible decision maker to reflect their preferences among different uncertain situations. Statistics such as as expected value, median and variance have historically been used to help provide valuation functions. With the consideration of risky alternatives the nature of the decision makers' preferences between different uncertain situations becomes more complex then can be captured by these simple statistics. In order to capture the decision makers preference in these situations we need more sophisticated structures for modeling the valuation functions.

One approach to modeling a decision makers preference structure, i.e..valuation function, is to use a rule based [6]. A rule base consists of a collection of statements, rules, each of which expresses the decision-makers valuation (attitude) about a particular uncertain situation. The totality of these individual components constitutes the decision makers preference function. The use of a rule base allows a decision maker to express their preferences in a modular fashion. We see how this rule base (knowledge base) is used, an alternative is presented to the rule base which then provides a value for the alternative. The value V is some score associated with the alternative. Fuzzy system modeling [6, 7] provides a well established framework for constructing these types of models used to capture the decision makers' valuation function in the form of a rule base. An individual component rule in the preference rule base is of the form If **antecedent** then V *is* S_i where the term **antecedent** describes some characterization of a risky alternative. An example could be "if an alternative has a very bad outcome with a substantial probability of occurrence then give it a very low value."

In this approach we use predicates to construct the antecedent. Here we use Pred_i to indicate a predicate corresponding to some property or feature of an alternative. For any alternative A we can calculate $\text{Pred}_i(A)$, the degree to which A satisfies the predicate. The antecedent of a rule may consist of a single predicate or a collection of predicates connected by some logical or other aggregation procedure. Typically the antecedent can be expressed in terms of properties associated with surrogate features of the uncertainty profile of an alternative. Things like variance, probability of particular situations, expected values are examples of these features. The consequent of the rule, V *is* S_i indicates a valuation of an alternative that satisfies this rule.

Given a collection of rules R_i: If Pred_i then V *is* S_i the general procedure for working with these rules is as follows. For the alternative A we calculate $\text{Pred}_i(A)$, the degree R_i is valid for this alternative. This gives us a collection of pairs ($\text{Pred}_i(A)$, S_i). We then aggregate these pairs to get an overall valuation for the alternative being valuated, $V(A) = \text{Agg}_i(\text{Pred}_i(A), S_i)$. The methodology used to aggregate these pairs depends upon the structure underlying the partitioning of the uncertainty profile space by the rules. We note in fuzzy systems modeling the most common aggregation is the weighted average

3 Valuation Functions and Uncertainty Profiles

Formally a risky alternative is characterized by an **uncertainty profile**. In part an uncertainty profile consists of a collection of possible outcomes (payoffs) that can occur as a result of selecting this alternative. We shall denote this collection of possible payoffs as X. In addition a uncertainty profile usually contains information about the realizability of each of the payoffs. In the following we assume that the uncertainty profile of an alternative is captured by a probability model, we are assuming that the payoff of a risky alternative is a random variable **R**. One of our concerns here is with the characterization of the features of this random variable that can be used as predicates in the antecedent of the rules used in the rule base definition of the valuation function. We must emphasize that the representation of the features used must be such that we can evaluate the degree of satisfaction of the associated predicate for an alternative given our knowledge of the uncertainty profile of the alternative. Well-established features associated with a random variable are expected value, variance, model and median. A typical example of the use of these features in a rule based is the form

If the expected payoff *is* ***high*** then V *is* ***good***"

Here the expected value is the feature being used. The predicate here is "the expected payoff *is* ***high.***" Thus for a given alternative we must determine the degree to which this is true. Specifically if we have the uncertainty profile of the alternative expressed in terms of a random variable with known probability distribution we can calculate the expected value. With ***high*** expressed as a fuzzy set we can calculate the degree to which the predicate is satisfied. Another example is a rule of the form

If the expected payoff is **high** and the variance is **small** then V is **very good.**

Here our antecedent consists of two predicates connected by an "and." The second predicate is the "variance is **small**". Here then for a given alternative we would calculate its expected value and its variance from its uncertainty profile. We then calculate the satisfaction of each of the two predicates and then take the "anding" of these two values. We could use the minimum of these values as the "and" [9]. It important to emphasize that with the use of predicates and these rules we have circumvented the issue of combining expected values and variances.

In making decisions in which we have risky alternatives the responsible decision maker's mental preference structure is generally more complex then that which can expressed simply using the basic features such as expected value and variance. Making decisions in risky environments require us to use more sophisticated features of an alternatives uncertainty profile.

One feature of an uncertainty profile that can play an important role in the formulating decision rules in the face of risky alternatives is the probability of some subset of payoffs. An example of a rule using this type of feature is

"If the probability of a severe loss *is* **low** then the value of the alternative *is* **high**".

In this case the feature used in the rule is "the alternative's probability of having a severe loss." The predicate here is the degree to which this feature attains a value that is considered as **low**. The process of evaluating this antecedent predicate involves the following. We represent the concept "low probability" as a fuzzy subset, **LOW**, of the unit interval. If Prob(S) is the probability of having a severe loss under the alternative then the degree to which the predicate is satisfied is **LOW**(Prob(S)), the membership grade of value Prob(S) in the fuzzy subset **LOW**.

The issue now becomes that of obtaining Prob(S), the probability of having a severe loss under the alternative. The determination of this depends upon our definition of severe loss and our knowledge about the uncertainty profile associated with the alternative. Initially we shall assume complete information about the probability associated with the random variable, the uncertainty profile of the alternative. If **R** is a continuous random variable, we assume the availability of the probability density function f. If the random variable is discrete we assume the availability of the probability mass function. In addition to our knowledge of the uncertainty profile we need a definition of the concept of "severe loss." Here we can use fuzzy sets to help in the definition.

Consider the payoff random variable whose uncertainty is captured by its probability density function f(x). Let us calculate the "probability of a severe loss". In order to obtain this we first need a definition of the term "severe loss". We define the concept of a severe loss as a fuzzy subset S on X such that S(x) is the degree to which an outcome x satisfies the concept of being a severe loss. Using this definition and the probability density function f(x) we obtain the probability of a severe loss as $\text{Prob}(S) = \int_R f(x)S(x)dx$ [10]. We note if S is a crisp subset then this becomes $\text{Prob}(S) = \int_{x \in S} f(x))dx$. If S is " less or equal a" then $\text{Prob}(S) = \int_{-\infty}^{a} f(x))dx$

In similar manner we can define a large payoff as the fuzzy subset L More generally if $\mathcal{E}$ is any linguistically expressed description of the payoff space which can be represented as a fuzzy subset E then we can obtain $\text{Prob}(\mathcal{E}) =. \int_r f(x)E(x)dx$ We emphasize the subjective nature of the concept $\mathcal{E}$ and the related fuzzy subset E.

Note: In the case in which the random variable describing the payoffs is discrete and captured by a probability mass P then Prob($\mathcal{E}$) = ΣP(x)E(x)

4 Perception Based Granular Probability Distributions

In the complex environment of decision-making the information needed to fully detail the probability measure associated with an alternative's uncertainty profile may only be partially or imprecisely available. Techniques such as the Dempster-Shafer theory of evidence [11] provide useful structures for representation of an alternative's uncertainty profile in the cases of lack of precise knowledge about the exact probability measure. Another approach recently developed by Zadeh [4] is rooted in the observation that much of the information appearing in an alternative's uncertainly profile is based upon the perceptions of the decision maker. In the light of this understanding Zadeh [4] has introduced the idea of **P**erception **B**ased **G**ranular (PBG) probability distributions to address situations in which we have less than perfect information about the uncertainty profile.

Zadeh [4] observed that the type of probability information associated with an uncertainty profile is generally a reflection of perceptions as well as measurements by the decision making entity He suggested that an appropriate way of representing this type of information is with a **P**erception **B**ased **G**ranular (PBG) probability distribution. With the aid of a PBG probability distribution the human can very naturally express their perceptions of an uncertainty profile As we shall see a PBG probability distributions generalize the idea of ordinary probability distribution.

Let **R** be a random variable whose domain X is a subset of the real line. A PBG probability distribution consists of a collection of tuples (A_i, Q_i). Within each tuple A_i is an imprecise element from the domain X of **R** represented as a fuzzy subset of X. Q_i is an amount of probability allocated to that range, generally having a imprecise linguistic nature and expressed as a fuzzy subset of the unit interval. If **R** takes its values in the interval X = [-10 to 10] an example of a PBG probability distribution is (low, about 0.5), (near zero, about 0.3), (near 10, about 0.2)

We now distinguish between two types of situations regarding the underlying domains. The first is when X is a continuous subset of the real line, X = [a, b], and the second is when X is discrete X = $\{x_1, ..., x_n\}$. We first consider the case in which X is discrete. Here the underlying measure is a probability distribution P, whose actual values are unknown. The PBG probability distribution is providing partial information about the underlying probability distribution. Let us look at this situation. First we recall with X = $\{x_1, ..., x_n\}$ then a valid probability distribution P on X is a collection $[p_1, .., p_n]$ such that Prob(x_i) = p_i and $p_i \in [0, 1]$ and the sum is one. If we let $\mathcal{P}_X$ be denote set of all valid probability distributions on X then formally a PBG probability distribution induces a possibility distribution over $\mathcal{P}_X$. Let $\mathcal{K} = \{(A_i, Q_i)|i = 1, ..., m\}$ be a PBG probability distribution on X. If $\Pi_{\mathcal{K}}$ is the induced possibility distribution then for each valid probability distribution, $P \in \mathcal{P}_X$, $\Pi_{\mathcal{K}}(P)$ indicates the possibility that P is the actual probability distribution on X.

With $P = [p_1, ..., p_n]$ in the following we describe one approach to determine $\Pi_{\mathcal{K}}(P)$ given $\mathcal{K} = ((A_i, Q_i)|i = 1, ..., m\}$.. (**1**) For each A_i calculate $\text{Prob}(A_i)$ using P: $\text{Prob}(A_i|P) = \sum_{j=1}^{n} A_i(x_j)\, p_j$ (**2**) For each i calculate, $\tau_i = Q_i(\text{Prob}(A_i|p))$. This is the compatibility of P with Q_i (**3**) $\Pi_{\mathcal{K}}(P) = \text{Min}_i[\tau_i]$

In the case in which X = [a, b], it is continuous, the random variable is characterized by a probability measure. Here the PBG probability distribution is only providing partial information about underlying probability measure. We note that a valid probability measure f associated with X is such that $f(x) \geq 0$ for all $x \in [a, b]$ and $\int_a^b f(x)dx = 1$. We let $\mathcal{F}_X$ be the collection of all valid probability measures on X. In this case a PBG probability induces a possibility distribution over the set $\mathcal{F}_X$. Again we shall assume $\mathcal{K} = ((A_i, Q_i), i = 1, ..., m)$ is the PBG probability distribution corresponding to the uncertainty profile. We let $\Pi_{\mathcal{K}}$ be the induced possibility distribution over $\mathcal{F}_X$. Here $\Pi_{\mathcal{K}}(f)$ indicates the possibility that f can be the actual probability measure given $\mathcal{K}$. We determine $\Pi_{\mathcal{K}}(f)$ as follows: (**1**) For each A_i we calculate $\text{Prob}(A_i|f) = \int_a^b f(x)A_i(x)dx$ (**2**) For each i calculate, $\tau_i = Q_i(\text{Prob}(A_i|p))$. (**3**) $\Pi_{\mathcal{K}}(f) = \text{Min}_i[t_i]$

Let us look at the nature of the PBG probability distribution in more detail. A PBG probability distribution is essentially a generalization of the idea of an ordinary probability distribution. Consider the PBG probability distribution $((A_i, Q_i), i = 1, ..., m)$. Each Q_i is a fuzzy number drawn from the unit interval I, it is normal and unimodal. In particular there exists an $r \in [0, 1]$ such that $Q_i(r) = 1$. In addition since it is unimodal, there exist two values a_i and $b_i \in$ I such that: **1.** $Q_i(r)$ is non-decreasing for $r \in [0, a_i]$, **2.** $Q_i(r) = 1$ for $r \in [a_i, b_i]$ and **3.** $Q_i(r)$ is non-increasing for $r \in [b_i, 1]$

One implication of the uni-modality of the granular probabilities is the interval nature of the associated level sets [12]. Thus if Q_i^α is the α-level set of Q_i, $Q_i^\alpha = \{r / Q_i(r) \geq \alpha\}$, then $Q_i^\alpha = [l_i(\alpha), u_i(\alpha)]$. It is also the case that the unimodality of Q_i implies that if $\alpha > \beta$ then $Q_i^\alpha \subseteq Q_i^\beta$ the level sets are nested.

We note two special cases of these granular probabilities. The first is the case when Q_i is a precise value q_i in I, $Q_i = \{q_i\}$. The second is when Q_i is an interval, $Q_i = [a_i, b_i]$. Here $Q_i(r) = 1$ for $r \in [a_i, b_i]$ and $Q_i(r) = 0$ for $r \notin [a_i, b_i]$

Generally the A_i are human comprehensible concepts associated with the space X. As discussed by Gardenfors [13] concepts on a domain are expressed as convex subsets. Thus formally the A_i are normal and unimodal, they are fuzzy numbers from the domain X. Two special cases of A_i are singletons and crisp intervals.

5 Evaluating Decision Functions with PBG Uncertainty Profiles

Previously we indicated the rules based approach for specifying the valuation function can involve rules in which we have antecedent terms of the form:

If Prob(Fuzzy Event) *is* Large then ... (I)

Here we shall investigate a method for evaluating the satisfaction of this type of antecedent by risky alternatives for this case in which an alternative's uncertainty profile is expressed in terms of a PBG probability distribution.

We first formalize the above antecedent. Let **R** indicate the payoff associated with the alternative being evaluated, it is a random variable on real line. In order to formalize the antecedent in I we let F be a fuzzy subset of the domain of **R**, this corresponds to a general fuzzy event. In addition we let Q be a fuzzy probability corresponding to what we generically denoted as Large in (I). Using these notations our rule becomes

If Prob(**R** *is* F) *is* Q then ....

Let us use W to indicate the variable corresponding to the "probability of the event **R** *is* F." Using this notation we can express our rule as "If W *is* Q then ....." The firing of this rule is determined by the compatibility of the value of W with the fuzzy subset Q.

Consider a risky alternative whose uncertainty profile is expressed using the PBG probability distribution $\mathcal{K} = ((A_i, Q_i), i = 1, ..., m)$. Here A_i is a fuzzy subset of X and Q_i is a fuzzy subset corresponding to amount of probability.

The task of evaluating the degree to which a risky alternative satisfies the rule is formulated as follows. We need to determine the compatibility of the value of W, the probability of the event **R** *is* F with Q, given that all we know about **R** *is* $\mathcal{K}$.

Consider the firing of the rule "If W *is* Q then .....". If we know that the probability of the event **R** *is* F is precisely equal to the value b, W = b, then the degree of firing τ is simply Q(b). More generally, if the value for W is a fuzzy probability B, then using the established procedure in fuzzy systems modeling we obtain as the firing level $\tau = \text{Max}_y[Q(y) \wedge B(y)]$, we take the maximum of the intersection of Q and B.

The situation we are faced with is slightly different than either of these. Instead of knowing the value of W, the probability that **R** *is* F, all we have is the PBG probability distribution $\mathcal{K}$ on **R**. In this case our task becomes to calculate the value of W from our information about **R**.

If instead of having a PBG probability distribution we had an ordinary probability distribution $P = [(x_i, p_i)]$, then to calculate W, probability that **R** *is* F, we use W $= \sum_{i=1}^{n} F(x_i)\, p_i$ We must extend this approach to our situation where we have the PBG probability distribution $\mathcal{K}$. With $\mathcal{K}$ we have that both A_i and Q_i are fuzzy subsets. The fact that A_i is not crisp conceptually provides more difficulty than the fuzziness of Q_i.

If we temporarily consider the situation in which Q_i is precise, $Q = q_i$ and A_i is an interval we can get some insight into how to proceed. Also for simplicity assume that F is a crisp subset. In calculating W we are essentially obtaining the sum of the probabilities of the possible values of **R** that lie in F. When A_i is an interval it is difficult to decide whether the probability is associated with element in F or not. To get around this problem we must obtain upper and lower bounds on W. The actual probability lies between these values. Using this idea for the more general situation where all the objects are fuzzy we obtain

$$\text{Upper}_F = \sum_{i=1}^{n} \text{Poss}[F/A_i]\, Q_i \text{ and } \text{Lower}_F = \sum_{i=1}^{n} (1 - \text{Poss}[\overline{F}/A_i])\, Q_i$$

where $\text{Poss}[F/A_i] = \text{Max}_x[F(x) \wedge A_i(x)]$ and $\text{Poss}[\overline{F}/A_i] = \text{Max}_x[(1 - F(x)) \wedge A_i(x)]$. $\text{Poss}[F/A_i]$ is the degree of intersection of A_i and F while $1 - \text{Poss}[\overline{F}/A_i]$ is the degree to which A_i is included in $\overline{F}$. There values are closely related to the measures of plausibility and belief in Dempster-Shafer theory [11].

At this point we must draw upon some of results from fuzzy arithmetic [14]. We recall if A and B are two fuzzy numbers then their sum $D = A \oplus B$ is also a fuzzy number such that $D(z) = \underset{\substack{x,\, y \text{ s.t.} \\ x + y = z}}{\text{Max}} [A(x) \wedge B(y)]$. We also note that if α is a scalar then αA is a fuzzy number D such that $D(z) = \underset{\substack{x \text{ s.t.} \\ \alpha x = z}}{\text{Max}} [A(x)]$. More generally if $D_1, ..., D_n$ are fuzzy numbers and $\alpha_1, ..., \alpha_n$ are nonnegative scalars then

$$D = \alpha_1 D_1 \oplus \alpha_2 D_2 \oplus .... \oplus \alpha_n D_n$$

is a fuzzy number such that $D(z) = \underset{\substack{x_i \text{ s.t.} \\ \Sigma_i \alpha_i x_i = z}}{\text{Max}} [Ai(x_i)]$

The point we can conclude from this digression is that we have available to us the facility to calculate the values Upper_F or Lower_F. More specifically if we denote $\lambda_i = \text{Poss}[F/A_i] \in [0,1]$ then Upper_F is a fuzzy number H defined on the unit interval such that for all $z \in [0, 1]$, $H(z) = \underset{\substack{z_i \text{ s.t.} \\ \Sigma_i \lambda_i z_i = z}}{\text{Max}} [\text{Min}_i[Q_i(z_i)]$.

If we denote $\gamma_i = 1 - \text{Poss}[\overline{F}/A_i] \in [0, 1]$ then Lower_F is a fuzzy number L defined on the unit interval such that for all $z \in [0,1]$, $L(z) = \underset{\substack{z_i \text{ s.t.} \\ \Sigma_i \gamma_i z_i = z}}{\text{Max}} [\text{Min}_i[Q_i(z_i)]$

We must now consider the relationship between the fuzzy subsets H and L. In anticipation of uncovering this we look at the relationship between $\lambda_i = \text{Poss}[F/A_i]$ and $\gamma = 1 - \text{Poss}[\overline{F}/A_i]$. Here we use the fact that F and A_i are normal, they have at least one element with membership grade 1. Assume $\gamma = \alpha$, then $\text{Max}_x[(1 - F(x)) \wedge A_i(x)] = 1 - \alpha$. Since A_i is normal there exists some x^* where $A_i(x^*) = 1$ and therefore $(1 - F(x^*)) \wedge 1 = (1 - F(x^*)) \leq 1 - \alpha$ hence $F(x^*) \geq \alpha$. Since $\lambda_i = \text{Max}_x[F(x) \wedge A_i(x)]$

$\geq F(x^*) \wedge A_i(x^*) \geq \alpha$. Hence we get $\lambda_i \geq \gamma_i$ for all i. Thus we see that $L = \sum_{i=1}^{n} \gamma_j Q_j$ and $H = \sum_{i=1}^{n} \lambda_j Q_j$ where $\lambda_j \geq \gamma_j$ for all j.

Before preceding we want to introduce a type of relationship between fuzzy numbers

Definition: Let G_1 and G_2 be two fuzzy numbers such that

$$G_j(x) \text{ is non-decreasing} \quad \text{for } x \leq a_j$$
$$G_j(x) = 1 \quad \text{for } x \in [a_j, b_j]$$
$$G_j(x) \text{ is non-increasing} \quad \text{for } x > b_j$$

where $a_1 \leq a_2$ and $b_2 \geq b_1$. If in addition we have $G_1(x) \geq G_2(x)$ for all $x \leq a_1$ and $G_2(x) \geq G_1(x)$. for all $x \geq a_2$we say G_2 is to **the right of** G_1 and denote $G_2 \geq_R G_1$

The relationship $G_2 \geq_R G_1$ can be also expressed in terms of level sets. If the α level set of G_i is $G_i(\alpha) = [a_i(\alpha), b_i(\alpha)]$, then the relationship $G_2 \geq_R G_1$ is equivalent to the condition that for each $\alpha \in [0,1]$ we have $a_1(\alpha) \leq a_2(\alpha)$ and $b_1(\alpha) \leq b_2(\alpha)$.

It can be shown that if $G_2 = \sum_{i=1}^{n} \lambda_i Q_i$ and $G_1 = \sum_{i=1}^{n} \gamma_i Q_i$ where $0 \leq \gamma_i \leq \lambda_i \leq 1$ for all i and the Q_i are non-negative fuzzy number then $G_2 \geq_R G_1$. From this it follows that $H \geq_R L$, the upper bound is always to the right of the lower bound.

Earlier we indicated that the value of W, the probability that **R** is F, lies between the H and L. In particular, we have the following constraints on the value of W:

W is greater than or equal L ***and*** W is less than or equal H.

If we let L* indicate the fuzzy subset *greater than or equal L* and let H* indicate the fuzzy subset *less than or equal H* then W *is* E where $E = L^* \cap H^*$. It is the intersection of the fuzzy subsets L* and H*.

Let us now calculate L* and H* from L and H. L* is obtained as

$$L^*(x) = \text{Max}_y[\text{GTE}(x, y) \wedge L(y)]$$

where GTE is the relationship "greater then or equal" defined on $[0, 1] \times [0, 1]$ by

$$\text{GTE}(x, y) = 1 \text{ if } x \geq y \text{ and } \text{GTE}(x, y) = 0 \text{ if } x < y$$

Here $L(x)$ is non-decreasing for $x \leq a_1$ and $L(x) = 1$ for $x \in [a_1, b_1]$ it is non-increasing for $x \geq b_1$. It is easy to show that in this case that L* is such that $L^*(x) = L(x)$ for $x \leq a_1$ and $L^*(x) = 1$ for $x \geq a_1$.

Similarly for H* we have $H^*(x) = \text{Max}_y[\text{LTE}(x, y) \wedge H(y)]$ LTE is the relationship "less then or equal" defined on $[0, 1] \times [0, 1]$ by

$$\text{LTE}(x, y) = 1 \text{ if } x \leq y \text{ and } \text{LTE}(x, y) = 0 \text{ if } x > y$$

If $H(x)$ is a fuzzy number with value one in the interval $[a_2, b_2]$ then H* is a fuzzy number such $H^*(x) = 1$ for $x \leq b_2$ and $H^*(x) = H(x)$ for $x > b_2$.

Combining L* and H* to get E, the possible values for W, we have $E = H^* \cap L^*$ hence $E(x) = H^*(x) \wedge L^*(x)$. From this we get

$E(x) = L(x)$ for $x \in [0, a_1]$, $E(x) = 1$ for $x \in [a_1, b_2]$, $E(x) = H(x)$ for $x \in [b_2, 1]$

Returning to our concern with determining the firing level of the rule

If W *is* Q then

when our input is $W = \mathcal{K}$ we now use this E to calculate the firing level of the rule as

$$\tau = \text{Max}_x[Q(x) \wedge E(x)]$$

6 Cumulative Distribution Functions

Here consider the situation where the information about the uncertainty profile of an alternative is available in terms of a cumulative distribution function and more generally a **P**erception **B**ased **G**ranular **C**umulative **D**istribution function, PBG-CD function.

If **R** is a random variable taking its value on the real line a cumulative distribution function F is such that F(x) is the probability that $\mathbf{R} \leq x$. Formally F is a function $F: [-\infty, \infty] \rightarrow [0, 1]$, it is monotonic, $F(x) \geq F(y)$ if $x > y$. We note F is available whether **R** is discrete or continuous. If **R** is discrete then $F(x) = \sum_{i \text{ s.t. } x_i \leq x} p_i$. If **R** is continuous with probability density f then $F(x) = \int_{-\infty}^{\infty} f(x)dx$. In many applications the domain of F is bounded, there exists some value x_* s.t. such that $F(x) = 0$ for $x \leq x_*$ and some x^* such that. $F(x) = 1$ for all $x \geq x^*$.

With the availability of the CDF we can easily provide the information needed to determine the firing level of a rule of the form "If Prob(A) *is* Q then ....." If A is a crisp subset, $A = \{x / a_1 \leq x \leq a_2]$ then $\text{Prob}(A) = F(a_2) - F(a_1)$ and the firing level is $Q(F(a_2) - F(a_1))$. If A is a fuzzy subset we must look a little more carefully at the situation. Here we shall assume A is a fuzzy number of the form

$A(x) = 0$ for $x \leq b_1$, $A(x) \geq A(y)$ for $b_1 \leq y < x \leq a_1$, $A(x) = 1$ for $a_1 < x \leq a_2$
$A(x) \leq A(y)$ for $a_2 \leq y \leq x \leq b_2$ and $A(x) = 0$for $x \geq b_2$

We define a fuzzy subset $\tilde{a}_1$ such that $\tilde{a}_1(x) = A(x)$ for $b_1 \leq x \leq a_1$and $\tilde{a}_1(x) = 0$ elsewhere, We also define the fuzzy subset $\tilde{a}_2$ such that $\tilde{a}_2(x) = A(x)$ for $a_2 \leq x \leq b_2$ and $\tilde{a}_2(x) = 0$ elsewhere. $\tilde{a}_1$ and $\tilde{a}_2$ are fuzzy numbers which allow us to express $\text{Prob}(A) = F(\tilde{a}_2) - F(\tilde{a}_1)$. In order to obtain Prob(A) we need to obtain $F(\tilde{a}_2)$ and $F(\tilde{a}_1)$. Since the processes needed to obtain these values are similar we shall only concentrate on $F(\tilde{a}_2)$. Using Zadeh's extension principle [15, 16], since $\tilde{a}_2$ is a fuzzy number of real line, then $F(\tilde{a}_2)$ is a fuzzy subset of the unit interval such that $F(\tilde{a}_2)$

$= \cup_x\{\frac{\tilde{a}_2(x)}{F(x)}\}$ and since $\tilde{a}_2(x) = A(x)$ for $x \in [a_2, b_2]$ and $\tilde{a}_2(x) = 0$ elsewhere then

$F(\tilde{a}_2) = \bigcup_{x \in [a_2, b_2]} \{\frac{A(x)}{F(x)}\}$. Here $F(\tilde{a}_2)$ is a fuzzy number. In this case the possibility that $F(\tilde{a}_2)$ takes the value z is $\underset{\substack{x \in [a_2, b_2] \\ F(x) = z}}{\text{Max}} [A(x)]$. The monotonic nature of the cumulative distribution function F and the special form of $\tilde{a}_2$ results in a form of $F(\tilde{a}_2)$ as shown in fig. 1. We emphasize that $F(\tilde{a}_2)$ is a fuzzy number of the unit interval such that its membership grade is one at the value $F(a_2)$, and monotonically decreases to zero at the value $F(b_2)$. In the range from zero to $F(a_2)$ and $F(b_2)$ to 1 its membership value is also zero.

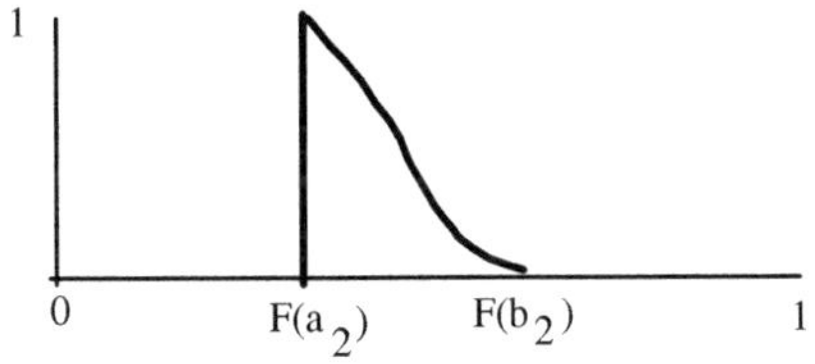

Fig. 1. Fuzzy subset $F(\tilde{a}_2)$

Some special situations are worth pointing out. If F is such that it is constant, F(x) = k, in the range $x \in [a_2, b_2]$ then $F(\tilde{a}_2)$ is a singleton set, $F(\tilde{a}_2) = \{\frac{1}{k}\}$

In a similar way we can show generally $F(\tilde{a}_1)$ is a fuzzy number of the unit interval such that $F(\tilde{a}_1) = \bigcup_{x \in [b_1, a_1]} \{\frac{A(x)}{F(x)}\}$, see fig. 2.

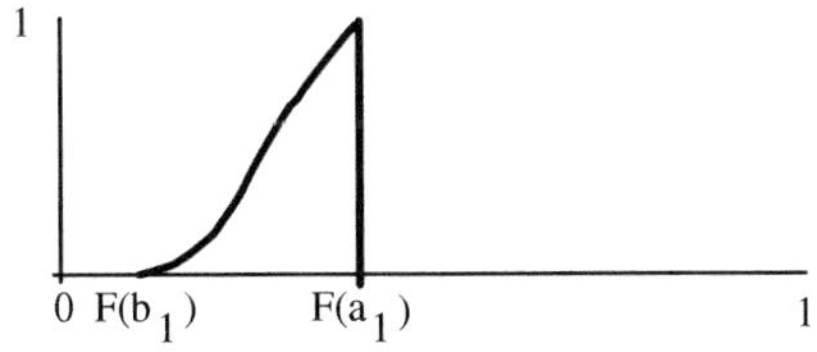

Fig. 2. Fuzzy subset $F(\tilde{a}_1)$

Using these fuzzy values for $F(\tilde{a}_2)$ and $F(\tilde{a}_1)$ we obtain $\text{Prob}(A) = F(\tilde{a}_2) - F(\tilde{a}_1)$ as a fuzzy number of unit interval having nonzero membership grade in the interval $(F(a_2) - F(a_1))$ to $(F(b_2) - F(b_1))$. Here if we let PA be the fuzzy subset denoting the value Prob(A) then

$$PA(z) = 0 \qquad z < F(a_2) - F(a_1)$$

$$\begin{array}{ll} PA(z) = 1 & z = F(a_2) - F(a_1) \\ PA(z) \text{ is decreasing} & F(a_2) - F(a_1) < z < F(b_2) - F(b_1) \\ PA(z) = 0 & z > F(b_2) - F(b_1) \end{array}$$

In some practical situations it may be much more efficient to defuzzify $F(\tilde{a}_1)$ and $F(\tilde{a}_2)$ and use these scalar values to obtain a scalar value for Prob(A).

Let us consider the defuzzification of $F(\tilde{a}_2)$ which we recall was $F(\tilde{a}_2) = \bigcup_{x \in [a_2, b_2]} \{\frac{A(x)}{F(x)}\}$. Letting d_2 denote the defuzzified value of $F(\tilde{a}_2)$ we get

$$d_2 = \frac{\int_{a_2}^{b_2} F(x)A(x)dx}{\int_{a_2}^{b_2} A(x)dx}$$

We observed that if F(x) is constant, F(x) = k in the range $[a_2, b_2]$, then $d_2 = k$. Actually as we have already pointed out if F(x) = k in the range a_2 to b_2 then $F(\tilde{a}_2)$ is itself a constant value k, no fuzziness exists.

In many real situations it may be difficult for a decision maker to obtain a precise manifestation of the cumulative distribution of the payoff of a risky alternative. In these cases a decision maker may be only able to obtain a imprecise characterization of the underlying cumulative distribution in the form of what we shall call a **P**erception **B**ased **G**ranular **C**umulative **D**istribution function, PBG-CD function. A PBG-CD is a granular description of the cumulative distribution function in a form that is widely used in fuzzy modeling [6]. When using a PBG-CD we partition the range R into fuzzy intervals $B_1, ..., B_n$. We then express the value of F in each one of these fuzzy ranges using a fuzzy subset of the unit interval F_i. With PBG-CD function we have a rule-based representation of the cumulative distribution function F

If U is B_i then F is F_i.

In working with the fuzzy rule based description of the underlying function we can draw upon the well established literature of fuzzy systems modeling.

In order to find the value of F at some value for U, *a*, we proceed as follows. We first obtain the firing level of each rule $\tau_i = B_i(a)$. We then calculate $\omega_i = \frac{\tau_i}{\sum_{i=1}^{n} \tau_i}$.

Using this we calculate F(a) as the fuzzy subset $\mathbf{F_a} = \sum_{=1}^{n} \omega_i F_i$. Here we get for F(a) a fuzzy subset of the unit interval such that $\mathbf{F_a}(y)$ is the possibility that F(a) assumes the value y. We can apply a defuzzification operation on $\mathbf{F_a}$ to obtain a scalar value.

In the following example we illustrate the generation of a perception based granular CD function

Example: We consider an investment alternative in which the investor has the following perceptions of the outcome of his investment.

He is certain that he won't lose more then $500 dollars
He believes his chances of losing more then $100 is about 10%
He believes his chances of losing any money is 20%
He feels that there is about a 90% chance that he will win at most $500
He is certain that he won't win more then a $1000

We can use this to construct a rule based description of the cumulative distribution function. In particular if F(U) = Prob(**R** ≤ U) with **R** being the random payoff then the rule base is

If U *is* **less then $500** then F *is* **zero**
If U *is* **"near -$100"** then F *is* **about 10%**
If U *is* **zero** then F *is* **about 20%**
If U *is* **about $500** then F *is* **about 90%**
If U *is* **greater then 1000** then F *is* **100%**

7 Conclusion

We focused on the issue of decision making in risky situations. We discussed the need for using decision functions to aid in capturing the decision maker's preference among these types of uncertain alternatives. The use of fuzzy rule based formulations to model these functions was investigated. We discussed the role of Zadeh's perception based granular probability distributions as a means of modeling the uncertainty profiles of the alternatives. We look at various properties of this method of describing uncertainty and showed how they induced possibility distributions of the space of probability distributions Tools for evaluating rule based decision functions in the face of perception based uncertainty profiles were presented. We considered the situation in which uncertainty profiles are expressed in terms of a cumulative distribution function. We introduced the idea of a perception based granular cumulative distribution and describe its representation in terms of a fuzzy rule based model.

References

[1] Zadeh, L.A.: From computing with numbers to computing with words-From manipulation of measurements to manipulations of perceptions. IEEE Transactions on Circuits and Systems 45, 105–119 (1999)
[2] Zadeh, L.A.: A new direction in AI - toward a computational theory of perceptions. AI Magazine 22(1), 73–84 (Spring 2001)
[3] Zadeh, L.A.: Toward a logic of perceptions based on fuzzy logic. In: Novak, W., Perfilieva, I. (eds.) Discovering the World with Fuzzy Logic, pp. 4–28. Physica-Verlag, Heidelberg (2001)
[4] Zadeh, L.A.: Toward a perception-based theory of probabilistic reasoning with imprecise probabilities. Journal of Statistical Planning and Inference 105, 233–264 (2002)
[5] Yager, R.R.: Using a notion of acceptable in uncertain ordinal decision making. International Journal of Uncertainty, Fuzziness and Knowledge-Based Systems 10, 241–256 (2002)
[6] Yager, R.R., Filev, D.P.: Essentials of Fuzzy Modeling and Control. John Wiley, New York (1994)

[7] Takagi, T., Sugeno, M.: Fuzzy identification of systems and its application to modeling and control. IEEE Transactions on Systems, Man and Cybernetics 15, 116–132 (1985)
[8] Klir, G.J.: Uncertainty and Information. John Wiley & Sons, New York (2006)
[9] Klir, G.J., Yuan, B.: Fuzzy Sets and Fuzzy Logic: Theory and Applications. Prentice-Hall, Upper Saddle River (1995)
[10] Zadeh, L.A.: Probability measures of fuzzy events. Journal of Mathematical Analysis and Applications 10, 421–427 (1968)
[11] Yager, R.R., Liu, L., Dempster, A. P., Shafer, G. (Advisory eds.) : Classic Works of the Dempster-Shafer Theory of Belief Functions. Springer, Heidelberg (to appear)
[12] Dubois, D., Prade, H.: Fuzzy Sets and Systems: Theory and Applications. Academic Press, New York (1980)
[13] Gardenfors, P.: Conceptual Spaces: the geometry of thought. MIT Press, Cambridge (2000)
[14] Dubois, D., Prade, H.: Fuzzy numbers: An overview. In: Bezdek, J.C. (ed.) Analysis of Fuzzy Information. Mathematics and Logic, vol. 1, pp. 3–39. CRC Press, Boca Raton (1987)
[15] Zadeh, L.A.: Fuzzy Sets. Information and Control 8, 338–353 (1965)
[16] Yager, R.R.: A characterization of the extension principle. Fuzzy Sets and Systems 18, 205–217 (1986)

Author Index

Afrati, Foto N. 28
Amgoud, Leila 42, 56
Assaghir, Zainab 70

Bell, David A. 361
Benferhat, Salem 3
Ben Hariz, Sarra 84
Ben Yaghlane, Boutheina 191
Besnard, Philippe 42
Bosc, Patrick 98
Bounhas, Myriam 112

Cholvy, Laurence 126

d'Amato, Claudia 137
de Keijzer, Ander 7
Denœux, Thierry 333
Destercke, Sebastien 151
Dhoedt, Bart 347
Dickerson, John P. 319
Dubois, Didier 11, 219

Elouedi, Zied 84

Fanizzi, Nicola 137
Fazzinga, Bettina 137
Flesca, Sergio 163
Furfaro, Filippo 163

Gottlob, Georg 137

Hadjali, Allel 98
Hansen, Clifford W. 177
Helton, Jon C. 177
Hüllermeier, Eyke 16

Kaytoue, Mehdi 70
Koch, Christoph 1

Laâmari, Wafa 191
Lawry, Jonathan 205

Liu, Weiru 236, 361
Loquin, Kevin 219
Lukasiewicz, Thomas 137

Magnani, Matteo 250
Ma, Jianbing 236
Martinez-Alvarez, Miguel 278
Martinez, Maria Vanina 264
Mellouli, Khaled 112
Miller, Paul 236
Montesi, Danilo 250

Papini, Odile 20
Parisi, Francesco 163, 264
Pivert, Olivier 98, 292
Prade, Henri 70, 112, 292
Pugliese, Andrea 264

Quost, Benjamin 333

Roelleke, Thomas 278

Saad, Emad 306
Sallaberry, Cédric J. 177
Schaub, Torsten 2
Schockaert, Steven 347
Serrurier, Mathieu 112
Simari, Gerardo I. 264, 319
Simon, Christophe 191
Smits, Grégory 98
Strauss, Olivier 24
Subrahmanian, V.S. 264, 319

Van Laere, Olivier 347
Vasilakopoulos, Angelos 28
Vesic, Srdjan 56

Wang, Ying 361

Yager, Ronald R. 375

Batch number: 10704693

Printed by Printforce, the Netherlands